Lecture Notes in Artificial Intelligence   4413

Edited by J. G. Carbonell and J. Siekmann

Subseries of Lecture Notes in Computer Science

T0191728

Marcin S. Szczuka   Daniel Howard
Dominik Ślęzak   Haeng-kon Kim
Tai-hoon Kim   Il-seok Ko   Geuk Lee
Peter M.A. Sloot (Eds.)

Advances in Hybrid Information Technology

First International Conference, ICHIT 2006
Jeju Island, Korea, November 9-11, 2006
Revised Selected Papers

Springer

Series Editors

Jaime G. Carbonell, Carnegie Mellon University, Pittsburgh, PA, USA
Jörg Siekmann, University of Saarland, Saarbrücken, Germany

Volume Editors

Marcin S. Szczuka
Warsaw University, Poland, szczuka@mimuw.edu.pl

Daniel Howard
QinetiQ Group PLC, Malvern, UK, dhoward@taz.qinetiq.com

Dominik Ślęzak
Infobright Inc, Toronto, ON, Canada, slezak@infobright.com

Haeng-kon Kim
Daegu Catholic University, Gyeongbuk, Korea, hangkon@cu.ac.kr

Tai-hoon Kim
Hannam University, Daejeon, Korea, taihoonn@hannam.ac.kr

Il-seok Ko
Dongguk University, Korea, isko@dongguk.edu

Geuk Lee
Hannam University, Daejeon, Korea, leegeuk@ce.hannam.ac.kr

Peter M.A. Sloot
University of Amsterdam, The Netherlands, sloot@science.uva.nl

Library of Congress Control Number: 2007941600

CR Subject Classification (1998): I.2, H.4, H.3, J.1, H.5, K.6, K.4

LNCS Sublibrary: SL 7 – Artificial Intelligence

ISSN   0302-9743
ISBN-10   3-540-77367-3 Springer Berlin Heidelberg New York
ISBN-13   978-3-540-77367-2 Springer Berlin Heidelberg New York

Springer is a part of Springer Science+Business Media

springer.com

© Springer-Verlag Berlin Heidelberg 2007
Printed in Germany

Typesetting: Camera-ready by author, data conversion by Scientific Publishing Services, Chennai, India
Printed on acid-free paper   SPIN: 12206988   06/3180   5 4 3 2 1 0

Preface

As information technologies (IT) become specialized and fragmented, it is easy to lose sight that many topics in IT have common threads and because of this, advances in one sub-discipline may transmit to another. The presentation of results between different sub-disciplines of IT encourages this interchange for the advancement of IT as a whole. Of particular interest is the hybrid approach or combining ideas from one discipline with those of another to achieve a result that is more significant than the sum of the individual parts. Through this hybrid philosophy, a new or common principle can be discovered which has the propensity to propagate throughout this multifaceted discipline.

This volume comprises the selection of extended versions of papers that were presented in their shortened form at the 2006 International Conference on Hybrid Information Technology (http://www.sersc.org/ICHIT2006/). Sixty-four papers out of the 235 that were published in ICHIT 2006 electronic proceedings were deemed suitable for inclusion in this volume, in a selection that was guided by technical quality and relevance to the balance of topics in hybrid information technology. The conference reflected a change in the thinking of scientists and practitioners, who now tend to join their efforts within multidisciplinary projects. As a consequence, the readers may observe that many papers might conceivably be classified into more than one chapter, given their interdisciplinary scope. The contributions in this monograph are clustered into six chapters: Data Analysis, Modeling, and Learning (11 papers); Imaging, Speech, and Complex Data (11 papers); Applications of Artificial Intelligence (11 papers); Hybrid, Smart, and Ubiquitous Systems (11 papers); Hardware and Software Engineering (9 papers); as well as Networking and Telecommunications (11 papers).

We would like to acknowledge the great effort of all in the ICHIT 2006 International Advisory Board and members of the International Program Committee of ICHIT 2006, as well as all organizations and individuals who supported the idea of publishing these advances in hybrid information technology, including SERSC (http://www.sersc.org/) and Springer. We strongly believe in the need for continuing this undertaking in the future, in the form of a conference, journal, or book series. In this respect we welcome any feedback.

October 2007

Marcin Szczuka
Daniel Howard
Dominik Ślęzak
Haeng-kon Kim
Tai-hoon Kim
Il-seok Ko
Geuk Lee
Peter Sloot

Table of Contents

Data Analysis, Modelling, and Learning

Imaging, Speech, and Complex Data

Applications of Artificial Intelligence

Hybrid, Smart, and Ubiquitous Systems

Hardware and Software Engineering

Networking and Telecommunications

Taking Class Importance into Account

José-Luis Polo, Fernando Berzal, and Juan-Carlos Cubero

Department of Computer Science and Artificial Intelligence,
University of Granada, 18071 Granada, Spain
{jlpolo,fberzal,JC.Cubero}@decsai.ugr.es

Abstract. In many classification problems, some classes are more important than others from the users' perspective. In this paper, we introduce a novel approach, *weighted classification*, to address this issue by modeling class importance through weights in the [0,1] interval. We also propose novel metrics to evaluate the performance of classifiers in a weighted classification context. In addition, we make some modifications to the ART[1] classification model [1] in order to deal with weighted classification.

1 Introduction

Classification is an extensively studied problem in Machine Learning research. Despite this, many classification problems exhibit specific features that render most classification models ineffective. Several models have been proposed to deal with class attribute peculiarities:

- In *imbalanced classification* [7] [8] problems, some classes are supported by a very low number of examples. Unfortunately, traditional classification models tend to ignore such classes, no matter what their importance is.
- *Cost-sensitive classification* [4] [5] [6] models take into account misclassification costs. These models are useful when the cost of a false positive is not the same for every class.
- *Subgroup discovery* [9]: In this case, there is only a class that is important for the expert. The aim of subgroup discovery is finding the most interesting subgroups of examples according to statistical criteria.

In this paper, we focus on a class attribute feature that is ignored by traditional classification models: the relative importance of each class.

2 The Weighted Classification Problem

Each class in a classification problem may have a different degree of importance. In some situations, the user might be interested in achieving the maximum possible accuracy for specific classes while keeping the classification model complexity

---

[1] ART in this paper stands for Association Rule Tree, not to be confused with Adaptive Resonance Theory - a paradigm commonly associated with Artificial Neural Networks.

M.S. Szczuka et al. (Eds.): ICHIT 2006, LNAI 4413, pp. 1–10, 2007.

to a minimum, even at the cost of lower accuracy for less important classes. She might also desire a minimum model complexity while preserving a reasonable accuracy level, even for the most important classes. Moreover, different users could attach different importance degrees to each class depending on their personal goals even for the same problem.

Therefore, we need classification inducers that take class importance into account when building classification models. In order to represent the relative importance of each class, we can resort to *relative weights* w_i for each class. For the sake of simplicity, we will assume that the weights w_i are floating-point values between 0 and 1. These values can always be normalized.

In this work, we focus on getting classification models as simple as possible for the important classes without penalizing classification accuracy. If we were only interested in classification accuracy, existing models could have been used. In particular, we could have used a cost-sensitive model [4] by defining a cost matrix, which would have reflected the relative importance of each class. However, classifier complexity is also a fundamental issue in supervised learning, since complexity is closely related to interpretability. A classifier might be useless from a practical point of view if it reaches a good accuracy level but is too complex to be understood by the decision makers who need a rationale behind their decisions.

In particular, *weighted classification models* (that is, classification models built by taking class weights into account) can be useful in situations such as the ones described by the following examples from the UCI Machine Learning repository [10]:

- **Extreme classes problems:** In some problems, experts could be specially interested in properly classifying 'extreme' classes, i.e. classes whose importance is paramount in the decision making process. For example, when dealing with the CAR data set, we could be interested in getting clear rules for good cars in order to recommend them (and for bad ones in order to avoid them).

  It should be noted that cost-sensitive classification could have also been used in this problem, since the expert might be interested in not mistaking a bad car for a good one, regardless of the required classifier complexity.
- **Two-class problems:** In binary classification problems, it is relatively common for the proper description of one class to be much more important than the other's for providing the rationale behind a given decision. For example, in the ADULT data set, where the class attribute is personal income, with values >50K and <=50K, a tax inspector might be more interested in people who earn more money in order to perform a financial investigation.
- **Classes and ontologies:** When the classes in a classification problem can be organized somehow, we can also resort to importance degrees in order to focus on related classes that might be specially relevant for the user. For example, a use hierarchy can be defined for the 6-class GLASS data set: three kinds of glass are used to make windows (one for vehicle windows, two for building windows), while the other three have other applications (containers,

tableware, and headlamps). If we were interested in identifying glass from a broken window, we could assign high importance degrees to all the kinds of glass used to make windows.

After the definition of weighted classification and the study of potential application areas, we face the problem of evaluating weighted classification models. In this paper, we propose two metrics that take into account class importance. They might be helpful when evaluating the accuracy and complexity of weighted classifiers:

- Classifier accuracy is the main goal of any classification system. In weighted classification problems, we recommend the use of the following **weighted accuracy** measure:

$$wAcc = \sum_{i=1}^{\#classes} w_i \cdot acc(i) \tag{1}$$

where $acc(i)$ is the average accuracy for the i-th class, w_i is the weight for the i-th class, and $\#classes$ is the number of classes.

- We also propose an analogous **weighted complexity** measure for evaluating classifier complexity, which is closely related to its understandability and interpretability:

$$wOpacity = \sum_{i=1}^{\#classes} w_i \cdot opacity(i) \tag{2}$$

where $opacity(i)$ is the value of the complexity measure for the i-th class. For instance, $opacity(i)$ might represent the average depth for nodes belonging to the i-th class in a decision tree. In this case, the complexity measure for i-th class can be defined as follows:

$$opacity(i) = depth(i) = \frac{\sum_{x \in class(i)} level(x)}{freq(i)} \tag{3}$$

where $level(x)$ is the depth of the leave corresponding to the example x, $freq(i)$ is the number of examples belonging to i-th class, and $class(i)$ is the set of examples belonging to the i-th class.

A weighted classifier should be evaluated according to these measures. An optimal classifier would optimize all of them at the same time, although this multi-objective optimization is not always possible, so we will usually have to achieve a trade-off between accuracy and complexity.

3 Adapting ART for Weighted Classification

In this paper, we show how standard classification models can be adapted for dealing with class weights. In particular, we focus on the ART classification model [1]. ART, which stands for Association Rule Tree, is a Separate and Conquer algorithm that is suitable for Data Mining applications because it makes use of efficient association rule mining techniques.

The special kind of decision list ART obtains can be considered as a degenerate, polythetic decision tree. The ART algorithm outline is shown in Figure 2. Unlike traditional TDIDT algorithms, ART branches the decision tree by simultaneously using several attributes.

Internally, ART makes use of association rules in order to find good descriptions of class values. When evaluating candidate rules, the classical confidence measure used in association rule mining is employed to rank the discovered rules, even though alternative criteria might be used [2].

Once ART discovers potentially useful classification rules, they are grouped according to the attributes in their antecedents, as shown in Figure 1. A rule selection mechanism is also necessary for choosing one of the resulting rule groups. The chosen group is used to branch the decision tree and the whole process is repeated for the remaining examples.

$$\{A_3\&B_2 \rightarrow C_1,\ B_4\&C_2 \rightarrow C_2,\ A_4\&B_1 \rightarrow C_2,\ B_0\&C_1 \rightarrow C_1,\ A_2\&C_2 \rightarrow C_1\}$$
$$\Downarrow$$
$$\{A_3\&B_2 \rightarrow C_1,\ A_4\&B_1 \rightarrow C_2\}$$
$$\{B_4\&C_2 \rightarrow C_2,\ B_0\&C_1 \rightarrow C_1\}$$
$$\{A_2\&C_2 \rightarrow C_1\}$$

Fig. 1. Grouping rules with compatible antecedents

In the following sections, we propose some modifications to the rule evaluation and rule selection criteria used by the ART algorithm in order to deal with weighted classification problems.

3.1   Weighted Selection Criterion

As we have mentioned before, a selection criterion is needed when alternative sets of rules are considered good enough to branch the tree. In ART, the criterion is based on the support of the rules belonging to the group. The best set of rules is the set that covers the maximum number of examples. However, this approach does not take the weights of the rules into account. We propose a modified criterion, which we call weighted coverage:

$$weightedCoverage(RuleSet) = \sum_{r \in RuleSet} support(r) \cdot w(r)$$

where $w(r)$ is the weight of the class in the consequent of the rule r and $support(r)$ is the number of examples supporting the rule r. In some sense, this is similar to the idea used in boosting algorithms such as AdaBoost [3].

By using weighted coverage, the set of rules that cover a larger number of more important classes is preferred over other sets. Since such a set will be selected as soon as possible, its level in the tree will tend to be lower and, therefore, the classifier opacity is expected to be reduced.

function ART **(data, MaxSize, MinSupp, MinEval): classifier;**
// **data**:   *Training dataset*
// **MaxSize**:   *Maximum LHS itemset size*
//   *(default value = 3)*
// **MinSupp**:   *Minimum support threshold*
//   *(default value = 0.05 = 5%)*
// **MinEval**:   *Minimum desirable rule evaluation value*
//   *(1.0 confidence threshold by default)*

$k = 1$;   // *LHS itemset size*
list = **null**;   // *Resulting decision list (degenerate tree)*

while ((list **is null**) **and** ($k \leq$ MaxSize))

    // *Rule mining*
    Find all the confident rules from input data with
    k items in the *LHS* and the class attribute in the *RHS*
    taking the rule evaluation tolerance into account
        e.g. $\{A_1.a_1 \ .. \ A_k.a_k\} \Rightarrow \{C.c_j\}$

    if there are candidate rules to grow the list

        // *Rule selection*
        Select the best set of rules with the same set of attributes
        $\{A_1..A_k\}$ in the *LHS* according to the selection criterion.

        // *Tree branching*
        list = List resulting from the selected rules
            $\{A_1.a_1 \ .. \ A_k.a_k\} \Rightarrow \{C.c_j\}$, where
            all training examples not covered by the selected
            association rules are grouped into an 'else' branch
            which is built calling the algorithm recursively:
            data = uncovered data // *Transaction trimming*
            MinEval = $max_{r \in list}$ evaluate(r)
            list.else = ART (data, MaxSize, MinSupp, MinEval);

    else
        $k = k + 1$;

if list **is null** // *no decision list has been built*
   list = default rule labelled with the most frequent class;

return list;

Fig. 2. ART Algorithm Outline

3.2   Weighted Rule Evaluation Criterion

ART uses confidence as rule evaluation criterion by default. A rule is a good or
valid rule when its confidence value is above a threshold that is determined by

the minimum desirable confidence value and a tolerance margin. To be precise, a rule r is valid when:

$$evaluate(r) \geq MinEval - Tolerance$$

where $evaluate(r)$ represents the rule quality according to the chosen evaluation criterion.

The key idea behind the adaptation of ART rule evaluation criterion to weighted classification is adjusting the tolerance margin according to the class of the rule we evaluate. Here we discuss three heuristics we have devised while trying to solve this problem. The experimental results we have obtained with them can be found in Section 4.

Tolerance Reduction (TR). If we only accept rules for the less important classes when their accuracy is very high, rules for the important classes will be more likely to be selected. This idea is expressed in the next formula:

$$tol_{TR} = tol \cdot \left(1 - \frac{MaxWeight - w}{MaxWeight}\right)$$

where w is the weight for the class of the rule being evaluated and *MaxWeight* is the maximum weight of the classes in the classification problem at hand. In the expression above, *tol* represents the original value for the tolerance in ART (0.1 by default) and tol_{TR} is the adjusted tolerance.

Using this heuristic criterion, when w equals the maximum weight, $tol_{TR} = tol$. That is, the original tolerance margin is preserved for the most important class. However, the tolerance margin will be reduced for any $w < MaxWeight$.

Relative Weight Premium (RWP). In this case, the tolerance margin will be increased according to the importance of the class. If the class is not important, the tolerance margin will be similar to the one used in ART. When the class is important, the tolerance margin will be higher, thus ensuring that more rules corresponding to important classes will be considered during the classifier construction.

$$tol_{RWP} = tol \cdot \left(1 + \frac{w - MinWeight}{MaxWeight - MinWeight}\right)$$

When w equals to the minimum weight, $tol_{RWP} = tol$. If w corresponds to the maximum weight, the fraction becomes 1 and $tol_{RWP} = 2 \cdot tol$. For any value between MinWeight and MaxWeight, the tolerance will be proportionally increased.

Weight Premium (WP). RWP increases the tolerance margin for important classes. However, it drives tolerance to the same values $[tol, 2 \cdot tol]$ regardless of the particular weights chosen by the user.

For instance, if there is a 2-class problem, the tolerance margin becomes $2 \cdot tol$ for the most important class and *tol* for the less important class using RWP. These margins will be the same no matter if our weights are $(0.9, 0.1)$ or $(0.6, 0.4)$.

We can easily modify the RWP criterion if we normalize with respect to the maximum weight in absolute terms:

$$tol_{WP} = tol \cdot \left(1 + \frac{w - MinWeight}{MaxWeight} \right)$$

Using this heuristics, the tolerance margin in the (0.9, 0.1) case becomes $(1.88 \cdot tol, tol)$ whereas in the (0.6, 0.4) case it becomes $(1.33 \cdot tol, tol)$.

4   Experimental Results

In this section, we evaluate the performance of our modified ART algorithm by performing standard ten-fold cross-validation experiments on several data sets available from the UCI Machine Learning repository [10]. The data sets we have used in our experimentation are summarized in Table 1.

Table 1. Data sets used to evaluate ART in weighted classification

Dataset	Size	Attributes	Classes
ADULT	48842	15	2
AUSTRALIAN	690	15	2
BREAST	699	9	2
CAR	1728	7	4
CHESS	3196	36	2
GLASS	214	9	6
HAYESROTH	160	5	3
HEART	270	14	2
IRIS	150	5	3
MUSHROOM	8124	23	2
NURSERY	12960	9	5
PIMA	768	9	2
SPLICE	3175	61	3
TICTACTOE	958	10	2
TITANIC	2201	4	2
VOTES	435	17	2
WAVEFORM	5000	22	3
WINE	178	14	3

We have used two kinds of weight distributions to determine the effect of a particular set of weights on classifier accuracy and complexity:

− First, we have performed experiments using extreme values for the class weights. We have selected a single class as the truly important class and we have made the others unimportant. We have assigned a 0.1 weight for all the unimportant classes and we have set the weight of the important class so that the sum of the weights equals 1. For instance, we have tested a weight distribution of $(0.9, 0.1)$ for 2-class problems, $(0.8, 0.1, 0.1)$ for 3-class problems, and so on.

– We have also tested a different kind of class weight distribution. Again, we have considered one class to be more important than the others but, in this case, the difference between the weight of the important class and the weight of the unimportant classes is not so extreme. Table 2 shows the particular weight distributions we have used.

For each data set in Table 1, we have performed $2 \cdot n_{ds}$ experiments, where n_{ds} is the number of classes in the data set. In each pair of experiments, we have selected a particular class as the important class and we have performed two cross-validation experiments, using extreme and non-extreme class weights.

Table 3 shows the experimental results we have obtained using the extreme weight distributions.

Each row in the table summarized the results obtained from a particular combination of heuristics. Each column in the table shows the overall average of each one of the measures we have used to evaluate the classifiers, as well as the number of times a particular heuristic combination matches or improves the standard ART algorithm (out of the 50 individual experiments performed for each heuristic combination).

The $wAcc$ and $wOpac$ measures correspond to the weighted accuracy and complexity metrics introduced in equations 1 and 2. The Acc_mic measure stands for the classifier accuracy with respect to the most important class, something that could be specially relevant to check the bias weights introduce in the learning algorithm. In a similar way, $Opac\_mic$ represents the average classifier opacity for the most important class (i.e. the average depth of the nodes corresponding to the important class in the resulting decision trees). Finally, Acc represents the

Table 2. Non-extreme values for the weights

Number of classes	Weights
2	$(0.7, 0.3)$
3	$(0.6, 0.2, 0.2)$
4	$(0.4, 0.2, 0.2, 0.2)$
5	$(0.5, 0.125, 0.125, 0.125, 0.125)$
6	$(0.4, 0.12, 0.12, 0.12, 0.12, 0.12)$

Table 3. Experimental results obtained using extreme values for the class weights

	wOpac	wAcc	Acc_mic	Opac_mic	Acc	Opac
Standard ART	5.67	78.67	78.67	5.67	**84.16**	5.16
Confidence + WCoverage	5.72 (37)	78.60 (42)	78.63 (43)	5.60 (38)	84.14 (45)	5.45 (22)
TR + Coverage	6.29 (32)	**81.08 (35)**	**81.87 (36)**	5.88 (39)	83.52 (33)	6.45 (25)
TR + WCoverage	6.42 (31)	80.26 (32)	80.84 (36)	5.78 (42)	83.16 (31)	7.20 (24)
RWP + Coverage	**4.54 (48)**	78.74 (37)	78.40 (36)	**4.32 (49)**	83.54 (37)	**4.44 (47)**
RWP + WCoverage	4.74 (46)	78.65 (36)	78.46 (37)	4.42 (48)	83.26 (34)	4.95 (41)
WP + Coverage	4.70 (48)	79.29 (40)	79.52 (39)	4.49 (49)	83.72 (38)	4.56 (47)
WP + WCoverage	4.86 (47)	79.07 (37)	79.45 (37)	4.57 (48)	83.56 (36)	5.01 (40)

Table 4. Experimental results obtained using non-extreme values for the class weights

	wOpac	wAcc	Acc_mic	Opac_mic	Acc	Opac
Standard ART	5.67	78.67	78.67	5.67	84.16	5.16
Confidence + WCoverage	5.81 (36)	78.65 (45)	78.66 (45)	5.62 (42)	**84.18 (47)**	5.41 (28)
TR + Coverage	5.13 (39)	**79.98 (33)**	**81.73 (43)**	4.77 (43)	83.61 (34)	4.84 (37)
TR + WCoverage	5.50 (37)	79.70 (33)	81.38 (42)	4.69 (44)	83.50 (32)	5.44 (33)
RWP + Coverage	**4.70 (46)**	78.18 (35)	78.40 (37)	**4.32 (49)**	83.65 (38)	**4.60 (46)**
RWP + WCoverage	4.96 (45)	78.42 (36)	78.79 (39)	4.37 (48)	83.73 (38)	5.01 (40)
WP + Coverage	5.14 (45)	78.90 (43)	79.11 (42)	4.81 (48)	84.01 (42)	4.80 (46)
WP + WCoverage	5.22 (44)	78.88 (41)	79.20 (41)	4.74 (49)	83.99 (43)	4.97 (39)

standard cross-validation classifier accuracy, while *Opac* is the average classifier opacity (i.e. the average decision tree depth), both without taking class weights into account.

The experiments show that the best results with respect to classifier complexity (about 20% improvement) are obtained by the *RWP* heuristics when combined with the standard coverage rule selection criterion.

It is important to emphasize that we achieve a 20% reduction in complexity without significantly penalizing the classical measures (*Acc* and *Opac*) nor their weighted counterparts (*wAcc* and *wOpac*).

As mentioned above, we have also performed some experiments using non-extreme weights distribution. Table 4 summarizes the results we have obtained. Once again, the classifier opacity (wOpac) is reduced by most of the heuristics we had devised. In addition, accuracy does not get worse and even improves in some cases.

Regarding to the heuristics we have developed to adapt ART to weighted classification, TR plus coverage tends to get the best results for accuracy measures, with minor improvements in classifier complexity. On the other hand, RWP plus coverage behaves better with respect to complexity measures without penalizing classifier accuracy, one of the main goals behind our work.

5   Conclusions and Future Work

Balancing complexity and accuracy in classification models is a difficult trade-off. We have devised a decision-list-oriented algorithm that finds shorter descriptions for the most important classes (as defined by a class weight assignment), and we have done so without penalizing classifier accuracy.

Several measures have been proposed to evaluate the behavior of supervised learning techniques in the context of weighted classification. In the experiments, we show that our method behaves well with respect to these novel measures, as well as with respect to the traditional measures of accuracy and complexity.

We expect to see a growing interest in complexity-oriented classification models in the near future. In particular, we intend to extend the work we have presented in this paper to general decision tree classifiers by introducing the appropriate complexity-oriented heuristics in the decision tree building process.

References

[1] Berzal, F., Cubero, J.C., Sánchez, D., Serrano, J.M.: ART: A hybrid classification model. Machine Learning 54(1), 67–92 (2004)

[2] Berzal, F., Cubero, J.C., Marín, N., Polo, J.L.: An overview of alternative rule evaluation criteria and their use in separate-and-conquer classifiers. In: Esposito, F., Raś, Z.W., Malerba, D., Semeraro, G. (eds.) ISMIS 2006. LNCS (LNAI), vol. 4203, Springer, Heidelberg (2006)

[3] Freund, Y., Schapire, R.E.: A decision-theoretic generalization of on-line learning and an application to boosting. Journal of Computer and System Sciences 55(1), 119–139 (1997)

[4] Elkan, C.: The foundations of Cost-Sensitive Learning. In: Proceedings of the Seventeenth International Joint Conference on Artificial Intelligence, pp. 973–978 (2001)

[5] Drummond, C., Holte, R.: Exploiting the Cost (In)sensitivity of Decision Tree Splitting Criteria. In: Proceedings of the Seventeenth International Conference on Machine Learning, pp. 239–246 (2000)

[6] Bradford, J., Kunz, C., Kohavi, R., Brunk, C., Brodley, C.: Pruning decision trees with misclassification costs. In: Proceedings of the European Conference on Machine Learning, pp. 131–136 (1998)

[7] Batista, G.E.A.P.A., Prati, R.C., Monard, M.C.: A Study of the Behavior of Several Methods for Balancing Machine Learning Training Data. Sigkdd Explorations 6(1), 20–29 (2004)

[8] Japkowicz, N., Stephen, S.: The Class Imbalance Problem: A Systematic Study. Intelligent Data Analysis Journal 6(5), 429–449 (2002)

[9] Lavrac, N., Kavsek, B., Flach, P., Todorovski, L.: Subgruop Discovery with CN2-SD. Journal of Machine Learning Research 5, 153–188 (2004)

[10] Blake, C., Merz, C.J.: UCI Repository of machine learning databases (1998), Available: http://www.ics.uci.edu/~mlearn/MLRepository.html

Tolerance Based Templates for Information Systems: Foundations and Perspectives

Piotr Synak[1,2] and Dominik Ślęzak[2]

[1] Polish-Japanese Institute of Information Technology
Koszykowa 86, 02-008 Warsaw, Poland
[2] Infobright Inc.
218 Adelaide St. W, Toronto, ON, M5H 1W8 Canada

Abstract. We discuss generalizations of the basic notion of a template defined over information systems using indiscernibility relation. Generalizations refer to the practical need of operating with more compound descriptors, over both symbolic and numeric attributes, as well as to a more entire extension from equivalence to tolerance relations between objects. We briefly show that the heuristic algorithms known from literature to search for templates in their classical indiscernibility-based form, can be easily adapted to the case of tolerance relations.

Keywords: Information Systems, Templates, Tolerance Relations.

1 Introduction

Information systems, studied in particular within the framework of the theory of rough sets [13,6], provide an efficient means for data representation and analysis. Basing on attribute-based indiscernibility relation between objects (records, rows), we can express various types of *patterns* satisfied, partially satisfied, or simply enough frequently occurring, within the real-world data tables.

Among very diversified definitions of patterns, *templates* are ones of the most basic constructions, based on conjunctions of single-attribute *descriptors* of various levels of complexity and generalization [9,10], widely applied in data mining [3,5,4]. To some extent, one can find analogies between templates and frequent itemsets, which are thoroughly studied in the literature [1,2]. Thus, potential area of application is very broad (see e.g. [11]).

In this particular paper, we discuss – besides providing a number of illustrative examples of templates and their generalizations – the need of extension of already known algorithmic framework for the template extraction onto the case of information systems with tolerance (similarity) relation instead of standard indiscernibility. Although tolerances have been already widely studied for information systems [7,16,22], we are the first to consider them in the context of efficient representation and search of generalized templates. In this way, we make an important step towards a wider application of templates in real-world data mining and knowledge discovery problems. It is worth to mention, that

M.S. Szczuka et al. (Eds.): ICHIT 2006, LNAI 4413, pp. 11–19, 2007.
© Springer-Verlag Berlin Heidelberg 2007

mining of complex patterns from data has been recognized as one of the ten most challenging problems in data mining [21]. The meaning of term "complex pattern" includes graph-based and structured patterns. We claim, that tolerance based templates are a milestone to templates based on graph-based descriptors.

The paper is an extension of [19]. It is organized as follows: In Section 2 we recall basic notions related to information systems and indiscernibility relations. In Section 3, we introduce the notion of a template for the most standard case of information systems. In Section 4, we generalize templates by building them from more sophisticated descriptors, definable for both symbolic and numeric attributes. In Section 5 we present temporal templates – patterns defined for information systems, where objects are linearly ordered. In Section 6, we discuss examples of tolerance relations in information systems with weakened requirements for indiscernibility relations. In Section 7, we consider the corresponding tolerance based templates. In Section 8, our main contribution in this paper, we explain how the most popular algorithms for efficient template generation can be extended towards extraction of tolerance based templates from data. Finally, in Section 9, we conclude the paper.

2   Preliminaries

In the paper, we use the notation of the theory of rough sets [13,6]. In particular, by $\mathbb{A} = (U, A)$ we denote an *information system* [12,14] with the universe U of *objects* and the attribute set A. Each *attribute* $a \in A$ is a function $a : U \to V_a$, where V_a is the *value set* of a. For a given set of attributes $B \subseteq A$, we define the *indiscernibility relation* $IND(B)$ on the universe U that partitions U into classes of indiscernible objects. We say that objects x and y are *indiscernible* with respect to B if and only if $a(x) = a(y)$ for each $a \in B$.

The values of attributes for a given object $x \in U$ form an elementary pattern generated by $\mathbb{A}$, where an *elementary pattern* (or *information signature*) $Inf_B(x)$ is a set $\{(a, a(x)) : a \in B\}$ of attribute-value pairs over $B \subseteq A$ consistent with a given object x. By

$$INF(A) = \{Inf_B(x) : x \in U, B \subseteq A\} \tag{1}$$

we denote the set of all signatures generated by A.

3   Templates in Information Systems

One of the main tasks of data mining process is searching for patterns in data [3,5,4]. There are several kinds of patterns considered in the literature. In the paper we consider one kind of such patterns, which we call templates (see, e.g., [9,10] and compare with [2]), that are defined by descriptors over the space of attributes of some information system.

Let $\mathbb{A} = (U, A)$ be an information system and $card(A) = m$. A *template* T of $\mathbb{A}$ is any propositional formula $\bigwedge(a_i = v_i)$, where $a_i \in A$, $a_i \neq a_j$ for $i \neq j$, and $v_i \in V_{a_i}$ [9,10]. Assuming $A = \{a_1, \ldots, a_m\}$ one can represent any template

$$T = (a_{i_1} = v_{i_1}) \wedge \ldots \wedge (a_{i_k} = v_{i_k}) \tag{2}$$

by the sequence $[u_1, \ldots, u_m]$ where on position p is either v_p if $p = i_1 \ldots i_k$ or "*" (don't care symbol) otherwise. We say that an object $x \in U$ *satisfies* the descriptor $(a = v)$ if $a(x) = v$. An object x satisfies (matches) the template T if it satisfies all the descriptors of T (i.e., if $x \in \|T\|_{\mathbb{A}}$). For any template T by $length(T)$ we denote the number of different descriptors $(a = v)$ occurring in T and by $supp_{\mathbb{A}}(T)$ we denote its *support*, i.e., the number of objects from the universe U satisfying T. If T consists of one descriptor $(a = v)$ only we also write $n_{\mathbb{A}}(a, v)$ (or $n(a, v)$) instead of $supp_{\mathbb{A}}(T)$. By the *quality* of template T we often understand the number $supp_{\mathbb{A}}(T) \times length(T)$. If s is an integer then by $Template_{\mathbb{A}}(s)$ we denote the set of all templates with support not less than s.

4   Generalized Templates

The idea of templates can be extended to so called *generalized templates*, i.e., templates of the form

$$T = \{(a \in V) : a \in B \subseteq A, V \subseteq V_a\}, \tag{3}$$

such that if $(a \in V) \in T$ and $(b \in W) \in T$ then $a \neq b$. The main difference is that instead of one-value we have many-valued descriptors. We say that an object x *satisfies* a generalized descriptor $(a \in V)$ iff $a(x) \in V$. An object x satisfies a generalized template T if it satisfies all the descriptors of T.

In the case of generalized templates, the definitions of length and support remain the same, but are related to the generalized descriptors. However, the quality function should take into account precision of particular descriptors, where intuitively by precise descriptor $(a \in V)$ we understand such that the cardinality of V is relatively low. Formally, the *precision* of descriptor $(a \in V)$ in $\mathbb{A}$ we define as

$$s_{\mathbb{A}}((a \in V)) = \begin{cases} \frac{card(V_a) - card(\{a(x):x \in U \wedge a(x) \in V\})}{card(V_a) - 1} & card(V_a) \neq 1 \\ 1 & card(V_a) = 1 \end{cases} \tag{4}$$

From this definition we can see that precision is one when there is only one value of attribute a among all objects satisfying the descriptor. And, if those values cover the whole domain of attribute a then the precision is zero.

The *precision* of a generalized template T we define as

$$S_{\mathbb{A}}(T) = \sum_{(a \in V) \in T} s_{\mathbb{A}}((a \in V)), \tag{5}$$

One can see that if precision of all descriptors of T is one then $S_{\mathbb{A}}(T)$ is equal to the length of T. If all the descriptors are very general then $S_{\mathbb{A}}(T)$ is close to zero.

5   Temporal Templates

Another class of patterns, defined for so called temporal information systems, $\mathbb{A} = (\{x_1, x_2, ..., x_n\}, A)$ ([18]), are temporal templates. They describe regularity in data occurring in a certain period of time. The basic assumption about data is that the objects are linearly ordered what gives the temporal interpretation.

Temporal template is a set of generalized descriptors involving any subset $B \subseteq A$, constrained in time:

$$\mathbf{T} = (T, t_s, t_e), \quad 1 \leq t_s \leq t_e \leq n, \tag{6}$$

where $T = \{(a \in V) : a \in B, V \subseteq V_a\}$ is a generalized template, and $[t_s, t_e]$ is a period of occurrence of T. By *width* of $\mathbf{T}$ we understand the length of period of occurrence, i.e., $width(\mathbf{T}) = t_e - t_s + 1$. *Support* is the number of objects from period $[t_s, t_e]$ matching all descriptors from T. Finally, *precision* of temporal template is defined as a sum of precisions of all descriptors from T. We consider quality of temporal template as a function of width, support and precision.

Two examples of temporal templates are presented in Figure 1. We have $\mathbf{T_1} = (T_1, 2, 8)$ and $\mathbf{T_2} = (T_2, 10, 13)$, where $T_1 = \{(a \in \{u\}), (c \in \{v\})\}$ and $T_2 = \{(b \in \{x\}), (d \in \{y\})\}$. We can see that $width(\mathbf{T_1}) = 7$, $width(\mathbf{T_2}) = 4$, $supp(\mathbf{T_1}) = 5$, $supp(\mathbf{T_2}) = 3$, $S_\mathbb{A}(T_1) = 1 + 1 = 2$, $S_\mathbb{A}(T_2) = 1 + 1 = 2$.

	A	a	b	c	d	e
	x_1	.	.	.	.	.
	x_2	u	.	v	.	.
	x_3	u	.	v	.	.
$\mathbf{T}_1$	x_4	.	.	.	.	.
	x_5	u	.	v	.	.
	x_6	u	.	v	.	.
	x_7	.	.	.	.	.
	x_8	u	.	v	.	.
	x_9	.	.	.	.	.
	x_{10}	.	x	.	y	.
$\mathbf{T}_2$	x_{11}	.	x	.	y	.
	x_{12}	.	.	.	y	.
	x_{13}	.	x	.	y	.
	x_{14}	.	x	.	.	.
	x_{15}	.	.	.	.	.

Fig. 1. Examples of temporal templates: $\mathbf{T_1} = (\{(a \in \{u\}), (c \in \{v\})\}, 2, 8)$ and $\mathbf{T_2} = (\{(b \in \{x\}), (d \in \{y\})\}, 10, 13)$

6 Tolerance Relations

The indiscernibility relation – the main driving tool of rough sets – in some cases can be insufficient to deal with data, for example, when we have real value attributes. In this case every object can potentially differ from the others. The equivalence classes of indiscernibility relation can be then one- or few-element sets and thus very specific. The standard rough set approach [13] can be extended by assuming any type of binary relation instead of the equivalence relation (see, e.g., [7,16,22]). In this paper, we consider reflexive and symmetric relations $\tau \subseteq U \times U$, usually called tolerance relations.

Any tolerance relation can extend the notion of indiscernibility of objects to their similarity. For a given information system $\mathbb{A} = (U, A)$, where $A = (a_1, \ldots, a_k)$, any object is characterized by its signature, i.e., a vector of attribute values. Similarity defined on attributes can be thus easily expressed as similarity between objects. Suppose we are given an attribute based tolerance $\tau_A \subseteq INF(A) \times INF(A)$. Then,

$$\forall_{x,y \in U} \{(x, y) \in \tau \Leftrightarrow (Inf_A(x), Inf_A(y)) \in \tau_A\}. \tag{7}$$

For any object $x \in U$ we can define a *tolerance class* with respect to τ:

$$\tau(x) = \{y \in U : (x, y) \in \tau\} \tag{8}$$

Let us emphasize that because tolerance relation is not transitive then one object may belong to two or more different tolerance classes.

There are several classes of tolerance relations considered in the literature (see, e.g., [17,16,10,8]). Any class is characterized by a first order formula and some parameters which are tuned up in the optimization process. One class of relations is based on some *similarity measures* defined on the attributes: $\delta_a : V_a \times V_a \rightarrow \mathbb{R}^+ \cup \{0\}$. Different measures can be used for different attributes. In the case of real attributes we can consider, for example, a distance between particular values. Symbolic attributes can be equipped with some preference order [17]. Sometimes, there is also a structure of attribute values given, making it possible to define a similarity measure.

Once the similarity measures are defined for each attribute there can be formulated several tolerance relations. Let us list here only a few examples [10]:

1. $(x, y) \in \tau_1(\varepsilon) \Leftrightarrow \max_{a_i \in A} \{\delta_{a_i}(x, y)\} \le \varepsilon$
2. $(x, y) \in \tau_2(\varepsilon_1, \ldots, \varepsilon_k) \Leftrightarrow \forall_{a_i \in A} [\delta_{a_i}(x, y) \le \varepsilon_i]$
3. $(x, y) \in \tau_3(w_1, \ldots, w_k, w) \Leftrightarrow \sum_{a_i \in A} w_i \cdot \delta_{a_i}(x, y) + w \le 0$
4. $(x, y) \in \tau_4(w) \Leftrightarrow \prod_{a_i \in A} \delta_{a_i}(x, y) \le w$

In the rest of the paper we are not going to construct such a global tolerance relations. We will rather consider tolerance to be defined for each attribute independently. In this case, if one wants to induce a global tolerance relation should it be defined in terms of satisfiability of all local relations. This can be done as follows:

$$(x, y) \in \tau \Leftrightarrow \forall_{a \in A}(a(x), a(y)) \in \tau_a. \tag{9}$$

7   Tolerance Based Templates

Let us assume that we are given an information system $\mathbb{A} = (U, A)$. Any template T for $\mathbb{A}$ determines some pattern in data by means of all the descriptors constituting T. More precisely, the pattern is determined by attributes forming descriptors and objects satisfying them. In the case when for each attribute $a \in A$ there is defined some local tolerance relation τ_a we can extend the semantics of a pattern by modifying definition of descriptor satisfaction by means of $(\tau_a)_{a \in A}$. We can assume that given descriptor is satisfied by an object if value from the descriptor is in the same tolerance class as the corresponding value of object. Thus, we say that an object $x \in U$ *satisfies* the descriptor $(a = v)$ *relatively to* τ_a if $(a(x), v) \in \tau_a$. An object x *satisfies* (matches) the template T if it satisfies all the descriptors of T relatively to the corresponding tolerance relations.

The same idea can be extended to generalized templates. A generalized descriptor $(a \in V)$ is satisfied by an object x relatively to tolerance relation τ if any value from descriptor is in the same tolerance class as the corresponding value of object, i.e., $\exists_{v \in V}(a(x), v) \in \tau_a$.

Let us emphasize that tolerance based templates may be extremely helpful when we have many different values in data, e.g., real attributes, and there are no straightforward patterns of values. Usually, in such cases some discretization methods were required to be applied before template generation process. Taking advantage from tolerance relations makes the solution more fitted to data and more accurate.

Tolerance relation can be also used to enhance expressiveness of temporal templates. In this case, we consider generalized templates which are based on tolerance, constrained in time. It means, that an object x_i supports temporal template $\mathbf{T} = (T, t_s, t_e)$ if $t_s \leq i \leq t_e$ and for each descriptor $(a \in V) \in T$ there exists $v \in V$ such that $(a(x_i), v) \in \tau_a$.

8   Tolerance Based Templates Generation

In this section we are going to show that most of the known methods of templates generation (see [9,10,8]) can be applied also to the case of tolerance based templates.

One of the methods is a greedy algorithm (see *Max I* and *Max II* method in [10]) iteratively adding optimal descriptors to initially empty template. The quality of a descriptor is measured by its support, i.e., the number of objects from the universe U satisfying it (see Section 3). In the tolerance based case the method remains the same but the support of a descriptor is measured by number of objects that satisfy this descriptor relatively to the corresponding tolerance. Thus, for any attribute $a \in A$ we search for a set of values V so the quality of descriptor $(a \in V)$ is maximal. In the case of attribute with some order defined on its domain we choose V from subsequences of a sorted list of all values. Once a descriptor is added to a template, all objects supporting it in the context of tolerance relation are removed from the universe used in the next iteration of the algorithm.

Another method is a heuristic based on random search of objects with respect to the attached weights (*Object weight* algorithm). Any found set of objects defines some template. The weights attached to objects are based on their average similarity to all the other objects (for details of similarity measure see [10]). In this method the similarity of objects is based on the equality of values. In the tolerance based case, the similarity can be based on the fact that the considered values belong to the same tolerance class. Most basic example of such a measure is as follows:

$$g(x, y) = \frac{card(\{a \in A : (a(x), a(y)) \in \tau_a\})}{card(A)}, \tag{10}$$

where $x, y \in U$.

In the *Attribute weight* method the weights are attached to the attributes as well as to their values and the random process tries to add iteratively descriptors to the initially empty set (or remove descriptors from the template generated from some selected object). In this method after some descriptor is chosen it is tested against improvement of template quality. In the tolerance based case the descriptor satisfied by the higher number of objects relatively to the corresponding tolerance relation will have higher chance to be chosen.

The genetic algorithm used for templates generation (see [20]) is based on finding an optimal permutation of attributes that for a selected base object x generates the best template. By combining single value descriptors obtained from pairs $(a, a(x))$ we incrementally construct a template, accordingly to the order of attributes defined by the permutation. By comparing template quality in each step we choose the optimal set of descriptors. One of the most important factors of template quality measure is its support. Thus, analogously to previous examples, this method can be easily adopted to the tolerance based case.

To generate temporal templates one can use a method based on time windows. The main idea is based on scanning the information system with time window of selected width. In each window there is generated the best generalized template (see [18]). There is a chance, that after a light shift of a time window, the previously found template is also the best one in a new window. By shifting the window we may discover the beginning (x_{t_s}) and the end (x_{t_e}) of the area where a given template is optimal or close to the optimal one. Shifting the time window through the whole information system generates a set of temporal templates.

9 Conclusions

We generalized the notion of a template onto the case of information systems equipped with tolerance, instead of standard indiscernibility equivalence relation. We provided examples illustrating the need of such generalization, referring to symbolic and numeric attributes, as well as to similarity functions defined both locally, for single attributes, and globally – for attribute sets. Then, we discussed possibilities of extension of the known algorithms extracting templates from data, which would enable to deal efficiently also with tolerance based patterns.

18   P. Synak and D. Ślęzak

Acknowledgements

The first author has been supported by the grants 8 T11C 025 19 and 3 T11C 007 28 from the Ministry of Scientific Research and Information Technology of the Republic of Poland and by the Research Center at the Polish-Japanese Institute of Information Technology, Warsaw, Poland.

References

1. Agrawal, R., Mannila, H., Srikant, R., Toivonen, H., Verkamo, A.I.: Fast discovery of association rules. In: Advances in Knowledge Discovery and Data Mining, pp. 307–328. AAAI Press/The MIT Press, Menlo Park, CA (1996)
2. Agrawal, R., Srikant, R.: Fast algorithms for mining association rules. In: Bocca, J.B., Jarke, M., Zaniolo, C. (eds.) Twentieth International Conference on Very Large Data Bases VLDB, pp. 487–499. Morgan Kaufmann, San Francisco (1994)
3. Fayyad, U.M., Piatetsky-Shapiro, G., Smyth, P., Uthurusamy, R.: Advances in Knowledge Discovery and Data Mining. The AAAI Press/The MIT Press, Cambridge, MA (1996)
4. Friedman, J.H., Hastie, T., Tibshirani, R.: The Elements of Statistical Learning: Data Mining, Inference, and Prediction. Springer, Heidelberg (2001)
5. Kloesgen, W., Żytkow, J.: Handbook of Knowledge Discovery and Data Mining. Oxford University Press, Oxford, UK (2002)
6. Komorowski, J., Polkowski, L., Skowron, A.: Rough sets: A tutorial. In: Pal, S.K., Skowron, A. (eds.) Rough Fuzzy Hybridization: A New Trend in Decision-Making, pp. 3–98. Springer, Singapore (1999)
7. Krawiec, K., Słowiński, R., Vanderpooten, D.: Learning decision rules from similarity based rough approximations. In: Polkowski, Skowron (eds.) [15], pp. 37–54
8. Nguyen, S.H.: Regularity Analysis and Its Applications in Data Mining. PhD thesis, Warsaw University, Warsaw, Poland (2000)
9. Nguyen, S.H., Skowron, A., Synak, P.: Rough sets in data mining: Approximate description of decision classes. In: Fourth European Congress on Intelligent Techniques and Soft Computing EUFIT, Aachen, Germany, September 2-5 1996, pp. 149–153. Verlag Mainz (1996)
10. Nguyen, S.H., Skowron, A., Synak, P.: Discovery of data patterns with applications to decomposition and classification problems (ch. 4). In: Polkowski, L., Skowron, A. (eds.) Rough Sets in Knowledge Discovery 2: Applications, Case Studies and Software Systems. Studies in Fuzziness and Soft Computing, ch. 4, vol. 19, pp. 55–97. Physica-Verlag, Heidelberg, Germany (1998)
11. Pasquier, N., Bastide, Y., Taouil, R., Lakhal, L.: Efficient mining of association rules using closed itemset lattices. J. Inf. Systems 24(1), 25–46 (1999)
12. Pawlak, Z.: Information systems - theoretical foundations. Information systems 6, 205–218 (1981)
13. Pawlak, Z.: Rough sets. International Journal of Computer and Information Sciences 11, 341–356 (1982)
14. Pawlak, Z.: Rough Sets: Theoretical Aspects of Reasoning about Data. D: System Theory, Knowledge Engineering and Problem Solving, vol. 9. Kluwer Academic Publishers, Dordrecht, The Netherlands (1991)
15. Polkowski, L., Skowron, A.: Rough Sets in Knowledge Discovery 2: Applications, Case Studies and Software Systems. Studies in Fuzziness and Soft Computing, vol. 19. Physica-Verlag, Heidelberg, Germany (1998)

16. Skowron, A., Stepaniuk, J.: Tolerance approximation spaces. Fundamenta Informaticae 27(2-3), 245–253 (1996)

17. Słowiński, R., Greco, S., Matarazzo, B.: Rough set analysis of preference-ordered data. In: Alpigini, J.J., Peters, J.F., Skowron, A., Zhong, N. (eds.) RSCTC 2002. LNCS (LNAI), vol. 2475, pp. 44–59. Springer, Heidelberg (2002)

18. Synak, P.: Temporal templates and analysis of time related data. In: Ziarko, W., Yao, Y. (eds.) RSCTC 2000. LNCS (LNAI), vol. 2005, pp. 420–427. Springer, Heidelberg (2001)

19. Synak, P., Ślęzak, D.: Tolerance based templates for information systems. In: Lee, G., Slezak, D., Kim, T.-h., Sloot, P. (eds.) International Conference on Hybrid Information Technology ICHIT, Cheju, Korea, November 9-11 2006, Science & Engineering Research Support Center (2006)

20. Wróblewski, J.: Genetic algorithms in decomposition and classification problem. In: Polkowski, Skowron. (ed.) [15], ch. 24, pp. 471–487

21. Yang, Q., Wu, X.: 10 challenging problems in data mining research. J. Inf. Technology & Decision Making 5(4), 597–604 (2006)

22. Yao, Y.Y., Wong, S.K.M., Lin, T.Y.: A review of rough set models. In: Lin, T.Y., Cercone, N. (eds.) Rough Sets and Data Mining. Analysis of Imprecise Data, pp. 47–75. Kluwer Academic Publishers, Boston, MA, USA (1997)

Reduction Based Symbolic Value Partition

Fan Min, Qihe Liu, Chunlan Fang, and Jianzhong Zhang

School of Computer Science and Engineering,
University of Electronic Science and Technology of China, Chengdu 610054, China
{minfan,qiheliu,fangcl,jianzhong}@uestc.edu.cn

Abstract. Theory of Rough Sets provides good foundations for the attribute reduction processes in data mining. For numeric attributes, it is enriched with appropriately designed discretization methods. However, not much has been done for symbolic attributes with large numbers of values. The paper presents a framework for the symbolic value partition problem, which is more general than the attribute reduction, and more complicated than the discretization problems. We demonstrate that such problem can be converted into a series of the attribute reduction phases. We propose an algorithm searching for a (sub)optimal attribute reduct coupled with attribute value domains partitions. Experimental results show that the algorithm can help in computing smaller rule sets with better coverage, comparing to the standard attribute reduction approaches.

Keywords: Symbolic Value Partition, Partition Reduct, Optimal Symbolic Value Partition Problem, Binarization.

1 Introduction

In the reduct problem and some other problems of Rough Sets [1], it is often assumed that the information systems only concern nominal attributes with appropriate domains. For continuous attributes or attributes with large domains, some preprocessing tools like *discretization* or *value grouping / partition* are needed to reduce attribute domains [2]. Moreover, both the reduct problem and the discretization problem can be viewed as special cases of the partition problem [3].

According to Nguyen H.S. [4] and Nguyen S.H. [2], the key issue of the partition problem is to find a mapping of the attribute domain to a new domain with minimal cardinality, which preserve semantics of decision table. This problem is referred to as the optimal symbolic value partition (OSVP) problem. A Decision Tree approach and a Rough Sets approach are proposed in [2]. The first approach partitions each attribute value set into two disjoint subsets in a top-down manner until some terminating condition holds. The second approach converts the partition problem into a reduct problem and a graph coloring problem. Although the second approach is quite interesting, it has two major drawbacks: first, the space complexity of the direct implementation is very high, second, the result might not be a partition reduct (see [3] for a counterexample).

M.S. Szczuka et al. (Eds.): ICHIT 2006, LNAI 4413, pp. 20–30, 2007.

In this paper, a framework for the symbolic value partition problem is proposed. The partition problem is converted into a series of the attribute reduction phases. Theoretical contributions are: (1) all possible outputs of the algorithm form the set of all partition reducts; (2) locally optimal solutions result in globally suboptimal solutions. Experiments are employed to show the advantages of our approach over the attribute reduction approach.

2   Preliminaries

In this section, we first enumerate some concepts of the partition problem introduced by Nguyen H.S. [4] with nonessential revisions, and propose the concept of partition reduct. Then we list the concept of M-relative reduct [5] that is useful in the algorithm. Finally we introduce the concept of decision table binarization. Rough Sets basic concepts such as positive region and attribute reduction should refer to [1].

2.1   The Partition Problem

Let $S = (U, C, \{d\})$ be a decision table where $C = \{a_i : U \rightarrow V_{a_i}\}$ for $i \in \{1, \ldots, |C|\}$. Any function $P_i = P_{a_i} : V_{a_i} \rightarrow W_{a_i} \cup V_{a_i}$ where $W_{a_i} \cap V_{a_i} = \emptyset$, $P_i(v) \in W_{a_i}$ or $P_i(v) = v$ is called a partition of V_{a_i}. The function P_i defines a new partition attribute $a_i^{P_i} = P_i \circ a_i$, i.e., $a_i^{P_i}(u) = P_i(a_i(u))$ for any object $u \in U$. The domain of $a_i^{P_i}$ is $V_{a_i}^{P_i} = \bigcup_{u \in U} \{a_i^{P_i}(u)\}$. Often P_i is also expressed by a set of value pairs, i.e., $P_i(v_1) = v_2 \Leftrightarrow (v_1, v_2) \in P_i$.

For example, for the decision table listed in Table 1, let $a_1 =$ Occupation, then $P_1 = \{(s, 1), (d, 2), (n, 2), (t, t), (1, 1)\}$ where s, d, $\ldots$, l stand for student, doctor, $\ldots$, low respectively is a partition of V_{a_1} where $W_{a_1} = \{1, 2\}$ and $V_{a_1}^{P_1} = \{1, 2, t\}$. That is, we do not distinguish between student and lawyer, or doctor and nurse, as will be shown in Table 8.

Any array of partition $P = [P_1, \ldots, P_{|C|}]$ is called a partition scheme of S. P defines from S a new decision table $S^P = (U, C^P, \{d\})$ where $C^P = \{a_1^{P_1}, \ldots, a_{|C|}^{P_{|C|}}\}$. Two partition schemes P', P are equivalent, i.e., $P' \equiv_S P$, iff $U/C^{P'_i} = U/C^{P_i}$ for $i \in \{1, \ldots, |C|\}$. The equivalence relation $\equiv_S$ has a finite number of equivalence classes. In the sequel we will not distinguish between equivalent families of partitions.

The rank of S is the value $\sum_{i=1}^{|C|} |V_{a_i}|$. P is consistent iff $POS_{C^P}(\{d\}) = POS_C(\{d\})$. S is unpartitionable iff there does not exist a consistent partition scheme P such that $rank(S^P) < rank(S)$. Similar with the definition of a reduct [6], we propose the following definition:

Definition 1. P is called a partition reduct of S iff P is consistent and S^P is unpartitionable.

The set of all partition reducts of S will be denoted by $PR(S)$. Any partition reduct P is optimal iff $rank(S^P)$ is minimal. The goal of the paper is to address the following problem which was proven NP-hard [2]:

PROBLEM: Optimal Symbolic Value Partition (OSVP)
Input: A decision table $S = (U, C, \{d\})$ where all attributes are symbolic.
Output: An optimal partition reduct P of S.

2.2   The M-Relative Reduct Problem

We have proposed the concept of M-relative reduct [5] to include a user specified attribute set M.

Definition 2. *Given a decision table* $S = (U, C, \{d\})$ *and a set of specified attributes* $M \subseteq C$, *any* $B \subseteq C$ *is called an* M-relative reduct of S iff:

1. $M \subseteq B$;
2. $POS_B(\{d\}) = POS_C(\{d\})$;
3. $\forall a \in (B - M), POS_{B-\{a\}}(\{d\}) \subset POS_C(\{d\})$.

Table 1. An exemplary decision table S

U	Occupation	Temperature	Cough	SARS
x_1	student	low	yes	suspicious
x_2	doctor	high	no	yes
x_3	nurse	high	yes	yes
x_4	nurse	normal	yes	yes
x_5	teacher	normal	no	suspicious
x_6	teacher	normal	yes	suspicious
x_7	lawyer	normal	yes	no
x_8	student	normal	no	no
x_9	student	high	no	no

Table 2. S_B, the binarized decision table of S

	(O, s)	(O, d)	(O, n)	(O, t)	(O, l)	(T, l)	(T, h)	(T, n)	(C, y)	(C, n)	d
x_1	1	0	0	0	0	1	0	0	0	1	suspicious
x_2	0	1	0	0	0	0	1	0	0	1	yes
x_3	0	0	1	0	0	0	1	0	1	0	yes
x_4	0	0	1	0	0	0	0	1	1	0	yes
x_5	0	0	0	1	0	0	0	1	0	1	suspicious
x_6	0	0	0	1	0	0	0	1	1	0	suspicious
x_7	0	0	0	0	1	0	0	1	1	0	no
x_8	1	0	0	0	0	0	0	1	0	1	no
x_9	1	0	0	0	0	0	1	0	0	1	no

2.3   Decision Table Binarization

Definition 3. *Given a decision table* $S = (U, C, \{d\})$, *the* binarized decision table *of S is*

$$S_B = (U, C_B, \{d\}) = (U, \{(a_i, v) | i \in \{1, \ldots, |C|\}, v \in V_{a_i}\}, \{d\}), \qquad (1)$$

where $(a_i, v) : U \rightarrow \{0, 1\}$ *and*

$$(a_i, v)(u) = \begin{cases} 1 & \text{if } a_i(u) = v; \\ 0 & \text{otherwise.} \end{cases} \qquad (2)$$

Tables 1 and 2 list a decision table and its binarized decision table, where (O, s), ..., (C, n) stands for (Occupation, student), ..., (Cough, no), respectively. Binarization is quite similar with scaling [7], and the only difference lies in that the former does not change the decision attribute.

3   The Reduction Based Symbolic Value Partition Algorithm

In this section we firstly analyze the example, then propose the optimal single group partition problem, and finally list and analyze the algorithm. Proofs of some theorems and properties will be omitted due to space limitation.

3.1   Example

We have obtained the binarized decision table as listed in Table 2. Next we come to the most important step of our approach, i.e., apply attribute reduction on S_B. According to Pawlak's definition [6], $R^1 = \{(O, d), (O, n), (O, t), (T, 1)\}$ is a reduct of S_B. Then we can convert $(U, R^1, \{d\})$ back to a "normal" decision table as listed in Table 3, where duplicated objects are removed.

In the new decision table, because (O, s) $\notin R^1$ and (O, 1) $\notin R^1$, we do not distinguish student from lawyer. From semantic point of view, we can replace student and lawyer with others, while here we used 1 instead. In fact, the new decision table can be constructed from S and a partition scheme $P^1 = [\{(s, 1),$

Table 3. S^{P^1}

U	O^{P^1}	T^{P^1}	C^{P^1}	d
x_1	1	low	1	suspicious
x_2	doctor	1	1	yes
x_3	nurse	1	1	yes
x_5	teacher	1	1	suspicious
x_7	1	1	1	no

Table 4. $S_B^{P^1}$

U	$(O^{P^1}, 1)$	(O^{P^1}, d)	(O^{P^1}, n)	(O^{P^1}, t)	$(T^{P^1}, 1)$	$(T^{P^1}, 1)$	$(C^{P^1}, 1)$	d
u_1	1	0	0	0	0	1	1	suspicious
u_2	0	1	0	0	1	0	1	yes
u_3	0	0	1	0	1	0	1	yes
u_5	0	0	0	1	1	0	1	suspicious
u_7	1	0	0	0	1	0	1	no

Table 5. S^{P^2}

U	O^{P^2}	T^{P^2}	C^{P^2}	d
u_1	1	2	1	suspicious
u_2	2	1	1	yes
u_5	teacher	1	1	suspicious
u_7	1	1	1	no

Table 6. $S_B^{P^2}$

U	$(O^{P^2}, 1)$	$(O^{P^2}, 2)$	(O^{P^2}, t)	$(T^{P^2}, 1)$	$(T^{P^2}, 2)$	$(C^{P^2}, 1)$	d
u_1	1	0	0	0	1	1	suspicious
u_2	0	1	0	1	0	1	yes
u_5	0	0	1	1	0	1	suspicious
u_7	1	0	0	1	0	1	no

Table 7. $S^P = S^{P^3}$

U	O^{P^3}	T^{P^3}	C^{P^3}	d
u_1	1	2	1	suspicious
u_2	2	1	1	yes
u_5	3	1	1	suspicious
u_7	1	1	1	no

Table 8. A more comprehensive version of S^{P^3}

U	O^{P^3}	T^{P^3}	C^{P^3}	d
u_1	{student, lawyer}	{low}	{yes, no}	suspicious
u_2	{doctor, nurse}	{normal, high}	{yes, no}	yes
u_5	{teacher}	{normal, high}	{yes, no}	suspicious
u_7	{student, lawyer}	{normal, high}	{yes, no}	no

(d, d), (n, n), (t, t), (1, 1)}, {(1, 1), (h, 1), (n, 1), {(y, 1), (n, 1)}] directly. Also note that $Rank(S^{P^1}) < Rank(S)$, indicating we have taken one step toward reducing the attribute domain.

Another key idea is to repeat this binarization, reduction and converting back process (where new values such as 2, 3, ... should be used) until the cardinality of any attribute cannot be reduced further. We obtain $S_B^{P^1}$ as listed in Table 4. $R^2 = \{(O, 1), (O, t), (T, 1)\}$ is a reduct of $S_B^{P^1}$ and we obtain S^{P^2} as listed in Table 5 where $P^2 = [\{(s, 1), (d, 2), (n, 2), (t, t), (1, 1)\}, \{(1, 2), (h, 1), (n, 1), \{(y, 1), (n, 1)\}]$. $S_B^{P^2}$ is listed in Table 6. $R^3 = \{(O, 1), (O, 2), (T, 1)\}$ is a reduct of $S_B^{P^1}$ and we obtain S^{P^3} as listed in Table 7. Since S^{P^3} is unpartitionable, the whole process terminates and the partition reduct is $P = P^3 = [\{(s, 1), (d, 2), (n, 2), (t, 3), (1, 1)\}, \{(1, 2), (h, 1), (n, 1)\}, \{(y, 1), (n, 1)\}]$. A more comprehensive version of S^{P^3} is listed in Table 8.

3.2   The Optimal Single Group Partition Problem

From the example we know that the whole process is essentially iterative, hence we shall focus on the first round of the process, i.e., the computation of P^1.

As listed in Table 3, any conditional attribute has exactly one new value, which corresponds to one or more initial values. For example, for Occupation, 1 corresponds to both **student** and **lawyer**; while for Cough, 1 corresponds to both **yes** and **no**. In other words, attribute values for each attribute form exactly one (single) new group. Hence we introduce here a special form of partition scheme.

Definition 4. *A partition scheme* $Q = [Q_1, \ldots, Q_{|C|}]$ *of S is called a* single group partition scheme *(SGPS) if for any* $i \in \{1, \ldots, |C|\}$, $|W_{a_i}| = 1$.

Given an SGPS $Q = [Q_1, \ldots, Q_{|C|}]$, for any $i \in \{1, \ldots, |C|\}$, Q_i essentially divides V_{a_i} into two disjoint subsets $V_{a_i}^F$ and $V_{a_i}^G$, and

$$Q_i(v) = \begin{cases} v & \text{if } v \in V_{a_i}^F; \\ k & \text{if } v \in V_{a_i}^G \end{cases} \tag{3}$$

where $k \notin V_{a_i}$. According to Definition 4, $V_{a_i}^G \neq \emptyset$. Hence any SGPS Q can be also represented by a set of attribute-value pairs, i.e., $Q = \{(a_i, v) | i \in \{1, \ldots, |C|\} \text{ and } v \in V_{a_i}^F\}$. With this form of SGPS we can define single group partition reducts as follows:

Definition 5. *Any SGPS* Q *is called a* single group partition reduct *(SGPR) of S iff* Q *is consistent and any* $Q' \subset Q$ *is not consistent.*

An SGPR Q is *optimal* iff $rank(S^Q)$ is minimal. We consider the optimal single group partition problem as listed in the next page:

Obviously, any SGPS $Q \subseteq C_B$ is as an attribute subset of S_B, and C^Q is the set of conditional attributes of S^Q. We have

$$Ind(Q) = Ind(C^Q). \tag{4}$$

And the following theorem is also obvious:

> **PROBLEM: Optimal Single Group Partition (OSGP)**
> **Input:** A decision table $S = (U, C, \{d\})$ where all attributes are symbolic.
> **Output:** An optimal SGPS Q of S.

Theorem 1. *Let $Red(S)$ and $SGR(S)$ denote the set of all relative reducts and the set of all single group partition reducts of decision table S, respectively,*

$$Red(S_B) = SGR(S). \tag{5}$$

This theorem indicates that the SGPS problem of S (constructing an SGPS scheme Q) is equivalent with the reduction problem of S_B (selecting an attribute subset Q from C_B). Moreover, from Equation (3) we know that

$$rank(S^Q) = \sum_{1 \leq i \leq |C|} (V_{a_i}^F + 1) = |Q| + |C|. \tag{6}$$

Finally, according to the definition of optimal metrics of reducts and SGPR, we have the following theorem:

Theorem 2. *The OSGP problem of S is equivalent with the optimal reduct (OR) problem of S_B.*

For example, since R^1 is an optimal reduct of S_B, according to Theorem 2, $Q^1 = R^1 = \{(O, d), (O, n), (O, t), (T, 1)\}$ is an optimal SGPS of S.

3.3   The Problem Conversion

Now we explain formally that the partition problem is converted into a series of reduct problem. Let $Q^1, Q^2, Q^3, \ldots, Q^K$ be SGPSs of $S^{P^0} = S, S^{P^1}, S^{P^2}, \ldots, S^{P^K}$, respectively where $P^i = Q^1 \circ Q^2 \circ \ldots \circ Q^i$ for $i = 1, 2, \ldots, K$. Obviously, P^K is a partition reduct of S if S^{P^K} is unpartitionable.

On the other hand, given a partition reduct P^K, it is straightforward for us to construct respective $Q^K, \ldots, Q^1$ in the reverse way. So we have the following theorem:

Theorem 3. *Let $P^K = Q^1 \circ Q^2 \circ \ldots \circ Q^K$, suppose that $S^{P^{K-1}}$ is partitionable and S^{P^K} is unpartitionable,*

$$\bigcup_{i \in \{1, \ldots, K\}, Q^i \in Red(S^{P^{i-1}})} \{P^K\} = PR(S). \tag{7}$$

3.4   The Algorithm

The Use of M-Relative Reduct. In the reduction step, we shall not just randomly choose a reduct for the partition purpose. In the example, R^1 is a reduct of S_B and we take it as an SGPR of S. It is easily seen that $Rank(S^{R^1}) < Rank(S)$. However, R^1 is also a reduct of $S_B^{P^1}$, if we take it again as an SGPR of S^{P^1}, $(S^{P^1})^{R^1}$ would be equivalent with S^{P^1}. In the worst case, the whole process would enter a dead loop if we always choose R^1 as the SGPR of the new decision table.

Hence we shall introduce the concept of M-relative reduct to control the computation of SGPRs and ensure quick converge of the algorithm. M should be deliberately set such that new attribute values introduced would never be replaced by others. In the example, since we have replaced **student** of Occupation with 1, we will never replace 1 again with any other new values. For this purpose we let $M^2 =\{(O, 1), (T, 1)\}$ for the computation of R^2 and $M^3 =\{(O, 1), (T, 1), (O, 2)\}$ for the computation of R^3.

Algorithm Description. The algorithm is listed in Fig. 1. We can compute $S_B^{P^i}$ without computing S^{P^i}, but for completeness the pseudo code is still listed.

$ReductionBasedSymbolicValuePartition$ $(S = (U, C, \{d\}))$
{**input:** A decision table S.}
{**output:** A partition reduct P.}
//Initialize. M^i is used for M-relative reduct.
Step 1. $M^1 = \emptyset$,
//The initial partition scheme P^0. In fact $S^{P^0} = S$.
Step 2. $P^0=[P_1^0, \ldots, P_{|C|}^0]$ where $P_i^0(v_i)=v_i$ for any $i \in \{1, \ldots, |C|\}$ and $v_i \in V_{a_i}$;
//Initialize unprocessed attribute-values pairs for each attribute.
//Now all attribute-values pairs are unprocessed.
Step 3. for $(i = 1; i \leq |C|; i + +)$ $H_i^0 = \{a_i\}_B$;
//Attack the OSVP-problem through attacking the OSGP-problem recursively.
Step 4. for $(i = 1; ; i + +)$ **begin**
//**Binarization.**
Step 4.1 compute $S_B^{P^{i-1}}$;
//**Reduction.**
Step 4.2 $R^i =$ an M-relative reduct of $S_B^{P^{i-1}}$ where $M = M^i$;
Step 4.3 $M^{i+1} = M^i$;//Initialize M^{i+1}.
Step 4.4 for $(j = 1; j \leq |C|; j + +)$ **begin**
//Compute P^i.
Step 4.4.1 $\forall (a_j, v) \notin H_j^{i-1} - R^i$, $P_j^i(v) = P_j^{i-1}(v)$;
Step 4.4.2 $\forall (a_j, v) \in H_j^{i-1} - R^i$, $P_j^i(v) = i$;
Step 4.4.3 $H_j^i = H_j^{i-1} \cap R^i$//Remove processed attribute-values pairs
//Compute M^{i+1}
Step 4.4.4 if $(H_j^i \neq \emptyset)$ $M^{i+1} = M^{i+1} \cup \{(a_j, i)\}$;
end//of for j;
//**Converting back to a "normal" decision table.**
Step 4.5 compute S^{P^i} where $P^i = [P_1^i, \ldots, P_{|C|}^i]$;
//See if all attribute-values pairs have been processed
Step 4.6 if $H^i = \bigcup_{j=1}^{|C|} H_j^i = \emptyset$ break; **end**;//of for i
Step 5. $P = P^i$, return P;

Fig. 1. The Reduction Based Symbolic Value Partition Algorithm

3.5   Algorithm Analysis

The introduction of M-relative reduct does not influence the essence of the whole process, and the set of all outputs of the algorithm is $PR(S)$, as indicated in Theorem 3.

Optimal Substructure. Since the goal of the OSVP-problem is to construct $P = P^K$ such that $rank(S^{P^K})$ is minimal, it is natural for us to choose locally optimal solutions, i.e., solutions of the OSGP-problem. Hence we need to require further in Step 4.2 that $R^i = $ an *optimal* M-relative reduct of $S_B^{P^{i-1}}$. Then the algorithm would be a *greedy* algorithm, and we need to explain why locally optimal solutions can result in globally suboptimal solutions. The following theorem gives partial reason.

Theorem 4. *The OSVP-problem has the optimal-substructure property.*

Proof. Let $P = [P_1, P_2, \ldots, P_{|C|}]$ be an optimal partition reduct of S, $P' = [P'_1, P'_2, \ldots, P'_{|C|}]$ where $P'_i = (P_i - \{(v,1)|(v,1) \in P_i\}) \cup \{(1,1)\}$ for $i \in \{1, \ldots, |C|\}$, we can see that $(S^{P^1})^{P'} = S^P$. We need to prove that P' is an optimal partition reduct of S^{P^1}.

Suppose that there is another partition reduct $P'' = [P''_1, P''_2, \ldots, P''_{|C|}]$ of S^{P^1} such that $rank((S^{P^1})^{P''}) < rank((S^{P^1})^{P'})$. We can then construct another partition scheme $P^x = [P^x_1, P^x_2, \ldots, P^x_{|C|}]$ where $P^x_i = (P^1_i - \{(v,v)|(v,v) \in P^1_i\}) \cup (P''_i - \{(1,1)\})$, and $S^{P^x} = (S^{P^1})^{P''}$. This in turn gives that

$$rank(S^{P^x}) = rank((S^{P^1})^{P''}) < rank((S^{P^1})^{P'}) = rank(S^P), \tag{8}$$

which means that P is not an optimal partition reduct and contradicts with the assumption.

Hence an optimal solution P of the OSVP-problem of S contains the optimal solution P' of the same problem of S^{P^1}, and the proof is completed.

Complexity Analysis. For most applications K is quite small, (e.g., $K = 3$ for the Australian dataset [8]) and $S_B^{P^{i-1}}$ has much less attributes and objects (duplicated objects are removed) than that of $S_B^{P^i}$, both the time and space complexities of the algorithm are determined by the reduct computation of S_B.

Any reduction algorithm could be employed, one very interesting approach is to borrow the idea of MD-heuristic [4] for the reduct computation since S_B contains only binary (boolean) conditional attributes. If we use this approach, the space complexity of our algorithm is

$$O(|U|rank(S)), \tag{9}$$

and the time complexity is

$$O(rank(S)|U|(|R| + \log|U|)), \tag{10}$$

where R is the reduct of S_B. In most applications $|R| \ll |U|$ and $Rank(S) \ll |U|$, hence this approach is applicable.

4   Experiments with Data

We experienced on five datasets from the UCI library [8] using RSES 2.2 [9] and our software called RDK (Rough sets Developer's Kit). The rule generation method was Exhaustive for Monks, LEM2 for Mushroom, and Australian where the cover parameter is set to 1. Moreover, for Mushroom and Australian datasets CV-5 was employed. Since Australian contains some continuous attributes, a discretization stage (using the *Global method*) was introduced before attribute reduction. This approach is called CLASSICAL. The other approach, called RBSVP, only differs from CLASSICAL in that attribute reduction was replaced by our symbolic value partition algorithm.

Experiment results are listed in Table 9. For the Monks datasets (i.e., Monk1, Monk2 and Monk3), RBSVP outperformed the attribute reduction approach in all three aspects, and the advatages in terms of the number of rules and rule coverage are quite obvious. For the Mushroom dataset, although two approaches performed the same in terms of rule coverage and rule accuracy, RBSVP helped to compute a smaller rule set. This indicate that symbolic value partition has a more general ability than attribute reduction.

Results on the Australian dataset are also interesting: Performances of two approaches are not quite distinguishable. The main reason lies in the attribute discretization stage.

Table 9. Experimental results

dataset	CLASSICAL			RBSVP		
	rules	coverage(%)	accuracy(%)	rules	coverage(%)	accuracy(%)
Monk1	48	74.3	99.1	10	100	100
Monk2	95	77.3	87.1	36	99.5	94.7
Monk3	46	75.2	90.8	39	90.3	91.8
Mushroom	24	100	100	14	100	100
Australian	212	65.8	88.1	209	68.2	87.4

5   Conclusions and Further Works

In this paper, we proposed an algorithm searching for a (sub)optimal partition reduct (see Theorem 4). With appropriate modification, we can also obtain an algorithm finding the set of all partition reducts (see Theorem 3). These algorithms are efficient (see equations (9) and (10) for the space and time complexities) thus applicable to many applications. In further works, we will extend these algorithms to suit mixed-mode data and/or decision tables with missing value, and apply them to the applications such as natural languages processing, image processing, etc.

Acknowledgement

Fan Min was supported by an information distribution project under grant No. 9140A06060106DZ223 and the Youth Foundation of UESTC.

This work was supported by National Natural Science Foundation of China, Grant No. 60702071.

The authors would like to thank Dr. Dominik Ślęzak and Dr. Mao Ye for their help in paper proofing.

References

1. Pawlak, Z.: Rough sets. International Journal of Computer and Information Sciences 11, 341–356 (1982)
2. Nguyen, S.H.: Regularity Analysis And Its Application In Data Mining. PhD thesis, Warsaw University, Warsaw, Poland (1999)
3. Min, F., Liu, Q., Fang, C.: Rough Sets Approach to Symbolic Value Partition. International Journal on Approximate Reasoning (Submitted 2007)
4. Nguyen, H.S.: Discretization of Real Value Attributes, Boolean Reasoning Approach. PhD thesis, Warsaw University, Warsaw, Poland (1997)
5. Min, F., Liu, Q., Tan, H., Chen, L.: The M-relative reduct problem. In: Wang, G.-Y., Peters, J.F., Skowron, A., Yao, Y. (eds.) RSKT 2006. LNCS (LNAI), vol. 4062, pp. 170–175. Springer, Heidelberg (2006)
6. Pawlak, Z.: Some issues on rough sets. In: Peters, J.F., Skowron, A., Grzymała-Busse, J.W., Kostek, B., Świniarski, R.W., Szczuka, M.S. (eds.) Transactions on Rough Sets I. LNCS, vol. 3100, pp. 1–58. Springer, Heidelberg (2004)
7. Ganter, B., Wille, R. (eds.): Formal Concept Analysis: Mathematical Foundations. Springer, New York (1999)
8. Blake, C.L., Merz, C.J.: UCI repository of machine learning databases (1998), http://www.ics.uci.edu/~mlearn/mlrepository.html
9. Bazan, J., Szczuka, M.: The RSES homepage (1994–2005), http://logic.mimuw.edu.pl/~rses

Investigative Data Mining for Counterterrorism

Muhammad Akram Shaikh, Jiaxin Wang, Hongbo Liu, and Yixu Song

State Key Lab of Intelligent Technology and Systems
Department of Computer Science & Technology
Tsinghua University, Beijing, P.R. China
Tel.: +86-10-87343432
alm04@mails.tsinghua.edu.cn

Abstract. After the tragic events of 9/11, the concern about national security has increased significantly. However, law enforcement agencies, particularly in view of current emphasis on terrorism, increasingly face the challenge of information overload and lack of advanced, automated techniques for the effective analysis of criminal and terrorism activities. Data mining applied in the context of law enforcement and intelligence analysis, called Investigative Data Mining (IDM), holds the promise of alleviating such problems. An important problem targeted by IDM is the identification of terror/crime networks, based on available intelligence and other information. In this paper, we present an understanding to show how IDM works and the importance of this approach in the context of terrorist network investigations and give particular emphasis on how to destabilize them by knowing the information about leaders and subgroups through hierarchical structure.

1 Introduction

Terrorists seldom operate in a vacuum but interact with one another to carry out various illegal activities. In particular, organized crimes such as terrorism, drug trafficking, gang-related offenses, frauds, and armed robberies require collaboration among criminals. Relationships between terrorists form the basis for organized crimes [1]; and are essential for smooth operation of a terrorist organization, which can be viewed as a network consisting of nodes (terrorists) and links (relationships). In terrorist networks, there may exist groups or teams, within which members have close relationships. One group also may interact with other groups to obtain or transfer illicit goods. Moreover, individuals play different roles in their groups [2]. For example, some key members may act as leaders to control activities of a group. Some others may serve as gatekeepers to ensure smooth flow of information or illicit goods and some act as outliers in a group. To analyze such terrorist networks, investigators must process large volumes of crime data gathered from multiple sources. This is a non-trivial process that consumes much human time and effort. Current practice of terrorist network analysis is primarily a manual process because of the lack of advanced, automated techniques. When there is a pressing need to untangle terrorist networks, manual approaches may fail to generate valuable knowledge in a timely

M.S. Szczuka et al. (Eds.): ICHIT 2006, LNAI 4413, pp. 31–41, 2007.

manner. Fighting against terrorist networks requires a more nimble intelligence apparatus that operates more actively and makes use of advanced information technology. Investigative Data-mining and automated data analysis techniques are powerful tools for intelligence and law enforcement officials fighting against such networks. The rest of the paper is organized as follows: Section 2 gives a brief overview of the IDM with specific reference to terrorist networks; Section 3 describes the process of IDM in terrorist networks and discuss in detail how to destabilize these Networks by finding the hierarchical structure with illustration; and section 4 concludes the paper and gives some future directions.

2   Investigative Data Mining (IDM)

The rapid growth of available data in all regions of society requires new computational methods. Besides traditional statistical techniques [3] and standard database approaches, current research known as Investigative Data Mining (IDM) uses modern methods that originate from research in Algorithms and Artificial Intelligence. The main goal is the quest for interesting and understandable patterns. This search has always been, and will always be, a critical task in law enforcement, especially for terrorist investigation, and more specific for the fight against terrorism. Examples are the discovery of interesting links between people (social networks, see, e.g., [4]) and other entities (means of transport, modus operandi, locations, communication channels like phone numbers, accounts, financial transactions and so on). There are many ways in which IDM can be defined, one of its approach is states as: "The technique which is used for the organization, sorting, visualization, determining associations and predicting criminal behavior in terrorist networks in order to destabilize them". IDM differs from traditional data mining applications in significant ways. Traditional data mining is generally applied against large transaction databases in order to classify people according to transaction characteristics and extra pattern in widespread applicability. The problem in IDM is to focus on smaller number of subjects within large background population and identify links and relationships from a far wider variety of activities.

3   The Process of IDM in Terrorist Network Investigations

The main focus of IDM approach is to identify important actors, crucial links, subgroups, roles, network characteristics, and so on, to answer substantive questions about terrorist organizational structures. There are three main levels of interest: the element, group, and network level. On the element level, one is interested in properties (both absolute and relative) of single actors, links, or incidences. Examples for this type of analyses are bottleneck identification and structural ranking of network items. On the group level, one is interested in classifying the elements of a network and properties of sub networks. Examples are actor equivalence classes, cluster identification and associations. Finally, on the network level, one is interested in properties of the overall network such as connectivity or balance. For illustration and understanding of the concept we will only be considering here the properties of terrorist networks at elemental level through out the paper. The process of IDM for terrorist network analysis includes mainly four phases: Link Analysis, Social Network Analysis, Network visualization and network destabilization as shown in figure 1. Explanation for each phase is given in the following sections.

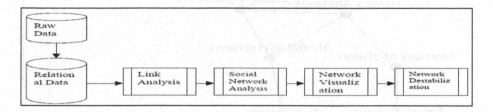

Fig. 1. Investigative Data Mining Process

3.1 Link Analysis

Link analysis is a technique of the data mining field concerned with extracting useful information from a large dataset of associations between entities. Based on graph theory, it is the process of building up networks of interconnected objects in order to explore patterns and trends [5]. In these graphs, nodes represent people, organizations, objects, or events. Edge represents relationship like interaction, ownership, or trust. Attributes store the details of each node and edge, like person's name, or interactions time of occurrence. Effectively combining multiple sources of data can lead law enforcement investigators to discover patterns to help them be proactive in their investigations. Link analysis is a good start in mapping terrorist activity and criminal intelligence by visualizing associations between entities and events [6]. Link analyses often involve seeing via a chart or a map the associations between suspects and locations, whether physical or on a network or the internet. The technique is often used to answer such questions as "who is who, who knows whom and when and where have they been in contact?" A critical first step in the mining of this data is viewing it in terms of relationships between people and organizations under investigation. One of the first tasks in Investigative data mining and criminal detection involves the visualization of these associations, which commonly involves the use of link analysis charts. Krebs [7] mapped a terrorist network comprised of the 19 hijackers in the September 11 attacks on the World Trade Center, using such an approach as shown in figure 2. How ever, the manual link analysis approach will become extremely ineffective and inefficient for large datasets.

3.2 Social Network Analysis (SNA)

Social Network Analysis (SNA), originating from social science research, is a set of analytical tools that can be used to map networks of relationships and

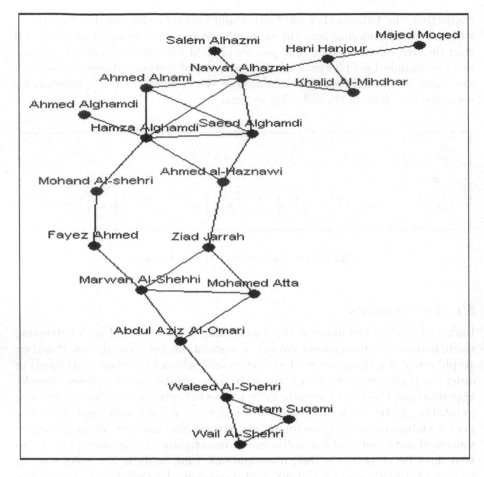

Fig. 2. Network of 9/11 Hijackers (Undirected Graph) [7]

provides an important means of assessing and promoting collaboration in strategically important groups [8]. SNA has recently been recognized as a promising technology for studying criminal and terrorist networks. SNA provides a set of measures and approaches for the investigation of terrorist networks. These techniques were originally designed to discover social structures in social networks [5] and are especially appropriate for studying criminal networks [1,9,10]. Social network analysis describes the roles of and interactions among nodes in a conceptual network. Investigators can use this technique to construct a network that illustrates criminals' roles, the flow of tangible and intangible goods and information, and associations among these entities. Further analysis can reveal critical roles and subgroups and vulnerabilities inside the network [10]. Specifically, in the literature the use of centrality and structural equivalence measures from SNA are used to measure the importance of each network member. Several centrality measures, such as degree, betweenness, closeness, and eigenvector can suggest the importance of a node in a network [5] and can automatically identify the leaders, gatekeepers, and outliers from a network. The degree of a particular node is its number of links; its betweenness is the number of geodesics (shortest paths between any two nodes) passing through it; and its closeness is the sum of all the geodesics between the particular node and every other node in the network whereas eigenvector centrality acknowledges that not all connections are equal. An individual's having a high degree, for instance, may imply his leadership; whereas an individual with a high betweenness may be a gatekeeper in the network [11]. Baker and Faulkner [12] employed these three measures, especially degree, to find the central individuals in a price-fixing conspiracy network in the electrical equipment industry.

3.3 Visualization

IDM process also includes visualization methods that present networks graphically. As we have already discussed link analysis charts are widely used by crime analysts but these were usually drawn manually. How ever, the manual link analysis approach will become extremely ineffective and inefficient for large datasets and it is very difficult to see the big picture in order to draw conclusions. A variety of commercial tools are available, including Analyst's Notebook [13], PolyAnalyst [14], Clementine [15], NetMap [16], VisualLinks [17], NetMiner [18], and Netdraw [19]. These types of commercial packages have proven valuable in extracting and visualizing data relating to criminal investigations, fraud detection, counterterrorism, national and security, and scientific research. These tools concentrate on showing summary properties of the data and exploring relationships between variables.

3.4 Destabilizing Terrorist Networks

Destabilizing techniques traditionally aim at neutralizing members of terrorist networks either through capture or death. The removal of a node from a network can make a cell less able to adapt, reduce its performance, and reduce its ability to communicate. These nodes are known as the 'critical' nodes within a network. The removal or isolation of these nodes ensures maximum damage to the network's ability to adapt, performance, and ability to communicate. In terrorist networks, there may exist groups or teams, within which members have close relationships. One group also may interact with other groups to obtain or transfer illicit goods. Moreover, individuals play different roles in their groups [2]. Network analysis methods support the study of structural properties in networks. One important method as described earlier is centrality analysis [5], which determines the relative importance of vertices (nodes) in a network based on their connectivity within the network structure. In general, the network studied in this paper can be represented by an undirected and un-weighted graph $G = (V, E)$, where V is the set of vertices (or nodes) and E is the set of edges (or links). Each edge connects exactly one pair of vertices, and a vertex pair can be connected by (a maximum of) one edge, i.e., multi-connection is not allowed. A terrorist network consists of V set of actors (nodes) and E relations (ties or edges) between these actors. The nodes may be individuals, groups (terrorist cells), organizations, or terrorist camps. The ties may fall within a level of analysis (e.g. individual to individual ties) or may cross-levels of analysis (individual-to-group analysis). A terrorist network can change in its nodes, links, groups, and even the overall structure. In this paper, we focus on detection and description of node level dynamics. Mathematically, a network can be represented by a matrix called the adjacency matrix A, which in the simplest case is an $N \times N$ symmetric matrix, where N is the number of vertices in the network. The adjacency matrix has elements.

$$A_{ij} = \begin{cases} 1 & \text{if } i \text{ and } j \text{ are connected} \\ 0 & \text{otherwise} \end{cases} \tag{1}$$

The matrix is symmetric since if there is an edge between i and j then clearly there is also an edge between j and i. Thus Turning to the analysis of network data, we start by looking at centrality measures, which are some of the most fundamental and frequently used measures of network structure. Centrality measures address the question, "Who is the most important or central person in this network?" There are many answers to this question, depending on what we mean by important. Perhaps the simplest of centrality measures is degree centrality, also called simply degree. The degree of a vertex in a network is the number of edges attached to it. In mathematical terms, the degree of a vertex i is [20]:

$$k_i = \sum_{j=1}^{n} A_{ij} \tag{2}$$

Though simple, degree is often a highly effective measure of the influence or importance of a node: in many social settings people with more connections tend to have more power. A more sophisticated version of the same idea is the so-called eigenvector centrality. Where degree centrality gives a simple count of the number of connections a vertex has, eigenvector centrality acknowledges that not all connections are equal. If we denote the centrality of vertex i by x_i, then we can allow for this effect by making x_i proportional to the average of the centralities of i's network neighbors [21]:

$$x_i = \frac{1}{\lambda} \sum_{j=1}^{n} A_{ij} x_j \tag{3}$$

Where λ is a constant. Defining the vector of centralities $x = (x1; x2; :::)$, we can rewrite this equation in matrix form as:

$$\lambda x = Ax \tag{4}$$

Hence we see that x is an eigenvector of the adjacency matrix with eigenvalue λ. Assuming that we wish the centralities to be non-negative, it can be shown that λ must be the largest eigenvalue of the adjacency matrix and x the corresponding eigenvector. The equation lends itself to the interpretation that a node that has

Table 1. Degree Centrality and Eigen-Vector Centrality of Terrorist network of 19 Alleged 9/11 hijackers

Node Label	Degree Centrality	Eigen Vector Centrality
Majed Moqed	1	3
Hani Hunjor	3	9
Khalid Mindhar	2	9
Nawaf Al-Hazmi	6	19
Saleem Al-Hazmi	1	6
Saeed Al-Ghamdi	4	18
Ahmed Al-Alnami	3	16
Ahmed Al-Ghamdi	1	6
Hamza Al-Hazmi	6	19
Ahmed Al-Haznawi	3	13
Mohand Al-Shehri	2	8
Fayez Ahmed	2	6
Zaid Jarrah	3	10
Marvan Al-Shehhi	4	11
Mohamed Atta	3	10
Abdul Aziz Al-Omari	3	10
Waleed Al-Shehri	3	7
Wail Al-Shehri	2	5
Satam Suqami	2	5

a high eigenvector score is one that is adjacent to nodes that are themselves high scorers. The idea is that even if a node influences just one other node, if that node influences many others (who themselves influence still more others), then the first node in that chain is highly influential. Hence, the eigenvector centrality measure is ideally suited for influence type processes. Using each node's degree centrality (DC) and Eigen Vector Centrality (EVC) as a proxy for its importance a method is proposed in this section to discover the hierarchy of a terrorist network (directed graph) into two stages as described below:

- Converting undirected graph intodirected graph.
- Converting directed graph into hierarchy.

This hierarchical structure gives an idea about the subgroups present in the network and also how information flows from higher ranks to lower (i.e. to give information about leaders and followers). This kind of information is also very useful in order to destabilize the terrorist networks by removing the important nodes from the network. Now consider the network of 19 alleged hijackers that prepared and executed September 11, 2001 attacks in the U.S. as shown in figure 2 [7].

The first stage is to convert undirected graph as shown in figure 2 into directed graph using degree and Eigen vector centrality in Table 1. The directed links are assigned to this network by a two-step process using degree and Eigen vector centrality in step 1 (column 2) and step 2 (column 3) respectively as shown in

Table 2. Undirected Links are converted in to Directed Links using degree centrality (DC) and Eigen vector centrality (EVC)

Undirected Links	Step1 (DC)	Step2 (EVC)	Directions
HaniHunjor- Majed Moqed	3 > 1		Out Ward Link
HaniHunjor- Nawaf Al-Hazmi	3 < 6		In Ward Link
HaniHunjor- Khalid Mindhar	3 > 2		Out Ward Link
Nawaf Al-Hazmi- Khalid Mindhar	6 > 2		Out Ward Link
Nawaf Al-Hazmi- Saleem Al-Hazmi	6 > 1		Out Ward Link
Nawaf Al-Hazmi- Ahmed Al-Alnami	6 > 3		Out Ward Link
Nawaf Al-Hazmi- Saeed Al-Ghamdi	6 > 4		Out Ward Link
Saeed Al-Ghamdi- Ahmed Al-Alnami	4 > 3		Out Ward Link
Ahmed Al-Ghamdi- Hamza Al-Hazmi	1 > 6		In Ward Link
Hamza Al-Hazmi- Ahmed Al-Alnami	6 > 3		Out Ward Link
Hamza Al-Hazmi- Saeed Al-Ghamdi	6 > 4		Out Ward Link
Hamza Al-Hazmi- Mohand Al-Shehri	6 > 2		Out Ward Link
Ahmed Al-Haznawi- Zaid Jarrah	3 = 3	13 > 10	Out Ward Link
Hamza Al-Hazmi- Ahmed Al-Haznawi	6 > 3		Out Ward Link
Ahmed Al-Haznawi- Saeed Al-Ghamdi	3 < 4		In Ward Link
Mohand Al-Shehri- Fayez Ahmed	2 = 2	8 > 6	Out Ward Link
Fayez Ahmed- Marvan Al-Shehhi	2 < 4		In Ward Link
Marvan Al-Shehhi- Zaid Jarrah	4 > 3		Out Ward Link
Marvan Al-Shehhi- Mohamed Atta	4 > 3		Out Ward Link
Marvan Al-Shehhi- Abdul Aziz Al-Omari	4 > 3		Out Ward Link
Zaid Jarrah- Mohamed Atta	3 = 3	10 = 10	Ignored
Abdul Aziz Al-Omari- Mohamed Atta	3 = 3	10 = 10	Ignored
Abdul Aziz Al-Omari- Waleed Al-Shehri	3 = 3	10 > 7	Out Ward Link
Waleed Al-Shehri- Wail Al-Shehri	3 > 2		Out Ward Link
Waleed Al-Shehri- Satam Suqami	3 > 2		Out Ward Link
Satam Suqami- Wail Al-Shehri	2 = 2	5 = 5	Ignored
Hamza Al-Hazmi- Nawaf Al-Hazmi	6 = 6	19 = 19	Ignored

Table 2. Step 2 has a condition that it is applied when step 1 fails i.e. the scores of adjacent nodes are equal. After the tabulation of links, we are in a position to draw the directed graph, which is not a difficult case if we consider the last column of Table 2 as shown in figure 3. Then we identify the parents and children pairs. For example, if we have two nodes, which are competing for being parent of a node, then we have to identify its correct parent. The correct parent will be the one, which is connected with maximum neighbors. This represents the fact that the true leader, with respect to a node, which is more influential on its neighborhood. When we identify parents, in such a way we traverse all the nodes. Then a tree structure is obtained, which we call hierarchy of network. The steps considering the influence assumptions follow as:

1. Identify the nodes from which one or more links are originating and repeat each of the steps from 2 to 4 for each node.

2. Taking the node with minimum numbers of links originating and traverse it's each link.
3. Every node adjacent to the current link will be placed under its predecessor, if no other link is pointing towards it.
4. If any other link is pointing towards it, then it will be placed under the node that has more links directing towards its neighborhood.

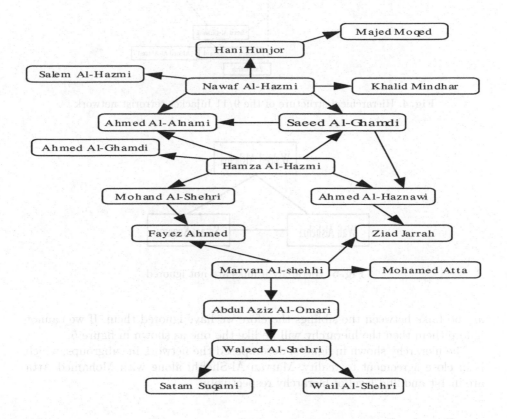

Fig. 3. Network of 9/11 Hijackers (Directed Graph)

By considering the power and influence of each node in the network of 9/11 (directed graph), the hierarchical structure is shown in figure 4. The important thing to discuss is that why we are ignoring few links as shown in table 2. Since higher degree node holds the power, or is influential, therefore equal degree denotes equal in that aspect. Similarly if Eigen-Vector Centrality becomes equal, it suggests that either the nodes have same neighborhood or the different neighborhood but equal in terms of power. So if the two nodes equals in both aspects, equilibrium of power establishes between them, and power stops flowing in between them, as they both are equally strong entities. Therefore in terms of hierarchy they usually are siblings, and in general hierarchical structure there

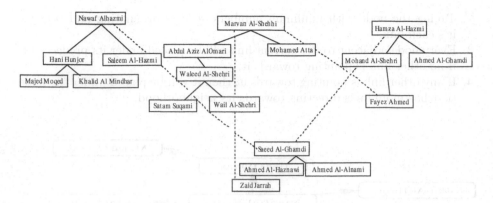

Fig. 4. Hierarchical structure of the 9/11 hijacker terrorist network

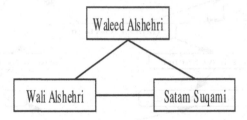

Fig. 5. Hierarchy if links are not ignored

are no links between the siblings therefore we have ignored them. If we cannot ignore them then the hierarchy will be like the one as shown in figure 5.

The hierarchy shown in figure 4 decomposed the network in subgroups, which is in close agreement to reality. Marvan Al-Shehhi along with Mohamed Atta are in 1st and 2nd level of hierarchy respectively.

4   Conclusion

It is believed that reliable data and sophisticated analytical techniques are critical for law enforcement and intelligence agencies to understand and possibly disrupt terrorist or criminal networks. In this paper, we have presented an overview of investigative data mining with its basic process and tried our best to shed some light on the issues. We believe that investigative data mining has a promising future for increasing the effectiveness and efficiency of counter terrorism and intelligence analysis. In addition to this we have also discussed the few available approaches for destabilizing terrorist networks specially finding the hierarchical structure in order to destabilize the terrorist networks by knowing the important nodes (leaders) and subgroups in the network. Many future directions can be explored in this still young field. For example, more visual and intuitive criminal and intelligence investigation techniques can be developed for counter-terrorism. Moreover we are also working on some practical approaches and algorithms for destabilizing terrorist networks, which are the integration of different techniques in which the concept is borrowed not only from SNA measures but also from mathematical order theory [21] and web structure analysis [22].

References

1. McIllwain, J.S.: Organized crime: A social network approach. Crime, Law & Social Change 32, 301–323 (1999)
2. Xu, J., Chen, H.: Criminal network analysis and visualization: A data mining perspective. Communications of the ACM 48(6), 101–107 (2005)
3. Heuer, R.: Psychology of intelligence analysis. Center for the study of Intelligence, Central Intelligence Agency (2005)
4. Pillar, P.R.: Counterterrorism after al qaeda. The Washington Quarterly 27(3), 101–113 (2004)
5. Wasserman, S., Faust, K.: Social Network Analysis: Methods and Applications. Cambridge University Press, Cambridge (1994)
6. Mena, J.: Investigative Data Mining for security and criminal Detection. Butterworth Heinemann, Elsevier Science (2003)
7. Krebs, V.E.: Mapping networks of terrorist cells. Connections 24(3), 43–52 (2001)
8. Chan, K., Liebowitz, J.: The synergy of social network analysis and knowledge mapping: a case study. Int. J. Management and Decision Making 7(1), 19–35 (2006)
9. McAndrew, D.: The structural analysis of criminal networks. In: Canter, D., Alison, L. (eds.) The Social Psychology of Crime: Groups, Teams, and Networks, Offender Profiling Series, III, Aldershot, Dartmouth, pp. 53–94 (1999)
10. Sparrow, M.K.: The application of network analysis to criminal intelligence: An assessment of the prospects. Social Networks 13, 251–274 (1991)
11. Chen, H., et al.: Crime data mining: A general framework and some examples, pp. 0018–9162. IEEE computer society, Los Alamitos (2004)
12. Baker, W.E., Faulkner, R.R.: The social organization of conspiracy: Illegal networks in the heavy electrical equipment industry. American Sociological Review 58(12), 837–860 (1993)
13. http://www.i2inc.com/Products/Analysts_Notebook/default.asp (Accessed on May 12, 2006)
14. http://www.megaputer.com/products/pa/index.php3 (Accessed on June 12, 2006)
15. http://www.spss.com/clementine/ (Accessed on June 26, 2006)
16. http://www.netmapanalytics.com (Accessed on June 27, 2006)
17. http://www.visualanalytics.com/Products/VL3-0Features.cfm (Accessed on June 26, 2006)
18. http://www.netminer.com/NetMiner/home_01.jsp (Accessed on May 9 2006)
19. http://www.analytictech.com/download_products.htm(Accessed on May 20, 2006)
20. Newman, M.E.J.: The structure and function of complex networks. SIAM Review 45, 167–256 (2003)
21. Farely, D.J.: Breaking al qaeda cells: A mathematical analysis of counterterrorism. operations. Studies in conflict terrorism 26, 399–411 (2003)
22. White, S., Smyth, P.: Algorithms for estimating relative importance in networks. In: SIGKDD 2003, ACM, New York (2003)

Data Integration Using Lazy Types

Fernando Berzal, Juan-Carlos Cubero, Nicolás Marín, and Maria Amparo Vila

Intelligent Databases and Information Systems Research Group
Department of Computer Science and
Artificial Intelligence
Escuela Técnica Superior de Ingenierías
Informática y de Telecomunicación
University of Granada, 18071, Granada, Spain
{fberzal,jc.cubero,nicm,vila}@decsai.ugr.es
http://frontdb.ugr.es

Abstract. The development of applications that use the different data sources available in organizations require to solve a data integration problem. Most of the methodologies and tools that simplify the task of finding an integrated schema propose conventional object-oriented solutions as the basis for building a global view of the system. As we will see in this work, the use of the conventional object-oriented data model is not as appropriate as we would like when dealing with data variability and we present a novel typing framework, lazy typing, that can be used for obtaining a global schema in the data integration process. This typing framework eases the transparent development of applications that use this integrated schema and reconcile data.

1 Introduction

Many organizations, specially those of significant size, use different data sources supporting their custom applications. The development of new applications usually require the use of those different data sources that are available at the organization (as, for example, it happens in the development of a Data Warehouse). In these situations, we have to find a common way to manage all data, independently of their source.

Data integration is one of the main problems in the development of this kind of systems[4]. Data integration consists of taking different data sources into account and producing a global schema with the mappings corresponding to all the original data sources, so that we can manage the available data through an integrated and reconciled view.

A variety of methodologies and tools [5] exist in the market that simplify the task of finding an integrated schema, but they tend to share the *shortcoming* of using conventional object-oriented solutions as the basis for building a global view of the system. As we will see, the use of the conventional object-oriented data model is not as suitable as we would like when dealing with data variability.

In this paper we present a novel typing framework, lazy typing, that can be used for obtaining a global schema in the data integration process. Moreover, it

M.S. Szczuka et al. (Eds.): ICHIT 2006, LNAI 4413, pp. 42–50, 2007.

eases the development of new applications that use this integrated schema and reconcile data.

Section 2 introduces the data integration process and presents a case study that will be used in the rest of this paper. Section 3 analyses the limitations of conventional object-oriented solutions for building integrated schemata. We discuss the benefits of using lazy typing in Section 4 and present a framework that allows the transparent use of lazy typing capabilities in current programming platforms like the Microsoft .NET framework.

2 Data Integration

We proceed through the following standard tasks [2,6] in order to integrate different data sources:

Pre-integration. Schema transformation for the sake of homogeneity both from a syntactic and a semantic point of view. The crucial issue in this phase is the selection of a common data model (usually the object-oriented model).

Correspondence identification. Identification of the relationships that hold between the different schemata. In this step, we have to focus on *what is represented* rather than *how it is represented*. Mappings can be described extensively (i.e., among instances) or, preferably, intensively (i.e., among types) [6]. This task is far from being trivial and it usually has to be carried out with human assistance.

Integration. Conflict resolution and generation of a global (integrated) schema.

A global schema is developed as the result of this process. This integrated schema is usually complex so that it can accommodate the heterogeneous data coming from the different data sources. This complexity, together with the rigidness of class hierarchies, makes its use for the implementation of new applications harder.

As we will see in next section, conventional object-oriented solutions can be improved with a more flexible typing system that eases both the design of the integrated schema and the management of the reconciled data obtained from the sources.

Case Study

Consider that we are involved in the development of a software system aimed to provide information about the farms in a certain region. The system will provide information such as the area for each plot of land devoted to a single crop, the number of trees in the plot, and, even, aerial images of the farms.

Plot data can be obtained from there different data sources:

- The Registry Database.
- An Aerial Image Repository.
- A Database with data obtained from a survey process.

That is, the information available for each plot at any moment might vary. There might be some basic registry data for all plots, aerial images for some, and detailed historical records for only a small fraction of them. Then, when the plot information system wants to compute a given plots area, the available information might range from a rough location of the plot to its exact perimeter. Suppose the property registry provides its actual area so that we do not need to estimate it. We could approximate the plots number of trees by using the plot area and the average tree density for a certain kind of crop. We could also automatically compute that number from an aerial image once we know the plots

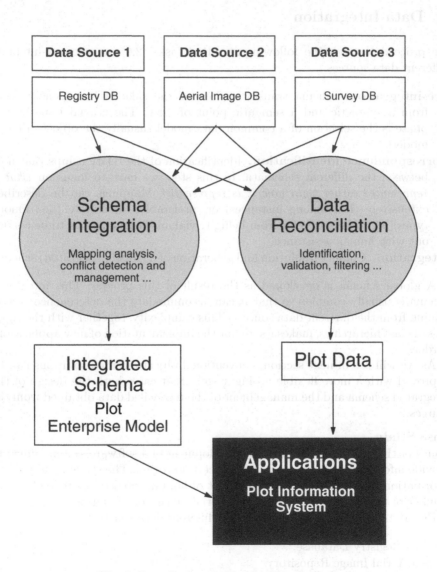

Fig. 1. Data Integration in the Plot Example

geographical limits. In short, we have alternative ways to compute what were interested in according to available data.

The integration process for this problem is depicted in Figure 1. The global schema for our plot representation problem has to adapt to the different plot configurations that we can obtain when reconciling data from our three different sources.

3   Designing a Suitable Integrated Schema

A global schema suitable for plot management is represented in Figure 2. This global schema has been designed using conventional object oriented modeling capabilities. As can be observed in the figure, we can create a class hierarchy and override method implementations when appropriate, thus providing the needed polymorphism.

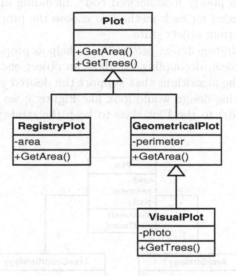

Fig. 2. A conventional Plot global schema using a static class hierarchy

Unfortunately, this global schema obtained using conventional inheritance hierarchies has important drawbacks:

- Even small changes in the data sources could cause huge reorganizations of the class hierarchy. For example, imagine that from the data sources we can obtain new plot configurations for which we could compute the number of trees from the aerial image and prefer to rely on registry data about the plot area, a situation our class hierarchy does not take into account.
- It is probably that a simple inheritance hierarchy will not fit our needs (this is the case with our last consideration in the plot example). And most programming languages today only accept single-implementation inheritance.

Multiple-interface inheritance would be useful here from a declarative point of view, although it wouldnt avoid the need to implement the needed functionality variants, which would probably include a lot of duplicated code.

– Another problem arises when objects evolve (for example, because we obtain new data about a particular plot): The object might need to change its behavior and, hence, its type in the hierarchy. Though the object would keep its external interface, it would lost its identity due to the change of type. This is a significant drawback, if we track object identities in the integrated system.

When its crucial to keep the object identity and when the problem domain gets so complicated that a class hierarchy becomes unmanageable, we can choose to not create such a hierarchy. We could include everything in a single monolithic class, but we have to include the conditional logic needed to select the suitable implementation for GetArea() and GetTrees(). However, the programmer must still maintain such a poorly modularized code, including all the embarrassing conditional logic needed for each method to choose the proper implementation depending on the current object state.

The well-known *strategy design pattern*[3] can help us properly modularize the solution to our problem, decoupling the data an object encapsulates from the implementation of the algorithms that support the desired variability in object behavior. The resulting design would look like Figure 3: we must add two new attributes (data fields) to the Plot class to keep the strategies responsible for

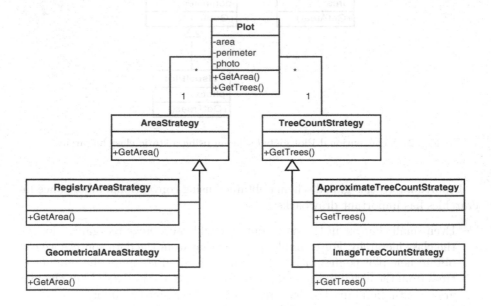

Fig. 3. A more flexible Plot global schema using strategies

implementing the GetArea() and GetTrees() methods. These methods implementations are now trivial because they just delegate to the corresponding strategy.

Unfortunately, those strategies must change when the object state changes. So, we must also include conditional logic to update the corresponding strategies every time the object state changes so that we invoke a suitable strategy given the available data (constructors, setters, ...). Thus, although the strategy design pattern elegantly solves some of the problems rigid inheritance hierarchies cause, it adds unnecessary implementation complexity and makes the programmer responsible for maintaining error-prone conditional logic to control the alternative implementation to use in each situation for each particular object.

4   Lazy Types for Coping with Data Heterogeneity

When we need to manage entities with differing precision levels or when entities present structural irregularities, we require more expressive and powerful modeling techniques to concisely define the type of a given class of objects.

Lazy typing techniques [1] offer a more flexible solution for coping with the (dynamic) data heterogeneity than the alternatives we have analyzed before.

By definition, a lazy type's structure and behavior dynamically adapt to the available data at the instance level. As with conventional types, a set of attributes determines its structure, and a set of method signatures defines its interface. A lazily typed object (lazy object, for short) encapsulates a set of attributes. However, this set is not immutable in lazy objects, even though its always a subset of the set of attributes defining the lazy type. This set of attributes is able to dynamically evolve during the objects lifetime.

Alternative method implementations describe lazy-type behavior for the different structural situations which can arise from the data integration process. These implementations share their signature, so the type maintains its external interface and the programmer can transparently use lazy objects.

Invoking a lazy method will *automatically* delegate to one of the alternative implementations according to the objects current state.

All this dynamism is obtained without requiring the programmer to add conditional logic nor increasing the design model complexity.

Applying Lazy Types in Our Case Study
Lets go back to our plot representation problem:

- First, we define the external plot interfacethat is, the part of a plot behavior that stays the same regardless of its internal structure (GetArea() and GetTrees() methods). This interface becomes the lazy types public interface.
- Next, we identify all the data we might collect about a plot, taking into account the different data sources we are integrating in the system (observed area -observedArea, geographical perimeter -perimeter, and an aerial photograph -photo).
- Finally, we design alternative method implementations and implement the lazy Lot class (see figure 4).

```
[Lazy] public class Lot {

    private float observedArea;
    private Polygon perimeter;
    private Image photo;

    . . .

    public float GetArea () {
        return observedArea;
        }

    [AlternativeImplementation(GetArea)]

    protected float GetAreaFromPerimeter () {
        return perimeter.GetArea();
        }

    public int GetTrees () {
        return (int) ( GetArea() * AverageTreeDensity );
        }

    [AlternativeImplementation(GetTrees)]

    protected int GetTreesFromPhoto () {
        return ImageMorphologyAnalyzer.GetObjectCount
            (photo, perimeter, AverageTreeSize);
        }


}
```

Fig. 4. Class Lot

The code in figure 4 shows how the lazy Lot class would look in C♯. As you can see in the source code, a [Lazy] metadata attribute indicates that the class corresponds to a lazy type. Another attribute, [AlternativeImplementation], marks the alternative method implementations that describe the lazy lot objects dynamically varying behavior.

As can be observed, our lazy class implementation looks like a standard class and avoids the need to create a class hierarchy without the artificial complexity of a strategy-based solution.

A reflective-object factory creates lazy objects implementing the public lot interface. This factory permits the flexible instantiation of plots and their dynamic evolution. To create a lazy lot object, we would type

```
Lot lotObject = (Lot) LazyFactory. Create(typeof(Lot));
```

Once the lazy-object factory creates a lazy object, setting object properties will make strategies change automatically without programmer intervention.

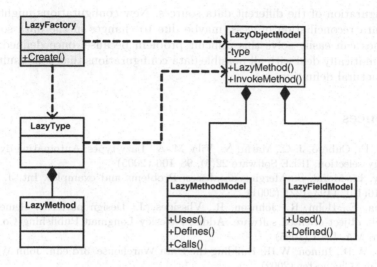

Fig. 5. A framework to support lazy typing

Implementation Issues

We have developed a reusable library that allows for the transparent use of lazy types within the .NET Framework[1].

Our implementation parses compiled intermediate code (Microsoft intermediate language, or MSIL, in the .NET Framework) to build a lazy-object model. We can then use a simple dataflow analysis of the alternative method implementations, as indicated by the metadata attributes, to determine when to invoke each alternative implementation. By using this information, implementation strategies can be dynamically changed in response to newly available data. Our lazy-object factory creates new types to represent lazy objects using the reflection capabilities included in the .NET Framework.

5 Conclusions

As we have seen in this paper, lazy types flexibility and their dynamic adaptation capabilities let us create a single type for any entity in the global schema resulting from a data integration process. This single type can be easily created as the (semiautomatic) union of the type descriptions provided by each data source. Our supporting framework allows the transparent use of this kind of types in conventional programming platforms like .NET or Java, thus freeing the programmer from having to deal with complicated conditional logic nor complex (and static) type hierarchies.

Moreover, the use of lazy types provides another advantage for the data integration process: when we build an integrated schema, we only have to consider

---

[1] Freely available from http://elvex.ugr.es/software/lazy

the configuration of the different data sources. New configurations might arise in the data reconciliation process, maybe due to changes in the data sources. Lazy types can easily solve this difficult problem because, once defined, they can automatically deal with all possible data configurations that can result from their structural definition.

References

1. Berzal, F., Cubero, J.-C., Marín, N., Vila, M.-A.: Lazy types: Automating dynamic strategy selection. IEEE Software 22(5), 98–106 (2005)
2. Cholvy, L., Moral, S.: Merging databases: Problems and examples. Int. J. Intell. Syst. 16(10), 1193–1221 (2001)
3. Gamma, E., Helm, R., Johnson, R., Vlissides, J.: Design patterns: elements of reusable object-oriented software. Addison-Wesley Longman Publishing Co., Inc., Boston, MA, USA (1995)
4. Inmon, W.H., Inmon, W.H.: Building the Data Warehouse, 3rd edn. John Wiley & Sons, Inc, Chichester (2002)
5. Jarke, M., Vassiliou, Y., Vassiliadis, P., Lenzerini, M.: Fundamentals of Data Warehouses. Springer, Heidelberg (1999)
6. Parent, C., Spaccapietra, S.: Issues and approaches of database integration. Commun. ACM 41(5), 166–178 (1998)

Data Generalization Algorithm for the Extraction of Road Horizontal Alignment Design Elements Using the GPS/INS Data

Sunhee Choi[1] and Junggon Sung[2]

[1] Researcher, Korea Institute of Construction Technology
[2] Research Fellow, Korea Institute of Construction Technology
2311, Daehwa-Dong, Ilsanseo-Gu, Goyang-Si,
Gyeonggi-Do, 411-712 Republic of Korea
{sunny,jgsung}@kict.re.kr

Abstract. This paper provides the methodologies to extract the road horizontal alignment design elements using the acquisition data from the Global Positioning System (GPS) and Inertial Navigation System (INS). For this study, highly accurate GPS/INS data from the RoSSAV (Road Safety Survey and Analysis Vehicle) were collected, and also extraction algorithm of road horizontal alignment design elements was proposed according to the statistical inference.

Keywords: GPS/INS, Data Generalization, Horizontal Curve, Highway.

1 Introduction

The development of advanced instruments and surveying techniques such as GPS, INS and laser scanners has aroused great interests to extract road alignment design elements from the real-world coordinates.

Specifically with issues related to ITS (Intelligent Transportation Systems) such as a pre-warning system for driver about hazardous sections of sharp turns or slopes, and an automatic steering control system of advanced vehicle itself, road alignment design elements are considered as an essential part in advanced safety technologies. This paper provides a scientific methodology to extract the road horizontal alignment design elements using the acquisition data from the GPS and INS.

For this study, highly accurate GPS/INS data from the RoSSAV (Road Safety Survey and Analysis Vehicle) were collected and the extraction algorithm of road horizontal alignment design elements was proposed and tested.

2 Algorithm

The data generalization algorithm is subdivided by three steps as follows:

- **1st step:** Data Grouping Algorithm by Tangent · Curve · Clothoid Section
- **2nd step:** Tangent · Curve · Clothoid Analysis Algorithm

M.S. Szczuka et al. (Eds.): ICHIT 2006, LNAI 4413, pp. 51–62, 2007.

- **3rd step:** Tangent · Curve · Clothoid Section Beginning · Ending Position Estimation Algorithm

The algorithm is shown in Fig.1.

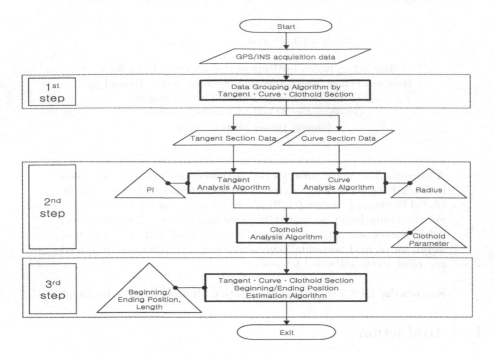

Fig. 1. Process of the Algorithm

2.1 Data Grouping Algorithm by Tangent · Curve · Clothoid Section

Since the acquisition data include tangent sections, curve sections, clothoid sections etc, it is impossible to analyze the acquisition data with only an algorithm for the whole. So, the data need to be grouped into each section as tangent, curve or clothoid.

The heading and roll data, which have different characteristics in both the tangent section and the curve section are acquired by the attitude of vehicle, and they are proper to be applied to the 1st step algorithm.

As a result, the acquisition data are grouped into tangent sections or curve sections.

Building Up Tangent Section Groups (lG)

$$lG_j = B_i \cup B_{i+1} \cdots \cup B_{i+(n^{lG_j}-1)} \quad if \ RV_{i\cdots i+(n^{lG_j}-1)} \leq 0.4. \quad (1)$$

Here $j = 1, 2, \ldots, n^{lG}$, $B_i = i^{th}$ data block(block size: 10m) when $i \in \{1, 2, \ldots, n^B\}$, RV (*Representative Value: data range of a block*) $= max\_Heading$

- $min\_Heading$, n^{lG_j} = number of data blocks in lG_j, and 0.4 is a RV boundary between tangent data and curve data.

Building Up Curve Section Groups (cG)

$$cG_k = B_i \cup B_{i+1} \cdots \cup B_{i+(n^{cG_k}-1)} \quad if \ RV_{i\cdots i+(n^{cG_k}-1)} > 0.4. \tag{2}$$

Here $k = 1, 2, \ldots, n^{cG}$, $B_i = i^{th}$ data block(block size: 10m) when $i \in \{1, 2, \ldots, n^B\}$, $RV(Representative\ Value:\ data\ range\ of\ a\ block) = max\_Heading$ - $min\_Heading$, n^{cG_j} = number of data blocks in cG_j, and 0.4 is a RV boundary between tangent data and curve data.

2.2 Tangent · Curve · Clothoid Analysis Algorithm

In this 2nd step algorithm, the Least Squares Method(LSM) is applied, and PI(Point of Intersection), radius, and clothoid parameter are extracted.

Tangent. PI is the intersection point of next two tangent sections. Each tangent section data group needs to be fit into an equation of the first degree in order to find the point. When an equation is expressed by $Y = aX + b$, the factor a and b can be calculated from the following numerical method in (3) and (4). These are general expressions of the least squares.

$$a = \frac{n \sum xy - \sum x \sum y}{n \sum x^2 - (\sum x)^2} = \frac{\sum (x - \bar{x})(y - \bar{y})}{\sum (x - \bar{x})^2}. \tag{3}$$

$$b = \frac{\sum y \sum x^2 - \sum x \sum xy}{n \sum x^2 - (\sum x)^2} = \bar{y} - a\bar{x}. \tag{4}$$

Suppose two equations $Y = a_1 X + b_1$ and $Y = a_2 X + b_2$, which are next to each other. Then, PI can be calculated by following numerical expressions in (5) and (6).

$$x = \frac{1}{-a_1 + a_2}(b_1 - b_2). \tag{5}$$

$$y = \frac{1}{-a_1 + a_2}(a_2 b_2 - a_1 b_2). \tag{6}$$

Curve. Radius can be calculated by the estimation of a *"Circular Equation"*. The calculating method is as follows:

$$(x - Xm_1)^2 + (y - Ym_1)^2 = R_1{}^2$$
$$(x - Xm_2)^2 + (y - Ym_2)^2 = R_2{}^2$$
$$\vdots$$
$$(x - Xm_{n-2})^2 + (y - Ym_{n-2})^2 = R_{n-2}{}^2.$$

As a result, each curve section data group might be fit into an equation of curve with the smallest SSE(Sum of Squares Error) as the following expression in (7).

$$(x - \overline{Xm})^2 + (y - \overline{Ym})^2 = \overline{R}^2,$$ (7)

where

$$\overline{Xm} = \frac{\sum\limits_{d=1}^{n-2} Xm_d}{n-2}, \quad \overline{Ym} = \frac{\sum\limits_{d=1}^{n-2} Ym_d}{n-2}, \quad and \quad \overline{R} = \frac{\sum\limits_{d=1}^{n-2} R_d}{n-2}.$$

Clothoid. Suppose a clothoid section between the tangent section which can be fit into $Y = a_1 X + b_1$ and the curve section which can be fit into $(X - Xm)^2 + (Y - Ym)^2 = R^2$. Then, clothoid parameter A can be calculated from the following expression in (8).

$$A = \sqrt{R \cdot L}.$$ (8)

Here R is radius and L represents the length of clothoid, which can be calculated from the following expression (9) when S in expression (10) means the tangent offset.

$$L = \sqrt{24 \cdot S \cdot R}$$ (9)

and

$$S = \sqrt{(x_1 - Xm)^2 + (y_1 - Ym)^2} - R.$$ (10)

where

$$x_1 = \frac{-a_1}{a_1^2 + 1}(b_1 - \frac{1}{a_1}Xm - Ym) \quad and \quad y_1 = \frac{-a_1}{a_1^2 + 1}(\frac{b_1}{a_1} + Xm + a_1 Ym).$$

2.3   Tangent · Curve · Clothoid Section Beginning · Ending Position Estimation Algorithm

Previously, each tangent and curve section data was fitted into a general expression with the smallest SSE(Sum of Squares Error) and also, the clothoid parameter was calculated from the tangent offset which would be regarded as the shortest distance between a extension line of tangent and a neighbor curve.

Now, the beginning or ending point of each section can be estimated by calculating the shortest distance between the mapping point on the generalized expression and the original point.

For tangent section, expressed as $Y = aX + b$, the shortest distance from the line to the original point (X_o, Y_o) is given as below:

$$d_l = \frac{|aX_o - Y_o + b|}{\sqrt{a_1^2 + 1^2}}.$$ (11)

For curve section, expressed by $(X - Xm)^2 + (Y - Ym)^2 = R^2$, the shortest distance from the curve to the original point (X_o, Y_o) is given as below:

$$d_c = \sqrt{(X_o - Xm)^2 + (Y_o - Ym)^2} - R. \tag{12}$$

Therefore, the beginning or ending point of each section is estimated to the point that the distances become increased.

3 Data Collection

3.1 Study Area

An overall 7.3km-long road section in Pocheon-Gun, GyeongGi-Do was selected for this study. The road has four-lane in both directions and the road section includes 5 tangent sections, 4 curve sections, and 8 clothoid sections between the tangent sections and curve sections. The investigation area has relatively good horizontal alignment and it is shown in Fig.2.

3.2 Acquisition Data

GPS/INS Data were collected from the first lane in both directions by kinematic survey method using the RoSSAV in April 2006. The details of the acquisition data are shown in Table 1.

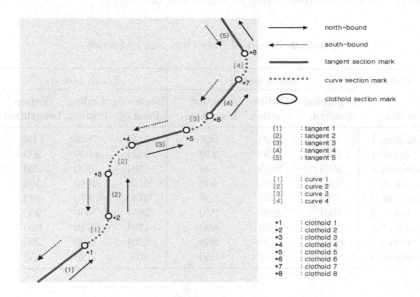

Fig. 2. Study Area

Table 1. RoSSAV Acquisition Data

	Data(unit)	Contents
GPS Time (sec)		Data acquisition time
	Easting(meter)	X
	Northing(meter)	Y
Coordinates	Height(meter)	Z
	Roll(deg)	X axis attitude
Attitude	Pitch(deg)	Y axis attitude
	Heading(deg)	Z axis attitude

* Note: the coordinates are converted to TM(Transverse Mercator) coordinates.

4   Data Analysis

4.1   Result of Data Grouping by Tangent · Curve · Clothoid Section

The data from the both directions were applied to the Data Grouping Algorithm (1st step) and the results were shown in Table 2. The beginning or ending position means accumulations by 1m from the head of acquisition data.

4.2   Result of the Tangent · Curve · Clothoid Analysis Algorithm

The output data from the Data Grouping Algorithm(1st step) were applied to Tangent · Curve · Clothoid Analysis Algorithm (2nd step).

Table 2. Results of Data Grouping by Section

Direction	North-bound			South-bound		
Section	Beginning Position	Ending Position	Section Length(m)	Beginning Position	Ending Position	Section Length(m)
tangent 1	41	1530	1490	6141	7630	1490
curve 1	1681	1890	210	5831	6030	200
tangent 2	2051	2870	820	4901	5720	820
curve 2	3101	3560	460	4151	4610	460
tangent 3	3741	4328	588	3141	3870	730
curve 3	4661	4880	220	2851	3050	200
tangent 4	5051	5950	890	1791	2710	920
curve 4	6131	6430	300	1301	1590	290
tangent 5	6631	7280	650	471	1120	650

Table 3. Results of Tangent Analysis

Direction	Section	$Y = aX + b$		PI(coordinates)	
		a	b	X	Y
North	tangent 1	1.0616	254686.7677		
South	tangent 1	1.0628	254419.0393		
Average of tangent 1 result		1.0622	254552.9035	229871.7008	498733.3946
North	tangent 2	52.8157	-11642215.2412		
South	tangent 2	55.2628	-12204502.1026		
Average of tangent 2 result		53.9818	-11910154.5506	229901.5784	500346.2439
North	tangent 3	0.3677	415809.8687		
South	tangent 3	0.3707	415131.0627		
Average of tangent 3 result		0.3692	415470.4657	231335.4166	500875.5927
North	tangent 4	1.6754	113292.3483		
South	tangent 4	1.6762	113122.7102		
Average of tangent 4 result		1.6758	113207.5293	232119.0782	502188.8399
North	tangent 5	-2.1715	1006248.9989		
South	tangent 5	-2.1697	1005819.2437		
Average of tangent 5 result		-2.1706	1006034.1213	-	-

4.3 Result of Tangent · Curve · Clothoid Section Begining · Ending Position Estimation Algorithm

As combining the result of Tangent · Curve · Clothoid Analysis Algorithm(2nd step), it might be possible to estimate the tangent, curve or clothoid sections'

Table 4. Results of Curve Analysis

Direction	Section	$(X - Xm)^2 + (Y - Ym)^2 = R^2$		
		Xm	Ym	R
North	curve 1	229369.1652	498937.8650	507.8158
South	curve 1	229382.3339	498933.0104	489.1261
Average of curve 1 result		229375.7496	498935.4377	498.4709
North	curve 2	230499.5741	499920.3389	599.8668
South	curve 2	230497.1088	499923.3480	601.5988
Average of curve 2 result		230498.3415	499921.8434	600.7328
North	curve 3	231004.4992	501277.1414	492.0094
South	curve 3	230997.2911	501283.5215	495.0524
Average of curve 3 result		231000.8952	501280.3315	493.5309
North	curve 4	231550.7069	502218.9909	504.0490
South	curve 4	231551.6779	502219.4590	498.1131
Average of curve 4 result		231551.1924	502219.2249	501.0810

Table 5. Results of Clothoid Analysis

Section	S	L	A
clothoid 1	1.1311	116.3240	240.7989
clothoid 2	1.1374	116.6496	241.1357
clothoid 3	3.7884	233.7092	374.6956
clothoid 4	4.0817	242.5875	381.7463
clothoid 5	2.0153	154.5019	276.1367
clothoid 6	1.1332	115.8579	239.1222
clothoid 7	2.1482	160.7312	283.7946
clothoid 8	1.9877	154.6092	278.3374

*Note: S is tangent offset, L is length of clothoid section, and A is clothoid parameter.

beginning · ending position, and also calculate the length of each section from the Tangent · Curve · Clothoid Section Beginning · Ending Position Estimation Algorithm(3rd step). The results are as follows:

Table 6. Results of Begining or Ending Position Estimation in Each Section

Section	Beginning·Ending Position Coordinates		Length(m)
	X	Y	
tangent 1	229665.2680	498514.1510	1452
clothoid 1	229760.4275	498619.7175	142
curve 1	229867.5190	498854.0245	260
clothoid 2	229876.0410	498965.1470	111
tangent 2	229889.9200	499713.4125	745
clothoid 3	229904.0940	500003.2975	289
curve 2	230226.4545	500457.5165	576
clothoid 4	230371.9150	500519.8240	158
tangent 3	231010.1020	500755.6605	677
clothoid 5	231205.3980	500831.1865	209
curve 3	231389.9690	500976.7295	236
clothoid 6	231432.2675	501037.8760	74
tangent 4	231920.8320	501856.5595	949
clothoid 7	232001.8640	502002.3390	166
curve 4	232038.6970	502334.0490	339
clothoid 8	231937.8100	502582.2375	267
tangent 5	231669.7440	503334.6850	810

4.4 Residuals Between Observed Values and Estimated Values

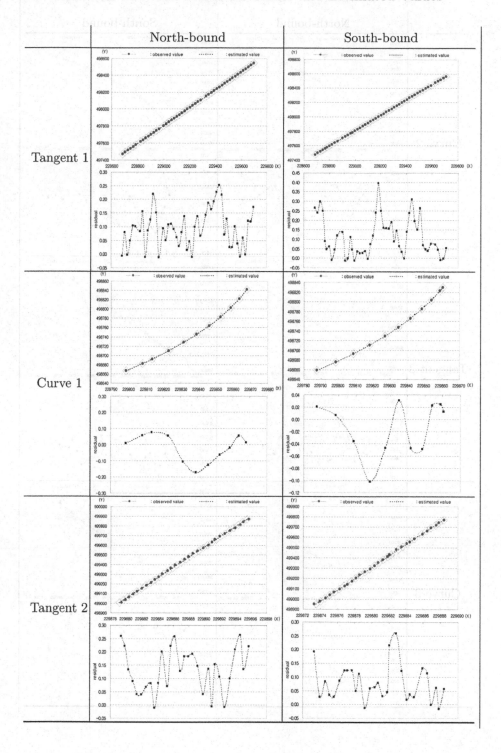

(continued)

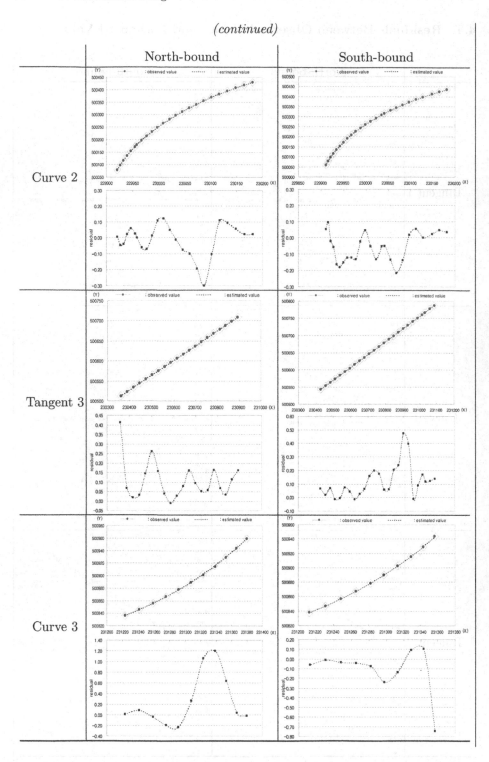

(continued)

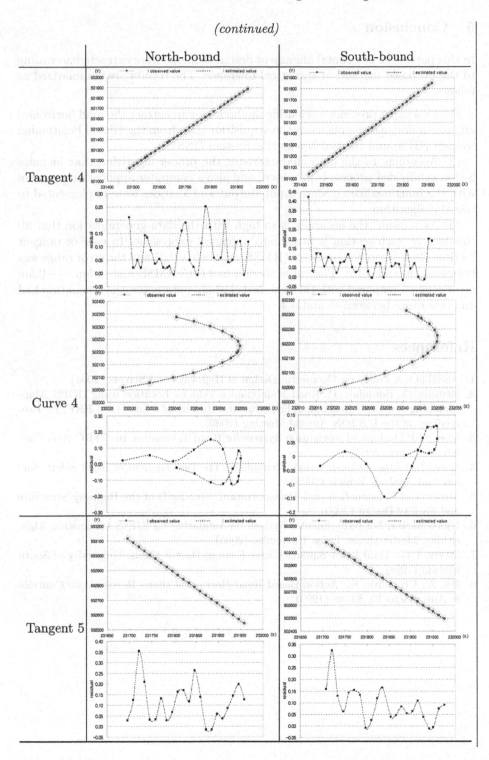

5   Conclusion

In this paper, the horizontal alignment design elements were extracted according to the proposed method of statistical inference. The results are summarized as follows:

(1) This paper provides a scientific methodology to extract the road horizontal alignment design elements using the acquisition data from the Global Positioning System (GPS) and Inertial Navigation System (INS).

(2) According to the data characteristic, the proper algorithm that includes 3-step subdivided process is proposed, and also a computer application program with the functionalities of data input/output and analyses were implemented to test the algorithm.

(3) As a result, the accuracy is so high with the data generalization that all the data by each section is fitted into a general expression. In case of tangent section, Root Mean Square Error (RMSE) was below 0.35 and the error range was between -1m and +1m. The curve section had 0.15 RMSE and -0.5m ∼ +0.5m error range. Compared with the designed clothoid parameter, the estimation had an error range between -5 and +4.

References

1. ASSHTO: A Policy on Geometric Design of Highways and Streets (2001)
2. Bétaille, D., Bonnifait, P.: Road Maintenance Vehicles Location using DGPS. Map-Matching and Dead-Reckoning, Experimental Results of a Smoothed EKF. In: Proceedings of the U.S.ION Annual Meeting (2000)
3. Yves, G.P.: Enhanced Navigation System for Road Telematics. In: STRC 2003 Conference Paper (2003)
4. Korea Institute of Construction Technology: The Development of Road Safety Survey and Analysis Vehicle (2005)
5. Ministry of Construction and Transportation: Standards of the Highway Structure and Facility Design (2000)
6. Hjortsmarker, N.: Experimental System for validating GPS/INS Integration Algorithms. Master thesis, Lulea University (2005)
7. Davis, T.G.: Total Least Squares Curve Fitting. Doctor thesis, University of South Florida (1998)
8. Hu, Z., Uchimura, K.: Action-Based Road Horizontal shape Recognition. Controle & Automacão 10, 83–88 (1999)

Personalized E-Learning Process Using Effective Assessment and Feedback

Cheonshik Kim[1], Myunghee Jung[1], Shaikh Muhammad Allayear[2], and Sung Soon Park[2]

[1] Major in Digital Media Engineering, Anyang University
[2] Computer Science, Anyang University
{mipsan,mhjung,allayear,sspark}@anyang.ac.kr

Abstract. The amount and quality of feedback provided to the learner has an impact on the learning process. Personalized feedback is particularly important to the effective delivery of e-learning courses. E-learning delivery methods such as web-based instruction are required to overcome the barriers to traditional-type classroom feedback. Thereby, the feedback for a learner should consist not only of adaptive information about his errors and performance, but also of adaptive hints for the improvement of his solution. Furthermore, the tutoring component is required to individually motivate the learners. In this paper, an adaptive assessment and feedback process model for personalized e-learning is proposed and developed for the purpose of maximizing the effects of learning.

Keywords: E-learning, Assessment, Feedback, IRT, ICT.

1 Introduction

Mastery Learning (ML) is an instructional strategy based on the principle that all students can learn a set of reasonable objectives with appropriate instruction and sufficient time to learn [1]. Using this strategy, the curriculum should be divided into relatively small learning units, each having their own objectives and assessment. Each unit is preceded by brief diagnostic tests, or formative assessments. The results are used to provide supplementary instruction, or corrective activities to help the learner overcome problems [2]. ML focuses on the process of mastering learning materials. A learner should not proceed to new material until basic prerequisite materials are mastered. The instructor should ensure that each learner can achieve the indicated objectives of a learning unit. An e-learning system can be fully developed to adapt and implement the concepts and theories of ML.

One of the most important factors in on-line learning is personalized feedback from an instructor to a learner which provides qualitative instruction and sufficient time to master learning materials. For this, correct evaluation of the level of learning and guide to the personalized learning environment (style) is required, which is an important component to improve the e-learning process. The tests and evaluations are a source of feedback for learners and instructors and

M.S. Szczuka et al. (Eds.): ICHIT 2006, LNAI 4413, pp. 63–72, 2007.
© Springer-Verlag Berlin Heidelberg 2007

thus, an integral part of the learning process. They are utilized to establish the learner's knowledge level and update learning profiles. Therefore, the evaluation is one of the first considerations for an online course [3]. A computerized test is usually evaluated by summing the score for each question and translating it in one scale. This method of evaluation does not take into account many aspects strictly inherent to some characteristics of the question such as the difficulty and discriminating power [4]. Recent testing models provide the ability to choose test items appropriate to the examinee's level of proficiency during the testing session, i.e., to tailor the test to the individual in real time. These models, known as the Item Response Theory (IRT), adapt mathematical characterizations of an examinee's test responses [5]. The IRT relates characteristics of items and individuals to the probability of a positive response and provides the most information for each examinee. The concept can be utilized for qualitative and exact assessment. The IRT also facilitates computer adaptive testing.

Each learner prefers different learning styles and techniques. A learner usually has a mix of learning styles. Some may find that they have a dominant style of learning, with far less use of other styles. In addition, different learners require a different interval of time in order to learn the same material. An e-learning system has the advantage of providing diverse learning environments (styles) with different techniques and enough time required to learn the materials, while adapting instructional strategies. An instructor should also provide feedback and guide learners to recognize and understand the learning environments and techniques best suited to the individual learner. This approach may improve the speed and quality of learning.

In this paper, a feedback process model, employing adaptive learner assessment, appropriate instruction and preferred learning styles based on the concepts of mastering learning, is proposed to maximize the learning effect. Here, the activity of assessment and feedback within an e-learning process is emphasized to determine how it helps to improve the learning process for all participants: the students, teachers, the designers of content, etc.

The paper is organized as follows. Section 2 describes some related works. Section 3 briefly describes problem assessment using the IRT. Section 4 explains how to use data mining to find student's learning styles from students' behavior. Section 5 presents a model for proposed adaptive assessment and feedback. Section 6 presents our conclusions.

2 Related Works

In the theory of Mastery Learning, the learning rate (LR) is demonstrated by the following relationship:

$$LR = T_s/T_n,$$

where, T_s and T_n are time spent for learning and time needed for learning, respectively. The theory implies that, given sufficient time and quality of instruction, nearly all students can learn. The basic model for effective instruction of each learning unit consists of introduction, presentation, structured practice, guided practice, and independent practice. Structured practice leads learners through practice examples working in a lock-step fashion while guided practice is based on monitoring the students' work and responding with corrective feedback. Independent practice is to be assigned when learners have achieved more than 85% accuracy. A learner proceeds to new material after basic prerequisite materials are mastered. When the contents are structured and the learning process is implemented, adopting the ideals of ML, the problems resulting from the lack of face-to-face teaching in e-learning are reduced. Following the instructions for ML materials, the material is to be divided into relatively small units, each with clear objectives and assessments. The result of testing is used to provide supplementary learning or corrective activities to help the learner overcome problems and master the materials. Assessment is directly connected to effective teaching and learning as it typically demonstrates understanding and achievement. Instruction based on assessment potentially diagnoses any misconceptions or challenges the student is facing. Assessment in an e-learning environment should include the ability to: [6]

- communicate achievement status for students,
- provide self-evaluation to the learner,
- identify student placement for educational paths or programs,
- motivate the learner, and
- evaluate the effectiveness of instructional programs.

Despite the apparent advantages of an online course of study, problems still exist. Without appropriate structure, online study could result in a sense of isolation and a lack of social interaction. Therefore, it is recommended that occasional face-to-face meetings are necessary in order for students to develop a sense of a learning community and social contact.

Interesting research data exist that in instances in which assessment does not include grades, providing qualitative feedback subsequently to work improves learning significantly [7]. Performance of learning refers to the level of learning and the ability to integrate acquired skills and knowledge. An example of online assessment is Computerized Adaptive Testing (CAT), which works to continually evaluate the ability of the student. CAT accomplishes this by compiling individual tests for users that avoid questions that are too easy or too difficult.

The amount and quality of feedback provided to the learner has an impact on the learner's satisfaction with the progress of learning. Feedback is particularly important to the effective delivery of e-learning courses because distance learners do not receive the day-to-day feedback available in traditional classroom settings [8]. Instructor-student feedback should be fostered in e-learning courses in order to provide supplementary instruction or corrective activities to help the learner overcome problems and discover the learning environments (styles) best suited to the learner.

3   Problem Assessment Using IRT

The IRT, or latent trait theory, is the most popular modern testing theory currently under active research [5]. Compared with the perspective of more traditional approaches, such as classical testing theories, an advantage of IRT is that it potentially provides information that enables a researcher to improve the reliability of an assessment. This is achieved through the extraction of more sophisticated information regarding psychometric properties of individual assessment items. IRT provides an estimate of the true score that is not based on the number of correct items. Therefore, the instructor is free to give different test items to different students but still place the students on the same scale.

The Rasch model is the only IRT model in which the total score across all items characterizes a student completely. The Rasch model is also the simplest model having the minimum number of parameters for the learner. One parameter corresponds to each category of an item. The item parameter is generically referred to as a threshold. One threshold exists in the case of a dichotomous item or two in the case of three ordered categories. The probability of a correct response to an item, i is defined using the formula in Eq. (1),

$$P_i(\theta) = \frac{1}{1 + e^{-a_i(\theta - b_i)}} \tag{1}$$

where θ is the learner parameter and a_i and b_i are item parameters. The parameter b_i represents the item location, which, in the case of attainment testing, is referred to as the item difficulty. The item parameter a_i represents the discrimination of the item, that is, the degree to which the item discriminates between learners in different regions on the latent continuum. θ refers to the learner's ability level. Therefore, the value of 0 is assigned to θ to calculate the item information. The item parameters, ai and bi are estimated using BILOG3[9]. Then, the learner parameter is obtained using the estimated item parameters.

The true score of the test is obtained using the following equation,

$$TrueScore = \sum_{i=1}^{N} P_i(\theta) \tag{2}$$

Where $P_i(\theta)$ is the probability of a correct response to an item, i.

4   Extracting E-Learning Style Using Web Mining

Web-usage mining is the task of discovering the activities of the users while they are browsing and navigating through the Web. The aim of understanding the navigation preferences of visitors is to enhance the quality of electronic services, to personalize the Web portals [10] or to improve the Web structure and Web server performance. The mined data consist of the log files, or the secondary data available on the web, while the documents accessible through the Web are considered as primary data. Web-usage mining applies data mining techniques for discovery of usage patterns from Web data in order to understand and better serve the needs of user's navigating on the Web. As in every data mining task, the process of Web-usage mining consists of three main steps: (i) preprocessing, (ii) pattern discovery and (iii) pattern analysis [11].

When a user visits a website, he leaves traces in the server's log file. Log information provides analytical data that can be utilized to understand the learning pattern for feedback process. A server log is a file (or several files) created automatically and maintained by a server of activities it performed. Fig. 1 shows a schematic of a process by which log information is recorded on a server computer while a learner is studying on a computer connected to the server. Log data are incomplete and must be preprocessed. Using web mining, a learner's tendencies are discerned for web personalization.

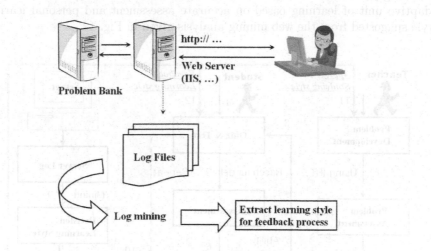

Fig. 1. Application process of a log mining technology

First, log files should be pre-processed and then learner's learning pattern is analyzed using the information such as user IP, URL, referenced web pages. For the summarization and pattern extraction to identify correlation between the reference pages, the association rule is utilized, which pick out patterns that occur in the original form through the database [12].

In association rule, given a set of transactions described by an unordered set of items, association of X and Y, which are conjunctions of items, is discovered from the data. The rule means that transactions which contain the items in X tend to also contain the items in Y. The association rule is based on two numbers that indicate the support and confidence of the rule.

- Support - percentage of transactions from the original database containing both X and Y.
- Confidence - percentage of transactions containing items in X that also contain items in Y.

Discovery algorithm in association rule investigate all possible patterns for rules that meet the user-specified support and confidence thresholds. First, discovering process finds all item sets appearing together in a transaction whose support is greater than the specified threshold. Item sets with minimum support are called frequent itemsets. Then, association rules are generated from the frequent item sets. Confidence is calculated as $support(XY)/support(X)$. All rules that meet the confidence threshold are reported as discoveries of the algorithm.

5   A Model for Adaptive Assessment and Feedback

The model for assessment and adaptive feedback is proposed as the following flow diagram shown in Figure 2. The expected flow is shown to develop an adaptive unit of learning based on accurate assessment and personal learning style suggested from the web mining analysis given in Fig. 1.

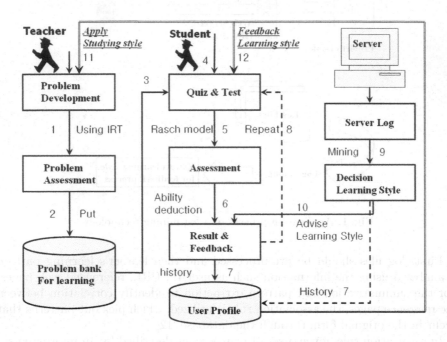

Fig. 2. Application process of a log mining technology

- **Problem development:** The teacher requires ascertaining the studying style of various students and the various levels of interaction and cooperation in the on-line classroom. Therefore, the teacher must generate suitable problems test students having various characteristics.
- **Problem bank:** Not only do test developers need to compose the test items, they must also determine the difficulty of each item in order to ensure that a

test will be neither too hard nor too easy. Using problem banks, test developers can develop a good test. Problem banks are files of suitable test items having a known instructional level and various pertinent item characteristics.

- **Assessment:** In the proposed model, assessment is in an online form in which a tester provides test items in a sequence that is dependent on the correctness of the response to the previous item. Through this process, each tester administers a unique set of test items that provides an accurate measure of the ability of the learner. Items are selected from an item bank of developed, reviewed and field-tested assessment items specific to a course and grade level.
- **Result and feedback:** The learner's process to solve the tests is stored in the user profile. The results are analyzed by evaluation management. For example, it is possible for the learner to check the number of points achieved at any time during an assessment. In addition to assessment, a feedback process is needed for efficient e-learning. Personalized feedback is an important part of individualized assessments. Another type of feedback is to provide 'hints'. In order to provide hints customized to the learner, the feedback manager must know the learner's history, which consists of information from the learner's profile and data from evaluation management. Thereby, advice, forecasts or warnings are displayed.
- **Decision Learning style:** Traditional schooling has used (and continues to use) primarily linguistic and logical teaching methods and a limited range of learning and teaching techniques. Many schools still rely on classroom- and book-based teaching, repetition, and pressured exams for reinforcement and review. Therefore, by recognizing and understanding a student's learning styles, the proposed e-learning model can apply techniques better suited to the individual student. This approach improves the speed and quality of learning. Learning style is designed based on log mining algorithms in order to achieve high scalability. Subsequently, a user profile, which stores user preferences and levels of expertise, is collected for personalized information using data mining.

In the proposed learning model, CAT plays an important role associating teaching and learning. CAT adjusts the degree of difficulty of the questions according to the learner's level of knowledge. Adjusting the degree of difficulty leads to individualization of the assessment with respect to the performance of the learner. The degree of difficulty of any future question is determined by the correctness of the answer to the previous question. The process terminates as soon as the learner's ability is estimated. In order to supply adequate questions, the corresponding degree of difficulty has to be determined in advance. Special care has to be taken in preparation of the questions. Furthermore, the total number of questions posed during an assessment can be reduced thus making the assessment more efficient.

6   Experiment and Discussion

The proposed prototype is implemented in Linux using an Apache Web Server. The Front-end script language, PHP 4.3, and MySQL server are used to implement the e-learning system. In this section, students are evaluated using the assessment and feedback model proposed in the previous sections. Additionally, the proposed model is shown to be effective in on-line learning while traditional assessment models evaluate the learning ability by a number of correct answers.

In the proposed assessment and feedback model, a 'true score' is computed using the IRT concept. The learner's ability level is determined using the 'true score'. Additionally, the assessment system determines the student's learning pattern using association rule mining and records the time required for learning. Table 1 shows learning levels categorized as 'low', 'middle' and 'high'. If a learner's level is 'low' or 'middle', the assessment model provides the learner with the proper feedback. In this case, an alternative learning style is suggested to learners who need to improve. After a learner completes the feedback learning process, he should solve the test again. If the learning level corresponds to 'high' after a test, the learning model allows a learner to proceed to study the following unit.

In an experiment implementing the assessment model, 10 students solved data-structured problems of a yes-no form. The learning system was used to evaluate the examination using Eq. (1) and Eq. (2). The results are given in Table 2. While tests generate equal scores in the traditional method if they have the same number of incorrectness, the learners' ability is evaluated more

Table 1. Criteria of assessment

Assessment Level	Score		Needs-Learing Syste
Low	0.1	0.4	GVA and Quiz and 1:1 Chatting
Middle	0.5	0.7	GVA and Quiz Bord
High	0.8	1.0	Board and Quiz

Table 2. The true score and learning style of testee

Testee ID	Number of correct answers	Ability	True score	Degree of assessment	Time of learning (hours)	Learning style
0001	6	0.617	5.565	Middle	1	GVA, Board
0002	5	0.279	4.935	Middle	1	Board
0003	6	0.570	5.480	Middle	2	Chatting
0004	1	-1.4475	1.848	Low	1	GVA, Board
0005	7	1.041	6.306	Middle	3	PPT, Board
0006	0	-1.770	1.429	Low	1	Board, Chatting
0007	4	0.083	4.236	Low	2	GVA, Chatting
0008	4	-0.296	3.823	Low	1	Chatting
0009	8	1.604	7.186	Middle	2	PPT, Chatting
0010	3	-0.630	3.192	Low	2	GVA

Table 3. The true score and learning style of testee

Testee ID	Number of correct answers	Ability	True score	Degree of assessment
0001	7	0.454	6.705	Middle
0002	7	0.564	6.888	Middle
0003	6	0.125	6.120	Middle
0004	1	-1.833	2.545	Low
0005	8	0.854	7.339	Middle
0006	4	-0.761	4.416	Low
0007	6	0.71	6.020	Middle
0008	5	-0.429	5.059	Middle
0009	9	1.418	8.081	High
0010	5	-0.381	5.153	Middle

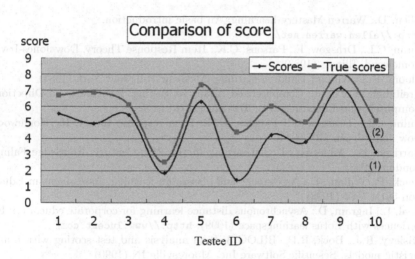

Fig. 3. Comparison of the score (1) before and (2) after feedback

accurately using the IRT. In the IRT-based system, each item is associated with a degree of difficulty. Also, the IRT reflects the weight of items in scores. Therefore, a 'true score' is calculated using Eq. 2. The resulting true score is shown in Table 2. "Time of learning" refers to the time that each learner needs to study and learn the content before test. "Learning pattern" defines the learning style of a learner in on-line learning. The data for the first testee in Table 2 displays that the learner used the GVA and BOARD for 1 hour. "Time of learning" and "Learning pattern" are utilized in providing personalized feedback to a learner.

The proposed model based on assessment and feedbacks allow a learner to obtain personalized feedback on his e-learning. In Table 3, the effects of feedback are displayed and the corresponding results are illustrated in Fig. 3. Learner's scores are improved through learning assessment using the IRT and the personalized feedback based on the analysis of learning pattern.

7 Conclusions

In this paper, an efficient assessment and feedback model is proposed for personalized e-learning environments. One of the most important factors in on-line learning is the personalized feedback of an instructor to a learner, which provides quality instruction and sufficient time to master learning materials. In order to achieve this goal, an accurate evaluation of the level of learning and guide of the personalized learning environment (pattern) is required, which are important components for improving the learning process. It is expected that an e-learning can be effectively and systematically improved using the proposed model based on accurate assessment and personalized feedback.

References

1. Allen, D.: Warren Mastery Learning: An basic introduction, http://allen.warren.net/
2. Hulin, C.L., Drasgow, F., Parsons, C.K.: Item Response Theory, Dow Jones-Irwin, Homewood, Illinois (1983)
3. Bloom, B.S.: All Our Children learning. McGraw-Hill, New York (1981)
4. Kreitzberg, C., et al.: Computerized Adaptive Testing: Principles and Directions. Computers and Education 2(4), 319–329 (1978)
5. Hulin, C.L., Drasgow, F., Parsons, C.K.: Item response theory, IL, Homewood. Dow Jones-Irwin (1983)
6. Garrison, D., Anderson, T.: E-Learning in the 21st Century. Routledge Falmer, London (2003)
7. Black, P., Wiliam, D.: Assessment and classroom learning. Assessment in Education 5(1), 7–74 (1998)
8. Neal, L., Ingram, D.: Asynchronous distance learning for corporate education: Experiences with Lotus learningspace (1999), http://www.lucent.com/
9. Mislevy, R.J., Bock, R.D.: BILOG 3: Item analysis and test scoring with binary logistic models. Scientific Software Inc, Mooresville IN (1990)
10. Lacher, M.S., Koch, M., Wndl, W.: A Framework for Personalizable Community Web Portals, Human-Computer Interaction International, New Orleans, LA, USA, August 5-10, 2001 (August 2001)
11. Xing, D., Shen, J.: Efficient data mining for web navigation patterns. Information and Software Technology 46(1), 55–63 (2004)
12. Kwan, I.S.Y., Fong, J., Wong, H.K.: An e-customer behavior model with online analytical mining for internet marketing planning. Decision Support Systems 41(1), 189–204 (2005)

Optimally Pricing European Options with Real Distributions

Chieh-Chung Sheng[1], Hsiao-Ya Chiu[2], and An-Pin Chen[3]

[1] The Department of Management Information, Yu Da College of Business No.168,
Hsueh-fu Rd, Chaochiao Township, Miaoli County, 361 Taiwan, ROC
[2] The Department of Management Information, Yu Da College of Business No.168,
Hsueh-fu Rd, Chaochiao Township, Miaoli County, 361 Taiwan, ROC
[3] Institute of Information Management, National Chiao Tung University,
1001 Ta Hsueh Road, Hsinchu, Taiwan 300, ROC

Abstract. Most option pricing methods use mathematical distributions to approximate underlying asset behavior. However, it is difficult to approximate the real distribution using pure mathematical distribution approaches. This study first introduces an innovative computational method of pricing European options based on the real distributions of the underlying asset. This computational approach can also be applied to expected value related applications that require real distributions rather than mathematical distributions. The contributions of this study include the following: a) it solves the risk neutral issue related to price options with real distributions, b) it proposes a simple method adjusting the standard deviation according to the practical need to apply short term volatility to real world applications and c) it demonstrates that modern databases are capable of handling large amounts of sample data to provide efficient execution speeds.

Keywords: Option pricing, real distribution, expected value application.

1 Introduction

Option pricing methods have been widely discussed since the development of the Black-Scholes model (BS model) in 1973 [2]. Numerous studies have focused on relaxing the restrictive assumptions in BS model by using various methodologies to approximate the real return distribution on assets in a risk-neutral manner to obtain the fair option price. Although it seems natural to obtain the option price based on real asset return distribution, this idea has rarely been implemented due to the fact that the real distribution never behaves in a risk-neutral manner. Furthermore, option value reduces considerably as it approaches its mature day but the high-frequency pricing methodology received little attention. This study proposes a computational model for pricing European options using the real return of the underlying asset and verifies the high-frequency pricing performance based on empirical study. Experimental results indicate not only that the real distribution pricing method outperforms the BS model, but also that modern

M.S. Szczuka et al. (Eds.): ICHIT 2006, LNAI 4413, pp. 73–82, 2007.

computational methods can be adopted to implement possibility distribution applications rather than using mathematical distributions to approximate the real distribution using the closed form formula. The rest of this paper is organized as follows. Section 2 briefly discusses the traditional option pricing methodologies. Section 3 then discusses observations of asset real return distribution and the feasibility of applying the real distribution map to price European options. Next, section 4 presents an empirical study for verifying the performance of applying real distribution to price European options. Finally, conclusions and future research directions are presented in section 5.

2   Preliminaries to Option Pricing Models

Cox and Ross (1976) established the option price as the expected payoff value discounted at the risk-free interest rate over the risk-neutral distribution of the underlying asset [1]. However, applying the real distribution rather than a mathematical risk-neutral distribution is difficult because the real distribution never behaves in a risk-neutral manner. This characteristic violates put-call parity rules [3] because the derived option price has arbitrage possibilities. A simple example is that if a distribution is risk-neutral then the mean value μ must equal the risk-free interest rate R. However, the μ in a real distribution rarely equals R. The other issue in the application of real distribution is that it varies consistently. On the nth day to maturity, an n-day distribution map must be available for option pricing. This leads to the unpleasant result that at least n different distribution maps are required to valuate the option price before maturity. Additionally, asset short term volatility is rarely consistent with the volatility implied by the real distribution map, leading to large pricing error. Consequently, real asset return distribution cannot be used to obtain the option price, encouraging researchers to apply mathematically risk-neutral distributions instead. The most classical among these approaches is the BS model that assumes that the payoff of the underlying asset follows the geometric Brownian motion, which has a lognormal distribution with constant volatility and risk-free interest rate prior to maturity [2]. Since the development of the BS model, improvements have been made in terms of developing more realistic option pricing methodologies. Examples include (a) the stochastic interest-rate/volatility option model [4][5][12]; (b) jump-diffusion related models [6][7]; (c) Markovian models [8][9]; and (d) stochastic-volatility jump-diffusion models [10][11]. However, all these models focus on identifying the "right" distributions and determining the option price using close form formulas. Consequently, the mathematical distribution will never perfectly fit any actual distribution of the underlying asset.

The field of computer science also tries to price options with artificial intelligence models. There are some alternative methods applying distinct computational approaches to improve the options pricing performance. The most popular one is neural network approach. In contrast to classical mathematical methodologies, a neural network is a non-parametric estimation technique which does not make any distributional assumption regarding the underlying asset. Instead, it builds up a model with sets of unknown parameters and let the optimization routine search for the best fitted parameters to obtain the desired results. For example, Hutchinson-Lo-Poggio (1994) showed that it is possible to use a neural network to price S&P future options [14]. Andrew Carverhill et al. (2003) followed the line of research and examined the best method to set up and train a multi-layer perceptron neural network for option pricing and hedging [13]. Meissner-Kawano (2001) also trained their neural networks with option prices [15]. All these works demonstrate that there exists alternative methodology to price options via modern computational theories. However, studies on using real distribution to price options are rare. Thus, this study focused on determining the option price based on "real" distribution obtained from a historical sample of the underlying asset.

3   Computational Method to Price European Options

This study proposes a computational method for pricing European options using high frequency time interval in time tick of one minute each. High frequency examples are used to obtain a larger sample for verification purposes if the execution efficiency of this computation method is found to be feasible for application to real world applications. However, the same concept can also be applied to price European options regardless of time interval.

Option value is the expected return on maturity. Thus, the option price can be estimated by totaling the historical distribution. Assume I days of sampling data, with each day containing J time ticks. Then for each sample of ith day and j^{th} time tick $X_{i,j}$, the one-day payoff rate $R_{i,j}$ is

$$
R_{i,j} = \begin{cases} \frac{X_{i+1,1}}{X_{i,j}}, & \text{if the final settlement price is determined by the opening} \\ \text{price on final settlement day} \\ \frac{X_{i+1,n}}{X_{i,j}}, & \text{if the final settlement price is determined by the closing} \\ \text{price on final settlement day} \end{cases}
$$

The payoff rate can be preprocessed and stored in a database table for further use in achieving a feasible execution speed when calculating option prices for practical use.

Assume that the option matures the next day and has strike price S, final settlement price S_t, exercise price K and current time-tick j. The call price C is the expected value of S_t is larger than K at maturity, $C = E(max(St - K, 0))$ where E denotes the expected value. Given m sampling days, the call price C can be approximated using the following formula:

$$
C(S,K,j) = E(\max(S_t - K,0)) = \frac{\sum_{i=1}^{m-1} S \times \max(R_{i,j} - \frac{K}{S},0)}{m-1}
$$

Similarly, the Put price P can be approximated as follows:

$$P(S, K, j) = E(\max(K - S_t, 0)) = \frac{\sum_{i=1}^{m-1} S \times \max(\frac{K}{S} - R_{i,j}, 0)}{m - 1}$$

Next, consider the riskless interest rate r. The option price is recalculated by replacing the exercise price K with its present value K', where time to maturity is τ

$$K' = e^{-r\tau} K$$

By introducing K' into the formula, the Call/Put price can be represented as

$$C(S, K, j, r, \tau) = \frac{\sum_{i=1}^{m-1} S \times \max(R_{i,j} - \frac{e^{-r\tau}K}{S}, 0)}{m - 1} \quad (1)$$

$$P(S, K, j, r, \tau) = \frac{\sum_{i=1}^{m-1} S \times \max(\frac{e^{-r\tau}K}{S} - R_{i,j}, 0)}{m - 1} \quad (2)$$

However, when attempting to determine the option price using (1) and (2), it quickly becomes obvious that the calculated price does not follow the put-call parity rule first identified by Stoll (1969) [3]. This is because the mean value μ of a real distribution does not equal zero, implying that the real distribution is not risk-neutral. Notably, arbitraging opportunities occur when the distribution is not risk-neutral. Another notable issue is that the real distribution has its own volatility that is difficult to change. For example, if a real distribution is formed from a 10 year period of sample data and has a standard deviation σ_1, but the forecasted volatility of the target option is σ_2, then the option must be priced using a distribution with a standard deviation σ_2 instead of σ_1. Furthermore, if the intrinsic volatility of the distribution cannot be transformed to fit the short term volatility, the pricing error will be too large for practical use. Given that changing the mean value without influencing the variance is extremely difficult, this study established a computational method for adjusting both the mean value and variance of an existing distribution to yield the desired values while maintaining a distribution as similar as possible to the original.

To obtain risk-neutral characteristics based on the real distribution, the mean value μ of sampling data must be adjusted to zero. By observing the real distribution, if the μ moves from the positive value to zero, the rightmost sampling data reduces while the leftmost sampling data increases. This observation encourages the derivation of a computational method for adjusting the mean value of real distributions.

The first step attaches a weighting factor wi to each sampled payoff rate $R_{i,j}$. Each w_i is assigned an original value 1.0 to indicate that it has a "sampling count" of 1. The Call and Put prices thus can be represented as

$$C(S, K, j, r, \tau) = \frac{\sum_{i=1}^{m-1} S \times \max(w_i(R_{i,j} - \frac{e^{-r\tau}K}{S}), 0)}{m - 1} \quad (3)$$

$$P(S,K,j,r,\tau) = \frac{\sum_{i=1}^{m-1} S \times \max(w_i(\frac{e^{-r\tau}K}{S} - R_{i,j}), 0)}{m-1} \tag{4}$$

For each set of sampling data, the mean value μ' and standard deviation σ' can be calculated as:

$$\begin{cases} \mu' = \dfrac{\sum_{i=1}^{n} R_{i,j} \times w_i}{\sum_{i=1}^{n} w_i} \\[4ex] \sigma' = \sqrt{\dfrac{\sum_{i=1}^{n} (R_{i,j} - \mu')^2 \times w_i}{\sum_{i=1}^{n} w_i}} \end{cases} \quad \text{for the } j^{th} \text{ tick to maturity}$$

The second step involves adjusting the weighting factors to transform the real distribution into a risk-neutral manner. To achieve this objective, the first process is to sort the sampled payoff rates and position them on the X-axis with weighting factor 1 in the direction of the Y-axis. If sample data is collected from the original distribution to obtain a smaller mean value, the result will be reduced counts of larger samples and higher counts of smaller ones. Assuming that the sample appearance probability changes in a linear manner, the weighting factors can be rotated to modify the distribution, as indicated in Fig. 1. Thus, by fixing the rotation point to $X = 0$, the weighting factors can be rotated clockwise to decrease or anti-clockwise to increase the mean values.

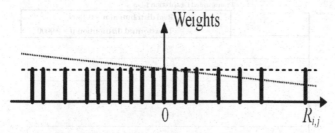

Fig. 1. Rotate the factor weights clockwise will decrease the mean value of a distribution

The weighting factors can be determined by solving the linear equations via the following steps:
Let

$$X_a = \sum_{i=1}^{n} R_{i,j} \big|_{R_{i,j} \geq 0}$$
$$X_b = \sum_{i=1}^{n} R_{i,j} \big|_{R_{i,j} < 0} \tag{5}$$

$$X_{a2} = \sum_{i=1}^{n} (R_{i,j})^2 \big|_{R_{i,j} \geq 0}$$
$$X_{b2} = \sum_{i=1}^{n} (R_{i,j})^2 \big|_{R_{i,j} < 0} \tag{6}$$

where m_a denotes the slope of weighting factors for $R_{i,j} \geq 0$, while m_b represents the slope of weighting factors for $R_{i,j} < 0$.

Solve $\begin{cases} X_a m_a = X_b m_b \\ \sum_{i=1}^{n} X_{i,j}(1 - m_a R_{i,j}) + \sum_{i=1}^{n} X_{i,j}(1 - m_b R_{i,j}) = 0 \end{cases}$

We have

$$\begin{cases} m_a = \frac{(X_a + X_b)X_b}{X_b X_{a2} - X_{b2} X_a} \\ m_b = \frac{(X_a + X_b)X_a}{X_b X_{a2} - X_{b2} X_a} \end{cases} \tag{7}$$

Thus, the weighting factor can be transformed as follows:

$$\begin{cases} w_i = 1 - m_a R_{i,j} \big|_{R_{ij} \geq 0} \\ w_i = 1 + m_b R_{i,j} \big|_{R_{ij} < 0} \end{cases} \tag{8}$$

Combining (7) and (8) produces the following weighting formula:

$$w_i = 1 - \frac{(X_a + X_b)X_b}{X_b X_{a2} - X_{b2} X_a} R_{i,j} \tag{9}$$

Using this computational method, any distribution can be transformed into a risk-neutral distribution while maintaining similar characteristics to the original, as shown in Fig 2.

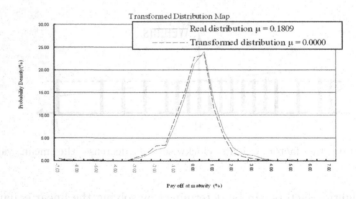

Fig. 2. The transformed distribution after rotating the weighting factors

After transforming the real distribution into a risk-neutral distribution, the second step is to adjust its intrinsic volatility to fit the forecast value. If the intrinsic volatility after applying formula (9) is v, the forecast volatility is v'; formula (9) can be rewritten as:

$$w_i = 1 - \frac{(X_a + X_b)X_b v'}{(X_b X_{a2} - X_{b2} X_a)v} R_{i,j} \tag{10}$$

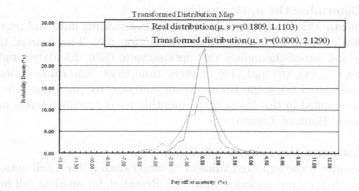

Fig. 3. Transformed distribution after applying formula (10)

Formula (10) can transform the real distribution into the desired volatility with risk-neutral characteristics while maintaining a shape similar to the original distribution.

The transformed distribution is illustrated in Fig 3. Thus, the option price can be determined via (3), (4), (9) and (10).

4   Empirical Tests

This study uses tick price data for the period from 03/01/2001 to 17/12/20031 to verify the feasibility of using computational methods to price Taiwan Stock Exchange Capitalization Weighted Stock Index (TAIEX) options using real distributions. The sampling count was 270 per day and the sample data totaled 216,810. Data for the period 03/01/2001 to 31/12/2002 were adopted as the initial distribution map and pricing errors of high frequency transactions were verified on the last trading day of each month during2003. The trading hours of the TAIEX run from 9:00 to 13:00. The final settlement price is calculated based on the opening price of the final settlement day. The detailed verification procedure is presented below:

Step 1: Generate the initial distribution map
Filter out the incorrect and duplicated data in the database, generate the distribution map, and store it in a database table in the form of (Transaction_date, Time_Ticks, Return_Rate) giving market price data for every minute between 03/01/2001 and 31/12/2002. Because the trading hours are 9:00 to 13:30, the first minute (9:01) is taken as Time_Ticks = 1 while the last (13:30) is Time_Ticks = 270. The Return_Rate Ri,j equals the sampled market price of TAIEX divided by the opening price for the following day:

$$R_{i,j} = \frac{X_{i+1,1}}{X_{i,j}}$$

Step 2: Determine the option price
For each month, this study first addressed the last trading date and regenerated the distribution map for the previous transaction day. The nearest three in-the-money and out-of-the-money call options were then calculated and priced by applying (3), (4), (9) and (10) of every time ticks. The same option prices were also calculated using the BS model for comparative purposes. The riskless interest rate applied in this study is the monthly fixed deposit interest rate used by the Central Bank of Taiwan.

Step 3: Estimate the pricing efficiency
The option price is the expected value of S_t larger than K for a call option, or S_t smaller than K for a put option at maturity. Restated, for an ideal call price $C = MAX(S_t - K, 0)$, the put price should be $P = MAX(K - S_t, 0)$. Consequently, if someone uses C dollars to purchase a call option, they should obtain C dollars by holding to the maturity. The returning ratio Rc and Rp were calculated for each option price to determine the pricing efficiency when the ideal value equals 1.0:

$$Rc = \frac{\sum \max(S_t - K, 0)}{\sum C} \text{ for call options,}$$

$$Rp = \frac{\sum \max(K - S_t, 0)}{\sum P} \text{ for put options.}$$

The final results are listed in Table 1. According to the empirical test, the computational method achieves better pricing performance than the traditional BS model.

Table 1. Pricing error

	Computational Method	Black-Scholes Method
Call Option, Rc	0.9290	0.9037
Pricing Error	7.10%	9.63%
Put Option, Rp	0.9874	0.9081
Pricing Error	1.26%	9.19%

Besides the pricing performance test, the execution speed has been tested by Microsoft Visual FoxPro. The computational option model in this study is sufficiently efficient to price 1,000 option prices in 16 seconds (approximately 0.02 seconds each) with a distribution map that contains 216,810 total sampling data running on an Intel Pentium4 2.6 GHz with 1GB Ram personal computer system. All analytical results indicate that this computational method offers a feasible method of providing good pricing performance and efficient execution speeds with modern personal computer systems.

5   Conclusions

Most modern option pricing models apply mathematical distributions to approximate underlying asset behavior and attempt to calculate the desired option price using close form formulas. This study first introduces a computational model for pricing European options with real distributions and then demonstrates its feasibility in real world applications. This study solves two main issues in applying real distribution to option pricing. First, this study uses weighting factors to adjust the mean value of a real distribution to zero while maintaining its distribution characteristics to follow the put-call parity rule. Second, this study scales the distribution to adjust its standard deviation to meet the needs of applying dynamic volatility to practical problems. The same concept can be easily applied to any expected value related application that cannot use mathematical distribution to obtain feasible solutions.

Although the proposed computational method is practical for real world application, room still exists for improvement. First, the weighting factor rotating method used to adjust the value of the distribution means can be improved. This study assumes that the weighting factors are altered linearly. Nonlinear modification methodologies still require further study. Second, this study use a simple method based on altering standard deviation that may be insufficient for complex applications. Third, the computational method must be simplified in execution speed critical applications.

References

1. Cox, J.C., Ross, S.A.: The valuation of options for alternative stochastic processes. Journal of Financial Economics 3, 145–166 (1976)
2. Black, F., Scholes, M.: The pricing of options and corporate liabilities. Journal of Political Economy 81(3), 637–654 (1973)
3. Stoll, H.R.: The relationship between put and call option prices. Journal of Finance 23, 801–824 (1969)
4. Merton, R.C.: Theory of Rational Option Pricing. Bell Journal of Economics and Management Science 4(1), 141–183 (1973)
5. Amin, K., Jarrow, R.: Pricing Options on Risky Assets in a Stochastic Interest Rate Economy. Mathematical Finance 2, 217–237 (1992)
6. Bates, D.S.: The Crash of 1987: Was It Expected? The Evidence from Options Markets. The Journal of Finance 46, 1009–1044 (1991)
7. Madan, D.B., Carr, P., Chang, E.C.: The Variance Gamma Process and Options Pricing. European Finance Review 2(1), 79–105 (1998)
8. Rubinstein, M.: Implied binomial trees. Journal of Finance 49, 771–818 (1994)
9. Ait-Sahalia, Y., Lo, A.W.: Nonparametric estimation of state-price densities implicit in financial asset prices. Journal of Finance 52, 499–548 (1996)
10. Bates, D.: Jumps and stochastic volatility: exchange rate processes implicit in Deutsch mark options. Review of financial studies 9, 69–107
11. Scott, L.O.: Pricing Stock Options in a Jump-Diffusion Model with Stochastic Volatility and Interest Rates: Applications of Fourier Inversion Methods. Mathematical Finance 7, 345–358 (1997)

12. Bates, D.S.: Jumps and Stochastic Volatility: Exchange Rate Processes Implicit in Deutsche Mark Options. Review of Financial Studies 9, 69–107 (1996)
13. Carverhill, A.P., Cheuk, T.H.F.: Alternative Neural Network Approach for Option Pricing and Hedging (December 2003), Available at SSRN:
    http://ssrn.com/abstract=480562orD0I:10.2139/ssrn.480562
14. Hutchinson, J.M., Lo, A.W., Poggio, T.: A nonparametric approach to Pricing and Hedging Derivatives Securities Via Learning Networks. Journal of Finance 49, 851–889 (1994)
15. Meissner, G., Kawano, N.: Capturing the volatility smile of options on high-tech stocks: A combined GARCH-Neural network approach. Journal of economics and finance 25, 276–292 (2001)

Applying Stated Preference Methods to Investigate Effects of Traffic Information on Route Choice

Hye-Jin Cho[1] and Kangsoo Kim[2]

[1] Korea Institute of Construction and Technology
2311, Daehwa-Dong, Ilsanseo-Gu, Goyang-Si,
Gyeonggi-Do, 411-712 Korea
hjcho@kict.re.kr
[2] Korea Development Institute
P.O.Box 113, Cheongnyang, Seoul 130-012, Korea
kskim@kdi.re.kr

Abstract. This research is exploring the extent to which providing traffic information on VMS affects drivers' route choice behaviour. The information include extra delay and charges. Three different charging regimes were tested. Stated preference(SP) surveys were conducted and route choice logit models were estimated. The results show that drivers' route choice is affected by length of delay and by road user charges on VMS. The fixed charges may be most likely to induce drivers to change their behaviour. Drivers value delay time more highly and they become increasingly sensitive to delay time as it increases.

Keywords: Traffic Information, Charging, Route Choice, SP.

1 Introduction

Traffic Information Systems aims to help and improve drivers' decision making processes and so achieve efficient use of the existing road network by providing drivers with fast and reliable traffic information [2]. Providing traffic information will achieve the user optimum equilibrium instead of the system optimum equilibrium and generate latent demand [9,10]. A road user charging imposes charges on drivers directly based on the amount of road use in order to promote efficient network use [17,25].

Traffic information systems gives information to achieve efficient use of the existing road network, while road user charges provides the incentive to follow the information toward system optimum equilibrium [1]. The latent demand generated by traffic information provision can remain suppressed when road user charges were introduced [10]. Therefore the combination of Traffic Information Systems and road user charging might be expected to maximize the benefits of each of them, as well as compensate for drawbacks and will lead to synergy effects on the cost side [7,9,10,25].

M.S. Szczuka et al. (Eds.): ICHIT 2006, LNAI 4413, pp. 83–92, 2007.

An important issue in the implementation of these systems is to develop an understanding of the way these systems affect driver's behaviour and this effect impact travel demand in the network. This study aims to investigate the effects of traffic information on route choice behaviour, to find out the value of delay time information in terms of normal travel time and to investigate the drivers' route choice behaviour in responses to different road user charging regimes.

The next section briefly reviews relevant literatures and section 3 summarises the survey design and data collection. Section 4 reports the process of estimating logit models and discussion about the results. Finally section 5 summarizes the results and suggests a further study.

2   Literature Review

The research concerning traffic information and drivers' responses to them has covered a wide range of aspects. The extent to which responses to traffic information vary also depends on the characteristics of traffic information. Quantitative information has more influence on route choice than qualitative information. Several researchers have found that drivers are more likely to divert as a result of information when the length of delay is reported on their usual route increase [12,13,14,15,18,19]. Previous studies [23,26,27] have also found that delay time was valued more highly than free flow time. The value of delay time indicates the extent to which drivers perceive delay time and in particular, the delay threshold is the point at which drivers change their behaviour. Estimating value of delay time help to understand drivers' behaviour in response to delay information and to predict their response to delays stated on VMS signs.Therefore, this study investigate the extent to which the amount of delay information affect drivers' route choice and the way in which drivers evaluate the value of information concerning delay time in detail.

There are various charging methods and a considerable amount of research about road user charges has been undertaken. However, previous studies concerning road user charges mainly considered cordon charges, or tolls, and many of them focused on the public acceptability, elasticity and influence on travel pattern. Schemes which were implemented or planned to be implemented has also been very limited to simple charges such as tolls or cordon charges [11,16,20].

Recent technical development related to road user charges may make this possible, whereby charge levels are decided depending on the current road conditions. For these variable systems, various methods have been studied to determine the charging levels including time-based charges, distance-based charges, and delay-based charges. Several researches about variable road user charges have been carried out, but most of them are focus on the acceptability. Several studies have investigated the effects of these variable road user charges on network conditions, which were based on assumption about drivers' responses. They have done this without any consideration of the possibility that the behavioural response might depend on the nature of the charging regime [20,21,24]. The time-based charge and delay-time based charges have been criticized because it would be difficult for drivers to predict the charge accurately [5,8,22]. This is one of reasons why they were excluded from the more in-dept study. There is hardly any literature about detail analysis about those charging particularly about drivers' response to them. There is also little literature about the value of time in terms of which was based on the drivers' responses rather than assumed values. Therefore, this study estimate value of time in terms of variable road user charges including the fixed charges, the time-based charge, and delay time based charges.

Recently, several authors tested the various variable road user charges to investigate which charging method are more effective or the way drivers perceived those methods. Smith et al. [24] compared the network effects of four road-user charging systems in the Cambridge network (including toll, distance-based charge, time-based charge, and delay-based charge). They concluded that delay-based charging reduces congestion substantially at comparatively low levels of charging and suggested delay-based charging would have the best effects in terms of relief of congestion. Bonsall et al. [6] asked drivers the most acceptable option of charges among fixed charges, distance-based charges, time-based charges and delay time-based charges. They found that the fixed charges were the most preferable and people were hostile to unpredictable charges. This study investigates best option to change drivers' behaviors among various charging regimes.

3 Methodology

3.1 Experimental Design

A stated preference(SP) approach was used with a hypothetical network and a choice -based sample is selected. It is assumed that a respondent is driving from home to work on a normal working day in the morning. There are two routes available. Route 1 is short, 4 miles as the crow flies, and goes through the city centre, while Route 2 is longer, 10 miles in total, but no delay is normally expected.

A self-completion survey was used and Variable Message Signs(VMS) were used as the source of traffic information in the SP survey, as well as observed traffic conditions at the decision point through the windscreen.

There are several kinds of road user charges depending on the way of measuring the levels of congestion and the amount of road use. Two road user charges, time-based charges and delay time-based charges are adopted in this study. With total time-based charges, the charge is directly proportion to the travel time spent in the charged area and with delay time-based charges, vehicles are charged directly in proportion to delay they experience within the charged area. Fixed charges are also used in this study. It is assumed that with the fixed charge, each vehicle is charged a fixed amount which might in practice be the same every day or might be designed to reflect traffic conditions in some previous time period, in which case, therefore the drivers might need information on the value of the charge.

In order to represent differences between charge types, the information for fixed charges was given as an accurate estimate of the charge, while the information for total time-based charges and delay time-based charges was given as an estimate of the charge with a range reflecting the uncertainty. The difference between estimates of total time-based charge and delay time-based charge is that the total time-based charge has a narrow range of estimates while delay time-based charge has a wide range of estimates, because delay time is more unpredictable than total travel time.

3.2   Data Collection and Overview

The main survey was conducted and total 281 were returned containing a total of 2626 choice observations. Among respondents, 57.5% are male. A half of respondents are between 35 and 54 years old. Most of them are full-time employee and 51% of respondents have flexible work start time. The below table summered the respondents' socio-economic characteristics and travel patterns.

The respondents were asked about their experience using traffic information system and their perception about usefulness of these systems and information. Results are shown in Fig 1 and Fig 2. About 74%of respondents have used Variable Message Signs before and 73% of them agreed that it was useful. This indicates that respondents' route choices in SP questions are based on their experience with VMS and good perception about its usefulness. 82%of respondents have experience with radio message and 82%of them say it is useful.

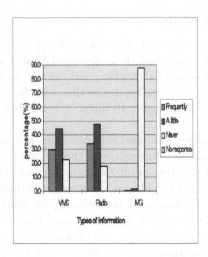

Fig. 1. Experience with Traffic Information

There are three main responses to road user charging; first paying a charge, secondly driving to the edge of the charge area, park and then walk or use public transport, and finally travelling before 7:30 am. The dominant choice for a charge

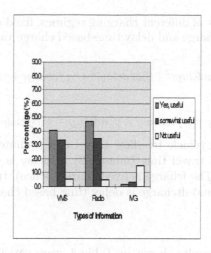

Fig. 2. Usefulness of Traffic Information

is to pay a charge but as levels of charge increase, the percentage of those who would pay a charge decreased from 37.8 to 9.4. The percentage of people who would travel before 7:30 a.m. to avoid a charge was 25.1% and as charge levels rise up, it increases to 33.1% and become the dominant choice for charges. The percentage of respondents who would drive to the edge of the charge area, park and then walk or use public transport, increased from 15.3% to 21.6%. These results indicated that the levels of road user charges increase, drivers would consider changing their departure time first and then consider park and ride to avoid a charge.

4   Modelling and Analysis

4.1   Modelling

In order to explain the influence of information system and road user charging on route choice behaviour, logit models were estimated. Logit model is the simplest and most popular practical discrete choice model and expresses the probability that an individual choose some alternative j as a function of the utilities(Uij) of the n alternatives in the choice set. Where, Pij is probability that individual I will choose route j and Uij is the utility of choosing route j.

$$P_{ij} = \frac{exp(U_{ij})}{\sum_{k=1}^{n} exp(U_k)} \qquad (1)$$

The utility for choosing routes consists of relevant variables representing individual's travel situations, traffic information characteristics and charging systems. To explore the effects of traffic information on route choice behaviour, free travel time variable, fttime, normal delay, ndelay, extra delay variable, exdelay for traffic information and dummy variable, htraffic for heavy traffic were used and for the effects of different charging regimes, fixed charge, fcharge, total time-based charge, ttcharge and delay time-based charge variable, dtcharge were used.

$$U_{route1} = \alpha fttime_1 + \beta_1 ndelay_1{}^2 + \beta_2 exdelay_1{}^2 + \gamma_1 fcharge + \gamma_2 ttcharge + \gamma_3 dtcharge \tag{2}$$

$$U_{route2} = \alpha fttime_1 + \beta_1 ndelay_1{}^2 + \beta_2 exdelay_1{}^2 \tag{3}$$

Where, fttime is In-vehicle free-flow travel time(minutes), ndelay is normal delay included normal travel time (minutes2), exdelay is extra delay reported on VMS (minutes2) . The fcharge is fixed charge(pence), ttcharge is total time-based charge (pence) and dtcharge is delay time-based charge(pence).

4.2   Discussion

The model estimate results shown in Table 1 were satisfactory. The value of rho-squared in estimated logit model was reasonably high and all parameters have intuitive signs and significant. A parameter, fttime for free-flow travel time indicates that drivers' are less likely to choose the route as normal travel time increases.

Table 1. Model Estimation Results

	Coefficient	T-ratios
fttime(min)	-0.103	-11.6
ndelay2(min)	-0.017	-8.8
exdelay2(min)	-0.007	-6.3
fcharge(pence)	-0.027	-14.8
ttcharge(pence)	-0.022	-13.4
dtcharge(pence)	-0.024	-14.5
Number of observation	2626	
Final Likelihood	-1500.781	
Rho-Squared w.r.t. zero	0.176	

The estimates for normal delay minutes, ndelay2 and extra delay, exdelay2 suggest that when the normal delay or extra delay increases, their preference of that route decreases. A number of research has found that drivers are more likely to divert when the length of delay reported on their usual route increase [3,12,13,14,15,18,19].

This model gave a good explanation of the value of the normal delay and value of the extra delay time. The value of normal delay time and extra delay time in terms of free travel time increased at an increasing rate as delay time increased. This produced the plausible delay threshold, which was 10 minutes of normal delay and 15 minutes of extra delay. These delay thresholds are also consistent with those from Khattak et al. [15].

Table 2. Value of Normal Delay and Extra Delay

Normal Delay(min)			Extra Delay(min)		
Normal Delay	Travel time	Ratio(%)	Normal Delay	Travel time	Ratio(%)
5 minutes	4	0.8	5 minutes	1.75	0.36
10 minutes	16	1.6	10 minutes	7	0.7
15 minutes	36	2.4	15 minutes	15.75	1.05
20 minutes	64	3.2	20 minutes	28	1.4
25 minutes	100	4	25 minutes	43.75	1.75

The value of normal delay and extra delay are expressed in units of normal travel time and summarized in Table 2 and shown in Figure 3. The estimated value of delay time are plausible and the curve of the value of normal delay and extra delay are non-linear and concave. This indicates that the value of normal delay time and extra delay time in terms of free travel time increase at an increasing rate as delay time increases. Therefore, drivers thus value delay time more highly and became increasingly sensitive to delay time as it increases. The derived non-linear curve of value of delay is consistent with the relevant studies [12,27]. Wardman et al. [27] also found that additional delays on the VMS were valued more highly than normally expected travel time due to the uncertainty, stress, frustration and the worse driving conditions involved.

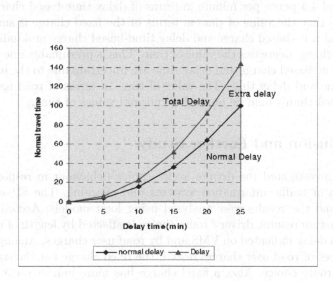

Fig. 3. Value of Normal Delay and Extra Delay per Minute of Normal Travel Time

According to the results, drivers' route choices are affected more by the normal delay time than by the amount of extra delay time stated on VMS. The reason may be because of the perceived unreliability of VMS information. Drivers may consider the extra delay information as a possible delay rather than a certain delay and therefore they may take it less seriously. This result differs from that of Bonsall et al. [4]. They assessed the impact of VMS information on drivers' propensity to divert to park and ride and reported that extra delays mentioned on VMS had a greater impact on drivers than did normal delay. This difference reflects the different contexts of the choices being investigated and does not give cause for concern.

The implication of all estimates for charges is that if charging is introduced on a route, it has influence on drivers' route choice and drivers become more likely to change their route as the level of charge increases on route. Among three charging parameters, fcharge is bigger than other charging parameters. Three parameters for charging regimes are significant. T-statistics tests results for the relevant difference between estimated coefficients show that fcharge coefficient estimate for the fixed charge is statistically significantly different from the other charge coefficient estimates, ttcharge for the time-based charge and dtcharge for the delay-based charge at the usual 5% level of significance. This suggests that a fixed charge has significantly more influence on route choice than do time-based charges or delay-based charges. The ttcharge and dtcharge coefficient estimates have similar values and are not statistically and significantly different from each other.

The values of time in terms of different charging regimes are estimated based on the model estimate results. The values of time are: 3.9 pence per minute in terms of fixed charging; 4.5 pence per minute in terms of total time-based charging; and 4.3 pence per minute in terms of delay time-based charging. This result shows that the value of time in terms of the fixed charge is smaller than those of total time-based charge and delay time-based charge and indicates that certainty of charge decreases the value of time. This is presumably due to the fact that total time based charge and delay time are uncertain due to the uncertainty of travel time and delay time. The value of time in terms of road user charges are less valued than would be implied by normal values of time.

5   Conclusion and Further Study

This study investigated the drivers' route choice behaviour in response to the combination of traffic information systems and charging. The SP survey was conducted and the results were analysed using logit models. According to the model estimation results, drivers' route choice is affected by length of normal delay, by extra delay indicated on VMS and by road user charges. Among the three different types of road user charges tested, the fixed charge has the strongest effect on the route choice. Also, a fixed charge has more impact on route choice than even the upper bound of a range of uncertain charges. This indicates that fixed charges may be most likely to induce drivers to change their behaviour. Also, as is well known, the technology required for the fixed charges is cheaper and less sophisticated than that required for variable charges. Therefore, fixed charges may be the best option for the purpose of the efficient use of network by changing drivers' behaviour and in terms of the easy and cheap implementation.

Drivers value delay time more highly and they become increasingly sensitive to delay time as it increases. The value of delay times indicates the extent to which drivers perceive delay time and in particular, the delay threshold is the point at which drivers change their behaviour. These values help to understand drivers' behaviour in response to delay and to predict their response to delays stated on VMS signs. There may also help to determine a strategy for providing delay time information.

References

1. Bonsall, P.W., Joint, M.: Drivers compliance with route guidance advice: the evidence and its implications. In: Proceedings of VNS conference (December 1991)
2. Bonsall, P.W, Perry, T.: Using an interactive route-choice simulator to investigate drivers' compliance with route guidance advice. Transportation Research Record 1306, 50–68 (1991)
3. Bonsall, P.W., Merrall, A.: Analysing and Modelling the Influence of Roadside Variable Message Displays on Drivers' Route Choice, presented at WCTR held at Sydney (July 1995)
4. Bonsall, P.W., Whelan, G.A., Page, M.: Stated preference experiments to determine the impact of information and guidance on drivers' route choice, PTRC 23rd 287–306 (1995)
5. Bonsall, P.W., Palmer, I.: Do Time-Based Road User Charges Induce Risk-Taking? - Results From A Driving Simulator. Traffic Engineering and Control 38(4), 200–204, 208 (1997)
6. Bonsall, P.W., Cho, H.J., Palmer, I., Thorpe, N.: Experiments to Determine Drivers' Response to Road-User Charges, presented at WCTR held in Antwerp (July 1998)
7. Cho, H.J.: The effects of combination of advanced traveller informaion systems with congestion pricing: Pilot stated preference survey. In: UTSG Conference (1997)
8. Collins, H., Inwood, H.: Attitudes to road pricing in the Bristol area. Traffic Engineering and Control (1996)
9. Emmerink, R., Nijkamp, P., Rietveld, P.: The integration of Road-pricing and motorist information systems, Tinbergen Institute Discussion paper TI-95-62 (1995)
10. Emmerink, R., Nijkamp, P., Rietveld, P., Ommeren, J.: Variable message signs and radio traffic information: An integrated imperical analysis of drivers' route choice behavior. Transpn Res.-A 30(2), 135–153 (1996)
11. Holland, E.P., Watson, E.P.: Traffic Restraint in Singapore. Traffic Engineering and Control 19(1), 14–22 (1978)
12. Huchingson, R.D, Dudek, C.: Delay, Time Saved, and Travel Time Information for Freeway Traffic Management. Transportation Research Record 722, 36–39 (1979)
13. Khattak, A., Schofer, J., Koppleman, F.S.: Factors Influencing Commuters' Enroute Decision Behaviour in Response to Delay. Transportation Research Record 1318, 125–136 (1991)
14. Khattak, A.J., Koppleman, F.S., Schofer, J.L.: Stated preferences for investigating commuters' diversion propensity. Transportation 20, 107–127 (1993a)

15. Khattak, A.J., Schofer, J.L., Koppleman, F.S.: Commuters' enroute diversion and return decisions: analysis and implications for advanced traveler information systems. Transportation Research-A 27A(2) (1993b)
16. Larsen, O.I.: The Toll Ring in Bergen Norway- The First Year of Operation. Traffic Engineering and Control 29(4), 216–222 (1988)
17. Lewis, N.C.: Road pricing-theory and practice. Thomas Telford, London (1993)
18. Mahmassani, H.S., Caplice, C., Walton, C.: Characteristics of Urban Commuter Behaviour Switching Propensity and Use of Infomation. Transportation Research Record 1285, 57–69 (1990)
19. Mannering, F.: Analysis of Commuter Flexibility in Changing Routes and Departure Times. Transportation Research 23B, 53–60 (1989)
20. Meland, S., Polak, J.: Impact of the trondheim toll ring on travel behaviour: Some preliminary findings, PTRC 21st, 103–115 (1993)
21. Milne, D.S., Van Vliet, D.: Implementing Road User Charging in Saturn, Working Paper 410, Institute for Transport Studies, University of Leeds (1993)
22. MVA: The London Congestion Charging Research Programme, Department of Transport, HMSO (1995)
23. Faber, O.: TPA, Setting Forth: Revealed Preference and Stated Preference Models, Technical Note 10, Prepared for Scottish Office(1993)
24. Smith, M.J., May, A.D., Wisten, M.B., Milne, D.S., Van Vliet, D., Ghali, M.O.: A comparision of the network effects of four road-user charging systems. Traffic Engineering and Control 35(5), 311–316 (1994)
25. Van Vuren, T., Smart, V.: Route guidance and road-pricing problems practicalities and possibilities. Transport Reviews 10(3), 269–283 (1990)
26. Wardman, M.: Stated Preference Methods and Travel Demand Forecasting: An Examination of the Scale on the Value of Time, Workingham, Berkshire (1991)
27. Wardman, M., Bonsall, P.W., Shires, J.: Stated Preference Analysis Of Driver Route Choice Reaction To Variable Message Sign Information, Institute for Transport Studies, Working Paper 475 (1997)

A Study on Determining the Priorities of ITS Services Using Analytic Hierarchy and Network Processes

Byung Doo Jung[1], Young-in Kwon[2], Hyun Kim[2], and Seon Woo Lee[3]

[1] Department of Transportation Engineering, Keimyung University
1000 Sindang-dong, Dalseo-gu, Daegu, Korea
[2] The Korea Transport Institute
2311, Daehwa-dong, Ilsan-gu, Koyang-city, Kyunggi-do, Korea
[3] The Transport Environmental Institute
Sindang-2dong, Jung-gu, Seoul, Korea

Abstract. Daegu Metropolitan City is currently in the process of implementing an Intelligent Transportation Systems (ITS) basic plan in order to establish these systems and the foundation of basic services, in addition to setting establishment goals based on the national basic plan of ITS. Some criteria have proven to be very effective at determining the priorities of ITS services, measuring their contribution to solving transportation problems, identifying the services preferred by users, and evaluating ITS systems and related technologies. In this study, the authors prioritize six ITS services using the Analytic Network Process (ANP), which considers mutual dependence between the evaluation items and alternatives. The Analytic Hierarchy Process (AHP), meanwhile, is a one-way process that does not consider the independence of feedback from the services. According to the results of the super decisions ratings, the Regional Traffic Information Center System was chosen to be the top priority project followed by the Urban Arterial Incident Management System and the Bus Information System.

Keywords: Analytic Network Process (ANP), Analytic Hierarchy Process (AHP), Intelligent Transportation Systems (ITS), Priorities, Multi Criteria Analysis.

1 Introduction

The use of intelligent transportation systems (ITS) is increasing, as further attempts are made to alleviate traffic problems by means other than expanding the physical capacity and size of roadways. ITS field trials have demonstrated various benefits associated with individual applications and integrated systems. In Korea, following the enactment of the 'Transport System Efficiency Promotion Act' in 1999, the 'National ITS Plan 21' was established in 2000. Since then, ITS projects have begun to be implemented at the city level.

M.S. Szczuka et al. (Eds.): ICHIT 2006, LNAI 4413, pp. 93–102, 2007.
© Springer-Verlag Berlin Heidelberg 2007

Daegu Metropolitan City, the 3rd largest city in Korea, is also establishing a basic ITS plan and considering the priority for detailed items to be implemented for each fields. Before setting priorities for ITS establishment, applicability and timing should be determined in consideration of technologies at home and abroad. In addition, users' preferences must also be considered to maximize the effect of the introduction of ITS services focusing on resolving the transportation problems that the citizens are really experiencing. Any city government with plans to introduce ITS services also intends to resolve the current transportation problems and to consider as much as possible the potential impact of the ITS service on the overall traffic flow, since it is something that can never be neglected.

The Analytic Hierarchy Process (AHP) is recommended as one of the Multi-Criteria Analysis methods to be used for the decision making or priority setting in the public and private sectors. The AHP has increased in use and popularity due to the fact that the process reflects the way people think and make decisions by simplifying complex decisions to a series of one-on-one comparisons. One of the axioms of AHP is that the elements of one level should be subordinated to the elements of the next upper level. However, in most of the prioritization issues, super elements and subordinate elements are interacting with each other, and the latter is dependent on the former. Thus, some are arguing that those issues cannot be explained simply with a hierarchical structure.

When ITS services are prioritized, interaction between evaluation items and alternatives as well as users' preferences should be considered. To this end, we used the Analytic Network Process (ANP), which considers mutual dependence between the evaluation items and alternatives, to decide which ITS services should be prioritized.

2   Outlines of ANP and Research Trend

2.1   Research Trends of ANP

AHP is a multiple-attribute decision-making tool developed by Thomas Saaty in the early 1970s. AHP is most commonly used as a supporting system for a group's decision-making, as it is easy to apply and highly respected for its process of measurement and weight calculation according to its hierarchical evaluation structure. In Korea, the Korea Development Institute (KDI) used AHP as a multi-criteria analysis method when conducting a comprehensive evaluation of the preliminary feasibility study. And the number of positive studies using AHP is increasing.

Conventional AHP established by Saaty is classified into an inner dependence method and an outer dependence method. When alternatives are dependent on one another, an inner dependence method is suggested. When alternatives and evaluation items are related, an outer dependence method is suggested. Also, ANP can represent the dependence between different levels. The development from a conventional AHP hierarchical structure to the recently suggested 4-phase ANP network structure is shown in (Figure 1). As such, its application areas have been expanded (Saaty, 2001).

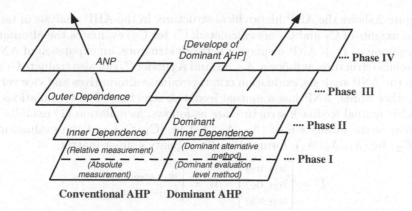

Fig. 1. Theoretical Development of AHP and ANP

In Korea, approximately ten theses have been published in local journals in relation to ANP. These days, the number of comparative studies on AHP and ANP are also increasing. Nevertheless, in the transportation area, positive studies using ANP have not been sufficiently conducted, unlike AHP.

2.2 Theoretical Consideration of ANP

In most cases, decision-making issues cannot be explained simply by a hierarchical structure, as the super elements and subordinate elements are interacting and dependent on one another. In other words, although the AHP basic structure is used under the assumption that the evaluation items and alternatives are not hierarchically related, dependence exists among the criteria and alternatives in many cases. To this end, T.L. Saaty suggested ANP, which allows feedback among alternatives or evaluation item groups to resolve issues with such a structural dependence.

ANP is a new method, an expansion of AHP, which can resolve issues related to inner as well as outer dependence and feedback by using a super matrix. A structure that considers feedback can efficiently demonstrate the complicated structure of interactions created during decision-making.

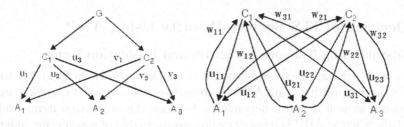

Fig. 2. Structure of AHP and ANP

Figure 2 shows the AHP hierarchical structure. In the AHP analysis of target G, the weights of C_1 and C_2 are calculated. C_1 (or C_2) evaluates the alternative A_j by giving u_j (v_j). ANP displays a network structure, an expansion of ANP's hierarchical structure, as follows: A_1, A_2, and A_3 under C_i can also evaluate C_1 and C_2. In the ANP analysis, evaluation criteria evaluate alternatives and vice versa.

In other words, ANP has a mutual feedback structure. The formulations for the ANP mutual feedback structure are as follows: formulation (1) exhibits the matrices of the evaluation of (A_1, A_2, A_3) by (C_1, C_2) and the evaluation of (C_1, C_2) by (A_1, A_2, A_3); formulation (2) is Saaty's super matrix.

$$U = \begin{bmatrix} u_{11} & u_{12} \\ u_{21} & u_{22} \\ u_{31} & u_{32} \end{bmatrix} \quad W = \begin{bmatrix} w_{11} & w_{12} & w_{13} \\ w_{21} & w_{22} & w_{23} \end{bmatrix} \tag{1}$$

$$S = \begin{bmatrix} 0 & W \\ U & 0 \end{bmatrix} = \begin{matrix} C_1 \\ C_2 \\ A_1 \\ A_2 \\ A_3 \end{matrix} \begin{matrix} C_1 & C_2 & A_1 & A_2 & A_3 \\ \begin{bmatrix} 0 & 0 & w_{11} & w_{12} & w_{13} \\ 0 & 0 & w_{21} & w_{22} & w_{23} \\ u_{11} & u_{12} & 0 & 0 & 0 \\ u_{21} & u_{22} & 0 & 0 & 0 \\ u_{31} & u_{32} & 0 & 0 & 0 \end{bmatrix} \end{matrix} \tag{2}$$

In the above formulations, the valuation criteria and alternatives (C_1, C_2, A_1, A_2, A_3) are all considered equally and the sum of the evaluation is one.

$$\sum_{i=1}^{3} u_{ij} = 1 (j = 1, 2) \quad \sum_{i=1}^{2} w_{ij} = 1 (j = 1, 2, 3) \tag{3}$$

The calculation of the network's graphical figure is conducted with super matrix W, an algebraic expression. The super matrix is two-dimensional, and its rows are effective factors. If each row is normalized and has a sum of one, the stochastic is made. This matrix becomes the first super matrix.

As explained above, ANP can measure the dominance of each element within the network structure. Such a structure can produce several networks according to different decision-making issues, and the networks, in this case, are created in a tree-like fashion. Therefore, it can be proven that the operation of ANP converges into a certain value depending on the degree of dominance based on the graph theory. This value is an outcome of a compound action of dependence among elements, a synergistic effect due to the interactions of elements, and a feedback effect from subordinate elements to super ordinate elements.

3   Decision of ITS Service Priority Using ANP

3.1   Establishing Network Structure and Evaluation Items

In this study, the authors established a network as shown in Figure 3 to set the ITS service priority using ANP. As mentioned above, ANP considers inner-dependence as well as outer-dependence between the evaluation items and alternatives, whereas AHP is a one-way process incapable of considering independence and feedback.

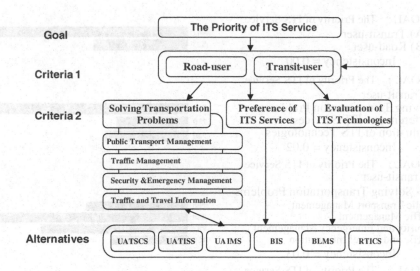

Fig. 3. Network Structure for the Prioritization of ITS Services

In this study, 'Super Decisions 1.4.1', which supports both relative and absolute measurement, was used and the authors created the evaluation items and alternatives as seen in Figure 3 and indicated each interrelation with an arrow. The cluster from which an arrow begins is the basis of the cluster at which the arrow arrives. In fact, the figure does not show the dependence between clusters but the dependence between nodes in the clusters.

In the analysis, ANP has the same hierarchical structure as AHP in terms of the goal to decide ITS service priority, criteria 1 (users of road and public transportation), criteria 2 (solving transportation problems and system preferences), the level and effect of technology, and the constitution of the ITS services (alternative systems). However, the ANP analysis is different in that it considers feedback on the six ITS services that have different influences over road users and public transportation users in resolving transportation problems.

3.2 Calculation of Weights

In this analysis, we used data from the Daegu area (508 public transportation users, 522 road users, 23 experts and city government workers) from the 2004 Korea Transport Institute (KOTI) survey was used. The survey involved eight ITS services and was carried out during the establishment of the Basic ITS Plan for the districts of Daejeon, Daegu and Gwangju. The intention of the survey was to measure the seriousness of transportation problems, gauge satisfaction with ITS services, and identify the preferred ITS service.

For the separate AHP and ANP analysis, the authors surveyed an additional twenty transportation experts. Figure 4 shows the weighted calculation results of the evaluation items of the public transportation users and road users. To

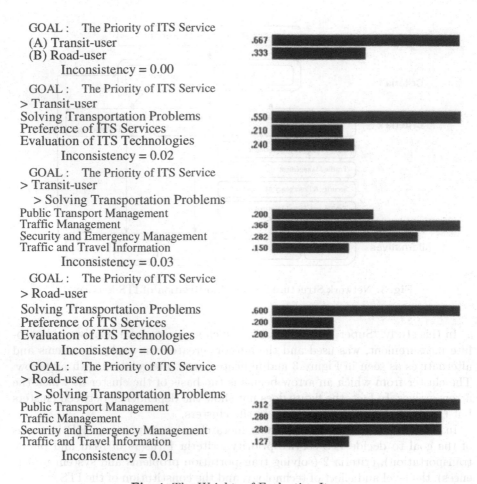

GOAL :   The Priority of ITS Service
(A) Transit-user   .667
(B) Road-user   .333
        Inconsistency = 0.00

GOAL :   The Priority of ITS Service
> Transit-user
Solving Transportation Problems   .550
Preference of ITS Services   .210
Evaluation of ITS Technologies   .240
        Inconsistency = 0.02

GOAL :   The Priority of ITS Service
> Transit-user
        > Solving Transportation Problems
Public Transport Management   .200
Traffic Management   .368
Security and Emergency Management   .282
Traffic and Travel Information   .150
        Inconsistency = 0.03

GOAL :   The Priority of ITS Service
> Road-user
Solving Transportation Problems   .600
Preference of ITS Services   .200
Evaluation of ITS Technologies   .200
        Inconsistency = 0.00

GOAL :   The Priority of ITS Service
> Road-user
        > Solving Transportation Problems
Public Transport Management   .312
Traffic Management   .280
Security and Emergency Management   .280
Traffic and Travel Information   .127
        Inconsistency = 0.01

Fig. 4. The Weights of Evaluation Items

carry out a comparative study on AHP and ANP analysis afterwards, the same criteria, alternatives, and hierarchical network structure were applied.

Unlike the AHP analysis, ANP reflects the impact of transportation issues, denoted as T1 through T4, for six alternative systems, which is called the 'Super Matrix'. The un-weighted Super Matrix is calculated by inputting the pairwise comparison between nodes and clusters excluding the pairwise comparison between alternatives, as in (Table 1).

And, from the Un-weighted Super Matrix calculated through the pairwise comparison, one can find the total of each row, calculate the Weighted Super Matrix (which again sums to one), and calculate the Limit Super Matrix by applying the unique characteristics of the Limit Super Matrix. Lastly, weights used for absolute measurement are selected among different weights produced from the Limit Super Matrix. Then, the weights were normalized as shown in (Table 2).

Table 1. Super Matrix of Un-weighted Priorities

	A1	A2	A3	A4	A5	A6	Priority of ITS	Pref. of ITS
A1	0	0	0	0	0	0	0	0.167
A2	0	0	0	0	0	0	0	0.167
A3	0	0	0	0	0	0	0	0.167
A4	0	0	0	0	0	0	0	0.167
A5	0	0	0	0	0	0	0	0.167
A6	0	0	0	0	0	0	0	0.167
Priority of ITS	0	0	0	0	0	0	0	0
Pref. of ITS	0	0	0	0	0	0	0	0
Road-User	1	1	1	1	1	1	1	0
Tech. of ITS	0	0	0	0	0	0	0	0
T1	0.241	0.106	0.087	0.585	0.626	0.073	0	0
T2	0.565	0.305	0.3	0.132	0.266	0.422	0	0
T3	0.133	0.061	0.396	0.132	0.054	0.214	0	0
T4	0.061	0.528	0.217	0.151	0.054	0.291	0	0
Transit-User	1	1	1	1	1	1	1	0

	Road-User	Tech. of ITS	T1	T2	T3	T4	Transit-User
A1	0	0.167	0.167	0.167	0.167	0.167	0
A2	0	0.167	0.167	0.167	0.167	0.167	0
A3	0	0.167	0.167	0.167	0.167	0.167	0
A4	0	0.167	0.167	0.167	0.167	0.167	0
A5	0	0.167	0.167	0.167	0.167	0.167	0
A6	0	0.167	0.167	0.167	0.167	0.167	0
Priority of ITS	0	0	0	0	0	0	0
Pref. of ITS	1	0	0	0	0	0	1
Road-User	0	0	0	0	0	0	0
Tech. of ITS	1	0	0	0	0	0	1
T1	0.312	0	0	0	0	0	0.2
T2	0.28	0	0	0	0	0	0.368
T3	0.28	0	0	0	0	0	0.282
T4	0.127	0	0	0	0	0	0.15
Transit-User	0	0	0	0	0	0	0

Notes) A1: Urban Arterial Traffic Signal Control System (UATSCS), A2: Urban Arterial Traffic Information Service System (UATISS), A3: Urban Arterial Incident Management System (UAIMS), A4: Bus Information System (BIS), A5: Bus Lane Management System (BLMS), A6: Regional Traffic Information Center System (RTICS), T1: Public Transport Management, T2: Traffic Management, T3: Security and Emergency Management, T4: Traffic and Travel Information

Table 2. Priorities Obtained from Limit Super Matrix

Elements in Components		Priorities from Limit Matrix	Normalized Priorities
Criteria 1	Road-User	0.125	0.20000
	Transit-User	0.125	0.20000
Criteria 2	Preference of ITS Service	0.051	0.08197
	Effects of ITS Technologies	0.055	0.08804
	Public Transport Management	0.073	0.11670
	Traffic Management	0.088	0.14038
	Security and Emergency Management	0.061	0.09766
	Traffic and Travel Information	0.047	0.07525

3.3 Analysis of Prioritization Results

The authors analyzed ANP programs with ratings of 'Super Decision' and evaluated the priority values applied to each evaluation item by applying the priority ratings of five scales of system preferences, three scales of technology level and efficiency, three scales of transportation problem solving, three scales of public transportation users, and three scales of road users.

According to the results of the prioritization through ANP analysis, the Regional Traffic Information Center System (A6) was selected as the top priority project followed by the Urban Arterial Incident Management System (A3), Bus Information System (A4), Bus Lane Management System (A5), Urban Arterial Traffic Signal Control System (A1), and Urban Arterial Traffic Information Service System (A2). The mutual network evaluation of ITS services on the preferred system of public transportation users as well as road users and the evaluation items of transportation problem solving showed different results from those of AHP analysis.

According to the mutual network evaluation, the Urban Arterial Traffic Signal Control System (A1), which was ranked third in the AHP analysis, fell to fifth, and the Bus Information System(A4) took the number three position. As such, the priority has changed. It clearly shows the effect of ANP analysis, which can

	Preference 0.08197	Technology 0.08804	T1 0.11670	T2 0.14038	T3 0.09766	T4 0.07525	R-user 0.20000	T-user 0.20000
A1	B	excellent	low	hi	low	medium	medium	medium
A2	B	medium	low	medium	low	hi	medium	medium
A3	B	medium	low	hi	hi	hi	hi	medium
A4	B	hi	hi	low	low	hi	medium	hi
A5	A	medium	hi	medium	low	low	medium	hi
A6	C	excellent	hi	hi	medium	hi	hi	medium

Fig. 5. Example of the input of priority values

Table 3. Comparison of ITS Priorities by AHP and ANP: Value and Ranking

	ITS Services	AHP	ANP
A1	Urban Arterial Traffic Signal Control System (UATSCS)	0.1762(3)	0.1463(5)
A2	Urban Arterial Traffic Information Service System (UATISS)	0.1214(6)	0.1232(6)
A3	Urban Arterial Incident Management System (UAIMS)	0.1811(2)	0.1872(2)
A4	Bus Information System (BIS)	0.1491(5)	0.1673(3)
A5	Bus Lane Management System (BLMS)	0.1546(4)	0.1661(4)
A6	Regional Traffic Information Center System (RTICS)	0.2177(1)	0.2099(1)

mutually evaluate the importance of transportation problem solving and the preferences of public transportation users as well as road users.

4   Conclusions and Further Research

Many decision problems cannot be structured hierarchically because they involve the interaction and dependence of higher-level elements on lower-level elements. Not only does the importance of the criteria (i.e., contribution to solving transportation problems) determine the importance of the priorities of ITS services as in a hierarchy, but also the importance of the ITS services themselves determine the importance of the criteria.

By using ANP, the authors first categorized public transportation users and drivers into the upper level. Solving transportation problems, users' preferences of ITS services, and the evaluation of system technologies and effects were placed in the lower level. Then, we evaluated six alternative systems were evaluated and each ITS service was prioritized.

Basically, the structure of ANP is similar to that of AHP in terms of purpose, criteria, and alternatives. However, given that each system contributes to the resolution of transportation issues to a different degree, the authors established a dependent relationship between systems (standards) and the settlement of transportation issues (alternatives).

According to the ANP analysis, the 'Regional Traffic Information Center System (A6)' and the 'Urban Arterial Incident Management System (A3)' took the first and second most important spots, respectively, duplicating the AHP analysis results.

However, the ranking of the 'Bus Information System (A4)' and the 'Urban Arterial Traffic Signal Control System (A1)' changed as 1) mutual evaluation criteria among evaluation factors are heavily considered and 2) ITS services have different degrees of impact on users that cannot be reflected through AHP.

All things considered, ANP does not always derive more accurate outcomes than AHP. ANP sometimes provides low reliability, and its method of analyzing networks is complex. Nevertheless, its method facilitates the comprehensive evaluation of a group's decision-making if structures of practical cases are fully factored into the ANP principle.

From now on, the reasonability of ANP analysis compared to AHP analysis should be further studied, testing the influence of evaluation results by sensitivity analysis. Additionally, different groups' opinions should be applied in prioritization and a second feedback procedure is also required.

References

1. Daegu Metropolitan City, Final Report on the Basic Plan of Daegu ITS Basic Plan (2002)
2. Byungdoo, J.: The Analysis of Priorities of ITS Services Using Analytic Hierarchy Process. Journal of Korea Planners Association 37(6), 137–147 (2002)
3. Youngchan, L.: A Model of Analytic Network Process for the evaluation of R&D. Journal of Industrial and Systems Engineering 25(5) (2002)
4. Kinosita: Introduction to AHP and Theory and Application of AHP, Japan Technology (2000)
5. Takahashi: Issues from AHP to ANP, Series I-VI, Operations Research, Japan Society of Operation Research (January-June 1998)
6. Saaty, T.L.: Decision Making - the Analytic Hierarchy and Network Processes (AHP/ANP). Journal of Systems Science and Systems Engineering 13(1) (2004)
7. Saaty, T.L.: Theory and applications of the analytic network process: Decision making with benefits, opportunities, costs and risks. RWS Publications (2005)
8. Saaty, T.L.: The Analytic Network Process: Decision Marking with Dependence and Feedback. RWS Publications (2001)
9. Takahashi, I.: Recent theoretical developments of AHP and ANP in Japan. In: Proceedings of Fifth Conference of International Symposium on Analytic Hierarchy Process, pp. 46–56 (1999)
10. Expert Choice, Inc.: Expert Choice 2000 - Quick Start Guide and Tutorials (2000)

An Introduction of Indicator Variables and Their Application to the Characteristics of Congested Traffic Flow at the Merge Area

Sang-Gu Kim[1], Youngho Kim[2], Taewan Kim[3], and YoungTae Son[4]

[1] Division of Transportation and Logistics, Chonnam National University,
San 96-1, Dundeok, Yeosu, 550-749, Korea
[2] Department of Highway Research, The Korea Transport Institute,
2311, Daehwa, Ilsan, Gyeonggido, 411-701, Korea
[3] Department of Urban Engineering, Chung-Ang Univeristy,
San 72-1, Daeduk, Ansung, Gyunggido, 456-756, Korea
twkim@cau.ac.kr
[4] Department of Transportation Engineering, MyongJi University,
San 38-2 Namdong, Yongin, Gyunggido, 449-728, Korea

Abstract. Research on the merge area has mainly dealt with free flow traffic and research on the congested traffic at the merge area is rare. This study investigates the relationship between mainline traffic and on-ramp traffic at three different segments of the merge area. For this purpose, new indicators based on traffic variables such as flow, speed, and density are used. The results show that a negative relationship exists between mainline and on-ramp flow. It is also found that the speed and the density of the right two lanes in the mainline traffic are significantly affected by the on-ramp flow. Based on the correlation analysis of the indicators, it is confirmed that the right two lanes of the freeway mainline are influenced by the ramp flow. The revealed relationships between mainline and on-ramp traffic may help to analyze the capacity of the downstream freeway segment of the merge area in congested traffic.

Keywords: Congested Traffic, Merge Area, Indicators, Correlation Analysis.

1 Introduction

Most traffic congestion on a freeway occurs at the merge area, where mainline traffic and on-ramp traffic compete for space. So far, the majority of research on traffic flow at merge areas has mainly dealt with free flow traffic and little research has focused on congested traffic at merge areas. It is generally considered that merging vehicles in congested traffic have a different effect on mainline traffic compared to those in free flow traffic. Thus, the analyzing method for congested traffic at merge areas needs to be different from that for free flow traffic. Existing analysis of traffic flow at merge areas has been conducted under the assumption that only the on-ramp traffic has an effect on the mainline traffic.

M.S. Szczuka et al. (Eds.): ICHIT 2006, LNAI 4413, pp. 103–113, 2007.
© Springer-Verlag Berlin Heidelberg 2007

However, data collected at various merge areas showed a close mutual relationship between the mainline traffic and the on-ramp traffic. So, for the analysis in merge area to be reasonably analyzed, interactions between the traffic are to be studied at first. This study divides the merge area into three segments, which are the upstream segment, merging segment, and downstream segment. The traffic conditions at the merge area, the most conflicted section of the freeway, are determined by conditions of mainline traffic and on-ramp traffic and are usually analyzed based on the empirical studies of Highway Capacity Manual (HCM) methodologies. The 1985 HCM edition [1] suggested a model for an outside lane flow based on regression analysis, using the mainline flow and the on-ramp flow to determine Level of Service. Then, HCM 2000 [2] used the density model estimated by the on-ramp flow, the flow entering the ramp influence area, and the length of acceleration lane, for determining LOS. However, the equations in HCM are produced based on the field data collected under free flow traffic conditions and are not suitable for applying to congested traffic conditions. This study investigates the relationships between the on-ramp traffic and the mainline traffic in each lane and at each station, by employing new variables called "*indicators*". Correlation analysis between indicators are performed using the traffic data collected from 3 different merge area sites.

2   Data Collection and Analysis

2.1   Data Collection and Sites

In order to analyze the relationships between mainline traffic and on-ramp traffic, traffic data are required which include traffic variables such as flow, speed, and density in each lane for a sufficiently long section of the merge influence area. In addition, the detection interval should be short enough for microscopic analysis. For the purpose of this analysis, aerial photographic data, which were originally obtained by FHWA [5], are used in this research. These data contain digitized vehicle locations at 18 observation sites, and from these data 3 merge area sites are used for the analysis as given in Table 1.

Table 1. Data Collection Sites

Site	Length of the section(ft)	Number of lanes*
Santa Monica Blvd.(I-405)	1,616	4/1
Roscoe Blvd.(I-405)	1,788	4/1
Backlick Rd.(I-95)	1,641	3/1

* :the number of lanes in the mainline/the number of outside lanes

The geometric configurations of these 3 merge sites are shown in Figure 1. The original data describe vehicle locations for every second during a span of about one hour and other information such as vehicle ID, type, length, travel speed are also included. The data are transformed into flow, density, and space mean speed for 30-sec intervals at every 100ft.

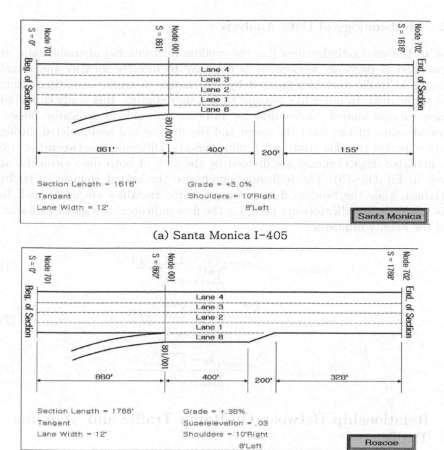

(a) Santa Monica I-405

(b) Roscoe I-405

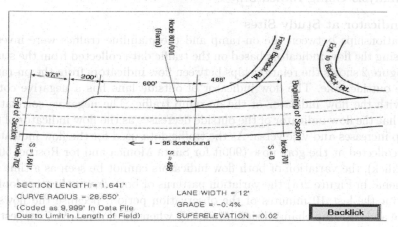

(c) Backlick I-95

Fig. 1. Geometric Configurations of Study Sites

2.2   Methodology of Data Analysis

Due to the stochastic features (i.e. the random fluctuations) of traffic data, as mentioned in previous sections, it is not easy to directly identify the actual changes in traffic variables induced by the change in traffic conditions using raw traffic data. In order to overcome such a shortcoming, this study employed a new variable named *"indicators"* [6]. In order to build the indicator, 30-sec. interval values of the flow, the speed and the density are accumulated during a given period (in this study a 2.5 min. period). Differences between the two accumulated data elements are divided by the sum of both data elements, as given in Eq. (1)~(3). The indicators emphasize the actual changes of traffic variables, while the random fluctuations of traffic variables are minimized. In this study, three indicators are used, i.e. the flow indicator, the speed indicator, and the density indicator.

$$q_{indicator} = \frac{\sum_{i=n-4}^{n} q_i - \sum_{i=n-9}^{n-5} q_i}{\sum_{i=n-9}^{n} q_i} \tag{1}$$

$$v_{indicator} = \frac{\sum_{i=n-4}^{n} v_i - \sum_{i=n-9}^{n-5} v_i}{\sum_{i=n-9}^{n} v_i} \tag{2}$$

$$k_{indicator} = \frac{\sum_{i=n-4}^{n} k_i - \sum_{i=n-9}^{n-5} k_i}{\sum_{i=n-9}^{n} k_i}, \tag{3}$$

3   Relationship Between On-Ramp Traffic and Mainline Traffic

3.1   Analysis Using Indicators

Flow Indicator at Study Sites

The relationships between the on-ramp and the mainline traffics were investigated using the flow indicators based on the traffic data collected from the study sites. Figure 2 shows the relationships between flow indicators from the on-ramp and the outside lane. The flow indicator of outside lane has a negative correlation with the flow indicator of the on-ramp traffic. The negative correlation means that the flow indicator of the outside decreases as the flow indicator of the on-ramp increases and vice versa. As the traffic data used for the flow indicators was all collected at the gore nose (900ft for Santa Monica and for Roscoe, 500ft for Backlick), the variation of both flow indicators cannot be seen as a time lag phenomena. In Figure 2(a) the variation patterns of both flow indicators looked similar for the last 10 minutes of the observation period, because the flows of mainline and on-ramp changed simultaneously when the downstream congestion of Santa Monica site dissipated.

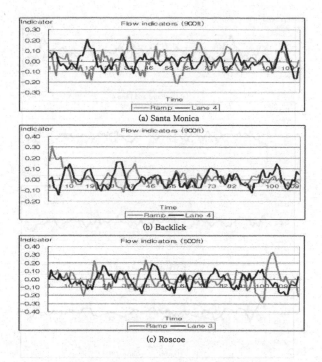

Fig. 2. Variations of Flow Indicators

Flow Indicator in Each Lane

Among the three study sites, the Santa Monica site is selected for the flow indicator dependant on mainline lanes, because this site shows congested traffic during the entire observation period. As shown in Figure 3(a), the most distinguishable negative correlation appears in the outside lane, which is lane 4. The amplitude of the flow indicator in lane 4 is also the largest. The analysis using the flow indicators explicitly shows the relationships between the mainline traffic and on-ramp traffic.

Flow indicator at Each Station

This section describes the variation patterns of the flow indicators in the outside lane and on-ramp at certain significant stations. These stations represent three freeway segments defined in the previous section. In Figure 4(a), the flow indicators of lane 4 show that the upstream segment of freeway traffic is influenced largely by the on-ramp traffic. The flow indicators of lane 4 for the merging and downstream segments of freeway traffic have a small fluctuation, as shown in Figures 4(b) and (c). The small fluctuation in the flow indicator of lane 4 in Figure 4(c) comes from the fact that the downstream freeway traffic is the sum of the on-ramp traffic and the mainline traffic. The change pattern of the on-ramp traffic shows a negative correlation with the mainline traffic from Figure 4(a) and therefore they compensate each other.

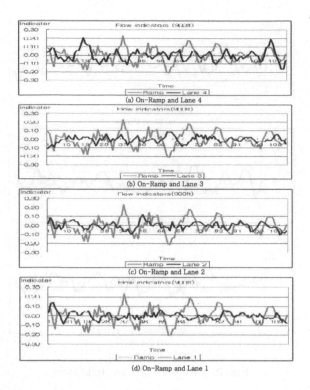

Fig. 3. Flow Indicators of the On-ramp and the Lanes

3.2 Correlation Analysis

Flow

In the previous section, the relationships between mainline traffic and on-ramp traffic are briefly described. In this section, the correlation analyses between two flow indicators at the Santa Monica site are conducted. Figure 5 shows correlation coefficients by 100ft intervals between the mainline flow indicator and the on-ramp flow indicator. Results show that a weak correlation exists between the flow indicators of the median lane and the on-ramp. However, it is shown that for the outside lane and the on-ramp, correlation coefficients are on the increase from 861ft, where the gore nose begins, and a positive correlation exists after 1,000ft. This means that the flow of the outside lane in the downstream segment of the freeway increases as the on-ramp flow increases.

Lane 3, or the lane next to the outside lane, has similar patterns to that of the outside lane, however the difference is that the positive correlation appears from 1,261ft, at which point the acceleration lane ends. In the correlation analysis between the flow indicators of the outside lane and the on-ramp, the values of correlation coefficient(ρ) show a range from -0.4253 to -0.2922 at the section prior to 900ft and from 0.097 to 0.2681 at the section after 1,000ft. To check if the correlation coefficients are statistically significant, a two-sided t-test is

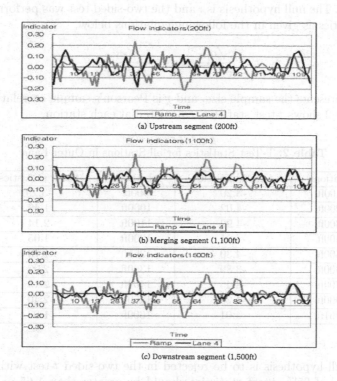

(a) Upstream segment (200ft)

(b) Merging segment (1,100ft)

(c) Downstream segment (1,500ft)

Fig. 4. Flow Indicators Depending on the Detection Stations

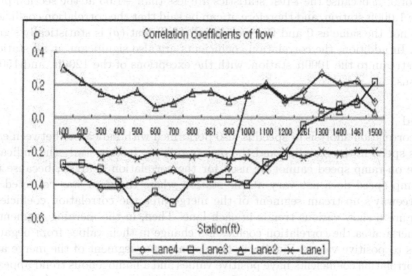

Fig. 5. Correlation Coefficients of the Flow in Each Lane

undertaken. The null hypothesis is : and the two-sided test was performed using
t-test statistics as given in the following equations below.

$$t = \frac{r\sqrt{n-2}}{\sqrt{1-r^2}}, r = \frac{s_{XY}}{\sqrt{s_{XX}s_{YY}}} \tag{4}$$

where n represents the sample size, and r is Pearson's sample correlation coefficient. Table 2 shows t-test statistics calculated at each station.

Table 2. T-Test Statistics for all Stations in Outer Lane

stations	t-test statistics	stations	t-test statistics
100ft	-3.20	900ft	-3.29
200ft	-4.02	1000ft	1.49
300ft	-4.92	1100ft	2.14
400ft	-4.81	1200ft	1.05
500ft	-4.30	1261ft	1.73
600ft	-3.89	1300ft	2.92
700ft	-3.45	1400ft	2.19
800ft	-4.25	1460ft	2.40
861ft	-4.01	1600ft	1.03

If the null hypothesis is to be rejected in the two-sided t-test with the confidence level of 95%, t-test statistics should be greater than 1.65 or less than
-1.65. As shown in Table 2, the null hypothesis is rejected with the confidence
level of 95% because the t-test statistics are less than -1.65 at the section prior
to the 1,000ft station, and therefore, it can be said that the correlation coefficient
(ρ) is not the same as 0 and the correlation coefficient (ρ) is statistically significant. In addition, the correlation coefficients are also significant at all stations
downstream to the 1000ft station, with the exceptions of the 1200ft, and 1500ft
stations.

Speed

The correlation analysis of speed is also performed with indicators between each
lane's speed in the mainline and the on-ramp flows. In this study the indicator
of the on-ramp speed cannot be used for the correlation analysis, because the
on-ramp speed does not vary at the station where the data was collected. In
the freeway's upstream segment of the merge area the correlation coefficients
in Figure 6 show various trends in each lane. Then, in the merging segment of
the merge area the correlation coefficients change in their values from negative
values to positive values. Finally, in the downstream segment of the merge area
the correlation coefficients have positive values and a homogenous trend appears.
Like the correlation analysis of the flow indicators in the previous subsection, in

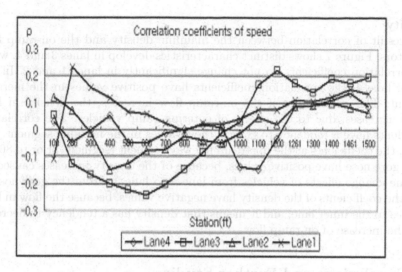

Fig. 6. Correlation Coefficients of Speed in Each Lane

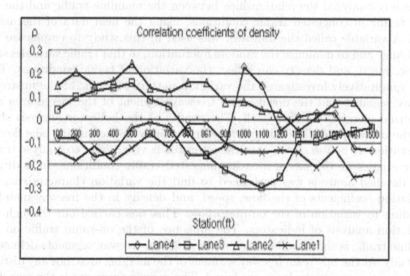

Fig. 7. Correlation Coefficients of Density by Each Lane

the case of lanes 1 and 2 of the freeway mainline, there are no distinct characteristics and the correlation coefficients in lanes 3 and 4 start to change their signs from negative to positive in the merging segment. In the case of lane 3, the correlation coefficients of the speed indicator have negative values to the indicator of the on-ramp flow at the upstream freeway segment prior to the 900ft station, because the mainline vehicles change their speeds in advance to avoid conflicts with the on-ramp vehicles.

Density

As a result of correlation between the mainline density and the on-ramp flow indicators, Figure 7 shows distinct characteristics develop in lanes 3 and 4, while the correlation coefficients do not change significantly in lanes 1 and 2. In the case of lane 4, the correlation coefficients have positive values in the merging segment. This means that if the on-ramp flow increases, the density in lane 4 also increases, due to the effects of entering ramp vehicles. The correlation coefficients have a tendency to change their values in the merging segment. For lane 3, the coefficients at the upstream freeway segment (segment prior to 861ft) of the gore nose have positive values, because of the density increases caused by the lane change effects of vehicles from lane 4 to lane 3. From the gore nose of 861ft the coefficients of the density have negative values, because the flow in lane 3 moves to the inner lane, and it means that density has a tendency to decrease with the increase of on-ramp flow.

4   Conclusions and Further Studies

This study analyzed the relationships between the mainline traffic and the on-ramp traffic in congested traffic conditions, using the field data of the merge areas. A variable called the indicator was used in this study to emphasize the real change and to minimize the random fluctuation, so that traffic variables such as flow, speed, and density can reflect the variation of traffic conditions. This study qualitatively investigated the variation pattern of traffic at the upstream freeway segment and the downstream freeway segment of the merge area and the actual merging segment itself. The results of the indicator analysis show that the on-ramp flow has a negative relationship with the outside lane flow at each station. It is also shown that the indicator is very useful to analyze traffic characteristics at the merge area in congested traffic conditions. In addition, more detailed analysis was performed to find the variation characteristics for correlation coefficients of the flow, speed, and density in the freeway mainline according to variation of the on-ramp flow. This was carried out through the correlation analysis of indicators. The influence of the on-ramp traffic on the mainline traffic is shown to be different for different freeway segments defined in this study. At the upstream freeway segment of the merging area, not any distinct characteristics across the lanes are found. The merging segment is the transient segment, where the correlation between mainline and on-ramp traffic changes. At the downstream freeway segment of the merge area, the traffic across the lanes takes on a stabilized pattern and shows homogenous traffic characteristics. As indicated in the Highway Capacity Manual, it is confirmed by the correlation analysis of the indicators that the ramp influence area is the right two lanes of the freeway mainline. The research results of this study can help to develop a model, in which the on-ramp flow rate could be determined depending on the state of mainline traffic. It can be also applied to current simulation models that have the shortcoming of not being able to determine the on-ramp flow realistically. Results of this research can be used to analyze capacity reduction of downstream traffic and provide a basis to develop the relationship between the mainline traffic and the merging flow from the on-ramp. Based on the relationships between the two traffic flows that are revealed in this study, further research is needed to develop a model that can determine the on-ramp flow under congested traffic, in real traffic situations, and to identify other variables that have an influence on the on-ramp flow.

References

1. TRB: Highway Capacity Manual (1985)
2. TRB: Highway Capacity Manual (2000)
3. Drew, D.R.: Traffic Flow Theory and Control, pp. 217–218. McGraw-Hill, New York (1968)
4. Kim, S.G., Son, Y.T.: Development of a New Merge Capacity Model and the Effects of Ramp Flow on the Merge Capacity. CD-ROM. Transportation Research Board, National Research Council, Washington, D.C. (2003)
5. Smith, S.A.: Freeway Data Collection for Studying Vehicle Interactions (Technical Report). Publication FHWA-RD-85-108. FHWA, U.S. Department of Transportation (1985)
6. Kim, Y.: Online Traffic Flow Model Applying Dynamic Flow-Density Relation. Ph.D. Thesis, Munich University of Technology (2003)

Image Resize Application of Novel Stochastic Methods of Function Recovery*

Daniel Howard[1] and Joseph Kolibal[2]

[1] Defence and Technology Division, QinetiQ plc, St Andrews Road,
Malvern WR14 3PS, UK
[2] Department of Mathematics, University of Southern Mississippi,
Hattiesburg, MS 39406, USA

Abstract. A novel family of stochastic methods developed for function recovery tasks is presented and its properties are discussed in some detail. A new image resize facility based on these new methods is applied to an image and this compares favorably in quality to the application to this image of an equivalent facility from a popular commercial graphics package.

1 Introduction

Stochastic Matrix Methods are a novel approach to function recovery [1,2,4,5,6]. since they combine many of the nice properties of Bernstein polynomials with the benefits that a spectral approximation enjoys over a polynomial approximation. Moreover, this 'hybridization' yields additional emergent benefits that exceed the 'sum' of its 'constituent parts' and which are summarized as:

1. The resulting spectral method can handle non uniformly distributed input data in a straightforward manner;
2. The resulting method possesses a free parameter σ that can be adjusted to give different response curves to the data;
3. As explained in the text there are two opportunities to set the σ parameter enabling: data interpolation ($\sigma_2 = \sigma_1$ causing the function to pass exactly through the data); data approximation (only σ_2 causing the function to pass close to the data); smoothing or peak sharpening (a quasi-interpolation with $\sigma_2 \neq \sigma_1$). Moreover, although σ is discussed here as constant, it is a function $\sigma(x)$ and could be used this way also.
4. The method becomes a useful generalization whereby the error function of the Bernstein polynomial (the mollifier of the new method) can be replaced by other functions, e.g. the arc tangent, resulting in a function recovery of a different character (use of different mollifiers and choices for σ results in a family of methods).
5. The method can be made responsive to changes in the value of data, but when these changes are very large we have developed: (a) a novel family of 'limiters'; (b) a weighting method; and (c) it can be applied to respect a discontinuity.

---

* This work was carried out as part of the Electronic Systems Domain of the Ministry of Defence Research Programme.

M.S. Szczuka et al. (Eds.): ICHIT 2006, LNAI 4413, pp. 114–127, 2007.

When the mollifier is a smooth function such as the error function (e.g. when the error function is chosen for the mollifier it is given the name: 'Bernstein Function') then any order of derivative is immediately available everywhere. This may be of interest to applications such as focusing of a camera or to using the method to reverse engineer the analytical solution to a differential equation [2].

Remaining sections of this introduction describe the method in more detail. There is a lot of scope for improvements with its further mathematical development and vast scope for applications particularly in an adaptive fashion. This paper developed the method into an image zooming algorithm that is shown to compete favorably with the image resize facility of a commercial graphics package.

1.1   Function Recovery with Stochastic Matrix Methods

Recovering the underlying characteristics of noisy data sets is imperative to understanding the behavior of many processes. Intrinsically, this involves finding the best functional representation of the discrete data. In essence the problem is one of reconstructing a function from a limited number of points, and so the task of testing function recovery methods is elegant and uncomplicated: sample a known function, then attempt to reconstruct the function from the sample. The task is made more difficult when there is noise that perturbs the sample data.

The method of stochastic approximation and interpolation [1] using mollifiers, typically those related to probabilistic kernels such as the Gaussian, is able to accomplish this task well even for very noisy data for which the noise approaches zero net deviation from the data on arbitrary subsets of the domain (that is, the noise statistically has a central tendency). For cases where the noise is markedly skewed away from this central limit, the results will reflect the skewing of the noise. The technique when using an error function as the mollifier can be seen to be an extension of Bernstein polynomial approximation (these are the Bernstein functions defined in [1][1]), although the technique has variable behaviour ranging from approximating, to interpolating to de-convolving, and thus it is not merely approximating as are the Bernstein polynomials.

Simplifying to one dimension, a function $f(x)$ is sampled at points $f(x_i) = y_i$ with $x_i \in [0, 1]$. The stochastic interpolant to $\{(x_j, y_j)\}, j = 1, \ldots n$ is given by

$$A_{mn} A_{nn}^{-1} \mathbf{y}. \tag{1}$$

The matrix A_{nn} is a row stochastic matrix whose coefficients consist of the $n \times n$ values a_{jk} containing the discrete cumulative probabilities in which

$$a_{jk} = P(x_k \to x_j), \tag{2}$$

that is to say, the probability that the datum at x_j is influenced by the datum at x_k. Similarly, in constructing A_{mn}, its coefficients consist of the $n \times n$ values $\tilde{a}_{jk}$ containing the discrete cumulative probabilities in which

---

[1] The technique is general and admits functions other than the error function as the mollifier, e.g. the arc tangent.

$$\tilde{a}_{jl} = P(x_l \to x_j), \tag{3}$$

that is to say, the probability that the datum at x_j at which the function is being evaluated is influenced by the data at datum at x_l. Interpolation requires that $a_{jk} = \tilde{a}_{jl}$ if $x_k = x_l$. Setting the partition values as $z_k = (x_k + x_{k+1})/2$, and choosing the generator of the row space of A_{nn} to be

$$a_{jk} = \frac{1}{2}\left[\text{erf}\left(\frac{z_{k+1} - x_j}{\sqrt{\sigma n}} \right) + \text{erf}\left(\frac{x_j - z_k}{\sqrt{\sigma n}} \right) \right], \tag{4}$$

and the row space of A_{mn} to be

$$a_{jl} = \frac{1}{2}\left[\text{erf}\left(\frac{z_{k+1} - x_l}{\sqrt{\sigma n}} \right) + \text{erf}\left(\frac{x_l - z_k}{\sqrt{\sigma n}} \right) \right]. \tag{5}$$

results in stochastic Bernstein interpolation. This provides an elementary example of stochastic function recovery that is extensible to more complex generating functions, particularly those involving limiters to accommodate highly irregularly structured data with rapid variations.

By choosing σ differently in generating the row space of A_{mn} from that of A_{nn} (considering the product $A_{mn}(\sigma_2) A_{nn}^{-1}(\sigma_1) \mathbf{y}$) it is possible to: filter, or smooth if $\sigma_2 > \sigma_1$; interpolate if $\sigma_2 = \sigma_1$; or de-convolve if $\sigma_2 < \sigma_1$ (see [6] for illustrations). By choosing $\sigma_1 = 0$, $A_{nn} = I_{nn}$ and the technique only approximates the data $\mathbf{y}$.

The extension of stochastic approximation and interpolation to the problem of function recovery provides advantages in that the method can be tuned to approximate or interpolate, and thus to represent with less or greater fidelity the data at each point as lying close or not so close to the recovered function. The method has excellent representational fidelity and appears to be a good candidate for function recovery from experimental measurements which contain significant random clutter. When used to achieve convolution-deconvolution using a Gaussian kernel, it provides the user with the ability to do peak sharpening [6].

The performance of the method on functional recovery tests performances on one dimensional data sets. The method can also be applied to "zoom up" an image, and the results are compared with the typical image resize function available in a popular commercial image handling package. Application on a test image gave excellent results (figure 6). For more complex images and to obtain higher quality still, the method can be implemented similar to function recovery, but using a more sophisticated choice for the transitions probabilities $P(x_k \to x_j)$ and $P(x_l \to x_j)$, incorporating peak sharpening as well as the introduction of limiters to correctly capture rapid transitions in intensity exhibited in image data. This would allow for edge preservation, while providing smooth interpolation of smoothly varying data.

1.2   Stochastic Bernstein Function Recovery

The Bernstein functions are the natural extension of the Bernstein polynomials and, in their most useful form, reduce to the familiar Gaussian probability density function (pdf) in which the terms in the stochastic matrix are generated using the error function.

While any smooth function will suffice that is normalized so that it sums to one, such that it is a pdf[2], the Gaussian is perhaps of greatest utility in function recovery for problems associated with random errors perturbing the data. Using the error function introduces a free parameter, $\sigma(x)$ as seen in (4) and (5), allowing the data recovery to be tuned to select from a family of smooth functions which approximate or interpolate the data. Moreover, the process of multiplying the vector of data values by A_{nn} can be shown to be equivalent to a discrete convolution process, and thus multiplying the data by A_{nn}^{-1} can be shown to be equivalent to deconvolution, giving rise to a deconvolution, convolution approach to data recovery.

Function recovery using stochastic Bernstein approximation or interpolation is simple and straight forward and can be achieved efficiently. The approach can handle non-uniformly distributed data (non uniformly spaced input data) while approximating or interpolating this data with a globally smooth function. Moreover, non-uniformly distributed output values can also be obtained. In the case that the mollifier has analytical derivatives, e.g. the error function, the derivatives of the approximant can also be obtained to any order of derivation, as the recovered function is infinitely differentiable. Derivative information is desirable in certain applications.

In [2] we compared these derivatives to those recovered by a localized discrete differences formula (a finite differences method) in the context of the solution of non-self-adjoint differential equations. The derivatives from the stochastic Bernstein matrices will not be identical in general to those that are recovered with finite difference and in a given application [2] the former may be smoother than the latter.

Computational costs are low usually involving matrix vector products only. The solution of a matrix system is required in the case of stochastic Bernstein interpolation but this can be achieved in the same order of operations as a matrix multiply. This is achieved because either a pseudo-inverse is available or the stochastic matrix can be factored into a non-stochastic Toeplitz matrix and a diagonal matrix. The Toeplitz matrix can be inverted with a fast Toeplitz matrix solver in $\mathcal{O}(n^2)$ operations. Moreover, the size of the matrix system can be minimized if one divides (in appropriate places) the data input into a number of rather small non-overlapping patches (small matrices). In such a case, the cost of even the full matrix inversion operation becomes trivial.

In summary the approach has the advantages that it is:

1. Robust, the technique will not fail regardless of irregularity or roughness of data;
2. Adaptable to any data ordering in any dimension;
3. Hierarchical, specifically, data can be added at any time to improve the results;
4. Global so that recovery of harmonic functions improves as additional data is accumulated in time; and,
5. Adaptive through:
   (a) Windowing;
   (b) Tuning the convolution coefficients which can be functions of the data; and,
   (c) choosing the probability density functions used to reconstruct or recover the function using statistical correlation.

---

[2] For example the arc tangent function.

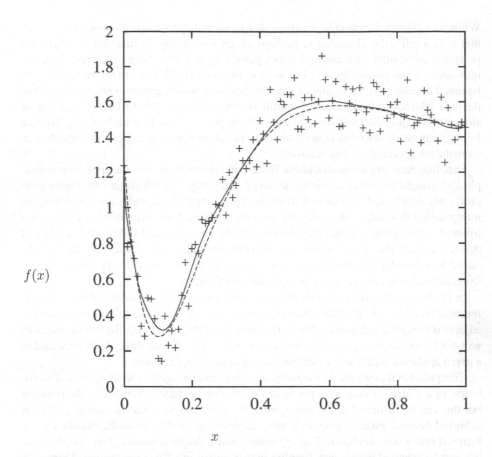

$f(x)$

Fig. 1. Illustrating function recovery using Bernstein functions, K_n, based on sampling the function $f(x) = u(x) + \epsilon(x)$ (crosses). The smooth function $u(x) = 4.26(e^{-3x} - 4e^{-6x} + 3e^{-9x}) + 1.28$ is shown using a dotted line and a solid line shows K_n recovered from the data, $\{(x_k, f_k)\}$. The perturbations are obtained using a normal distribution $N(0, 1)$ scaled by 0.2 to perturb $u(x)$.

1.3   Illustrating Data Recovery Using Stochastic Bernstein Approximation

Consider Fig. 1. Starting with an underlying smooth function, sampling the function and adding noise at each sample location, the stochastic Bernstein function recovery obtains a smooth representative function for the data based on the noisy data. In this case because of the underlying assumption that the data was noisy, the approach taken was to approximate the data in order to achieve the function recovery.

The following questions and answers are offered with respect to Fig. 1:

- **Question:** It looks a reasonable fit. Is it? *Answer:* A score of methods may give similar results. The method is useful because it gives good results in many cases without tweaking it, but it can be tweaked if needed.

- **Question:** About the data, is the x-coordinate uniformly sampled on $[0,1]$? *Answer:* It was uniformly sampled, however it need not be (the method works on any arbitrary grid).
- **Question:** About the data, is the variance constant across $[0,1]$? *Answer:* The random perturbations generated were applied uniformly. Other models are possible but that was the simplest.
- **Question:** About the fitted curve, what are the boundary conditions? The fit near $x = 0$ looks good, but it does not seem to be supported by the data in that region. Some standard methods (e.g. a standard kernel-based method) may perform poorly at the boundaries, but there exist modifications to handle that. You could probably get a similar fit with splines with appropriate boundary conditions. *Answer:* There are no boundary conditions, and the results are from the standard linear model, that is, no attempt was made to introduce any corrections using feedback from any additional information although doing so could have enhanced the results. The method of convolution does not come from solution of a differential equation.
- **Question:** About the fitted curve, how do you determine model order? The wiggle towards $x = 1$ seems odd, but who is to say that it is not supported by the data? *Answer:* The model does not have order since it is non-polynomial. Instead, there is a measure of the distance that the curve is away from the original data. The process only convolves the data with the mollifier, and thus smooths it further away or closer to the original curve.
- **Question:** About the method, can the method be used to obtain confidence bounds on the curve? *Answer:* The model by itself has no statistical components, and since it is not an $L2$ fit, there are no best fits, only families of functions which will be appropriate.

1.4   Recovery of Noisy Data: Families of Functions

Figures 2 to 4 illustrate equivalent approximations used to achieve function recovery from numerical data which is dominated by random noise.

The data are shown as red (dark gray) crosses, and the blue (dark gray) line shows the data connected with straight line interpolation from point to point. The green (light gray) curve shows the recovered function. The underlying function used to generate these data is

$$\cos(3x) \exp\left(|\cos\left(\exp((x+5)/3)\right)|\right).$$

The recovered function is similarly close to the data in all cases in terms of the evolution of the function, and the only difference in these cases is the values of σ_2, the parameter associated with the matrix A_{mn}. Note that choosing σ_2 very small results in a model in which the noise is increasingly regarded as data, so that as $\sigma_2 \to 0$, the approximating curve attempts to approximate the oscillatory data so closely that it nearly interpolates the data. In contrast, as σ_2 gets large, the behaviour of the recovered curve becomes more nearly the same, and while in the limit as $\sigma_2 \to \infty$, the function would approach a linear function attempting to fit all of the data, the behaviour for most large values of σ_2 is only slowly varying.

For a large range of values, as shown by this example, with values of σ going from the hundreds to the thousands, the qualitative features of the recovered function are quite

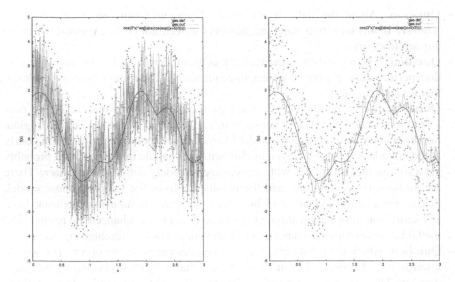

Fig. 2. Function recovery with a parameter value $\sigma = 1$ (left) and $\sigma = 10$ (right). As $\sigma \to 0$, the approximating curve approaches a piecewise constant curve which interpolates the data at the midpoint of each step.

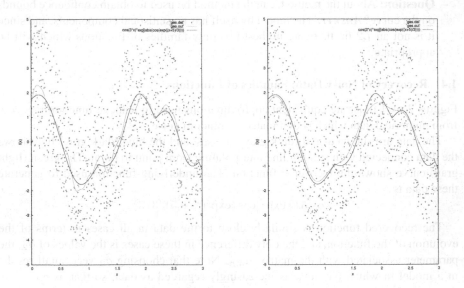

Fig. 3. Function recovery with a parameter value $\sigma = 100$ (left) and $\sigma = 1,000$ (right)

similar, tending to be more responsive for smaller values of σ, and less responsive, resulting in some over-smoothing for values of σ that are too large. It is recognized that since the recovered function is not unique there can be difficulties in finding an optimal solution to achieve the recovery, and this process is made even more complicated if the

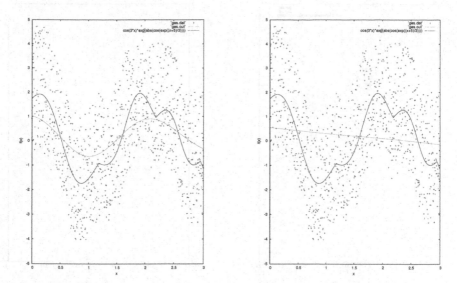

Fig. 4. Function recovery with a parameter value $\sigma = 10,000$ (left) and $\sigma = 100,000$ (right). As $\sigma \to \infty$, the approximating curve approaches a straight line going through the data.

parameter depends on the location or the data. However, the advantage of the approach is its flexibility, particularly when combined with other algorithms which can test the fitness of the results. This appears to be particularly the case when combining the approach with genetic algorithms to test the fitness of the recovered underlying function.

1.5 Iterative Feature Extraction

Figure 5 (Top Left) illustrates this feature extraction in the region from $x = 1.7$ to $x = 2.3$ (approximately). The purpose is to enhance the recovery in the region of Fig. 3 on the right which because $\sigma = 1000$, has slightly over-smoothed the data in the region of the double hump. The iterative recovery is the result of the following procedure:

1. Take the data from the function recovery that used 1,000 as the deconvolution parameter (right plot of Fig. 3).
2. Extract the smooth function on a dense set (about 1,000 points) in the region of interest.
3. Re-approximate using the previous approximation.

The resulting curve in Fig. 5 (Top Left) does a much better job of approximating the function in this region. The procedure amounts to cloning this segment of the curve, intensifying its attributes and then extracting the smooth underlying feature from this curve. This may be compared with the response to function recovery using $\sigma = 1,000$ as the parameter (right plot of Fig. 3) from $x = 1.7$ to $x = 2.3$. It illustrates a "windowing properties" feature in which attention is restricted to a subset of the data.

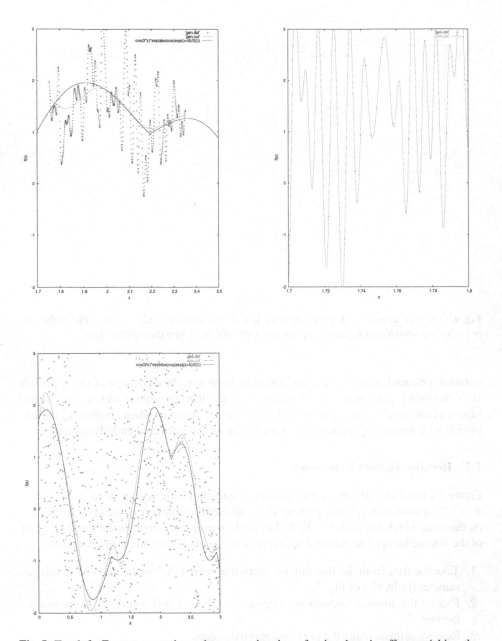

Fig. 5. Top left: Feature extraction using approximation of noisy data, in effect enriching the amount of information in the windowed region. Top right: A more detailed look at the interval [1.7, 1.8] showing stochastic interpolation of the noisy data. The purpose is to enrich the information available for subsequent approximation of the data. Bottom left: Function recovery using interpolated enrichment. The method finds all of the features of the underlying curve.

An important consideration of any method which aims to represent functions is how well it can do that task. Ultimately, what matters is the representational fidelity of the method. When it gets that right, everything else follows. Thus, a revisit to the case used in Figs. 2 to 4 is warranted, this time to interpolate the data along with the noise. The purpose is to improve the feature extraction, following the methodology used in highlighting the features of the data in Fig. 3, however this time using stochastic interpolation on the data instead of stochastic approximation. Whether approximation or interpolation is used, the motivation is to glean additional information about the noise, however the difference between these approaches is that interpolation will enhance the oscillatory behavior of a noisy function, while approximation only smooths it.

The first experiment takes the 1,000 points and interpolates these to 10,000 points uniformly spaced on the interval [0,3]. This gives a richer characterization of the data, finding a representative smooth function which fits the data and the noise. This is a lot of data. Thus, Fig. 5 (Top Right), provides a closer look in the interval [1.7,1.8] showing representative details from the interpolated function. From these figures, and in particular from Fig. 5 (Top Right), it is noticeable that the interpolant has peaks and valleys where the 1,000 data set did not. In effect, the method is predicting that the representative smooth function which interpolates the noisy data would be shaped as shown in these two figures for the value of $\sigma = 1.0$.

Instead of using only the given data for feature extraction, it is now possible to use the interpolated data containing 10,000 additional points to achieve feature extraction. Figure 5 (Bottom Left), displays the results that speak for themselves. The method finds all of the features of the underlying curve, except for the sudden drop-off at $x = 3$. However, there is not enough information at this point to discern that from the data. It slightly overshoots and smooths out the features, but given the level of noise, the extra step of interpolating the data has done credibly well at feature extraction. It has correctly predicted all of the features of the underlying smooth function.

2 Applications of the Method: Image Zoom

So far the method has these applications: (a) for image compression (forthcoming); (b) as a simpler and tunable alternative to the complex polynomial fit methods for removing the baseline shift often produced by a Matrix Assisted Laser Desorption Ionization-Time of Flight-Mass Spectrometry (MALDI-TOF) instrument [4]; (c) as progression to an earlier effort [3] to solve the convection-diffusion equations by a new method that circumvents the lack of ellipticity that characterizes solution of non-self-adjoint differential equations by traditional weighted residual methods [2]; (d) as an alternative to the Nonuniform rational B-spline (NURBS) for rendering of surfaces in computer graphics [7], where sometimes it provides either higher quality or more efficiency than the polynomial Bezier curves approach.

One application of the technique is for zooming up an image. This can be achieved in a straightforward manner, for example using a pixel zoom in which each pixel is doubled in size, or using more complex algorithms which attempt to correctly recover the shape of the two-dimensional image surface. Many algorithms simply engage in a pixel zoom and then smooth the resulting image using a Gaussian blur applied to the zoomed pixels. For computational efficiency, the pixels of the image are taken as a uniformly distributed

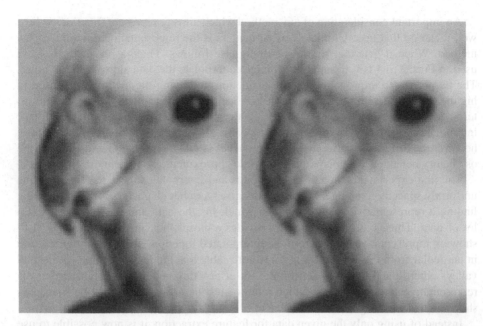

Fig. 6. Left: Commercial package image resize. Right: stochastic function recovery image resize.

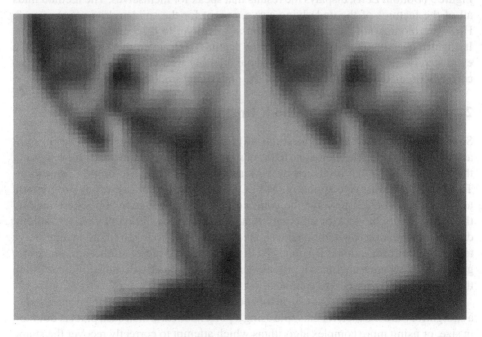

Fig. 7. Detail. Left: Commercial package image resize. Right: stochastic function recovery image resize.

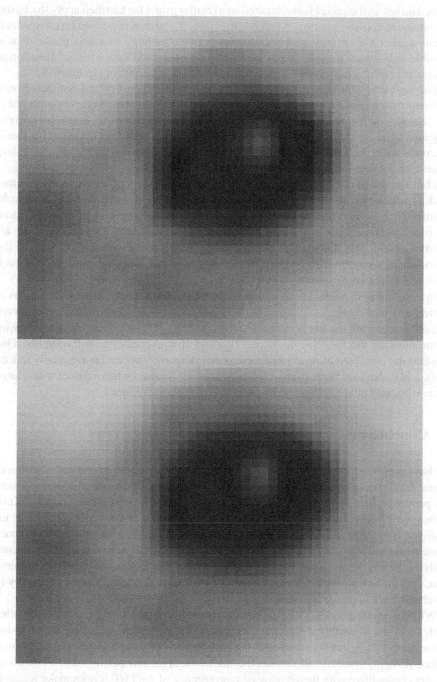

Fig. 8. Detail. Top: Commercial package image resize. Bottom: stochastic function recovery image resize.

input data set, so the image is constructed on a regular grid. One can then apply stochastic Bernstein interpolation in two alternative directions x and then in y to obtain interpolated values again on a uniform grid but one that is twice as large as a new image. It was possible to compare the quality of the result to the standard "image resize" function in a popular commercially available desktop application as shown in Figures 6–8.

The problem with image zooming is one of function recovery and not merely one of interpolation, since there is the requirement that edges be preserved. Smoothly varying regions require the usual interpolation. In this case stochastic function recovery may also be achieved using transition probabilities which reduce to the usual Bernstein functions when the data is smooth, but which are non-linear when the gridding is widely spaced, or the data values vary rapidly.

Thus the use of the stochastic formulation for interpolation provides a robust framework in which to construct a variety of kernel functions to develop the transition probabilities $P(x_k \rightarrow x_j)$ and $P(x_l \rightarrow x_j)$. Some of these functions are ad-hoc, and some are based on functional shape matching, however regardless of the structure, the algorithm assures that the result is interpolating so long as the row sums in each of the matrices are normalized to one and the transition probabilities are the same for each point in the matrix A_{mn} as in A_{nn}.

In developing these functions to handle two dimensional data, the use of limiters is particularly useful because it provides the required non-linearity within the framework of the linear model: the linear matrix product, $A_{mn}A_{nn}^{-1}$. Limiters are introduced easily by suppressing the range of the kernel function if it exceeds preset thresholds. This prevents the over-weighting of large regions that frequently occurs on irregularly spaced grids, and it also avoids the over-weighting of function values when a datum with a large value is located next to others with values that are relatively small.

3   Conclusions

We have presented a simple linear model for function recovery from one and two dimensional data, specializing our interest in noisy data. Using stochastic function recovery has proven to be a robust method for achieving excellent computational results. Much of this work is preliminary, however it provides a framework for further study and investigation. Clearly, the proposed method is not built within a standard Hilbert space framework, and the usual uniqueness that comes with L_2, i.e., least squares and spectral analysis, is lost. In its place, though, is a technique that in the limit tends to preserve area, and thus provides a mechanism for conveniently achieving function recovery in which measure is preserved.

While we have not elaborated on the computational details, nor included any detailed analysis of the algorithmic efficiency, in general the approach, including the inversion of the matrix A_{nn}, can be achieved at the cost of matrix-vector and matrix-matrix multiplies. Roughly speaking, this makes the cost of the algorithms $\mathcal{O}(n^2)$, and while this is not yet competitive with the $\mathcal{O}(n \log n)$ convergence of the FFT, it does make it usable for many applications. This is particularly the case when applying kernel functions with implicit functional non-linearities, such as the limiters, which then allow a non-linear algorithm to run in the same operation count as the linear algorithm.

References

1. Method and apparatus for approximating, deconvolving and interpolating data using Berstein functions. U.S. Provisional Patent Application No. 60544,975 #20050203982 (2004)
2. Howard, D., Kolibal, J.: Solution of differential equations with genetic programming and the stochastic bernstein interpolation. In: Biocoumputing-Developmental Systems Group, University of Limerick Technical Report No. BDS-TR-2005-001. University of Limerick, Limerick, Ireland (June 2005),
   http://www.genetic-programming.org/hc2005/bds.pdf
3. Howard, D., Roberts, S.C.: Genetic programming solution of the convection-diffusion equation. In: Spector, L., Goodman, E.D., Wu, A., Langdon, W.B., Voigt, H.-M., Gen, M., Sen, S., Dorigo, M., Pezeshk, S., Garzon, M.H., Burke, E. (eds.) Proceedings of the Genetic and Evolutionary Computation Conference (GECCO-2001), 7-11 July 2001, pp. 34-41. Morgan Kaufmann, San Francisco, California, USA (2001)
4. Kolibal, J., Howard, D.: Maldi-tof baseline drift removal using stochastic Bernstein approximation. EURASIP Journal on Applied Signal Processing, Special Issue on Advanced Signal Processing Techniques for Bioinformatics 2006, 1-9 (2006)
5. Kolibal, J., Saltiel, C.: Data regularization using stochastic methods. Submitted to SIAM (2005)
6. Kolibal, J., Howard, D.: The novel stochastic Bernstein method of functional approximation. In: Adaptive Hardware and Systems, 2006. AHS 2006. First NASA/ESA Conference on, Istanbul, Turkey, 15-18 June 2006, pp. 97-100. IEEE Computer Society Press, Los Alamitos (2006)
7. Seyfarth, R., Kolibal, J., Howard, D.: New mathematical method for computer graphics. In: Proceedings of 2006 International Conference on Hybrid Information Technology (ICHIT 2006), 11-13 November 2006, pp. 8-12. IEEE Computer Society Press, Los Alamitos (2006)

Automatic Face Analysis System Based on Face Recognition and Facial Physiognomy

Eung-Joo Lee[1] and Ki-Ryong Kwon[2],*

[1] Dep. of Information Communication Eng., TongMyong University, Korea
Ejlee@tu.ac.kr
[2] Div. of Electronic, Computer&Telecomm. Eng.,
Pukyong National Univ., Korea
krkwon@pknu.ac.kr

Abstract. An automatic face analysis system is proposed which uses face recognition and facial physiognomy. It first detects human's face, extracts its features, and classifies the shape of facial features. It will analyze the person's facial physiognomy and then automatically make an avatar drawing using the facial features. The face analysis method of the proposed algorithm can recognize face at real-time and analyze facial physiognomy which is composed of inherent physiological characteristics of humans, orientalism, and fortunes with regard to human's life. The proposed algorithm can draw the person's avatar automatically based on face recognition. We conform that the proposed algorithm could contribute to the scientific and quantitative on-line face analysis fields as well as the biometrics.

Keywords: Face Recognition, Facial Physiognomy, Face Analysis, Avatar Drawing.

1 Introduction

Due to increasing demands for computer users security, authentication, and biometric is a rapidly growing research area. Many on-going studies on recognition systems are based on biometrics that use characteristics of human body such as face, fingerprint, gesture, signature, voice, iris, and vein. Among them, Face recognition is an important research part of image processing with a large number of possible application areas such as security, multimedia contents and so on[1].

Generally, face recognition method uses a three-step approach. First face is detected from input image and then a set of facial features such as eyes, nose, and mouth is extracted. Lastly, face recognition is employed by the proposed measures. However, human face may change its appearance because of external distortions such as the scale, lighting condition, and tilting as well as internal variations such as make-up, hairstyle, glasses, etc. And many researches have been conducted to improve face detection and recognition efficiency as well as

---

* Corresponding author.

M.S. Szczuka et al. (Eds.): ICHIT 2006, LNAI 4413, pp. 128–138, 2007.

overcome these problems[2-5]. Among them, it is difficult to analyze face with a complex background and distortion in an image. Therefore, for the progress of multimedia content field as well as security, it is necessary to study the analysis of the face as well as improve the robustness of the face recognition system. Thus, we proposed an automatic face analysis system based on face recognition and facial physiognomy in multimedia environments such as internet, mo-bile-phone and PDA. The goal of our study is to detect person's face, extract facial features such as eyes, eyebrows, nose, mouth and facial shape, classify the shape of facial features, analyze facial physiognomy based on facial recognition, and draw avatar automatically which is resembled to person's face.

To do our work from face image, we first calculate Cr and I component which is component of YCbCr and YIQ color model, respectively. And also we convert these components into binary image and detect face using logical product of Cr and I component. Then we extract facial features and classify it finely. Namely, it is first to find a face in which the facial feature is guaranteed to be and then use a proposed facial feature extraction algorithm. For the extraction of facial features, we use face geometry based approach which is the use of vertical and horizontal projections, labeling, relative distances between features, and geometrical characteristic of face. And also, we suggest facial feature classification method particularly to find finely the shape of facial features for face analysis and avatar drawing. In the facial feature classification part, classification of eyebrows, mouth, eyes, facial shape, and hairstyle depends on the geometrical feature of face, pixel gray levels, and distribution of pixels in feature. Lastly, we recognize user's face using face feature vector function and analyze facial physiognomy, and make avatar drawing automatically using these extracted and classified facial features. In the facial physiognomy analysis step of proposed algorithm, we can recognize the person's face at real-time and analyze facial physiognomy which is composed of inherent physiological characteristics of humans, orientalism, and fortunes with regard to human's life.

This paper is organized as follows. First, we describe how the face is detected in a real input image by using multi-color model component and how the facial feature is extracted by facial geometry. And also face tracking, correction, and recognition are explained. In section 3, the analysis of facial physiognomy and drawing of facial avatar is also explained. In section 4, experimental results on a proposed algorithm are given.

2 Face Detection, Tracking, Tilted Face Correction, and Recognition

2.1 Face Detection Using Multi-color Model

There are several approaches to detect face such as color based method, ellipse fitting method, etc. Face detection based on the skin color is a very popular method because the human face has much skin color component than others. Ellipse fitting method is to ap-proximate the shape of the face since human face

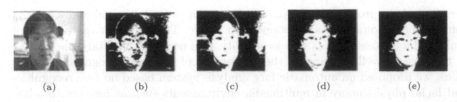

(a)   (b)   (c)   (d)   (e)

Fig. 1. Facial skin color detection in daylights; (a) Original image, (b) resulting image using H component, (c) Cr component, (d) I component, (e) multi-color model

resembles to an ellipse. However, color information is not sufficient alone with external distortion such as complex background and also ellipse based method is sensitive to internal variations[2]. This paper proposes an algorithm which can detect face accurately by multi-color model, regardless of the conditions of lighting and backgrounds. It also enables recognizing automatically tilted faces by correcting face rotation. In the face detection part, we follow the multi-color based approach using YCbCr and YIQ color model which is robust to the complex background or multi-faces in an image. In this step, we calculate Cr and I component which is component of YCbCr and YIQ color model, respectively. And then, we convert these components into binary image and then detect face using logical product of Cr and I component. One way of detecting face from the captured image by CCD camera is either to use skin color information after transforming it to HSI, YCbCr, and YIQ color space[6]. Each space has corresponding skin color region. Since in HSI color space, H contains the component of face color, Cr for YCbCr, and I for YIQ, detecting facial region can be possible. Still, hue component is sensitive to illumination and can be a cause of heavy background noise because face is detected from the skin color. And Cr component can help detect the clear image of face, and yet can not eliminate the noise. Lastly, I component plays a role in accurate face detection with no background noise. And it is luminance-resistant as well. Fig. 1 shows the results of facial skin color detection, using each color model. As shown in Fig. 1(b), the image, which is a result of using H component to detect the face color with daylight, has background noise and some other colors according to the illuminants. And also, Fig. 1(c), the resulting image using Cr component has clear face area itself but it also has background noise. At the same time, the resulting image with I component, as can be seen Fig. 1(d), simply shows strong face color, whether with day-light or not, which is different from the Fig. 1(b). However, I component covers a restricted area and sometimes can't detect face accurately when the object has not movement in the face tracking step. That is why we used multi-color model to detect only face. The face candidate is segmented by labeling, which connects pixels that have the same component in the image. And then, we use facial geometry ratio(horizontal to vertical length of face; 1:1.4 1:1.8) as well as vertical horizontal projection to detect face itself and eliminate background noise. The resulting equation for multi color modeling is as follows.

$$I_{mc} = I_{Cr} \cdot I_{In-Phase} \tag{1}$$

2.2   Face Tracking and Tilted Face Correction Using Eyes Feature

In this paper, a face tracking and recognition algorithm has two stages, motion information is used for tracking and face feature vector function is used for recognition. Tracking algorithm of the proposed algorithm first sets an object selected from the previous frame as a target object and calculates Euclidean distance between the objects from the frame and those from the previous to find one with smallest distance. The face tracking algorithm in this paper tracks the face with the least Euclidean distance from the area detected by using the multi-color model to each detected face from the next frame. If it fails to find the object to track, it sets the area detected from the current frame as a target object again and tracks the face real-time. Following Equation is to detect the target object to track. Here, $F\_Tracking_{object}$ represent target to tracking. Oj_{t-1} and Oj_t represent target object to previous frame and the frame.

$$F\_Tracking_{object} = min \sum ||Oj_{t-1} - Oj_t|| \tag{2}$$

Generally, human face may change its appearance because of external distortions. Especially, it is difficult to recognize and analyze person's face with a background noise and a tilting problem. Thus, we employed face correction method using eyes feature to solve the tilting distortion as well as face detection method based on multi-color model to improve the background noise and illumination problem as mentioned above. In the facial feature extraction step of proposed algorithm, we use face geometry information for the extraction of facial features such as eyes, nose and mouth. Face geometry can describe the geometrical shape of face from the coordinates of features. It is consists of each feature coordinate, center coordinate of eyes, length of left and right eyes, distance between eyes and mouth, and relative distance and ratio between features. The distance between two features can be approximated by their horizontal and vertical distance. In this case, the facial coordinates will be invalid when the slope of face is changed. Namely, if the face is tilted, face geometry is no longer valid, and a correction of the orientation of the face is necessary. Thus, we correct the tilted face with the correction of eyes slope. For a face almost frontal and upright, the slope between two eyes can be approximated by their horizontal angle. In this paper, in order

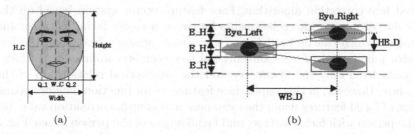

(a)   (b)

Fig. 2. Facial feature coordinate system; (a) Facial coordinate and (b) rotation angle of eyes feature

to recognize the images of both the rotated and the front faces, we extracted one of the facial features, eyes and corrected the facial rotation. We employed the horizontal Sobel edge operation within the detected face and extracted features applying labeling to the objects crowd in a certain range. As shown in Fig. 2, the left eye and its eyebrow are positioned at Q_1 that points to 1/4 of the width of the whole face, and the right eye and the eyebrow are at Q_1, 3/4 of the width. Each eyes object has the feature information that it is located at H_C, 1/2 of the height of the face, in the range of the width of the eyebrow objects. Although we can use this information to extract eye feature, it is better to expand the range as wide as the height of the eye object itself, E_H, as can be seen in Fig. 2(b). Rotation angle ($Angle_{Face}$) for the rotation correction is computed by dividing the width value (WE_D) between the two eye objects by the height value between the objects (HE_D) and applying tan^{-1} to the result. Fig. 3 is the resulting image of the eyes detection in various tilted images. Fig. 4 is the resulting image after the facial rotation was revised by the rotation angle.

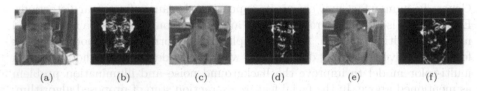

(a)   (b)   (c)   (d)   (e)   (f)

Fig. 3. Eyes feature detection; (a) Eyes detection in frontal face image, (b) vertical histogram of (a), (c) eyes detection in rightward tilted image(-15^0), (d) vertical histogram of (c), (e) eyes detection in leftward tilted image($+30^0$), and (f) vertical histogram of (e)

2.3   Face Recognition Using Face Feature Vector Function

Generally, face recognition method uses a three-step approach. First face is detected from input image and then a set of facial features is extracted. Lastly, face recognition is employed by recognition approach. This chapter mainly deals with the part of recognition, recognizing the face image by face feature vector function, which has been detected and corrected using both multi-color model and tilted face correction algorithm. Face feature vector system based on the person's face geometry, interrelationship be-tween person's facial features, and facial angles, which can be used to recognize each person's face using the feature vector found on the face. Conventional face geometry approaches of facial feature considered only the symmetry and the geometrical relationship of the person's face. However, in this paper, face feature vector function includes geometry value of facial features using the eyes, nose and mouth, correlation value between the person with facial feature, and facial angles of the person's face. Fig. 5 shows the face feature vector function for face recognition. If one considers the characteristics of the person with facial characteristics on Fig. 5, Equation GD_1

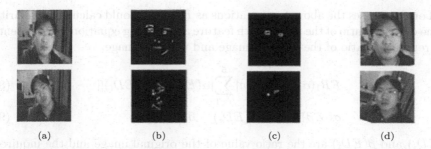

(a)   (b)   (c)   (d)

Fig. 4. Tilted face correction; (a) Tilted face image, (b) facial feature extraction, (c) eyes extraction, and (d) tilted face correction

is the distance between both eyes using the eyes' symmetry, GD_2 is the distance between the mid point between the eyes and the center of the nose(P_{NC}), GD_3 and GD_4 are the distance between the center of the nose(P_{NC}) and the center of the mouth(P_{MC}), and the distance between the center of the mouth(P_{MC})and the center of the chin(P_{CC}), respectively.

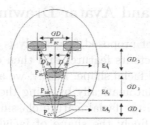

Fig. 5. Face feature vector function

And also, facial angle EA_1, EA_2, and EA_3 with facial characteristic can be extracted from using the cosine function property. Following equation represent the relative distance between the person with facial feature vector.

$$ED_1 = Abs[1.0 - \{0.7 \times \frac{GD_2}{GD_1}\}] \tag{3}$$

$$ED_2 = Abs[1.0 - \{1.1 \times \frac{GD_1}{(GD_2 + GD_3)}\}] \tag{4}$$

$$ED_3 = Abs[1.0 - \{1.6 \times \frac{GD_1}{(GD_2 + GD_3 + GD_4)}\}] \tag{5}$$

$$ED_2 = Abs[1.0 - \{1.7 \times \frac{GD_3 + GD_4}{(GD_2 + GD_3 + GD_4)}\}] \tag{6}$$

$$ED_5 = Abs[1.0 - \{1.5 \times \frac{GD_4}{GD_3}\}] \tag{7}$$

If one expresses the above five equations as ED_i, one could calculate similarity of the distance ratio of the person with feature as following equation. It represents the repetition ratio of the original image and the test image.

$$FR_1(\alpha, \beta) = Min| \sum_{i=1}^{5} [\alpha(ED_i) - \beta(ED_i)]| \qquad (8)$$

$$\sigma(\alpha, \beta) = 1 - |\alpha(ED_i) - \beta(ED_i)| \qquad (9)$$

$\alpha(ED_i)$ and $\beta(ED_i)$ are the ratio value of the original image and the inquired test image value and $\sigma(\alpha, \beta)$ is the correlation similarity of the two images. And also, following equation represent facial angle repetition rate and facial angle correlation similarity. Where, if $\sigma(\alpha, \beta)$ is close to 0, it corresponds more to the feature.

$$FR_2(\alpha, \beta) = Min| \sum_{i=1}^{3} [\alpha(EA_i) - \beta(EA_i)]| \qquad (10)$$

$$\sigma(\alpha, \beta) = 1 - |\alpha(EA_i) - \beta(EA_i)| \qquad (11)$$

3   A Face Analysis and Avatar Drawing System

The whole block-diagram of proposed face analysis algorithm is as follows; it first detects face and extracts its features and then classifies the shapes of the eyes, eyebrows, mouth, and facial shape. Lastly, we analyze facial physiognomy and make avatar drawing automatically by using these extracted and classified facial features. In this part, the facial feature classification method is also proposed particularly to find finely the shape of facial features and classify the physiognomy of face. Namely, it is first find a facial in which the facial feature is guaranteed to be and then use a proposed facial feature extraction algorithm to extract the shape. In the facial feature classification part of proposed algorithm, the detection of eyebrows, mouth, eyes, and facial shape depends on the face geometry, the pixel gray levels, and the distribution of pixels. In the facial physiognomy analysis step, we analyze the person's facial physiognomy which is composed of inherent physiological characteristics of humans, orientalism, and fortunes with regard to human's life. And also, in the facial avatar drawing step, the proposed algorithm can draw user's avatar automatically based on face recognition. It can be used and applied as a content of internet entertainment, mobile phone and PDA, and criminal prevention system such as making a montage picture and searching a criminal. In the facial feature classification step, we first extract eyebrows from the detected face image. Generally, eyebrows are on the upper side of eye, thus we define that it has same height as eyes and 1.5 times as length as eyes. For the classification of eyebrows, we used the thickness and corner of eyebrows in 3×3 sub region. For the classification of eyebrows thickness, we calculate the number of pixels in the left side and the right side horizontal area of eyebrows. We classify the thickness of eyebrows into left side, equal, and right side type, respectively. In

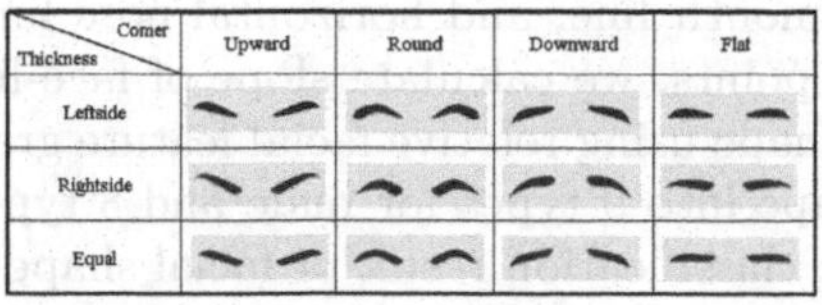

Fig. 6. Classification result of eyebrows

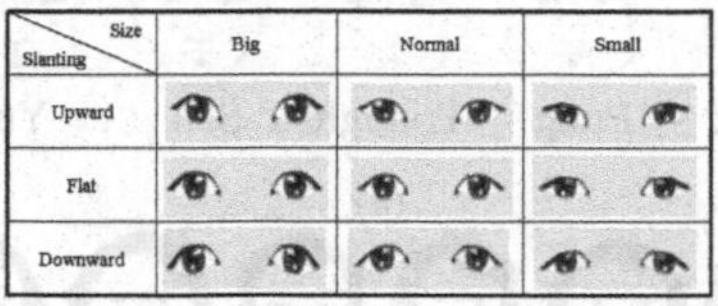

Fig. 7. Classification result of eyes

the corner of eyebrows classification step, we calculate each pixel distributions in the eyebrows sub-areas. Based upon the pixel distributions, we classify it into upward, round, downward, and flat corner of eyebrows, respectively. From these information, we classify eyebrows into twelve types as following Fig. 6. In the eyes classification part, we calculate the slant and size of eyes. First, we calculate the slant from the major axis end point of eyes. We classify it into upward, downward, and flat type, respectively. And also, we classify the size of eyes into small, normal, and big eyes based on the relative length between height and width of eyes. Fig. 7 shows the classification result of eyes using proposed algorithm. The mouth classification method of proposed algorithm depends on the slant of mouth end-point and the thickness of lip. To do this, we divide mouth into 4×3 sub-regions and estimate the position of mouth end-point using horizontally projection in the vertical sub-regions. From these projected pixel distribution, we classify the slant of mouth into end-point upward, downward, and flat type, respectively. We also compare the thickness of upper lip with the lower lip to estimate mouth thickness. So, we classify it into upper lip, lower lip, and equal lip type, respectively. Fig. 8 shows the classification result of mouth. In the proposed facial shape classification part, we first extract face-outline points which are crossing points between vertical

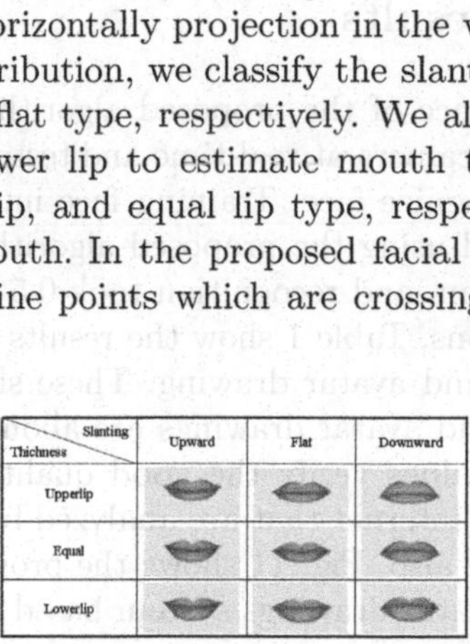

Fig. 8. Classification result of mouth

center line, horizontal mouth line, and horizontal nose line, respectively. From the extracted face-line points, we calculate slope of face-line in each areas and can be estimate facial shape using relative facial feature gradients. In this paper, we classify the face shape into 9 types for male and 8 types for female, respectively. Fig. 9 shows the classification result of facial shape for automatic avatar drawing.

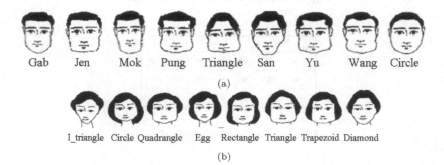

Gab   Jen   Mok   Pung   Triangle   San   Yu   Wang   Circle

(a)

I_triangle Circle Quadrangle   Egg   Rectangle Triangle Trapezoid Diamond

(b)

Fig. 9. Classification result of facial shape; (a) Facial shape for male and (b) facial shape for female

Table 1. Percentage of the proposed face analysis and avatar drawings

Face Image	Detection	Extraction	Analysis	Avatar Drawing
Number(320)	309	302	293	293
Rate[%]	96.6	94.4	91.6	91.6

4   Experimental Results

To evaluate the performance of the proposed algorithm, we used 320 training face images taken by web-camera at real-time and tested it to draw facial avatar as well as detect and recognize face. Training face images including tilted and frontal images were tested using the proposed algorithm, the result processing of face detection, correction, and recognition took 0.5 frame/sec, which is suitable for real-time operations. Table 1 show the results of face detection, feature extraction, face analysis and avatar drawing. These show that face analysis of the proposed algorithm and avatar drawings are about 91.5% according to the number of faces. These values verify the good quality. The automatic avatar drawings system based on internet that are analyzed by the proposed algorithm are shown in Fig. 10. And also, Fig. 11 shows the proposed facial physiognomy analysis and automatic avatar drawings system based on PDA.

(a)   (b)   (c)

Fig. 10. GUI of the proposed face analysis and automatic facial avatar drawing system based on internet; (a) Face analysis and avatar for male, (b) face analysis and avatar for female, and (c) face analysis, avatar drawing and life partner news in physiognomy DB

(a)   (b)   (c)

Fig. 11. Proposed face analysis and automatic facial avatar drawing system based on PDA; (a) Initial page, (b) face detection, and (c) facial physiognomy analysis

5 Conclusions

This paper presented an automatic face analysis system based on face recognition and facial physiognomy. The proposed system can detect user's face, extract facial features, classify shape of facial features, analyze facial physiognomy, and draw avatar automatically which is resembled to user's face. The proposed algorithm could contribute to the development of scientific and quantitative facial physiognomy system which can be ap-plied to the on-line facial application service as well as biometrics area. And also, it offers oriental facial physiognomy database and automatic avatar drawing scheme based on face recognition.

Acknowledgement

This research was supported by MIC, Korea, under the ITRC support program supervised by the IITA.

References

1. Selin Baskan, M., Baskan, S., Bulut, M.M., Atalay, V.: Projection Based Method for Segmentation of Human Face and its Evaluation. Pattern Recognition Letters 23, 1623–1629 (2002)

2. Kim, Y.G., Lee, O., Lee, C., Oh, H.: Facial Caricaturing System with Correction of Facial Decline. Proceeding of Korea Information Processing Society 8(1), 887–890 (2001)
3. Brunelli, P., et al.: Face Recognition: Feature Versus Templates. IEEE Transaction on Pattern Analysis Machine Intelligence 15, 1042–1052 (1993)
4. Samal, A., Iyengar., P.A.: Automatic Recognition and Analysis of Human Faces and Facial Expression: A Survey. Pattern Recognition 25, 65–77 (1992)
5. Jang, K.S.: Facial Feature Detection Using Heuristic Cost Function. Journal of Korea Information Processing Society 8(2), 183–188 (2001)
6. Lee, E.J.: Favorite Color Correction for Reference Color. IEEE Transaction on Consumer Electronics 44, 10–15 (1998)
7. Liu, Z., Wang, Y.: Face Detection and Tracking in Video Using Dynamic Program. In: IEEE International Conference on Image Processing, vol. 2, pp. 53–56 (2000)
8. Cardinaux, F., Sanderson, C., Bengio, S.: User Authentication via Adapted Statistical Models of Face Images. IEEE Transaction on Signal Processing 54(1), 361–373 (2005)
9. Yang, M.H., Kriegman, D., Ahuja, N.: Detecting Faces in Images: A Survey. IEEE Transaction on Pattern Analysis Machine Intelligence 24(1), 34–58 (2002)

Moving Cast Shadow Elimination Algorithm Using Principal Component Analysis in Vehicle Surveillance Video

Wooksun Shin[1], Jongseok Um[2], Doo Heon Song[3], and Changhoon Lee[1]

[1] Dept. of Computer Science, Konkuk University, Seoul, Korea
[2] Dept. of Multimedia Engineering, Hansung University, Seoul, Korea
[3] Dept. of Computer Game & Information, Yong-in SongDam College, Yongin, Korea
{wsshin,chlee}@konkuk.ac.kr
jsum@hansung.ac.kr, dsong@ysc.ac.kr

Abstract. Moving cast shadows on object distort figures which causes serious detection deficiency and analysis problems in ITS related applications. Thus, shadow removal plays an important role for robust object extraction from surveillance videos. In this paper, we propose an algorithm to eliminate moving cast shadow that uses features of color information about foreground and background figures. The significant information among the features of shadow, background and object is extracted by PCA transformation and tilting coordinates system. By appropriate analyses of the information, we found distributive characteristics of colors from the tilted PCA space. With this new color space, we can detect moving cast shadow and remove them effectively.

Keywords: Shadow Elimination, PCA, ITS, Color Shadow Model.

1 Introduction

The characteristics of a moving object extracted from a surveillance video could be applied to video scene analysis and/or Intelligent Traffic Monitoring System. Moving objects can be extracted through frame differences or background subtractions. However, if the shadow is not removed correctly, it will distort the shape of the objects which causes inaccurate recognitions and analyses. Thus, shadow removing is particularly important for the Intelligent Transportation System (ITS) related applications. The shadow may cause an occlusion of two adjacent objects or the misclassified shadow recognized as a part of a moving object can induce an inaccurate car type classification [2], [10], [11].

Researches of detecting and removing shadows could be classified into the statistical approaches [3],[4],[6] and the deterministic approaches [1], [9], [10] according to the treatment of uncertainty [8]. There are also many different principles in detecting shadows such as heuristic-based [9], [11], histogram-based [6], color constancy model [1], [4] depending on the assumptions they made [7].

In this paper, we propose an effective algorithm for detecting and removing the moving cast shadow from a fixed road surveillance camera. In such circumstances, we do not have to worry about the movement of people and the scene

M.S. Szczuka et al. (Eds.): ICHIT 2006, LNAI 4413, pp. 139–148, 2007.

has usually moderate noise level. According to the taxonomies mentioned above, our approach belongs to the statistical method while using the color information as the spectral characteristic and the region level information rather than the pixels as the spatial characteristics [1], [4], [6]. However, we do not use the temporal feature. We also have similar assumptions of color constancy model. The specialty of our algorithm lies to the development of the new color model that is able to discriminate moving shadows from moving object effectively.

We assume that the distributional characteristics of background figure and shadow have similar directional distributive pattern with locational difference. To find the directional distributive pattern, we first transform the color space from RGB to PCA space and then tilt the PCA transformed coordinate system. After all, we found two main axes representing those distributive patterns respectively. The shadow removing algorithm is developed based on that directional distributive characteristic. We also consider various weather conditions and car darkness which affect distributive direction.

The structure of this paper is as following. We discuss the shadow model in section 2, the algorithm to detect and remove shadows within object regions in section 3, the experimental results in section 4, and the conclusions in section 5.

2   Shadow Model

2.1   Foreground and Background Figure

By applying background subtraction, we can extract a moving object from the surveillance video. We obtain the foreground figures and the background figures through the current frame and the background images as shown in Figure 1.

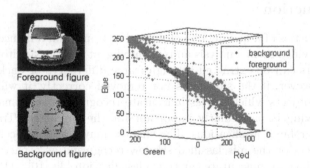

Fig. 1. Foreground figure and Background figure(left), RGB color space of foreground figure, Background figure(right)

The light gray dot displays the RGB value of the foreground figure which contains the shadow and the car object meanwhile the dark gray dot displays the background image. The background figure shows a narrow and short bar while the foreground figure shows a wider and longer one. This shows that the road in the background figure is brighter than the shadow. Thus, it is expected that the foreground figure has a wider distribution of the RGB value than that of background figure.

Table 1. Means and Variances of Input figures

	Mean(R,G,B)	Variance(R,G,B)
Foreground figure	158.2,170.1,165.9	7865.9,6517.4,6827.7
Background figure	169.3,182.3,164.9	60.8,61.1,56.1

2.2 Shadow Feature

Shadows can be classified into the self-shadow, which is created as a part of the object, and the cast shadow, which is created outside of the object. Self-shadows can become recognized as a part of the object but cast shadows are not a part of the object. Figure 2 shows the distributional characteristics of the background and the shadows with different sunlight intensities. With less bright sunlight, we can see that the background and the shadows are distributed in a straight line in the same direction. But with bright sunlight, each directional line starts to separate each other. As the sunlight increases, the degree of separation between two directional lines increases.

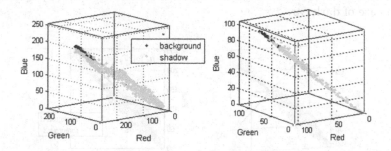

Fig. 2. RGB color space of Background and Shadow, bright sunlight (left), less bright sunlight (right)

3 Algorithm Description

3.1 Shadow Region Detection

For an easier analysis, we therefore need to use the PCA transformation to extract the characteristics of each RGB elements and then perform a linear conversion for the purpose of creating a new color space. Let X_i^F and X_i^B be the foreground figure and the background figure in respectively. Let μ_X^B and C_X^B be the mean and covariance of the background figure as in Equation (1).

$$X_i^F = [\, R_i^F \; G_i^F \; B_i^F \,]^T, \quad X_i^B = [\, R_i^B \; G_i^B \; B_i^B \,]^T \quad for \;\; i = 1, \; ... \,, \; M$$

$$\mu_X^B = [\, \mu_R \; \mu_G \; \mu_B \,]^T = \tfrac{1}{M} \sum_{i=1}^{M} X_i^B \,, \quad C_X^B = Covariance(X^B) \tag{1}$$

Let each eigenvalue of the covariance C_X be λ_1, λ_2, λ_3 in descending order. Then we find the transform matrix A^B that is composed of the eigenvectors e_1, e_2, and e_3 as shown in Equation (2).

$$Y^B = A^B(X^B - \mu_X^B)\,, \quad A^B = [\, e_1 \; e_2 \; e_3 \,]^T \tag{2}$$

Then we transform each value of the foreground figure using Equation (3). The left one in Figure 3 shows the distribution of Y^B and the distribution of Y^F where Y^F is in Equation (3)

$$Y^F = A^B(X^F - \mu_X^B) \tag{3}$$

The center of the background figure becomes the origin of the PCA space. Since shadows are darker than the road, the location of shadow area can be traced along the principal axis. As shown in Figure 3(right), the shadow is located to the left of Y_1 which is the transformed value on the principal axis of the background. However, the predicted shadow area may include a part of car. Dark sections of a car such as windshields might recognize as a shadow. Similar problem may arise in case of dark car.

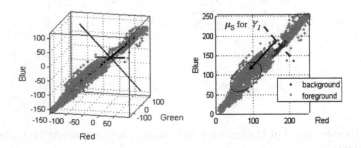

Fig. 3. RGB color space and Principal components vectors of foreground figure and background figure(left), candidate region of shadow(right)

3.2   Principal Components Modification

Spatial data of the foreground figure and the background figure have a specific directional pattern as shown in Figure 4. Let (Y_1, Y_2, Y_3) be the coordinates of the PCA space. The distribution of the background figure is densely accumulated in a narrow area along the principal axis. Like a background distribution, the foreground figure is distributed along the principal axis of the background with wider and thicker dispersion. So, it is highly probable that two distributions have identical distributive direction. When we project two distributions onto the Y_2Y_3 plane, there is a locational difference due to the different distributional shapes.

The Y_1 axis, the principal axis of the background figure, and the direction of the foreground pattern are similar but not identical. Even though the background figure and foreground figure are distributed along the principal axis, they are not aligned to the principal direction. Therefore, we tilted the coordinate system to find the direction of alignment by using the hill-climbing binary search which minimizes the number of projected points. If an accurate pattern direction is found through a tilting of the coordinate system, each direction of the pattern will become accurately aligned to the tilted Y_2Y_3 plane. Figure 4 shows the tilted result.

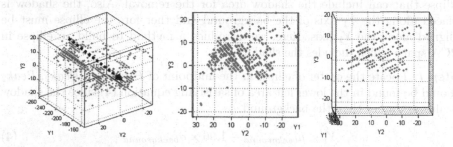

Fig. 4. Foreground pattern (left). Projecting foreground onto Y_2Y_3 plane (mid), Projecting foreground onto the tilted Y_2Y_3 plane (right).

By observing the characteristics of the shadow in the foreground figure, we can see that the distribution of the projected points onto the tilted Y_2Y_3 plane is different depending on the color of the car as shown in Figure 5. Since the distribution of shadow has a different shape according to the darkness of the car, it is difficult to estimate the distributional location of the shadow in the tilted Y_2Y_3 plane.

As shown in Figure 5, we find two locations having highest number of projected points ($Peak_1$, $Peak_2$). $Peak_1$ is almost identical to the origin that is

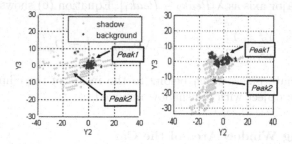

Fig. 5. Projected points and peaks in tilted Y_2Y_3 plane, bright-color car(left), dark color car(right)

the center of the background. $Peak_2$ is located in the center area of a shadow and its value differs according to the darkness of the vehicle. To simplify the computation, we align the line connecting $Peak_1$ and $Peak_2$ to Y_2 axis.

3.3   Removing Shadow Using Ellipse

The distribution of shadows and the distribution of backgrounds depend on the weather condition. Those distributions have the same distributive direction and are almost aligned to each other when there is little sunlight. However, they show different distributional characteristics as the sunlight increase according to the car darkness. The distribution of shadows in the tilted Y_2Y_3 plane shows a bigger dispersion in the Y_2 direction than Y_3. Therefore, we use an appropriate ellipse that can include the shadow area for the removal. Also, the shadow is placed along the Y_1 axis of the background and therefore the ellipse must be aligned along the Y_1 axis. Here, we introduce a method to build an ellipse in PCA space which includes the shadow area.

Step 1. Locate the center of ellipse at the midpoint of $Peak_1$ and $Peak_2$. $Peak_2$ should be found in the lower left area of satisfying equation (4) since the shadow is always darker than the background.

$$Y_1 < \mu_{background} - 1.96 \times \sigma_{background} \tag{4}$$

Here, $\mu_{background}$ is the mean of the background and $\sigma_{background}$ is the standard deviation of the background.

Step 2. Determine the length of the minor axis of the ellipse. Use the length of minor axis of the ellipse as $|Peak_2 - Peak_1|$. In case of little sunlight, $|Peak_2 - Peak_1|$ becomes so small that we cannot build an ellipse with an appropriate size. In such cases, we use the standard deviation of Y_2 instead of $|Peak_2 - Peak_1|$ as in equation (5).

$$D_{peak} = \max(\,|Peak_2 - Peak_1|,\ \tau\,)\,. \quad \tau = 1.5\,\sigma_{Y_2}. \tag{5}$$

Step 3. Determine the length of the major axis of the ellipse. Let be the ratio of the standard deviation of Y_2 and the standard deviation of Y_3. Then we set the length of major axis as $\lambda\,|Peak_2 - Peak_1|$. Equation (6) shows the resulting ellipse.

$$\frac{Y_2^2}{a^2} + \frac{Y_3^2}{b^2} = 1\,, \quad a = \lambda D_{peak}\,, \quad b = D_{peak}. \tag{6}$$

Step 4. Eliminate pixels which belong to the inside of the ellipse. Resulting image shows the object with the shadow removed.

3.4   Searching Window Area of the Car

Since the proposed algorithm is based on the distributional characteristics of color information, darker area in the car may be classified as the shadow. Since the window area of a car is usually darker than any other part of the car, the window area may be contained in the eliminating area. To increase shadow discrimination rate while keeping high shadow detection rate, we detect window from foreground images using template based searching. Constructing the template similar to window using rectangle, we search the window area in the car. If an appropriate area is found as the window by template masking, we protect pixels in that area in the shadow removal process.

4   Experimental Results

We used fifteen 320×240 24 bit RGB color videos that were recorded on the road during daytime with different sunlight intensities and locations. Data was obtained by using background subtraction to extract the vehicle object including the shadows. The extracted results were the foreground figure including the vehicle and the shadow, and the background figure including the image of the road. We used the proposed algorithm to use these background/foreground figures for the detection and the removal of shadows. Figure 6 shows the removed area which is predicted to be the shadows in the $Y_1Y_2Y_3$ space. We then need to remove the shadow area and must retransform to the original RGB space X^F in the $Y_1Y_2Y_3$ space by equation (7).

$$X^F = (A^B)^T Y^F + \mu_X^B \tag{7}$$

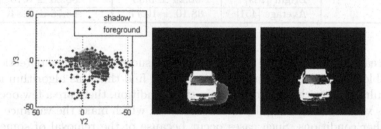

Fig. 6. Shadow in the Foreground(left), input figure(mid), result figure(right)

We need a quantitative evaluation on the efficiency of our shadow removal algorithm. Evaluation criteria are the shadow detection rate (η : how much dots which are part of a shadow can be removed) and the shadow discrimination rate (ξ : how accurately shadows are classified) used in a comparative study [8]. The shadow pixels that are correctly found are represented as True Positive (TP), the shadow pixels that are not found are represented as False Positive (FP), and the falsely found shadow pixels are represented as False Negative (FN). Also, the number of pixels of shadows and vehicle objects are then compared to the ground-truth to use for evaluation.

$$\eta = \frac{TP_{Shadow}}{TP_{Shadow} + FN_{Shadow}}, \quad \xi = \frac{\overline{TP}_{Foreground}}{TP_{Foreground} + FN_{Foreground}} \tag{8}$$

$\overline{TP}_{Foreground}$ is obtained by subtracting the number of dots that are detected as shadows(including the foreground object) from the number of ground-truth dots on the foreground object. The following table shows the result of our experiment. This experiment used 451 car objects extracted from fifteen videos with varying sunlight intensity and darkness of the moving object.

Table 2. Experiment Result (σ : Standard Deviation)

Light Intensity	Car Darkness [# of objects]	Shadow Detection Rate (% $\pm$ σ)	Shadow Discrimination Rate (% $\pm$ σ)
Sunny (6 videos)	Dark [44]	82.46 $\pm$ 12.39	83.61 $\pm$ 10.86
	Mid [51]	88.89 $\pm$ 8.05	81.90 $\pm$ 8.47
	Bright [169]	91.83 $\pm$ 7.57	90.21 $\pm$ 4.89
	Average [264]	89.70 $\pm$ 9.30	87.50 $\pm$ 7.88
Sunny&Cloudy (2 videos)	Dark [6]	76.76 $\pm$ 8.27	84.21 $\pm$ 5.75
	Mid [14]	85.53 $\pm$ 5.77	79.16 $\pm$ 5.35
	Bright [21]	88.61 $\pm$ 5.58	84.74 $\pm$ 5.62
	Average [41]	85.83 $\pm$ 7.30	82.76 $\pm$ 6.13
Cloudy (5 videos)	Dark [30]	79.64 $\pm$ 15.25	77.15 $\pm$ 16.64
	Mid [17]	85.03 $\pm$ 14.00	81.61 $\pm$ 6.99
	Bright [99]	87.89 $\pm$ 11.81	85.65 $\pm$ 6.73
	Average [146]	85.86 $\pm$ 13.26	83.44 $\pm$ 10.25
Summary	Dark [80]	80.97 $\pm$ 13.41	81.24 $\pm$ 13.46
	Mid [82]	87.51 $\pm$ 9.47	81.37 $\pm$ 7.79
	Bright [289]	90.24 $\pm$ 9.33	88.25 $\pm$ 6.10
	Average [451]	88.10 $\pm$ 10.77	85.76 $\pm$ 8.84

Whereas the experimental result shows very successful shadow detection and acceptable shadow discrimination rate, one can find that this algorithm is relatively vulnerable to the dark objects. In that condition, there are a few occasions that has very low detection/discrimination rate, which make the variance bigger than other conditions. Such cases occur because of the removal of some parts of the vehicle that are recognized as shadows such as the windshields. For the windshields, we protect a part of it by template masking in shadow removal procedure. With this compensation, the discrimination rate is up about 5% in average without losing the detection rate.

The encouraging part of the experiment is high shadow detection rate. Since this algorithm is designed for the ITS applications, as long as we do not need image restoration process by over-removing shadow pixels such as [10] equipped, this side effect may not be as harmful as the number in the table shows in real applications.

Regarding the performance comparison with other related algorithms, one may argue that any comparison is not fair enough unless all algorithms are tested with the same set of videos. However, the Table 3 could be at least reasonable evidence that our approach is meaningful. We found that the Highway II video comparison of [8] has the most similar environments to our experiment.

Table 3. Comparitive Result from Literature

Algorithms	Shadow Detection Accuracy (%)	Shadow Discrimination Accuracy (%)
SNP[6]*	51.20	78.92
SP[4]*	46.93	91.49
DNM1[1]*	54.07	78.93
DNM2[9]*	60.24	72.50
Our Algorithm	88.10	85.76

* Highway of [8].

Key: SNP (Statistical non-parametric approach), SP(Statistical parametric approach), DNM(Deterministic non-model based approach).

5 Conclusion

In this paper, we propose an algorithm which creates a new color model and uses this to detect and remove shadows through a series of statistical analysis on the background and shadows. It is possible to remove shadows regardless of the intensity of the sunlight and the darkness of the vehicle. The power of our approach comes from the separability of the shadow and the object in the new space based on their distributional patterns. It does not add any new strong constraints or assumptions about the position of light source, preprocessing of model fitting to forecast the object shape [11], or image restoration process [10]. In the experiment, the algorithm is particularly strong in shadow detection measure. Since our main target applications are ITS-related such as vehicle recognition, vehicle analysis, and tracking, our approach can be effective.

However, as many statistical approaches with color constancy model share the same problem, our new color space does not completely separate the moving shadow and the moving object. Also, in current implementation, we do not utilize the temporal feature. We should look at the color space and the algorithm deeper and try to generalize it for the more noisy scenes than the static background as we used here.

Acknowledgments. This research was supported by the MIC(Ministry of Information and Communication), Korea, under the ITRC(Information Technology Research Center) support program supervised by the IITA(Institute of Information Technology Assessment) and this research was partially supported by Hansung University in the year of 2006.

References

1. Cucchiara, R., Grana, C., Piccardi, M., Prati, A.: Detecting objects, shadows and ghosts in video streams by exploiting color and motion information. In: Proceedings of the IEEE Int'l Conference on Image Analysis and Processing, pp. 360–365 (2001)

2. Cucchiara, R., Grana, C., Prati, A., Piccardi, M.: Effective detection of moving objects, shadows and ghosts in surveillance videos, Australia-Japan Advanced Workshop on Computer Vision (2003)
3. Haritaoglu, I., Harwood, D., Davis, L.S.: W4: real-time surveillance of people and their activities. IEEE Transactions on Pattern Analysis and Machine Intelligence 22(8), 809–830 (2000)
4. Horprasert, T., Harwood, D., Davis, L.S.: A statistical approach for real-time robust background subtraction and shadow detection. In: Proceedings of IEEE ICCV 1999 FRAME-RATE Workshop (1999)
5. Jolliffe, I.T.: Principal component analysis. Springer, New York (1996)
6. Mikic, I., Cosman, P., Kogut, G., Trivedi, M.M.: Moving shadow and object detection in traffic scenes. In: Proceedings of Int'l Conference on Pattern Recognition, pp. 321–324 (2000)
7. Nadimi, S., Bhanu, B.: Physical models for moving shadow and object detection in video, Pattern Analysis and Machine Intelligence. IEEE Transactions 26(8), 1079–1087 (2004)
8. Prati, A., Cucchiara, R., Mikic, I., Trivedi, M.M.: Analysis and Detection of Shadows in Video Streams: A Comparative Evaluation. In: Proc. of Third Workshop on Empirical Evaluation Methods in Computer Vision - IEEE Int'l Conference on Computer Vision and Pattern Recognition, pp. 571–576 (2001)
9. Strauder, J., Mech, R., Ostermann, J.: Detection of moving cast shadow for object Segmentation. IEEE Transaction on Multimedia 1(1), 65–76 (1999)
10. Wang, J.M., Chung, Y.C., Chang, C.L, Chen, S.W.: Shadow Detection and Removal for Traffic Image. In: Proceedings of the IEEE Int'l Conference on Networking, Sensing & Control, pp. 649–654 (2004)
11. Yoneyama, A., Yeh, C.H., Kuo, C.-C.J.: Moving cast shadow elimination for robust vehicle extraction based on 2D joint vehicle/shadow models. In: Proceedings. IEEE Conference on Advanced Video and Signal Based Surveillance, pp. 229–236 (2003)

Automatic Marker-Driven Three Dimensional Watershed Transform for Tumor Volume Measurement

Yong-su Chae and Desok Kim

School of Engineering, Information and Communications University, Daejeon, Korea
{yschae,kimdesok}@icu.ac.kr

Abstract. Molecular imaging can detect abnormal functions of living tissue. Functional abnormality in gene expression or metabolism can be represented as altered volume or probe intensity. Accurate measurement of volume and probe intensity in tissue mainly relies on image segmentation techniques. Thus, segmentation is a critical technique in quantitative analysis. We developed an automatic object marker-driven three dimensional(3D) watershed transform for quantitative analysis of functional images. To reduce the discretization error in volume measurement less than 5%, the size criteria for digital spheres were investigated to provide the minimum volume. When applied to SPECT images, our segmentation technique produced 89% or higher accuracy in the volume and intensity of tumors and also showed high correlation with the ground truth segmentation ($\rho > 0.93$). The developed 3D method did not require interactive object marking and offered higher accuracy than a 2D watershed approach. Furthermore, it computed faster than the segmentation technique based on the marker-driven gradient modification.

Keywords: Image segmentation, mathematical morphology, marker-driven watershed segmentation, automatic object marker, 3D measurement.

1 Introduction

The 3D segmentation techniques are applied to different biomedical image analyses, for example, the detection of abnormal tissue such as tumors, visualization or measurement of body organs such as brains and hearts. Molecular images can be obtained from PET, SPECT, and optical imaging of live cells, small animals, and human. Quantitative analysis of these images enables us to investigate functional abnormalities of diseased tissue more objectively. Abnormal gene expression and metabolic rates of target molecules are represented as altered probe intensity. Recently, advanced fluorescent or radioisotopic probes for molecular imaging have been developed so that their sensitivity and specificity can be quantitatively studied [1].

Watershed is a term used in topography that divides the topographic surface into many meaningful regions and can be explained by a "drainage" analogy:

M.S. Szczuka et al. (Eds.): ICHIT 2006, LNAI 4413, pp. 149–158, 2007.

the raindrops falling on one of topographic regions flow into the same minimum point and the region is called a catchment basin. The topographic surface consists of many catchment basins and the lines separating them can be defined as watershed (or watershed line). When this analogy is implemented by mathematical morphology, it is called watershed transform. In watershed transform, the gradient image is used to represent the altitude of the topographic surface. Watershed transform was first proposed by Digabel and Lantuéjoul [2]. Later, it was applied to the grayscale images by Lantuéjoul and Beucher [3]. The speed and accuracy were greatly improved by Vincent and Soille using immersion-based fast watershed transform [4].

Numerous local minima in random noisy structures cause over-segmentation by basic watershed transform. To avoid this problem, previous studies suggested multi-resolutional filtering [5], wavelet analysis, adaptive anisotropic diffusion filter, and a marker-driven gradient modification [6,7,13,15]. In this study, we developed a new 3D marker-driven watershed transform based on fast watershed transform to measure the volume and probe intensity from 3D medical images. Object markers were automatically extracted through grayscale reconstruction and skeletonization. Upon applying our segmentation technique onto the differently sized digital sphere images, we estimated the minimum volume that reduced the discretization error to less than 5%. The segmented image, volume, and probe intensity were compared to the manually segmented ground truth to obtain the accuracy.

2   Methods

2.1   Three Dimensional Marker-Driven Watershed Transform

Based on the fast watershed algorithm [4], the 3D fast watershed transform algorithm was developed by simply adding an indexing mechanism that retrieves and manipulates neighboring pixels along the z-axis. Neighborhood operation was performed on 6 neighboring pixels instead of 26 neighboring pixels to speed up the computation. For a marker-driven approach, a user can interactively place a background marker and multiple object markers onto 2D gradient images or, alternatively these markers can be automatically extracted. These markers were imposed onto the 3D gradient image as only minima that were given the initial labels: in the immersion analogy, the water emerged from each minimum were uniquely labeled. However, in our marker-driven approach, the water emerged only from the interactively imposed marker was uniquely labeled. The water emerged from other irrelevant minima were not labeled, instead, designated as the unlabeled water.

During our version of immersion process, a new labeling event occurred when the labeled water met the unlabeled water (Fig. 1). The water in each catchment basin is shown initially labeled at the gradient intensity level, X_h (Fig. 1.a). Two catchment basins imposed by the object markers are labeled as "1" and "2"; the background is labeled as "0"; the catchment basins that contain irrelevant minima are labeled as "U". At the next gradient intensity level, X_{h+1}, the water

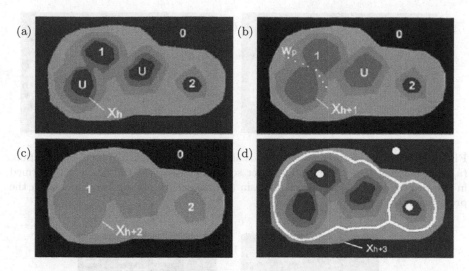

Fig. 1. Illustration of immersion process and labeling event. (a) At X_h, object labels 1 and 2, background label 0, unlabeled catchment basins U are shown; (b) at X_{h+1}, the label propagates into the lower left catchment basin through the minimum of the provisional watershed, w_p; (c) at X_{h+2}, the middle catchment basin labeled as 1; (d) at X_{h+3}, the provisional watershed (shown as thick solid line) along with originally imposed markers (shown as solid circles).

initially labeled as "1" encounters the water emerged from one of unlabeled catchment basins at the minimum of the provisional watershed (marked as "w_p" in Fig. 1.b) between the two corresponding catchment basins. When this occurs, two catchment basins become merged through the unlabeled water receiving the label "1" (Fig. 1.b). At X_{h+2}, another catchment basin becomes merged with the catchment basin 1 (Fig. 1.c). Finally at X_{h+3}, the provisional watershed can be found between the labeled catchment basins and the background (Fig. 1.d).

However, the propagation of labels may become arbitrary when more than one minimum are possessed by a catchment basin (marked with an arrowhead in Fig. 2.a) on the provisional watershed. Since the propagation of the label is performed in the raster scanning order, the labeled water whose position takes the precedence in the raster scanning order always meets the unlabeled water earlier. Thus, the labeling is basically decided by the minimum whose location takes the precedence in the raster scanning order (Fig. 2). To avoid the dependence of watershed on the marker position, the gradient image can be slightly blurred so that each minimum becomes unique in terms of its pixel intensity [8].

To validate our marker-driven watershed segmentation, the 3D segmentation was performed with a cubic model that contained a brighter inner cubicle (Fig. 3.a). The 3D gradient image of the cubic model was calculated first (Fig. 3.b). Without imposing a marker, basic watershed segmentation resulted in an over-segmentation (Fig. 3.c). The object marker and the background marker were imposed onto the gradient image (Fig. 3.b). At the end of the immersion

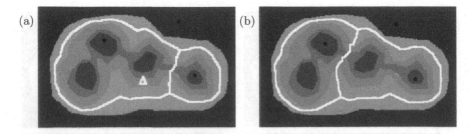

Fig. 2. Illustration of different watersheds showing a dependence of label propagation. (a) The labeling performed in the raster scanning order; (b) the labeling performed in the reverse order. The catchment basin (arrowhead) possesses two minima on the provisional watershed.

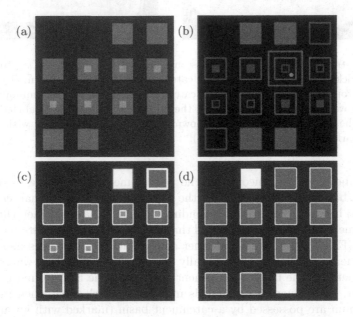

Fig. 3. Illustration of 3D watershed segmentation of a cubic model with a small cubicle at the center. (a) 3D image; (b) 3D gradient image with an object marker and a background marker; (c) basic 3D watershed showing over-segmentation; (d) marker-driven 3D watershed resulting in a sensitive segmentation.

process, watershed line was detected if a pixel in the background had any of its 26 neighboring pixels labeled differently to the background: the marker-driven 3D segmentation produced a sensitive result (Fig. 3.d).

2.2   Volume Measurement of 3D Object Models

The digital model of a sphere was constructed by stacking up circles with different radii drawn on the equally spaced z planes using Bresenham's arc algorithm

[10,11]. The total number of pixels in the digital sphere was first counted to produce a rough volume of the sphere. The calculated radius of the digital sphere was obtained from the rough volume to generate the calculated volume. The measured volume of the digital sphere was obtained by applying the 3D gradient operator to the digital sphere image and applying the marker-driven 3D watershed segmentation. Finally, all data were translated into the millimeter units, assuming the dimension of a pixel as $0.5 \times 0.5 \times 0.5$ mm. The relative errors between two kinds of volumetric data were investigated in many differently sized digital spheres. The relative errors were also calculated for digital spheres with different numbers of z planes, simulating the pixel distance along the z axis.

2.3 Volume and Intensity Measurement of SPECT Brain Images

The minimum radii of the digital sphere for different z-axis resolutions were determined based on the relative errors explained in the Section 2.2. Based on these size criteria, a set of seven SPECT images were obtained from the Whole Brain Atlas [12]. Brain tumors visualized by high Thallium uptake were larger than average 6mm in radius. The pixel distances along the x and y axes were 0.5mm and the one along the z axis was 1.5mm. Ground truth segmentation images were made by manual following the boundary of 2D tumor images shown in the pseudo color. The 3D grayscale images consisted of 20, 24, or 28 two-dimensional images of 128×128 pixels with grayscale intensity ranging from 0 to 255 (byte data) (Fig. 4.a). The 3D images were first filtered by a 3D closing operation and a 3D opening using a $3 \times 3 \times 3$ cubic structuring element. This morphological filtering filled small gaps and suppressed irrelevant noises in the 3D images. Especially, the 3D closing operation helped to define the 3D gradient more accurately by increasing the spatial correlation of the intensity between the successive 2D images (thus, suppressing abnormally high gradient component along the z axis). The 3D gradient image was obtained from the filtered image, f, using the following 3D gradient operator, $grad_{3d}$ (Fig. 4.b):

$$grad_{3d}(f) = \max\left(f \oplus S_x - f \ominus S_x, f \oplus S_y - f \ominus S_y, f \oplus S_z - f \ominus S_z\right) \quad (1)$$

where $\oplus$ and $\ominus$ were dilation and erosion operators, respectively; S_i was a linear structuring element three pixels wide along the i_{th} axis; **max** was the maximum operator. Then, the 3D gradient image was blurred with a Gaussian function with a radius of 1 pixel to reduce the dependence to the marker position (image not shown).

A 2D frame with the largest gradient intensity value was selected to detect object and background markers automatically within it. The boundary of the 2D frame was recognized as the background marker (Fig. 4.b). To locate the object markers, 2D grayscale reconstruction was applied using an estimated intensity difference between object and background (55 in our example). After subtracted from the original grayscale image, a binary image was obtained by a thresholding at the intensity value slightly lower than the intensity difference (53 in our example). Binary objects were individually labeled and small irrelevant objects

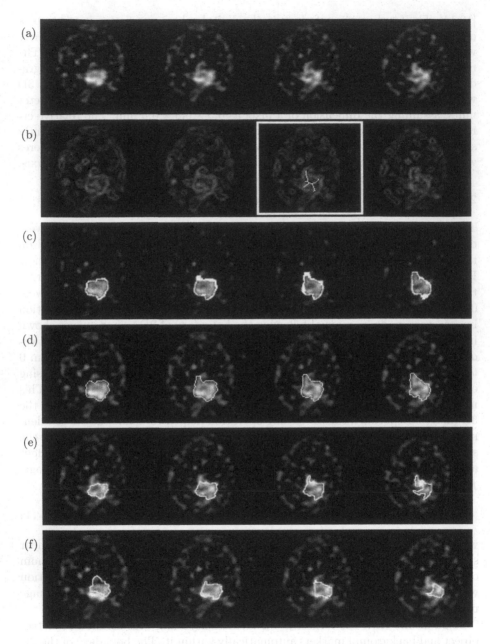

Fig. 4. The sequence of automatically detected marker-driven 3D watershed transform. (a) 3D gray scale image; (b) 3D gradient image with a background marker (square bracket) and an object marker (skeleton); (c) thick watershed segmentation result; (d) final segmentation result; (e) ground truth segmentation; (f) segmentation results of 2D marker-driven watershed transform.

were removed based on the area. Then, 2D skeletons were obtained from the binary image. The skeletons working as the object markers often contained long branches so that they were pruned two times [9]. The object marker and background marker were imposed onto the gradient image of the selected 2D frame (Fig. 4.b). The marker-driven 3D watershed transform was performed and the 3D watershed was accurately localized to the boundary of the tumor area that showed increased probe intensity compared to the rest of the brain (Fig. 4.c). The resultant watershed was often many pixels thick so that only the watershed one pixel distant from the background was selected as final watershed pixels (Fig. 4.c and Fig. 4.d). The volume and the total intensity of the segmented tumors were obtained by summing up the pixels and their intensity values, respectively. The measured values were compared to the ones measured from the manually obtained ground truth segmentation (Fig. 4.e). In addition, the 2D marker-driven watershed segmentation was performed on each (x, y) plane of the 3D gradient image to compare the performance (Fig. 4.f).

3   Results

3.1   Size Criteria of 3D Object Models

Digital spheres were segmented using the developed segmentation method. The number of pixels in the segmented sphere was counted and transformed to the volumetric unit (mm^3). The relative errors between the calculated volume and the measured volume were summarized in Table 1. When the pixel distance along the z axis was 0.5mm, the minimum volume that showed less than 5% error was $129.4mm^3$ with the radius of 3mm. At the pixel distance of 1.0mm, the error less than 5% first occurred when the minimum volume was $302.8mm^3$ with the radius of 4mm. Finally, at the pixel distance of 1.5mm, the minimum volume was $951.4mm^3$ with the radius of 6mm.

3.2   Accuracy and Speed of Volume and Intensity Measurement of 3D SPECT Images

To assess the accuracy of the segmentation result, the volume and total intensity were compared to those obtained from the ground truth segmentation. The volume, total intensity and their % errors measured in SPECT images are summarized in Table 2. The maximum % errors of the volume and the total intensity measurement were 10.6% and 8.7%, respectively. Thus, the volume measurement was at least 89% accurate (Dice coefficient for the volume measurement was 81.2% in average). The obtained volume and the total intensity were not significantly different to the ground truth results (Mann Whitney Wilcoxon test, $p > 0.1$) and highly correlated (Pearson's correlation coefficients, $\rho = 0.94$ for the volume and 0.99 for the probe intensity, $p < 0.01$). The 2D method produced much less volume and total intensity mainly due to the incomplete segmentation of tumors (Fig. 4.f), resulting in more than 50.0% error in the volume measurement (data not shown). The inaccurate 2D contouring of the tumor occurred

Table 1. The % errors between the measured volume and the calculated volume of digital spheres

Pixel Distance along the z Axis (mm)	Radius of Bresenham's Circle (mm)	Radius by Calculation (mm)	Volume by Measurement (mm³)	Volume by Calculation (mm³)	% Error
	1	1.2	6.1	7.9	22.37
	2	2.1	38.4	41.3	7.17
0.5	*3*	*3.1*	*129.4*	*129.7*	*0.24*
	4	4.1	296.9	298.3	0.48
	5	5.2	589.9	590.7	0.14
	2	2.1	33.3	43.4	23.36
	3	3.1	118.8	129.7	8.43
1.0	*4*	*4.1*	*302.8*	*298.3*	*1.49*
	5	5.2	571.3	590.7	3.29
	6	6.2	972.8	1,000.7	2.80
	3	3.1	121.1	134.7	10.06
	4.5	4.6	400.9	429.3	6.62
1.5	*6*	*6.2*	*951.4*	*1,000.7*	*4.93*
	7	7.2	1,524.4	1,557.0	2.09
	8	8.2	2,238.8	2,275.9	1.63

Table 2. The comparison of the % errors of the volume and total intensity measured by the manual method and the marker-driven 3D watershed transform

Image No.	Volume (mm³)			Total Intensity		
	Manual	3D Watershed	% Error	Manual	3D Watershed	% Error
1	1,496.3	1,399.9	6.4	809,916	783,093	3.3
2	1,241.6	1,295.3	4.3	844,038	808,509	4.2
3	2,041.1	2,154.8	5.5	1,791,894	1,670,505	6.7
4	2,166.8	2,197.5	1.4	3,130,800	2,977,221	4.9
5	2,237.6	2,000.3	10.6	2,273,496	2,075,361	8.7
6	1,717.9	1,870.9	8.9	2,118,189	2,135,214	0.8
7	1,497.8	1,404.0	6.2	1,948,716	1,787,643	8.2

mainly by missing the gradient information along the z axis that was essential to define the watershed of the 3D objects.

The 3D marker-driven watershed segmentation was completed in 39.9 sec, whereas the 2D watershed segmentation took 6.7 sec for 128×128×28 size images in 8 bit grayscale. The computation time was measured on the personal computer with a 3.0 GHz Intel Pentium 4 processor and 1 GB of memory running Windows

XP operating system. Although the 2D method took much less computation time, it could potentially lead to incomplete segmentation when the automatic detection of object markers fails in some 2D frames, especially in a 2D frame that happens to represent relatively small cross section of a 3D object. Finally, the maker-driven gradient modification took 350.4 sec on the same image followed by the 3D basic watershed segmentation completed in 34.4 sec [4], [6] (image not shown).

4   Conclusions and Discussion

In this study, we estimated the size criteria for the reliable volume measurement using digital sphere images. We obtained the SPECT images that contained brain tumors that met the size criteria and applied the developed 3D marker-driven watershed segmentation. The measurement accuracy for the volume and the probe intensity of brain tumors was approximately 89%, comparable to the results from previous studies [13,14,15]. Automatic detection of object and background markers removed user interaction. Compared to the 2D approach, the 3D approach resulted in more accurate segmentation since it took advantage of gradient information along the z axis. The developed 3D marker-driven watershed segmentation ran much faster than the 3D watershed segmentation based on the marker-driven gradient reconstruction that required a time consuming iterative morphology operation such as ultimate erosion [6]. Although the previous marker-driven approach provides a very powerful scheme for image segmentation, its gradient reconstruction may be replaced with less time consuming methods that still produced similarly accurate segmentation results [5]. We are planning to further investigate the size criteria for non-isotropic digital objects that should provide more general guideline for 3D volume measurement in medical images.

Acknowledgment. We are grateful to Drs. Keith A. Johnson and J. Alex Becker for providing the SPECT images. This work was supported by the Korea Science and Engineering Foundation (KOSEF) grant funded by the Korea government(MOST) (No. 2006-03764). Correspondence should be directed to Desok Kim, PhD, Information and Communications University, 103-6 Munji-dong, Yusung-gu, Daejeon, Korea.

References

1. Kim, K., Lee, M., Park, H., Kim, J.H., Kim, S., Chung, H., Choi, K., Kim, I.S., Seong, B.L., Kwon, I.C.: Cell-Permeable and Biocompatible Polymeric Nanoparticles for Apoptosis Imaging. Journal of the American Chemical Society 128, 3490–3491 (2006)
2. Digabel, H., Lantuéjoul, C.: Iterative Algorithms. In: Chermant, J.L. (ed.) Proc. 2nd European Symp. on Quantitative Anal. Microstructures in Material Science, Biology and Medicine, West Germany, pp. 85–99 (October 1977)

3. Beucher, S., Lantuéjoul, C.: Use of Watersheds in Contour Detection. In: Proc. Int. Workshop on Image Processing, Real-Time Edge and Motion Detection/Estimation, Rennes, France, pp. 17–21 (September 1979)
4. Vincent, L., Soille, P.: Watersheds in Digital Spaces: An Efficient Algorithm Based on Immersion Simulations. IEEE Trans. Patt. Anal. Mach. Int. 13, 583–598 (1991)
5. Kim, D.: Multiresolutional Watershed Segmentation with User-Guided Grouping. The International Society for Optical Engineering (SPIE), Medical Imaging 1998, San Diego, CA, pp. 1087–1095, (February 1998)
6. Serra, J., Vincent, L.: An Overview of Morphological Filtering. Circuits Systems Signal Process 11, 47–108 (1992)
7. Roerdink, J.B.T.M., Meijster, A.: The Watershed Transform: Definitions, Algorithms and Parallelization. Fundamenta Informaticae 41, 187–228 (2000)
8. Yim, P.J., Kim, D., Lucas, C.: A High-Resolution Four-Dimensional Surface Reconstruction of the Right Heart and Pulmonary Arteries. The International Society for Optical Engineering (SPIE), Medical Imaging, San Diego, CA, pp. 726–738 (February 1998)
9. Gonzalez, R.C, Woods, R.E.: Digital Image Processing, 2nd edn. Prentice Hall, Englewood Cliffs (2002)
10. Bresenham, J.: A Linear Algorithm for Incremental Display of Circular Arcs. Communications of the ACM 20, 100–106 (1977)
11. Kennedy, J.: A Fast Bresenham Type Algorithm for Drawing Circles. Available at: http://homepage.smc.edu/kennedy_john/BCIRCLE.PDF
12. Available at: http://www.med.harvard.edu/AANLIB/cases/case1/case.html
13. Sijbers, J., Scheunders, P., Verhoye, M., van der Linden, A., van Dyck, D., Raman, E.: Watershed-Based Segmentation of 3D MR Data for Volume Quantization. Magnetic Resonance Imaging 15(6), 679–688 (1997)
14. Grau, V., Kikinis, R., Alcaniz, M., Warfield, S.K.: Cortical Gray Matter Segmentation Using an Improved Watershed-Transform. Engineering in Medicine and Biology Society, 2003. In: Proceedings of the 25th Annual International Conference of the IEEE, vol. 1, pp. 618–621 (September 2003)
15. Lapeer, R.J., Tan, A.C., Aldridge, R.V.: A Combined Approach to 3D Medical Image Segmentation Using Marker-based Watersheds and Active Contours: The Active Watershed Method. In: Medical Image Understanding - MIUA 2002, Proceedings, pp. 165–168 (2002)

A Study on the Medical Image Transmission Service Based on IEEE 802.15.4a

Yang-Sun Lee[1], Jae-Min Kwak[2], Sung-Eon Cho[3,*], Ji-Woong Kim[4],
and Heau-Jo Kang[1]

[1] Div. of Computer Eng., Mokwon University,
800, Doan-dong, Seo-gu, Daejeon, 302-729, Korea
{yslee,hjkang}@mokwon.ac.kr
[2] SoC Research Center, Korea Electronics Technology Institute
68, Yatap-dong, Bundang-gu, Seongnam-shi, Gyeonggi-do, 463-816, Korea
kjm@keti.re.kr
[3] Dept. of Inform. & Comm. Eng., Sunchon National University,
Sunchon-shi, Chonnam, 540-472, Korea
chose@sunchon.ac.kr
[4] Dept. of Hospital Biomedical Eng.,Dongshin University,
Naju-shi, Chonnam, 520-714, Korea
kjwcomm@korea.com

Abstract. In this paper, the transmission service for medical image is proposed via IEEE 802.15.4a on WPAN environment. Also, transmission and receiving performance of medical image using TH UWB-IR system is evaluated on indoor multi-path fading environment. On the results, the proposed scheme can solve the problem of interference from the medical equipment in same frequency band, and minimize the loss due to the indoor multi-path fading environment. Therefore, the transmission with low power usage is possible.

Keywords: Ultra Wideband, IEEE 802.15.4a, Medical Image Transmission.

1 Introduction

The conventional medical treatment service is possible only when and where doctor is located in the medical institute. Therefore, the limitation of time and space exists. Most of medical institute has limited number of doctor and fixed work time. In case of emergency, the status of patient can take a turn for the worse when relevant doctor is absent[1].

In modern society, the advance of medical equipment and technology enables the mobile type of equipments. These equipments only show limited information to the user, and transmit the limited diagnosis information to online medical personnel. Recently, via the emergence of high speed internet and wireless network environment, vast amount of multimedia data can be transmitted, and

---

* Corresponding author.

M.S. Szczuka et al. (Eds.): ICHIT 2006, LNAI 4413, pp. 159–167, 2007.

efficient diagnosis and prevention of disease is possible, which is customized for the status and need of the patient. By the miniaturization, digitalization, and making wireless, low power of diagnosis probes, the user can carry them all the time, and monitor the status of disease and health whenever and wherever possible. Also the emergency case can be expected and transmitted to near medical facility. Therefore, the high quality of medical service can be provided on the road.

However, mobile medical equipment using wireless network has the interference problem with existing medical equipments, which use ISM spectrum, and larger battery power usage. High quality of transmission is not possible[2].

In this paper, the medical image transmission service via IEEE 802.15.4a specification on WPAN network is proposed. The proposed scheme can solve the problem of interference from the medical equipment on the same frequency band, and minimize the loss of medical image on indoor multi-path fading environment. Transmission with low power usage is possible.

2   TH UWB-IR System

2.1   Frequency Trend of IEEE 802.15.4a

Currently low rate WPAN area that the standardization is undergoing by IEEE 802.15.4a group enables the low cost/power, reliable data transmission and multiple device on network. This is expected to apply in much possible area[3].

15 channels are assigned from the physical class of IEEE 802.15.4a draft version.

Table 1. The plan of IEEE 802.15.4a frequency assignment

Channel Number	Center Frequency	Band Width (3dB)	Mandatory/Optional
1	3458	494	Optional
2	3952	494	Mandatory
3	4446	494	Optional
4	3952	1482	Optional
5	6337.5	507	Optional
6	7098	507	Optional
7	7605	507	Optional
8	8112	507	Mandatory
9	8619	507	Optional
10	9126	507	Optional
11	9633	507	Optional
12	10140	507	Optional
13	6591	1318.2	Optional
14	8112	1352	Optional
15	8961.75	1342.5	Optional

Frequency range of UWB technology is divided into low frequency range and high frequency range in 3.1 10.6 GHz as shown in Table 1. Channel 2 is used in low frequency range, and Channel 8 is employed in high frequency. Fig. 1 shows allocated frequency range in total UWB range.

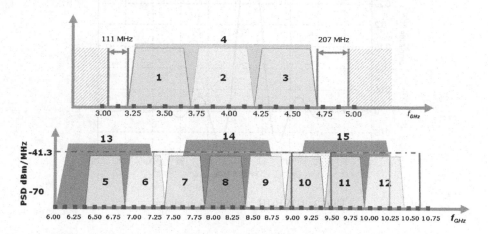

Fig. 1. The plan of IEEE 802.15.4a frequency assignment

2.2 IEEE 802.15.4a TH UWB-IR System

Pulse $p(t)$, of TH UWB-IR has a period of T_p and energy, $E_p = \int_{-\infty}^{\infty}[p(t)]^2dt$

$$p(t) = t\exp\left(-2\pi\left[\frac{t}{t_n}\right]^2\right) \tag{1}$$

Where, t_n is parameter for deciding pulse width, and uses under nanosecond value. When the pulse of Gaussian mono cycle like equation (1) is introduced to receiver, input wave form, $P_{RX}(t)$, is expressed as equation(2)[4].

$$P_{RX}(t) = \left(1 - 4\pi\left[\frac{t}{t_n}\right]^2\right)\exp\left(-2\pi\left[\frac{t}{t_n}\right]^2\right) \tag{2}$$

Where, spectrum of signal and structure on time domain is determined by t_n.

In TH UWB-IR system, demodulation use the correlator, and normalized signal correlation function, $\gamma_p(\tau)$, is defined as follows[5].

$$\gamma_p(\tau) = \int_{-\infty}^{+\infty} p_{RX}(t)p_{RX}(t+\tau)dt \tag{3}$$

$$= \left[1 - 4\pi\left[\frac{\tau}{t_n}\right]^2 + \frac{4\pi^2}{3}\left[\frac{\tau}{t_n}\right]^4\right]\exp\left(-\pi\left[\frac{\tau}{t_n}\right]^2\right)$$

Transmission signal of UWB-IR system is shown in Fig. 2, and transmission signal via considering the multiple accesses is shown in Fig. 3.

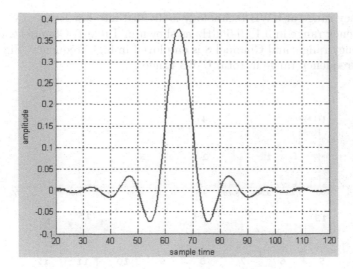

Fig. 2. UWB-IR transmission signal

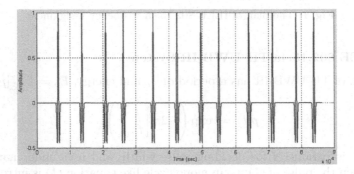

Fig. 3. TH UWB-IR transmission signal (TH code word=4)

2.3 Model of Indoor Multi-path Channel

UWB system directly transmit via pulse row, and received signal is classified depending on value of δ. Therefore, in case that time delay δ, which is received via indirect path, the signal detection performance is effected at correlator. Time delay between direct wave and indirect wave is calculated using following equation, where c stands for the velocity of wave.

$$\Delta t = (R - D)/c \tag{4}$$

Fig. 4 shows the effect of direct wave due to the path delay time.

In case of path 4 in Fig. 4, indirect pulse, which has larger time delay, Δt, than the width of direct pulse, has no effect on the received direct pulse wave.

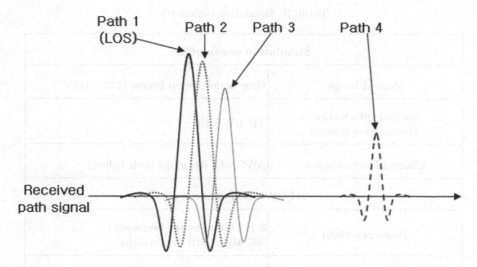

Fig. 4. The effect of between direct wave and path wave due to the path delay time

However, path 2 and 3, which has smaller time delay, Δt, than width of direct pulse, can generate critical interference in demodulation of reception signal.

In this study, to model the channel generated from the multi-path delay, the multi-path parameter of interior office environment, which is recommended by ITU-R M.1225, is applied to analyze the system performance [6]. In Table 2, the parameters suggested by ITU-R M.1225 are shown, Suggested wireless channel environment consists of channel A, which has relatively smaller delay spread, and channel B, which has intermediate delay spread.

Table 2. Parameter of TDL in indoor office environment

	ITU-R M.1225		Modeling Parameter	
Tap	Channel A	Channel B	Channel A	Channel
	Delay (ns)		Tap Weight	
1	0	0	0.6172	0.5784
2	50	100	0.3093	0.2525
3	110	200	0.0617	0.1102

3   Performance Analysis of System

In Table 3, the simulation parameters for medical picture transmission system are shown via TH UWB-IR system.

Table 3. Simulation parameter

Simulation parameter	
Medical Image	Gray scale bitmap Image (128 ×128)
Medical Information Transmission System	TH UWB-IR
Channel Environments	AWGN+3-ray multi-path fading
TH UWB-IR Parameter	
Frequency Band	8.112 GHz (center frequency) 507 MHz (3dB bandwidth)
Data rate	1 Mbps
Modulation index	2 (binary PPM)

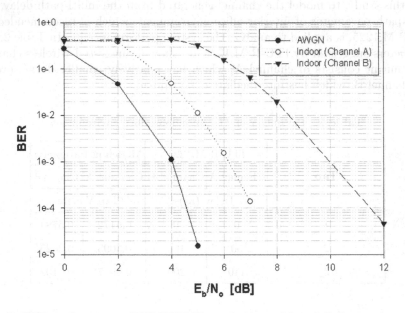

Fig. 5. BER performance of TH UWB-IR system in multi-path fading environment

Fig. 5 shows BER performance graph for 1Mbps data transmission on multi-path fading channel environment in TH UWB-IR medical picture transmission system.

In case of channel B, which has larger delay spread than channel A, 5dB of SNR performance difference is achieved to meet 10^{-3}BER like channel A in interior office environment, Fig. 6 shows the comparison on PSNR of received picture for multi-path fading channel environment in TH UWB-IR medical picture transmission system.

Quality of picture should be evaluated by person, but for the objective evaluation criteria S/N ratio is frequently used. This is different from the conventional transmission S/N ratio. Following equation is the definition of PSNR. Numerator terms repre-sent the maximum 255 signal of original image in case of 8bit/pixel picture. Denomi-nator terms represent the noise, which use the difference between original and recon-structed image.

$$PSNR(a,b) = 10\log_{10}\left[\frac{255^2}{\frac{1}{N\times M}\sum_{x=0}^{N-1}\sum_{y=0}^{M-1}[a(x,y)-b(x,y)]^2}\right] [dB] \qquad (5)$$

In above equation, a is original picture, and b is reconstructed picture. (x, y) is combination of pixel.

In case of image, over 30dB of PSNR is supposed for no deterioration. In Fig. 6, reception SNR is about 6dB for 30dB of PSNR. However, for channel B, higher SNR (over 11dB) is required. Table 4 visually shows the performances of received picture for varying status of multi-path fading channel in TH UWB-IR medical image transmission system.

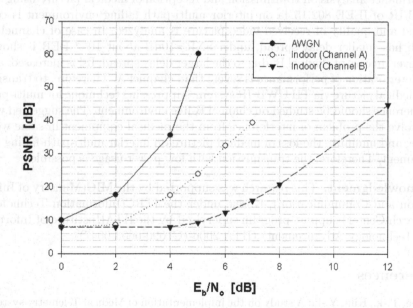

Fig. 6. Variation of PSNR Medical Image Transmission System according to channel environment

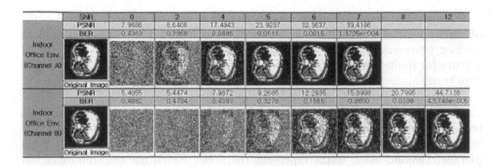

Fig. 7. Comparison of received image to TH UWB-IR Medical Image Transmission System in multi-path fading environment

In case of channel A, visible analysis of medical picture is possible for over 4dB of SNR. In case of channel B, visible analysis is possible for over 8dB of SNR. Therefore, to transmit medical picture, which requires minimum distortion, error correction technique should be applied for low power consumption and high quality transmission.

4 Conclusion

In this paper, the medical picture transmission service using IEEE 802.15.4a specification for low-rate transmission on WPAN environment is suggested. Also, the performance analysis on transmission and reception of medical picture using TH UWB-IR of IEEE 802.15.4a on interior multi-path fading environment is conducted and quality of received video picture is analyzed. In case of channel A, which has smaller delay spread, under SNR=10dB about 10^{-4} BER is shown. However, in case of channel B, due to the larger delay spread, ISI is increased, and the deterioration of performance is severe than channel A. Therefore, to transmit the medical picture via TH UWB-IR system, suppressing technique for multi-path interference should be studied on channel B-like environment. The suggested way can solve the problem of interference from the medical equipment in same work space, and minimize the loss of medical picture on interior multi-path fading environment. Therefore, the transmission with low power usage is possible.

Acknowledgment. This research was supported by the MIC(Ministry of Information and Communication), Korea, under the ITRC(Information Technology Research Center) support program supervised by the IITA(Institute of Information Technology Advancement) (IITA-2006-(C1090-0603-0047)).

References

1. Lee, T.-k., Kim, Y.-k.: A study on the implementation of Medical Telemetry systems using wireless public data network. In: KIMICS 2000 Conference, vol. 4(2), pp. 278–283 (2000)

2. Seo, I.-H., Kang, H.-J.: A Study on the OFDM Wireless Transmission System for Medical Information transmission in Mobile Multi-path Environment. In: KIMICS 2006 Conference, vol. 10, pp. 153–156 (May 2006)
3. IEEE 802.15.4a TG4a drafting, IEEE p802.15-05/0733r2 (December 2005)
4. Win, M.Z., Scholtz, R.A.: Impulse Radio:How It Works. IEEE Comm. Lett. 2, 36–38 (1998)
5. Lee, Y., Kang, H., Lee, M., Kim, T.: A Study on the Effect of Interference on Time Hopping Binary PPM Impulse Radio System. In: Huang, D.-S., Zhang, X.-P., Huang, G.-B. (eds.) ICIC 2005. LNCS, vol. 3645, pp. 665–674. Springer, Heidelberg (2005)
6. REC ITU-M.1225, Guidelines for Evaluation of Radio Transmission Technologies for IMT-2000
7. Ngan, K.N., Yap, C.W., Tan, K.T.: Video Coding for Wireless Communication Systems. Marcel Dekker (2001)

Detecting Image Based Spam Email*

Wanli Ma, Dat Tran, and Dharmendra Sharma

School of Information Sciences and Engineering, University of Canberra, Australia
{Wanli.Ma,Dat.Tran,Dharmendra.Sharma}@canberra.edu.au

Abstract. Image based spam email can easily circumvent widely used text based spam email filters. More and more spammers are adapting the technology. Being able to detect the nature of email from its image content is urgently needed. We propose to use OCR (optical character recognition) technology to extract the embedded text from the images and then assess the nature of the email by the extracted text using the same text based engine. This approach avoids maintaining an extra image based detection engine and also takes the benefit of the strong and reasonably mature text based engine. The success of this approach relies on the accuracy of the OCR. However, regardless of how good an OCR is, misrecognition is unavoidable. Therefore, a Markov model which has the ability to tolerate misspells is also proposed. The solution proposed in this paper can be integrated smoothly into existing spam email filters.

1 Introduction

Spam email is the common name for "unsolicited bulk email". It is one of the cyber nuisances we have to put up with in our daily life. Spam email does not just waste resources, but also poses a serious security threat. There are 2 types of spam email: unsolicited commercial email and the email used as delivery agents for malware (malicious software). The former uses email for commercial advertisement purposes and costs staff time and IT resources. It is estimated that unsolicited commercial email "costs the world US$50 billion in lost productivity and other expenses" a year [1]. The latter has more a sinister intention. For any type of malware, be it virus, worm or SpyWare, after being developed, it has to find a way to infect host computers. An easy and effective way to deliver malware is through unsolicited bulk email. LoveBug [2] and Slammer [3] are prominent examples. Slammer alone, "caused between US$950 million and US$1.2 billion" in its first five days worldwide and the LoveBug has cost the world US$8.8 billion [4].

There are many spam email filter products and they are all far from foolproof. SpamAssassin [5] is the one of the most mature and most widely used spam filters. Many organizations run spam email filters at their incoming mail servers yet all of us have and are continuously having first hand experience with the frustration of spam email. The research community has put significant effort

---

* This research work is supported by the divisional grants from the Division of Business, Law and Information Sciences, University of Canberra, Australia, and the university grants from University of Canberra, Australia.

M.S. Szczuka et al. (Eds.): ICHIT 2006, LNAI 4413, pp. 168–177, 2007.

into detecting spam email. Many different types of mathematic models have been proposed, for example, Na Bayes classifier [6], instance based learning - memory based approach [7], boosted decision tree [8], Maximum Entropy [9], Support Vector Machines [10], LVQ-based neural network [11] and practical entropy coding theory [12].

While the research community is improving the effectiveness of different mathematic models in detecting spam email, the spammers swiftly invented new techniques to circumvent spam filters. There are two popular ways of hoodwinking the spam filters:

1. Using deliberately misspelled words [13]: for example, spelling virus as "virrus", "vi-rus" or "viruus".
2. Using images as email attachments as shown in Fig 1. Aradhye et al. [15] estimated that 25% of spam email contains images. C.-T. Wu et al's count is 38% [14]. One of the authors counted spam email received from his working email address. Among the 256 spam emails received within 15 days, 91 or 36% of emails were image based. Given the fact that image based spam can successfully circumvent spam filters, the situation can only get worse in the future.

Fig. 1. SPAM Email by Image Attachments

To the best of our knowledge, there are only two proposals for detecting image based spam email. In [15], Aradhye et al. suggested to isolate text regions from the images. They extract five features from the regions and then train Support Vector Machines (SVMs) to detect spam images. In [14,16], C.-T. Wu et al. proposed a similar solution. Based on their observation that most spam images are "artificially generated and contain embedded images" [14], they extract three features from the image. They also use SVMs for the detection.

In this paper, we propose to extract the text from the spam images by OCR (optical character recognition) technology and then assess the nature of the email by the extracted text using the same text based detecting engine. The idea of OCR first emerged in 1950's. Eikvil has a good survey paper on early history and technology [17]. Nowadays, commercial grade OCR products, such as FineReader and OmniPage, are widely in use.

The advantage of our approach is that the same engine can be used to process both text and attachment images. Therefore, we only need to conduct the expensive operations of training and updating one engine. In addition, the solution can be easily and seamlessly integrated into the existing spam filters.

The success of the proposal relies on the accuracy and efficiency of the OCR program used. However, regardless of how accurate an OCR product is, it will make mistakes in extracting words from a spam image and produce misspells. Coincidentally, misspell is also a method used by spammers to deceive spam filters. We proposed a statistical framework which has the ability to detect misspelled keywords [18].

The rest of the paper is organized as follows: Section 2 discusses image based spam and the proposed spam detection system, and Section 3 examines the text extracted from spam images. Section 4 works through the Markov Keyword Model that is used to detect misspelled words. Section 5 discusses the experimental results obtained from the Markov Model. Section 6 concludes the paper.

2   Spam Email and Its Detection

At the very beginning, email was in ASCII text format only [19]. To be able to convey rich presentation styles, it was extended with multimedia abilities [20]. Image based spam email takes the advantage of using MIME multipart/alternative directive, which is designed to accommodate multiple displays of an email, such as plaintext format and HTML format. The directive suggests that the enclosed parts are the same in semantics, but with different presentation styles. Only one of them will be chosen to display, and a mailer "must place the body parts in increasing order of preference, that is with the preferred format last" [20].

Figure 2 is an example of spam email. The email has three alternative parts: part one is a plain text paragraph cut from a book, part two is a HTML formatted paragraph cut from a book, and part three is a JPEG format picture of Fig 1 (a). A mailer believes that these three parts are semantically identical and will only display one part. But in this email, the first two parts have nothing to do with the third part. These parts are purposely included in the email to deceive text based spam filters and they will not be displayed by the mailer.

The information reveals that the spam nature of these emails is the text in the images. Instead of treating the text as special image regions, we use OCR to extract the text from the images and then feed it into the spam email detection engine.

For an incoming email, the head and the text body of the email are fed into the email detection engine for assessment as usual. In addition, the attachments are unpacked. The text attachments are sent to the engine as they are, and the image attachments are piped through an OCR program and then sent to the engine, as shown in Fig. fig:DetectingSpamEmail. The OCR program inputs images and outputs the extracted embedded text from the images. In Fig. fig:DetectingSpamEmail `ripmime` [21] is the software tool which unpacks MIME attachments into temporary files, and `gocr` [22] is an open source OCR program.

```
From: spammer <faked_email address>
To: recipient_email_address
Content-type: multipart/alternative;

--Boundary_(ID_fkG49yFmM6kAJ0sBSY0dzg)
##### Part 1: plain text format #####
Langdon looked again at the fax an ancient myth confirmed in black and white.

--Boundary_(ID_fkG49yFmM6kAJ0sBSY0dzg)
##### Part 2: HTML format #####
<textarea style="visibility: hidden;">Stan Planton for being my</textarea>

--Boundary_(ID_fkG49yFmM6kAJ0sBSY0dzg)
##### Part 3: picture format. It has nothing to do with Part 1 or 2 #####
Content-type: image/jpeg; name=image001.jpg
Content-disposition: attachment; filename=image001.jpg

/9j/4AAQSkZJRgABAgAAZABkAAD/7AARRHVja3kAAQAEAAAAHgAA/
+4ADkFkb2JIAGTAAAAAAf/bXFxcXHx4XGhoaGhceHiMIJyUjHi8vMzML
```

Fig. 2. A spam email sample

This system design has a few advantages. First, the same detection engine is used to process the text part and the image part of email. Training and updating the detection engine are very expensive operations and in most cases, human intervention is needed. From an operational point of view, keeping one engine saves on cost, and from a system point of view, one engine preserves the integrity of the data and logic. Second, Bayesian filters of text based detection engines are the working horses on the field [23]. Our proposal takes full benefit of current text based detection engines. Finally, the system is ready to implement on real mail servers. The image process part only adds an extra branch to the data flow of an existing spam detection setup. In addition, to pipe the email text into the detecting engine, the email is also piped through `ripmime` and `gocr` and then back to the detecting engine. Both `ripmime` and `gocr` are open source software and are easy to install. The extra branch of data flow can be easily and seamlessly integrated into the existing spam filters.

3 Extracting Text from Images by OCR

Spam images always contain text, and the text is the key to identify the nature of the email - being spam or not. The same observation was also made in [14,15,16]. The success of our proposal relies on the accuracy and efficiency of the OCR program used.

We group the text embedded in the image into two categories: image text and text generated image text (tegit). Table 1, Row 1, contains image text. The text is designed together with the whole image and blended smoothly into the background. The text in Table 1, Row 2, is different. It is actually an image generated from text. We call it text generated image text, or "tegit" for short.

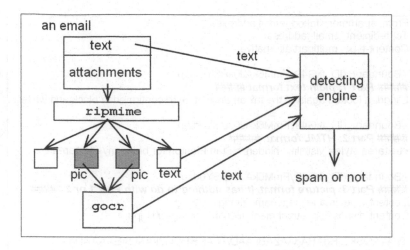

Fig. 3. Detecting Spam Email

Tegit uses images to circumvent spam filters and displays the same visual text appearance to mislead human readers. The text like display makes it appear less like spam to human readers.

There are many OCR products available, ranging from expensive commercial products to free open source software. The accuracy and efficiency of each

Table 1. Images and the Recognised Text

	Picture	Recognized Text
1	VIAGRAsoft $ 3.66 per 100 mg CIALISsoft $ 3.78 per 20 mg LEVITRA $ 4.90 per 20 mg VIAGRA $ 3.00 per 100 mg	VIAGRAsoft ___ ___ CIALISsoft _ $ 3.66 _ $ 3.78 per l * mg per * mg LEVITRA ___ VIAGRA m: $ 4.90 t $ 3.00 per * mg per l* mg
2	☆nab Dear National Australia Bank client. The National Australia Bank Technical Department is performing a scheduled software upgrade to improve the quality of the banking services. By clicking on the link below you will begin the procedure of the user details confirmation. http://ib.national.com.au/conferencepage/logon/creditcardlogin.asp These instructions are to be sent to and followed by all National Australia Bank clients. We apologize for any inconvenience and thank you for cooperation. National Australia Bank Technical Service ©National Australia Bank Limited.	___ ___ nab Dear NatIonal AustralIa Bank client The NatIonal AustralIa Bank TechnIcal Depa_ment Is performI ng a scheduled so_ware upgrade to Improve the qualIty of the bankIng services By clIckIng on the II nk
3	Enlarge Patch AMAZING MEDICAL BREAKTHROUGH IN MEN'S ENLARGEMENT!	_,,,,,,··,_,, ,,,,,,_,,·,, g·,,,,,,_;,_,,gu,H ,,·, ,,·,,,,··, _' Q·,,,,,,a,,,,,,·,,, /,M, g_

product varies. However, regardless of how accurate an OCR product is, it always has difficulties in recognizing similar characters, such as "l" (el) and "1" (one), "u" and "v", and "i" and "I".

In our experience, OCR provides good outcomes, in terms of being accepted by our proposed Markov model detecting engine, on tegit and most image text. Out of the 33 spam images we recently collected, 30 yielded good outcomes and only 3 were unacceptable. Table 1 gives three examples of images and the text extracted by gocr. The first two rows show good results, and the third row shows an unacceptable result. The image text of Row 3 is blended into the background in a way that gocr cannot handle. We will discuss a possible solution to the problem in Section 6.

Coincidentally, misspelling is also a method used by spammers to deceive spam filters. We proposed to use Markov keyword models to detect misspelled keywords. In the framework, a keyword is represented as a Markov chain where letters are states. A Markov model is then built for the keyword. In order to decide an unknown word as a misspelling of a given keyword, a statistical hypothesis test is used.

4   Markov Keyword Models

Let $X = \{X^{(1)}, X^{(2)}, \cdots, X^{(k)},\}$ be a set of K random variable sequences, where $X^{(k)} = \{X_1^{(1)}, X_2^{(2)}, \cdots, X_{T_k}^{(k)}\}$ is a sequence of T_k random variables, $k = 1, 2, \cdots, K$ and $T_k > 0$. Let $V = \{V_1, V_2, \cdots, V_M\}$ be the set of M states in a Markov chain. Consider the conditional probabilities

$$P(X_t^{(k)} = x_t^{(k)} \mid X_{t-1}^{(k)} = x_{t-1}^{(k)}, \cdots, X_1^{(k)} = x_1^{(k)}) \tag{1}$$

where $x_t^{(k)}$, $k = 1, 2, \cdots, K$ and $t = 1, 2, \cdots, T_k$ are values taken by the corresponding variables $X_t^{(k)}$. These probabilities are very complicated for calculation, so the Markov assumption is applied to reduce the complexity

$$P(X_t^{(k)} = x_t^{(k)} \mid X_{t-1}^{(k)} = x_{t-1}^{(k)}, \cdots, X_1^{(k)} = x_1^{(k)}) = P(X_t^{(k)} = x_t^{(k)} \mid X_{t-1}^{(k)} = x_{t-1}^{(k)}) \tag{2}$$

where $x_t^{(k)}$, $k = 1, 2, \cdots, K$ and $t = 1, 2, \cdots, T_k$. This means that the event at time t depends only on the immediately preceding event at time $t - 1$. The stochastic process based on the Markov assumption is called the Markov process. In order to restrict the variables $X_t^{(k)}$ taking values $x_t^{(k)}$ in the finite set V, the time-invariant assumption is applied

$$P(X_1^{(k)} = x_1^{(k)}) = P(X_1^{(k)} = V_i) \tag{3}$$

$$P(X_t^{(k)} = x_t^{(k)} \mid X_{t-1}^{(k)} = x_{t-1}^{(k)}) = P(X_t^{(k)} = V_j \mid X_{t-1}^{(k)} = V_i) \tag{4}$$

where $k = 1, 2, \cdots, K$, $i = 1, 2, \cdots, M$, $j = 1, 2, \cdots, M$. Such the Markov process is called Markov chain.

Define the following parameters

$$q = [q(i)], q(i) = P(X_1^{(k)} = V_i) \tag{5}$$

$$p = [p(i,j)], p(i,j) = P(X_t^{(k)} = V_j \mid X_{t-1}^{(k)} = V_i) \tag{6}$$

where $k = 1, 2, \cdots, K$, K is the number of words in the training document, $t = 1, 2, \cdots, T_k$, T_k is the word length, $i = 1, 2, \cdots, M$, $j = 1, 2, \cdots, M$, M is the number of alphabetical letters. The set $\lambda = (q, p)$ is called a Markov language model that represents the words in the training document as the Markov chains. A method to calculate the model set $\lambda = (q, p)$ is calculated as follows [24].

$$q(i) = \frac{n_i}{\sum_{s=1}^{M} n_s} \qquad\qquad p(i,j) = \frac{n_{ij}}{\sum_{s=1}^{M} n_{is}} \tag{7}$$

The equations in (7) are used to determine the Markov keyword model for each keyword and the Markov keyword list model for the entire keyword list.

5   Experimental Results

The misspelling word detection problem is formulated as a problem of statistical hypothesis testing. For an unknown word X and a claimed keyword W, the task is to determine if X is a misspelling of W. This task can be regarded as a basic hypothesis test between the null hypothesis H_0: X is a misspelling of the claimed keyword W against the alternative hypothesis H: X is not a misspelling of the claimed keyword W. A parametric form of the distribution under each hypothesis is assumed to estimate these probability density functions. Let λ_c be the claimed keyword model and λ be a model representing all other words, i.e. impostors. Let $P(X \mid \lambda_c)$ and $P(X \mid \lambda)$ be the likelihood functions of the claimed keyword and impostors, respectively. The similarity score is calculated as follows

$$S(X) = \frac{P(X \mid \lambda_c)}{P(X \mid \lambda)} \begin{cases} > \theta & \text{accept } H_0 \\ \leq \theta & \text{reject } H_0 \end{cases} \tag{8}$$

The denominator $P(X \mid \lambda)$ is called the normalization term and requires calculation of all impostors' likelihood functions. However it is impossible to do the calculation for all words in the dictionary, hence a subset of B "background" words is used to represent the population close to the claimed keyword [25]. The training and detection procedures of the proposed misspelling word detection tool are summarized as follows.

- Training:
  1. Given K keywords in the blacklist.
  2. Train K Markov keyword models using (7) where X is the sequence of letters in the keyword used to calculate n_i and n_{ij}.
  3. Train a Markov keyword list model using (7) where X is the sequence of letters of all keywords in the keyword list used to calculate n_i and n_{ij}.

– Detection:

1. Given an unknown word regarded as a sequence of letters X, i.e., $X = (x_1, x_2, \cdots, x_T)$, a claimed keyword and a predefined threshold θ.
2. Calculate the probabilities $P(X \mid \lambda_c)$ and $P(X \mid \lambda)$ using (9) when λ_c is the claimed keyword model and λ is the keyword list model.
3. Calculate the similarity score $S(X)$ and compare it with the threshold θ using (8).
4. If the score is greater than the threshold, the unknown word is referred to as a misspelling of the claimed keyword.

An email is regarded as a spam email if it contains a predefined number of keywords and/or their misspellings.

We collected a set of 50 keywords from our emails classified as spam emails in the banking, pornography and advertisement categories. This set was used to build the keyword list model. A keyword model was built for each word in this set, so the number of models in this experiment was 51.

We test the models against both deliberately misspelled email keywords and OCR text.

For deliberately misspelled email keywords, in addition to the misspelled words we collected from spam email, we also produced 10 misspellings for testing. The test set therefore contained 500 misspellings. The minimum word length was set to three for all words in the keyword list and the test set. It was first pre-processed to remove all special, common characters and punctuation marks such as commas, columns, semi-columns, quotes, stops, exclamation marks, question marks and signs. The next step was to convert all the characters into lowercase. The probability is calculated as follows:

$$P(X \mid \lambda) = q(x_1) + \sum_{t=2}^{T} p(x_{t-1}, x_t) \tag{9}$$

This probability calculation allows pairs of letters which do not appear in the keyword but appear in its misspellings to contribute nothing to the probability. The scores obtained for misspellings of a keyword are not very different from the score for the keyword. Therefore, it is possible to set a threshold to detect the possible misspellings of a given keyword with a minimum false acceptance of other words. Experiments showed that the proposed system could detect keywords and their misspellings in the test set with the equal error rate (false rejection error = false acceptance error) of 0.1%.

For OCR text, we extract the text from spam email image attachments by an OCR program (`gocr`). OCR text always comes with misspells - OCR noise. In our experience, the OCR program provides good outcomes in terms of being accepted by our proposed Markov method detecting engine on all tegits and most image text.

Out of the 33 spam images we recently collected, 23 (70%) are tegits, and 10 (30%) are of image text. Among the 23 files which are produced by the OCR program from tegits, we achieved 100% detection rate. And, for the 10 files from image text, we achieved 70% detection rate. Therefore, the overall weighted detection rate is 91%. Note that the result for OCR text is the same as using trigram method to recover the keyword [26], due to the small sample size, but the speed of using the Markov chain based method is faster than the trigram based method [27].

6   Conclusion and Future Work

In this paper, we proposed a new methodology to detect spam email based on its image content. Our proposal involved the use of optical character recognition technology to extract the embedded text from the images and then assess the nature of the email by examining the text with a mature text based engine. We also introduced the Keyword Markov Model which has the ability to tolerate both purposely misspelled words and words that are misrecognized by the OCR due to the similarity between certain characters. Our proposal can be smoothly integrated into existing spam email filters to increase the chances of catching spam emails.

Our preliminary test was very encouraging. In the future, we are going to build a mail server on a Linux host with `ripmime`, `gocr` and the proposed Markov text engine to perform a large scale real life test. For these images, which `gocr` does not yield meaningful outcomes, we are considering to use the OCR output text, whatever it is, as the signature of the image. The large scale test will be conducted to determine if the yielded text is stable with different sizes, brightness and color saturation.

References

1. Keizer, G.: Spam Could Cost Businesses Worldwide $50 Billion (accessed 09 October 2005), http://www.informationweek.com/story/showArticle.jhtml?articleID=60403649
2. Symantec: Love Letter Worm (accessed October 2005), http://securityresponse.symantec.com/avcenter/venc/data/vbs.loveletter.a.html
3. Symantec: Slammer Virus, (accessed October 2005), http://securityresponse.symantec.com/avcenter/venc/data/w32.sqlexp.worm.html
4. Lemos, R.: Counting the cost of Slammer (accessed 11 October 2005), http://news.com.com/2102-1001_3-982955.html?tag=st.util.print
5. SpamAssassin: The Apache SpamAssassin Project, http://spamassassin.apache.org/
6. Sahami, M., et al.: A Bayesian Approach to Filtering Junk E-mail. In: AAAI- 1998 Workshop on Learning for Text Categorization (1998)
7. Sakkis, G., et al.: A Memory-Based Approach to Anti-Spam Filtering for Mailing Lists. INFORMATION RETRIEVAL 6(1), 49–73 (2003)
8. Carreras, X., Marquez, L.: Boosting Trees for Anti-Spam Email Filtering. In: 4th International Conference on Recent Advances in Natural Language Processing (RANLP-2001) (2001)
9. Zhang, L., Yao, T.-S.: Filtering Junk Mail with A Maximum Entropy Model. In: 20th International Conference on Computer Processing of Oriental Languages (IC-CPOL 2003) (2003)

10. Drucker, H., Wu, D., Vapnik, V.N.: Support vector machines for spam categorization. IEEE Transactions on Neural Networks 10(5), 1048–1054 (1999)
11. Chuan, Z., et al.: A LVQ-based neural network anti-spam email approach. ACM SIGOPS Operating Systems Review 39(1), 34–39 (2005)
12. Zhou, Y., Mulekar, M.S., Nerellapalli, P.: Adaptive Spam Filtering Using Dynamic Feature Space. In: 17th IEEE International Conference on Tools with Artificial Intelligence (ICTAI 2005) (2005)
13. Graham-Cumming, J.: The Spammers' Compendium, http://www.jgc.org/tsc/
14. Wu, C.-T., et al.: Using visual features for anti-spam filtering. In: IEEE International Conference on Image Processing (ICIP 2005) (2005)
15. Aradhye, H.B., Myers, G.K., Herson, J.A.: Image Analysis for Efficient Categorization of Image-based Spam E-mail. In: Eighth International Conference on Document Analysis and Recognition (ICDAR 2005), IEEE, Los Alamitos (2005)
16. Wu, C.-T.: Embedded-Text Detection and Its Application to Anti-Spam Filtering. University of California, Santa Barbara: Santa Barbra, CA, USA (2005)
17. Eikvil, L.O.: Optical Character Recognition. Oslo, Norway, Norwegian Computing Center (1993)
18. Tran, D., et al.: A Proposed Statistical Model for Spam Email Detection (submitted for publishing 2006)
19. Postel, J.B.: Simple Mail Transfer Protocol, http://www.ietf.org/rfc/rfc0821.txt
20. Freed, N., Borenstein, N.: Multipurpose Internet Mail Extensions (MIME) Part Two: Media Types (accessed May 2006), http://www.ietf.org/rfc/rfc2046.txt
21. ripmime (accessed May 2006), http://www.pldaniels.com/ripmime
22. gocr (accessed May 2006), http://jocr.sourceforge.net
23. Pelletier, L., Almhana, J., Choulakian, V.: Adaptive filtering of spam. In: (CNSR 2004) Second Annual Conference on Communication Networks and Services Research (2004)
24. Tran, D., Sharma, D.: Markov Modeling Method for Written Language Identification and Verification. In: the Sixth International Conference on Intelligent Technologies InTech 2005, Thailand (2005)
25. Tran, D.: New Background Modeling for Speaker Verification. In: INTERSPEECH, ICSLP Conference, Korea (2004)
26. Ma, W., Tran, D., Sharma, D.: Detecting image based spam email by using OCR and trigram methods. In: Proceedings of Asia-Pacific Workshop on Visual Information Processing (VIP 2006), Beijing, China (November 2006)
27. Tran, D., Markov, D.S.: Models for Written Language Identification. In: The 12th International Conference on Neural Information Processing, Taiwan, pp. 67–70 (30 October-2 November 2005)

Efficient Fixed Codebook Search Method for ACELP Speech Codecs

Eung-Don Lee[1] and Jae-Min Ahn[2]

[1] Electronics and Telecommunication Research Institute, 161 Gajeong-Dong,
Yuseong-Gu, Daejeon, 305-700, Korea
edlee@etri.re.kr
[2] Div. of Electrical & Computer Eng., Chungnam National University, 161
Gung-Dong, Yuseong-Gu, Daejeon, 305-764, Korea
jmahn@cnu.ac.kr

Abstract. There are several sub-optimal search techniques for fast algebraic codebook search of ACELP speech codecs. Focused search method,
depth-first tree search method and pulse replacement methods are used
to reduce computational complexity of algebraic codebook search. In previous pulse replacement methods, the computational load is increased
as the pulse replacement procedure is repeated. In this paper, we propose a fast algebraic codebook search method based on iteration-free
pulse replacement. The proposed method is composed of two stages. At
the first stage, an initial codevector is determined by the backward filtered target vector or the pulse-position likelihood-estimate vector. At
the second stage, after computing pulse contributions for every track
the pulse replacement is performed to maximize the search criterion Q_k
over all combination replacing the pulses of the initial codevector with
the most important pulses for every track. The performance of the proposed algebraic codebook search method is measured in terms of the
segmental signal to noise ratio (SNRseg) and PESQ (Perceptual Evaluation of Speech Quality) using various speech data. Experimental results show that the proposed method is very efficient in computational
complexity and speech quality comparing to previous pulse replacement
methods.

Keywords: ACELP, PESQ, algebraic codebook.

1 Introduction

ACELP (Algebraic Code-Excited Linear-Prediction) speech codecs[2, 3] are most
widely used in digital speech communications due to their high performance. For
example, ITU G.729[4, 5] and G.723.1[6] for VoIP, 3GPP EFR[7] for GSM and
AMR[8, 9] for W-CDMA, and 3GPP2 EVRC[10] for PCS and SMV[11] for IMT-
2000 are adopted as the standard speech codecs. An ACELP algebraic codebook
is a set of codevectors, each consisting of several pulses with position constraints,

M.S. Szczuka et al. (Eds.): ICHIT 2006, LNAI 4413, pp. 178–187, 2007.
© Springer-Verlag Berlin Heidelberg 2007

and the optimal codevector from the algebraic codebook is searched for each subframe. Due to the enormous number of codevector candidates, it is desirable to develop efficient algebraic codebook search methods with no degradation in overall performance. The conventional algebraic codebook search method recommended in each ACELP codec uses various sub-optimal search techniques. For example, "focused search method" in G.729 and G.723.1 5.3 kbps and "depth-first tree search method" in G.729A, AMR, and SMV are used to reduce computational complexity.

Although the sub-optimal algebraic codebook search methods can reduce the computational complexity of algebraic codebook search, the algebraic codebook search is still the major reason for the low efficiency of ACELP codec. Recently, efficient algebraic codebook search methods based on pulse replacement were proposed, which are "least important pulse replacement method" and "global pulse replacement method." And also the global pulse replacement method was adopted as the standard of the algebraic codebook search method of ITU-T G.729.1[] 8kbps mode. However, the least important pulse replacement method is that its procedure is terminated without finding the optimal codevector even in increasing of the number of iteration of the pulse replacement procedure. Moreover in previous pulse replacement methods, the computational load is increased as the pulse replacement procedure is repeated.

In this paper, we propose iteration-free pulse replacement method eliminating the repetition procedure of the pulse replacement and apply the proposed method to G.729A. The performance of proposed algebraic codebook search method is measured in terms of the computational load and the speech quality, and compared with previous pulse replacement methods. In the pulse replacement method, the computational load is increased as the pulse replacement procedure is repeated.

2 Algebraic Codebook

2.1 Algebraic Codebook Structure

In the codebook of G.729A, each codevector contains 4 non -zero pulses. Each pulse can have either the amplitudes +1 or -1, and can assume the positions given in Table 1.

Table 1. Potential positions of individual pulses in the algebraic codebook

Track	Pulse	Sign	Positions
T_0	i_0	$s_0=\pm1$	m_0 : 0, 5, 10, 15, 20, 25, 30, 35
T_1	i_1	$s_1=\pm1$	m_1 : 1, 6, 11, 16, 21, 26, 31, 36
T_2	i_2	$s_2=\pm1$	m_2 : 2, 7, 12, 17, 22, 27, 32, 37
T_3	i_3	$s_3=\pm1$	m_3 : 3, 8, 13, 18, 23, 28, 33, 38
			4, 9, 14, 19, 24, 29, 34, 39

The codevector $c(n)$ is constructed by taking a zero vector of dimension 40, and putting the four unit pulses at the found locations, multiplied with their corresponding sign.

$$c(n) = s_0\delta(n-m_0) + s_1\delta(n-m_1) + s_2\delta(n-m_2) + s_3\delta(n-m_3), n = 0, ..., 39 \quad (1)$$

2.2 Algebraic Codebook Search

The algebraic codebook is searched by minimizing the mean-squared error between the weighted input speech and the weighted synthesis speech. The mean squared weighted error is defined by

$$E_k = \|x - g_c H c_k\|^2 \quad (2)$$

where, x is the target vector given by the weighted input speech after subtracting the zero-input response of the weighted synthesis filter and the adaptive codebook contribution, g_c is a scaling gain factor, and H is a lower triangular convolution matrix constructed from the impulse response of the weighted synthesis filter, and c_k is the codevector with index k. Minimizing Eq. (1) gives

$$E_k = \acute{x}x - \frac{(x^t H c_k)^2}{c_k^t H^t H c_k} \quad (3)$$

From Eq. (2), the optimum algebraic codevector is determined by maximizing

$$Q_k = \frac{(d^t c_k)^2}{c_k^t \Phi c_k} \quad (4)$$

where, $d(= H^t x)$ is the backward filtered target vector representing the correlation between the target signal and the impulse response of weighted synthesis filter, and $\Phi(= H^t H)$ is the autocorrelation matrix of the impulse response of weighted synthesis filter. The vector d and the matrix Φ are computed prior to the codebook search. The elements of the vector d are computed by

$$d(n) = \sum_{i=n}^{39} x_2(n)h(i-n), n = 0, ..., 39 \quad (5)$$

and the elements of the symmetric matrix Φ are computed by

$$\phi(i,j) = \sum_{n=j}^{39} h(n-i)h(n-j), (j \geq i) \quad (6)$$

The algebraic structure of the codebooks allows for very fast search procedures since the innovation vector c_k contains only a few nonzero pulses. The correlation in the numerator of Eq. (3) is given by

$$C = \sum_{i=0}^{3} s_i d(m_i) \quad (7)$$

where m_i is the position of the ith pulse, s_i is its amplitude. The energy in the denominator of Eq. (3) is given by

$$E = \sum_{i=0}^{3} \phi(m_i, m_i) + 2 \sum_{i=0}^{2} \sum_{j=i+1}^{3} s_i s_j \phi(m_i, m_j) \tag{8}$$

To simplify the search procedure, the pulse amplitudes are preset by the mere quantization of an appropriate signal as follows

$$s_i = \begin{cases} sign\{d(i)\} \\ sign\{b(i)\} \end{cases} \tag{9}$$

where, the elements of the pulse-position likelihood-estimator vector b is given by[5, 8]

$$b(n) = \frac{d(n)}{\sqrt{E_d}} + \frac{r_{LTP}(n)}{\sqrt{E_r}}, n = 0, ..., 39 \tag{10}$$

where, E_d is the energy of the backward filtered target vector d and E_r is the energy of the long-term prediction residual vector r_{LTP}.

3   Pulse Replacement Methods

The pulse replacement methods are based on two-stage structure. At the first stage, an initial codevector is searched with minimal search load. Then at the second stage, pulse replacement procedure is applied to the initial codevector to enhance the performance of codevector.

In order to reduce the computational load for finding an initial codevector, the initial codevector is determined by the pulse positions with maximum absolute values of "backward filtered target vector" or by the pulse positions with maximum absolute values of "pulse-position likelihood-estimate vector" due to high probability that one of them might be a member of the optimal codevector[14].

3.1   The Least Important Pulse Replacement Method

In the least important pulse replacement method, it is necessary to measure the contribution of each pulse and replace the least important pulse with a new one. The contribution of a pulse is measured by the value of Q_k of codevector after removing the corresponding pulse one by one from the initial codevector. Since the change of Q_k is caused by the removal of a pulse, the pulse resulting in codevector with the largest Q_k is the least important pulse. After removing the least important pulse, a new pulse is searched from the track with the least important pulse so that Q_k of new codevector is increased. Therefore, the codevector approaches to the optimal solution steadily as this procedure is repeated. If Q_k does not change, which is the case when the removed pulse is selected again, the pulse replacement procedure is terminated.

The computational complexity of the least important pulse replacement method is considerably low because the search is performed sequentially on

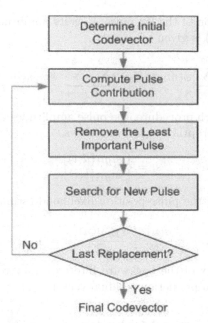

Fig. 1. Block diagram of the least important pulse replacement method

one pulse basis. Applying the least important pulse replacement method to G.729A, at the first stage, initial Q_k is computed from an initial codevector with 1 search. At the second stage, Q_k is computed 4 times for contribution measure of each pulse, and without duplicate computation, the new pulse search requires $7 \times 3/4 + 15 \times 1/4 = 9$ searches on the average. Therefore, the overall computational load with an initial codevector is $1 + (4+9) \times N$, where N is the number of iteration of the pulse replacement procedure.

3.2 Global Pulse Replacement Method

In order to prevent the termination of the pulse replacement procedure without finding the optimal codevector, in global pulse replacement method, a new pulse is searched for all tracks instead of the only track that has the least important pulse. That is, a new pulse is searched by replacing each pulse in each track with a new one so that Q_k of new codevector is maximized. Because the change of Q_k is always maximized in each procedure, the codevector approaches to the optimal solution rapidly as this procedure is repeated. If Q_k does not change, which is the case when the replaced pulse is selected again, the pulse replacement procedure is terminated.

At the first pulse replacement procedure, the computation of Q_k is performed on each track to search for a new pulse, and the computational load can be reduced for one of the tracks by removing duplicate computation at the other pulse replacement procedure. Applying global pulse replacement method to G.729A, at the first stage, initial Q_k is computed from an initial codevector with 1 search.

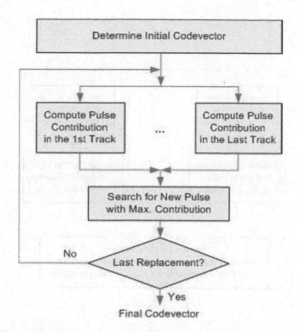

Fig. 2. Block diagram of global pulse replacement method

At the second stage without duplicate computation, the new pulse search requires 7+7+7+15 = 36 searches at the first pulse replacement procedure, and the new pulse search requires $3/4\times(7+7+7+15) = 27$ on the average from the second pulse replacement procedure. Therefore, the overall computational load is 1+36+27(N-1) for N≥1, where N is the number of iteration of the pulse replacement procedure.

3.3   Proposed Iteration-Free Pulse Replacement Method

In previous pulse replacement methods, the computational load is increased as the pulse replacement procedure is repeated. Therefore, in order to eliminate the repetition procedure of the pulse replacement, in the proposed method, new pulses are searched by several pulse replacements at one time after computing pulse contributions for every track to maximize the search criterion over all combination replacing the pulses of the initial codevector with the most important pulses for every track. That is, final codevector is searched from the combination replacing one pulse of the initial codevector with one pulse of the most important pulses, replacing two pulses of the initial codevector with two pulses of the most important pulses, replacing three pulses of the initial codevector with three pulses of the most important pulses, and replacing four pulses of the initial codevector with four pulses of the most important pulses.

Applying the proposed method to G.729A, assume that the pulses of the initial codevector are determined as (30, 31, 32, 4). At the first stage, initial Q_k is computed from an initial codevector with 1 search. At the second stage

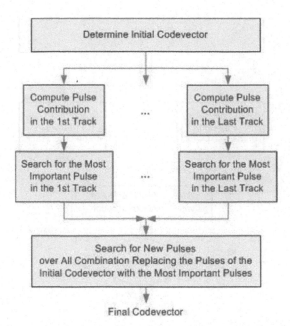

Fig. 3. Block diagram of proposed iteration-free pulse-replacement method

without duplicate computation, the computation of Q_k is performed for pulse contributions measure for every track with $7+7+7+15 = 36$ searches. Assume that the most important pulses are determined as $(5, 21, 17, 19)$.

To find the final codevector, at the first, the computation of Q_k is performed for the combination replacing one pulse of the initial codevector with one pulse of the most important pulses as follows.

$(\mathbf{5}, 31, 32, 4), (30, \mathbf{21}, 32, 4), (30, 31, \mathbf{17}, 4), (30, 31, 32, \mathbf{19})$

Note that the computation of Q_k is already included in the pulse contributions measure.

Next, the computation of Q_k is performed with $6(=4C2)$ searches for the combination replacing two pulses of the initial codevector with two pulses of the most important pulses as follows.

$(\mathbf{5}, \mathbf{21}, 32, 4), (\mathbf{5}, 31, \mathbf{17}, 4), (\mathbf{5}, 31, 32, \mathbf{19}),$
$(30, 21, 17, 4), (30, 21, 32, 19), (30, 31, 17, 19)$

And, the computation of Q_k is performed with $4(=4C3)$ searches for the combination replacing three pulses of the initial codevector with three pulses of the most important pulses as follows.

$(5, 21, 17, 4), (5, 21, 32, 19), (5, 31, 17, 19), (30, 21, 17, 19)$

Lastly, the computation of Q_k is performed with 1(=4C4) searche for the combination replacing four pulses of the initial codevector with four pulses of the most important pulses, which is the case of (5, 21, 17, 19).

Therefore, the overall computational load is 1+36+11 = 48.

4   Experimental Results

The speech quality is measured in terms of segmental SNR(SNR_{seg}) and PESQ (Perceptual Evaluation of Speech Quality) using the test speech data in Table. 2. The algebraic codebook computational load, SNR_{seg}, and PESQ are summarized in Table 3, Table 4, and Table 5, respectively. It can be seen that the proposed method reduces the computational complexity by 10% compared with the least important pulse replacement method with the improved speech quality and reduces the computational complexity by 85% and 60% compared with depth-first search method and global pulse replacement with the slight degradation of speech quality, respectively.

Table 2. Test speech data

Type of speech data	Level of noise
Origin	-
Origin + Music Noise	25dB
Origin + Office Noise	20dB
Origin + Babble Noise	30dB
Origin + Interfering Talker	15dB

∗ Origin consists of 3 males & 3 females speech data, 5 samples each totally 30 samples. Hence the test speech data consist of 150 speech samples.

Table 3. Comparison of computational load, where N is the number of iteration of the pulse replacement procedure

Method		Load
Depth-first Search		320
The Least	N=1	14
Important	N=2	27
Pulse	N=3	40
Replacement	N=4	53
	N=1	37
Global Pulse	N=2	64
Replacement	N=3	91
	N=4	118
Propose		48

Table 4. Comparison of SNR_{seg}(dB)

Method	N=2	N=4
Depth-first Search	10.17	
The Least Important Pulse Replacement	9.63	9.65
Global Pulse Replacement	10.00	10.07
Proposed	9.90	

Table 5. Comparison of PESQ

Method	N=2	N=4
Depth-first Search	3.76	
The Least Important Pulse Replacement	3.72	3.72
Global Pulse Replacement	3.75	3.76
Proposed	3.75	

5 Conclusion

In this paper, iteration-free pulse replacement method for algebraic codebook search is proposed and applied to G.729A. New pulses are searched by several pulse replacements at one time over all combination replacing the pulses of the initial codevector with the most important pulses. The experimental results show that the proposed method provides the speech quality equivalent to that of global pulse replacement method with considerably low complexity, and provides better speech quality than the least important pulse replacement method. The proposed algorithm can be applied to other ACELP-based speech codes to enhance the overall performance.

References

1. Schroeder, M.R., Atal, B.S.: Code Excited Linear Prediction(CELP) High Quality Speech at Very Low Bit Rates. In: IEEE Int. Conf. on Acoust. Speech Signal Process., pp. 937–940 (1985)
2. Adoul, J-P., Mabilleau, P., Delprat, M., Morissette, S.: Fast CELP coding based on algebraic codes. In: IEEE Int. Conf. on Acoust. Speech Signal Process., pp. 1957–1960 (1987)
3. Laflamme, C., Adoul, J-P., Salami, R., Morissette, S., Mabilleau, P.: 16 kbps Wideband Speech Coding Technique Based on Algebraic CELP. In: IEEE Int. Conf. on Acoust. Speech Signal Process., Toronto, CA, pp. 13–16 (May 1991)
4. ITU-T Rec. G.729, Coding of Speech at 8 kbit/s Using Conjugate-Structure Algebraic-Code-Excited Linear-Prediction (CS-ACELP) (March 1996)
5. ITU-T Rec. G.729 Annex A, Reduced Complexity 8 kbit/s CS-ACELP Speech Codec (November 1996)

6. ITU-T Rec. G.723.1, Dual Rate Speech Coder for Multimedia Communications Transmitting at 5.3 and 6.3 kbit/s (March 1996)
7. GSM 06.60, Enhanced Full Rate(EFR) Speech Transcoding (1996)
8. 3GPP TS 26.090, AMR Speech Codec; Transcoding Functions (Release 4) (March 2001)
9. 3GPP TS 26.190, AMR Wideband Speech Codec; Transcoding Functions (Release 5) (March 2001)
10. TIA/EIA/IS-127, Enhanced Variable Rate Codec, Speech Service Option 3 for Wideband Spread Spectrum Digital Systems (1997)
11. 3GPP2 C.S0030-0 Ver.2.0, Selectable Mode Vocoder Service Option for Wideband Spread Spectrum Communication Systems (December 2001)
12. ITU-T Rec. G.729.1, G.729 based Embedded Variable Bit-Rate Coder: An 8-32 kbit/s Scalable Wideband Coder Bitstream Interoperable with G.729 (May 2006)
13. Park, H.C., Choi, Y.C., Lee, D.Y.: Efficient Codebook Search Method for ACELP Speech Codes. In: 2002 IEEE Speech Coding Workshop Proceedings, pp. 17–19 (2002)
14. Ha, N.K.: A Fast Search Method of Algebraic Codebook by Reordering Search Sequence. In: IEEE Int. Conf. on Acoust. Speech Signal Process., vol. 1, pp. 21–24 (1999)
15. Lee, E.D., Lee, M.S., Kim, D.Y.: Global Pulse Replacement Method for Fixed Codebook Search of ACELP Speech Codec. In: CIIT2003. Proceedings of the 2nd IASTED International Conference, pp. 372–375 (2003)

Conventional Beamformer Using Post-filter for Speech Enhancement

Soojeong Lee, Kiho Choi, and Soonhyob Kim

Department of Computer Engineering, Kwangwoon University
Wolgye-dong, Nowon-gu, Seoul 447-1, Korea
{leesoo86,khchoi,kimsh}@kw.ac.kr

Abstract. This paper presents a combined handsfree speech enhancement method based on a spatialpost-filter. The scheme uses a linear microphone array to capture a speech signal that has been corrupted by babble noise, car noise, and interference signals. Simulation results for real environment show that the proposed structure achieves a maximum interference suppression of 12 dB, an improvement of 6 dB over the delay and sum beamformer. Furthermore, the system is robust in the presence of distortion as opposed to the generalized sidelobe canceller. The subjective evaluation has shown that the combined system of delay and sum with the minimum mean square error estimator using a noncausal signal to noise ratio (SNR) estimator obtained 3.8 points on a fivepoint.

Keywords: Speech enhancement, postfilter, spatial-temporal filer.

1 Introduction

Today, hands-free systems have become a popular method of communication due to their less restrictive nature. However, the use of hands-free systems introduces some new problems. Like any communication system, the hands-free interface suffers from unwanted interference. In a typical office environment, the signal is subjected to background noise and interference effects. Background noise is generally due to machinery, such as computers and air conditioning, as well as other ambient noise sources. Interference includes such highly non-stationary interferers as speech, music, and far-end echo. Further, reflective surfaces such as ceilings, walls and furniture cause reverberation. A sound wave has a constant propagation speed in air. Therefore, the pressure wave produced by a speaker takes a certain time to reach the microphone. If there is a group of microphones covering a region of space, then there is a predetermined amount of delay for the wavefront arrival at each microphone. Beamforming theory refers to a class of algorithms that combine the signals from each microphone to pass or block radiation in a specific direction [1], making it possible to distinguish signals based on their spatial origin.

Adaptive beamformers extract the signal from the direction of arrival as specified by the steering vector, which is a beamforming parameter. However, with a classical adaptive beamformer (GSC) [2],[3],[4],[7], target signal cancellation

M.S. Szczuka et al. (Eds.): ICHIT 2006, LNAI 4413, pp. 188–197, 2007.
© Springer-Verlag Berlin Heidelberg 2007

occurs in the presence of steering vector errors. The steering vector errors are caused by errors in microphone position, microphone gain, reverberation, and target direction. Therefore, steering vector errors are inevitable with actual microphone arrays, and target signal cancellation is a serious problem. Many signal-processing schemes have been proposed to avoid signal cancellation. These techniques are called robust adaptive beamforming as their performance is robust against errors [5]. However, they are not free from degradation in interference rejection or an increase in the number of microphones; moreover, in a complex acoustic environment, the length of the filter typically required is on the order of hundreds. Furthermore, the adaptation process suffers from a large convergence period that will degrade performance in a non-stationary environment [6].

This paper uses a spatial temporal scheme using nonadaptive subband beamforming with a post-filtering method structure to surmount the problems associated with hands-free applications. Subband processing allows the wideband speech signal to be partitioned into a number of narrowband signals [8]. These narrowband signals are processed collectively using data independent beamforming algorithms. There are several advantages to a subband beamformer, as compared to a fullband beamformer: shorter filter length, better interference suppression, a simple optional weighting in each frequency band, faster convergence, and smaller computational load. Therefore, post filtering is used to improve the beamforming performance in nondirectional and nonstationary noise environments. Further, the proposed scheme can be applied to reduce computational complications by using a non-adaptive process. This yields a more effective and efficient processing system.

The following Section describes conventional beamformers based on delay and summing. In Section 3, the structure of the proposed method is described. Section 4 contains simulations demonstrating the evaluation of the proposed system. Finally, Section 5 summarizes, draws conclusions, and proposes further ideas.

2   The Conventional Beamformer

A beamforming signal-processor uses the output of an array of microphones to implement some form of spatial filtering. The general purpose of a beamformer is to improve the signal-to-noise ratio (SNR). That is, to retrieve a desired input signal in the presence of noise and directional interference.

The signal processor sums the microphone outputs and produces the beamformer output [3]. The ability of a beamformer to improve the SNR is due to the increased sensitivity to the signals arriving at $0°$ to the array. Therefore,interference will not be detected as strongly if it arrives at another angle. The ability of a beamformer to improve the SNR is explained below. If X_k is received by the k_{th} microphone in the array and $Noise_1$, $Noise_2$, $Noise_i$, are independent, $X_1 = S + Noise_1$ and $X_k = S + Noise_k$. Therefore, the power output of the array $Y = 2S + Noise$ if the two microphone outputs are summed. This implies that $SNR_{output} = 2S/Noise$, which corresponds to a 3 dB increase in the SNR.

The conventional beamformer is also called a delay-and-sum(DAS) beam-former because of this characteristic. It is often impossible to steer the array of microphones physically, so the array is steered using time delay functions, with the delays corresponding to the differences in time that a signal arrives at the different array microphones. For example, a signal arrives at X_1, X_2 and X_i at the same time t_1, but we want to simulate a steered array where the signal arrives at X_2 at a later time t_2. To do this, we define $\tau_2 = t_2 - t_1$. Time domain beamforming can then be summarized using the following equation:

This is known as delay-and-sum beamforming, where a_i is a shading constant scaled to decrease the magnitudes of the sidelobes. Eq. (1) can be converted to the frequency domain by applying the Fourier Transform:

$$Y(t, \theta) = \sum_{i=1}^{m} a_i X_i (t - \tau_i) \ . \tag{1}$$

$$Y(\omega, \theta) = \sum_{i=1}^{m} a_i X_i (\omega) \, e^{-j\omega\tau_1(\theta)} \ . \tag{2}$$

$$X_r = [X_1(\omega) X_2(\omega) \cdots X_i(\omega)]^T \ . \tag{3}$$

$$W = \left[a_1 e^{-j\omega\tau_1} a_2 e^{-j\omega\tau_2} \cdots a_i e^{-j\omega\tau_i} \right]^T \ . \tag{4}$$

Eq.(3) is the Fourier Transform X_i and Eq. (4) is the weighting vector, which is the same as the steering vector, but includes the shading coefficients. Now, the output spectrum, Eq. (2) can be redefined in terms of Eq.(3) and Eq.(4).

$$P_y = E[YY^*] = W^H E\left[X_r X_r^k\right] W \ . \tag{5}$$

Eq. (5) is the power spectrum. Note that the weighting vector has a determin-istic nature and can be brought out of the expected value. H represents the trans-pose and complex conjugate. As $R = E\left[X_r X_r^k\right]$ is defined as the cross-spectral matrix, P_y can be redefined as $P_y = W^H RW$. For delay-and-sum beamformer, we assume $a_i = 1$. The delay-and-sum is the simplest beamformer; electronically aiming the array such that the direct arrivals of the source of interest sum in a coherent fashion. A drawback with the delay-and-sum(DAS) beamformer is that a strong interferer arriving from a different direction [1].

3   Combined Non-adaptive Beamforming and Post Filtering

The major system block diagram for a subband beamforming system is shown in Fig. 1. The analysis filter bank splits each input into a number of subbands. Each subband block is processed in parallel by a beamforming steering unit with delays t_1, t_2 and t_i which are adjusted such that the desired speech signal S arrives simultaneously at the i microphones. In the beamforming scheme, the

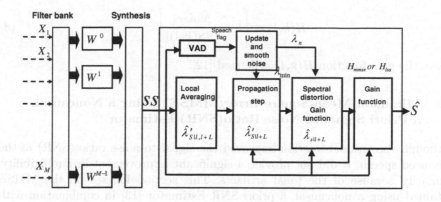

Fig. 1. Combined non-adaptive beamforming and post filtering

effect of the noise components n_i is reduced by computing the averaged signal. After beamforming, subbands are recombined by a synthesis bank to yield the system output. At first glance, it appears that extra complexity and computational burden is being introduced. However, by splitting the data into subbands, shorter filter lengths can be used, which actually improves the computational speed.

3.1 Post-filtering Schemes Using Spectral Subtraction

In the post-filtering scheme, the background noise is further decreased using different noise reduction techniques. Several techniques were developed based on Boll's spectral subtraction (SS) method [9],[11].

Taking the short-time discrete Fourier transform, we find the output of the step one of the Fig. 1. where $SS(k, l)$ is the noisy speech spectrum, $S(k, l)$ is the clean speech spectrum, $Noise(k, l)$ is the residue noise spectrum, k is the frequency bin index and l is the time frame index. The clean speech spectrum can be represented as,

$$|SS(k,l)| = |S(k,l)| + |Noise(k,l)| \,. \tag{6}$$

In equation (7), the spectrum of noise signal, $|Noise(k, l)|$, cannot be obtained directly and it is approximated by the expectation value $E\,[Noise(k, l)]$. Typically $E\,[Noise(k, l)]$ is estimated during non-speech period by using voice activity detection (VAD), it is denoted by $|\hat{N}oise|$. Where $|\hat{N}oise|$ represents the estimated clean speech spectrum.

$$\hat{S}(k,l) = \left[1 - \frac{|\hat{N}oise|}{|SS(k,l)|}\right] * |SS(k,l)| \,. \tag{7}$$

$$H(k,l) = \left[1 - \frac{\hat{N}oise|}{|SS(k,l)|}\right] .$$

(8)

where the gain function $H(k,l)$ is defined [12].

3.2   Minimum Mean-Square Error(MMSE) Using a Noncausal, a Priori Signal-to-Noise Ratio(SNR) Estimator

Although spectral subtraction improved the signal-to-noise ratio (SNR) of the enhanced speech, it did not provide a significant improvement in intelligibility, primarily because of the tonal artifacts. This section focuses on the MMSE obtained using a noncausal, a priori SNR Estimator [13] in combination with other widely used Wiener based estimators [14].

To obtain an estimate for $\hat{S}$, a spectral gain function can be applied to the corresponding noisy spectral component [13], [14] SS:

$$\hat{S} = H\left(\xi, \gamma\right) \cdot SS ,$$

(9)

where $\xi(k,l)$ and $\gamma(k,l)$ are defined by

$$\xi(k,l) = \frac{\lambda_s(k,l)}{\lambda_n(k,l)}, \gamma(k,l) = \frac{|SS(k,l)|^2}{\lambda_n(k,l)} .$$

(10)

$\lambda_s(k,l) = E\{|S(k,l)|^2\}$ is the variance of the spectral component of speech, and $\lambda_n(k,l) = E\{|Noise(k,l)|^2\}$, that of the noise. The variables, $\xi(k,l)$ and $\gamma(k,l)$, are also interpreted as the a priori and a posteriori signal-to-noise ratios.

The spectral power distortion, the log-spectral amplitude, and the MMSE, or gain function, can be written as [13],[14],[15]:

$$H_{spd} = \sqrt{(\frac{1}{\gamma} + \frac{\xi}{1+\gamma})\frac{\xi}{1+\xi}} .$$

(11)

$$H_{lsa} = \frac{\xi}{1+\xi}exp(0.5 \int_{\nu}^{\infty} \frac{e^{-t}}{t} dt) .$$

(12)

$$H_{mmse} = \frac{\sqrt{\pi\nu}}{2\gamma} \exp(\frac{-\nu}{2})[(1+\nu) \cdot I_0(\frac{\nu}{2}) + \nu \cdot I_1(\frac{\nu}{2})] .$$

(13)

where I_n is the Bessel function of order n, and $\nu = \frac{\xi\gamma}{1+\xi}$.

Now let $d(S, \hat{S})$ be a given distortion measure between S and $\hat{S}$, and let SS_0^{l+L} represent the set of spectral measurements up to frame $l + L$, where L denotes a time delay in frames [13]. Then, given SS_0^{l+L}, a noncausal estimator $S(\hat{k}, l)$, which minimizes the conditional expected value of the distortion measure, can be written as

$$\hat{S}(k,l) = Min_{\hat{s}}E\left[d(S, \hat{S})|SS_0^{l+L}\right] .$$

(14)

Given the noisy measurements SS_0^{l+L}, let $\lambda_{s|l+L} = E\{|S|^2|SS_0^{l+L}\}$ denote the conditional variance of S. Then, given SS_0^{l+L} and excluding the noisy measurements SS, $\lambda'_{s|l+L} = E\{|S|^2|SS_0^{l+L}\backslash SS\}$ denotes the conditional variance of S Given the noisy measurements SS_l^{l+L}, let $\lambda'_{s|l,l+L} = E\{|S|^2|SS_l^{l+L}\backslash SS\}$ denote the conditional variance of S. An estimate for $\lambda_{s|l+L}$ can then be obtained from the conditional variance of S given SS and $\hat{\lambda}'_{s|l+L}$, by applying the gain function $H_{spd}(\xi,\gamma)$ to SS, with $\xi = \hat{\lambda}'_{s|l+L}/\lambda_{noise}$ [13].

$$\lambda_{s|l+L} = H_{spd}^2\Big(\frac{\hat{\lambda}'_{s|l+L}}{\lambda_n},\gamma\Big)|SS|^2 = \frac{\hat{\lambda}'_{s|l+L}}{\lambda_n + \hat{\lambda}'_{s|l+L}}\Big(\lambda_{noise} + \frac{\hat{\lambda}'_{s|l+L}|SS|^2}{\lambda_n + \hat{\lambda}'_{s|l+L}}\Big). \quad (15)$$

4   Evaluation and Results

Thirty-one listeners participated in the comparison tests. Enhanced speech signals were played through loudspeakers and the listeners were asked to compare the samples from the perspectives of the amount of noise reduction, the nature of the residual noise, and the distortion of the speech signal itself. Mean opinion score (MOS), the most commonly used subjective quality measure [10], [14], was used to evaluate.

Case one: Using the generalized sidelobe canceller (GSC) with SS and 0 dB SNR, white noise slightly enhanced the speech quality. Little noise reduction was perceived, and musical noise was introduced. The speech signal distortion decreased as input SNR increased, but was still very large, and the enhanced speech was unintelligible at 0 dB SNR.

Case two: Enhanced speech was obtained using the delay and sum with SS, although it suffered from musical noise. It was especially prominent at the high input SNR values of 10 and 15 dB, but could not be perceived at the low SNR values of 5 and 0 dB.

Case three: Using the delay and sum with SS and tracking an a priori SNR using a decision-directed estimator [16] resulted in a great reduction of noise, and provided enhanced speech with only white residual noise. This white residual noise was much less annoying and disturbing than the musical noise.

Case four: Using a minimum mean-square error (MMSE) log-spectral estimator, or an MMSE amplitude estimator with a noncausal a priori SNR estimator, yielded very similar results: a great reduction of the overall noise, but the introduction of a little musical noise [14], [17].

To calibrate the results, the original male speech, without interference and noise, was set as 5. The subjects were told that the desired signal cancellation should have a lower score. The delay-and-sum beamformer therefore enhanced the desired signal through a direct path. The evaluation results are shown in Table 1. The conventional beamformer with SS was rated at 1.4 points because the interference suppression ratio (ISR) was a low 4 dB. The GSC with SS

Table 1. Mean opinion score(clean speech = 5 points)

SNR(dB)	GSC	MMSE	DAS + SS	mulSS	ScalSS	logSNR	priSNR
0	0.31	0.53	1.43	2.42	3.15	3.25	3.79
5	0.58	0.81	1.85	2.74	3.58	3.87	4.21
10	0.88	0.56	1.94	2.97	3.72	4.09	4.58
15	0.90	0.77	2.26	3.18	3.83	4.08	4.56

reduced interference considerably, but also cancelled much of the target signal, resulting in a poor score of 0.3 points. The delay and sum with MMSE using a noncausal a priori SNR estimator scored 3.8 points at 0 dB SNR.

4.1 Experimental Method

The speech quality noise and interference effects were investigated at input SNRs of 0, 5, 10 and 15 dB. Car, babble, and white noise were added to the clean speech signal to generate and test three different types of noisy speech signals using two objective measures. The first, noise suppression, provides an indication of how much noise is reduced in enhanced speech during nonspeech periods. A higher noise suppression value indicates more noise suppression in the enhanced speech signal. Second, the Itakura-Saito distance was obtained. This objective measure was implemented over the entire speech and nonspeech signal. A smaller value indicates lower spectral distortion in the enhanced speech signal. The two objective measures were then compared using different noise reduction algorithms [10],[11].

4.2 Performance

The proposed system was evaluated against two GSC beamformer schemes: a GSC subband beamformer targeted on the desired source location, and a subband beamformer using an MMSE beamformer. This simulation environment was an unrealistic example as the desired signal was drowned out by two jammers containing considerable diffuse white, babble, and car noise of similar power to the desired signal. The performance was compared in an environment consisting of a nonstationary jammer signal placed at 120°, with babble noise 0 dB, and a jammer signal placed at 150°, with car noise 0 dB, plus a white noise. All were placed 1.0 m from the pivot of the array, and the desired signal originated at 70°. The signals were sampled at 8 kHz using 16-bit quantization. The proposed system utilizes 32 subbands, which adequately suppress noise while maintaining a moderate level of distortion. The microphone array consisted of 4 and 5 elements spaced at 0.04m and 0.06m intervals. The simulated room was 5m × 4m × 3m, with the array assumed to be in the center of the room.

An arbitrary male voice, selected from the noisy speech corpus (NOIZEUS), was developed to facilitate the comparison of the speech enhancement algorithms among different research groups.

Table 2. Subband Delay and Sum beamformer using post-filtering vs. subband GSC using post-filtering, with seven microphones and a 4 cm microphone distance: Delay and Sum(DAS) beamformer vs GSC and DAS using SS vs GSC using SS

SNR(dB)	$noise_{supp}$	$inter_{supp}$	distort	$noise_{supp}$	$inter_{supp}$	distort
0	0.2/6.8	4.0/10.6	1.5/3.3	7.8/12.7	10.6/16.0	3.2/13.8
5	0.2/6.2	5.3/11.5	1.0/3.5	3.4/8.6	11.5/16.7	1.6/10.8
10	0.3/5.9	5.9/12.0	0.7/3.8	0.8/6.9	12.9/13.0	0.3/9.8
15	1.5/5.7	4.8/12.1	0.5/2.1	0.3/6.2	6.1/12.6	0.2/3.8

The proposed system has a much better noise and interference suppression capability than the standard delay-and-sum structure, albeit at the expense of a slight increase in signal distortion. As the SNR increases, the amounts of noise, jammer suppression, and distortion decrease rapidly for the proposed structures. The signal cancellation problem of the conventional GSC is apparent for high SNR; the noise suppression is actually negative. This is because the high amount of signal leakage degrades the output; many desired signal components have been removed by the adaptive filter. The adaptive filter does not distinguish the desired signal from interference, and consequently attempts to suppress it, resulting in rapidly increasing distortion levels for high SNR.

Table 3. Subband Delay and Sum beamformer using post-filtering

SNR(0/5/10dB)	SS	SSmulti	SSscal	logSNR	priSNR
$inter_{supp}$	15.7/12.3/8.1	9.3/8.4/6.8	6.8/6.0/5.6	10/7.3/6.6	8.1/6.2/5.4
$noise_{supp}$	11.8/6.9/1.9	5.5/3.0/0.6	2.9/0.5/-0.6	6.2/1.8/0.4	4.2/0.7/-0.7
distortion	4.4/2.4/0.8	1.0/0.7/0.4	0.7/0.5/0.3	5.8/1.8/0.3	1.6/1.2/0.2

The delay and sum with SS suffered from musical noise. Table 3 shows the results of five methods of post-filtering. The Spectral Subtraction using multi-band [18] had a little better noise and interference suppression capability than the Spectral Subtraction of Scalart [16]. However, when the delay and sum with SS, based on tracking an a priori SNR using a decision-directed estimator [16], greatly reduced the noise, the residual noise was much less annoying and disturbing than the musical noise. Using the MMSE log-spectral estimator or the MMSE amplitude estimator with a noncausal a priori SNR estimator, similarly resulted in a great reduction of the noise and the introduction of a little musical noise. The MMSE log-spectral estimator also resulted in distortion that steeply increased at 0 dB SNR [13],[14],[17].

5 Conclusion

This paper has presented a hands-free system that is applicable to office and car environments. By using a microphone array, unwanted interference sources and noise effects can be suppressed with a spatiotemporal filter. The design combines a conventional beamformer with spectral subtraction and an MMSE, noncausal SNR estimator. The combined system provides a similar suppression to the conventional GSC. In addition, the target signal cancellation problems of GSC systems are alleviated. However, the combined system of a conventional beamformer with SS suffered from musical noise. Using the conventional beamformer with an MMSE log-spectral estimator or an MMSE amplitude estimator with a noncausal a priori SNR estimator, similarly resulted in a great reduction in the noise and interference, albeit with a little musical noise and a small echo. The MOS evaluation has shown that the combined system of delay and sum with the MMSE amplitude estimator following a noncausal a priori SNR estimator obtained 3.8 points on a five-point scale.

Acknowledgments. This work was supported by the Realistic 3D-IT Research Program of Kwangwoon University under the National Fund from the Ministry of Education and Human Resources Development (2005).

References

1. Van Veen, B.D., Buckley, K.M.: Beamforming: A Versatile Approach to Spatial Filtering. IEEE ASSP Magazine, 4–22 (1988)
2. Capon, J.: High-resolution frequency-wavenumber spectrum analysis. Proc. of the IEEE 57(8), 1408–1419 (1969)
3. Applebaum, S.P., Chapman, D.J.: Adaptive arrays with main beam constraints. IEEE Trans. Antennas and Propagate AP–24, 650–662 (1976)
4. Griffiths, L.J., Jim, C.W.: An Alternative Approach to Linearly Constrained Adaptive Beamforming. IEEE Transactions on Antennas and Propagation AP–30, 27–34 (1982)
5. Hoshuyama, O., Sugiyama, A., Hirano, A.: A Robust Adaptive Beamformer for Microphone Arrays with a Blocking Matrix Using Constrained Adaptive Filters. IEEE Transactions on Signal Processing 47(10) (1999)
6. Denholm, J.: Hands-free Voice Interface using Multiple Microphones, A report submitted for the degree of Bachelor of Engineering in Electrical and Computer, Curtin University of Technology (2004)
7. Low, S.Y., Grbic, N., Nordholm, S.: Robust Microphone Array using Subband Adaptive Beamformer and Spectral Subtraction. In: ICCS 2002. The 8th International Conference on Communication Systems, vol. 2(25-28), pp. 1020–1024 (2002)

8. Nordholm, S., Claesson, I., Grbi, N.: Structures and Performance Limits in Subband Beamforming. IEEE transactions on Speech and Audio Processing (submitted April 2001)
9. Boll, S.F.: Suppression of Acoustic Noise in Speech using Spectral Subtraction. IEEE Transactions on Acoustics, Speech and Signal Processing ASSP-27(2), 113–120 (1979)
10. Davis, A.: Voice Activity Detectors and Spectral Subtraction, Curtin University of Technology, School of Electrical and Computer Engineering (2002)
11. Lee, H.W.: Post-filtering Noise-reduction Techniques for Hands-free Communication Systems, Curtin University of Technology, School of Electrical and Computer Engineering (2005)
12. Deller, J.R., Hansen, J.H.L., Proakis, J.G.: Discrete Time Processing of Speech Signals. IEEE Press, New York, USA (2000)
13. Cohen, I.: Speech Enhancement Using a Non causal A Priori SNR Estimator. IEEE Signal Processing Letters 11(9), 725–728 (2004)
14. Ephraim, Y., Malah, D.: Speech enhancement using a minimum mean-square error short-time spectral amplitude estimator. IEEE Transactions on Acoustics, Speech, and SignalProcessing ASSP-32(6), 1109–1121 (1984)
15. Wolfe, P.J., Godsill, S.J.: Simple alternatives to the Ephraim and the Malah suppression rule for speech enhancement. In: Proc. 11th, IEEE Workshop Statist. Signal Processing, Singapore, August 6–8, pp. 496–499 (2001)
16. Scalart, P., Vieira-Filho, J.: Speech enhancement based on a priori signal to noise estimation. In: Proc. 21st IEEE Int. Conf. Acoust. SpeechSignal Processing, Atlanta, GA, pp. 629–632 (1996)
17. Ephraim, Y., Malah, D.: Speech enhancement using a minimum mean square error log-spectral amplitude estimator. IEEE Trans. on Acoust., Speech, Signal Processing ASSP-33, 443–445 (1985)
18. Kamath, S., Loizou, P.: A multi-band spectral subtraction method for enhancing speech corrupted by colored noise. In: Proceedings International Conference on Acoustics, Speech and Signal Processing (2002)

Bandwidth Extension of a Narrowband Speech Coder for Music Delivery over IP

Young Han Lee[1], Hong Kook Kim[1], Mi Suk Lee[2], and Do Young Kim[2]

[1] Dept. of Information and Communications
Gwangju Institute of Science and Technology (GIST), Gwangju 500-712, Korea
{cpumaker,hongkook}@gist.ac.kr
[2] BcN Service Research Group, BcN Research Division ETRI
161 Gajeong-dong, Yuseong-gu, Daejeon, 305-350, Korea
{lms,dyk}@etri.re.kr

Abstract. In this paper, we propose a bandwidth extension (BWE) algorithm of a narrowband speech coder for music delivery services over IP networks. The proposed BWE algorithm is based on an embedded structure of using a baseline coder followed by an enhancement layer. To minimize the bit-rate increase by the enhancement layer, the proposed algorithm shares spectral envelope and excitation parameters between the baseline coder and the enhancement layer. In this paper, we choose the iLBC as the baseline coder and mel-frequency cepstral coefficients (MFCCs) are used to reconstruct higher frequency components at the enhancement layer. By doing this, the bit-rate of the proposed BWE coder is 15.45 kbit/s which is just 0.25 kbit/s higher than the iLBC. We compare the quality of the proposed BWE coder with that of the iLBC, and it is shown from an informal listening test that the proposed BWE coder provides significantly better quality than the iLBC for all four different kinds of music genres such as pop, classical, jazz and rock.

1 Introduction

A voice over Internet protocol (VoIP) service has been receiving a great attention in recent years and it is provided for free or very reasonable price compared with the legacy public switched telephone network (PSTN) service. In IP networks, the VoIP service is provided through speech coders such as ITU-T G.729 [1], ITU-T G.723.1 [2], ITU-T G.728 [3], and the Internet low bit-rate codec (iLBC) which was released on 2004 by the Internet engineering task force (IETF) [4]. In the early days of this service, these speech coders were good enough to process voice signals efficiently. However, it is greatly agreed that they are not enough to satisfy users' expectation of sound quality, especially when music signals are used. In order to launch an improved VoIP service, several limitations of the existing service should be overcome.

One of the limitations is the bandwidth of signals processed by existing VoIP speech coders. Speech coders in the VoIP system compress input signals with a low bit-rate around $5.3 \sim 16$ kbit/s under the assumption that the input signals

M.S. Szczuka et al. (Eds.): ICHIT 2006, LNAI 4413, pp. 198–208, 2007.
© Springer-Verlag Berlin Heidelberg 2007

are bandlimited up to 3.4 kHz, while music signals occupy at greater than 4 kHz. In order to improve the music quality in the VoIP system, the audio bandwidth up to 22.0 kHz [5] should be covered by the coder. Another limitation making degraded service quality is due to packet loss caused by a traffic congestion. In IP networks, packet loss makes sound clipping and skips which degrade voice or music quality [6]. This implies that audio coders are not suitable for a VoIP coder. Although audio coders are able to deal with the audio bandwidth, they do not have any packet loss concealment (PLC) algorithm typically.

In this paper, we propose a bandwidth extension of a narrowband speech coder for improving quality of music delivery over IP networks. The proposed coder is based on an embedded structure of using a baseline coder followed by an enhancement layer. The iLBC is used as a baseline coder because the iLBC provides very robust sound quality against packet loss [7]. To prevent the bit-rate from being increased by the enhancement layer, the proposed BWE coder shares spectral envelope and excitation parameters between the baseline coder and the enhancement layer. For this end, mel-frequency cepstral coefficients (MFCCs) [8] are used to represent spectral envelope and converted to linear prediction coefficients (LPCs) for both the baseline coder and the enhancement layer. In addition, the excitation signals for the enhancement layer are expanded from those decoded by the baseline coder, which results in no more additional bit increase for the enhancement layer. As a result, the proposed BWE coder operates in 15.45 kbit/s which is 0.25 kbit/s higher than the iLBC. In order to evaluate the performance of the proposed BWE coder, we perform a spectrum comparison and a preference listening test with the iLBC and the proposed BWE coder.

The rest of this paper is organized as follows. Following this Introduction, the encoder and decoder structures of the proposed BWE coder are described in Sections 2 and 3, respectively. In Section 4, the quality of the proposed BWE coder is evaluated and compared with that of the baseline coder. Finally, we conclude the paper in Section 5.

2   Proposed BWE Encoder

In this section, we describe the proposed BWE coder for music delivery services. The proposed BWE coder is based on an embedded structure of using a baseline coder followed by an enhancement layer. Figure 1(a) shows a basic principle of a conventional embedded audio coder, where the enhancement layer is devised to increase the bandwidth without regarding to the baseline coder [9]. Therefore, any parameters between the baseline coder and the enhancement layer are not related. On the other hand, the structure of the proposed BWE coder as shown in Figure 1(b) shares information on spectral envelope and excitation of the enhancement layer with the baseline coder. By doing this, we can develop the enhancement layer with a smaller number of bit increase than the conventional structure shown in Figure 1(a).

Figure 2 shows a block diagram of the proposed BWE encoder. The iLBC, which is used as a baseline coder of the proposed BWE coder, is an LPC-based

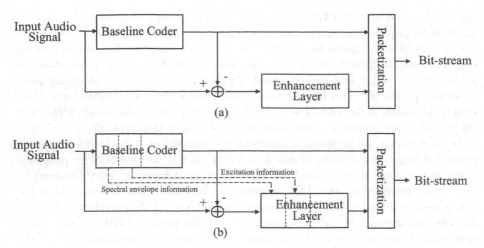

Fig. 1. Comparison of embedded encoding structures between (a) a conventional BWE coder and (b) the proposed BWE coder

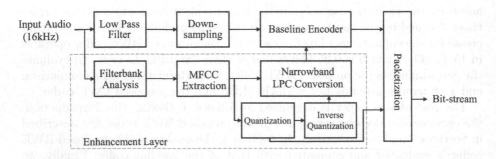

Fig. 2. A block diagram of the proposed BWE encoder

coder and operates in two different bit-rates: 15.2 kbit/s and 13.33 kbit/s with a frame size of 20 msec and 30 msec, respectively. Because many VoIP systems use 20 msec packets [6], 15.2 kbit/s iLBC is a candidate for the baseline coder. As described above, we first modify the iLBC to incorporate the sharing parts with the enhancement layer.

One of the sharing parts is the representation of spectral envelope. In the proposed BWE coder, MFCCs are used as spectral envelope parameters while the most conventional speech coders including iLBC use LPCs as shown in Fig. 4. A filterbank analysis is applied to audio signals sampled at 16 kHz. The filterbank analysis begins with a DC removal followed by a pre-emphasis filtering. After that, a Hamming window is applied to the pre-emphasized audio signals. In order to obtain MFCCs, a 512-point fast Fourier transform (FFT) is first performed to compute the magnitude spectrum. The magnitude spectrum is filtered by 23 triangular-shaped mel-filterbanks. The mel-filtered signal is transformed into a logarithmic scale. Finally, a discrete cosine transform (DCT) is

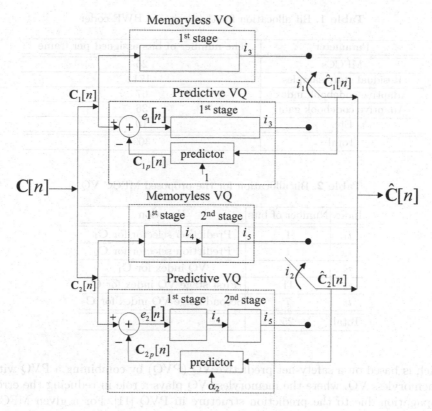

Fig. 3. The proposed MFCC quantization by combining a predictive VQ with a memoryless VQ

applied to obtain 23 MFCCs. Among them, only 12 MFCCs are quantized with a 25 bits/frame [10].

On the other hand, we have to obtain 10 LPCs for the baseline coder. In order to obtain LPCs, an inverse DCT is applied to the 12 quantized MFCCs and 23 filterbank energies are estimated as shown in Fig. 5. Next, the power density spectrum whose frequency range is from 0 to 8 kHz is also estimated by linearly interpolating these 23 filterbank energies. In order to obtain LPCs used for the baseline encoder, we only take a half of the power density spectrum, which results in the frequency range from 0 to 4 kHz. And then, a 256-point inverse FFT is applied to compute the autocorrelation coefficients. The autocorrelation coefficients are smoothed by applying a lag window. Finally, we obtain 10 LPCs using the Levinson-Durbin recursion. These 10 LPCs are used in the baseline coder to model speech signals.

The bit allocation for the proposed BWE coder is shown in Table 1. The main difference of the bit allocation for the proposed BWE coder from that of the iLBC is that 25 bits/frame are assigned to MFCCs, while 20 bits are used for LSF quantization in the iLBC. Figure 3 shows the proposed MFCC quantization

Table 1. Bit allocation for the proposed BWE coder

Parameter	The number of bits assigned per frame
MFCCs	25
Residual state samples	171
Adaptive codebook index	67
Adaptive codebook gain	36
Etc.	10
Total	309

Table 2. Bit allocation for the proposed MFCC VQ

Index	Number of bits	Function
i_1	1	Prediction selector for $\mathbf{C}_1$
i_2	1	Prediction selector for $\mathbf{C}_2$
i_3	5	VQ index for $\mathbf{C}_1$
i_4	11	First stage VQ index for $\mathbf{C}_2$
i_5	7	Second stage VQ index for $\mathbf{C}_2$
Total	25	

which is based on a safety-net predictive VQ (PVQ) by combining a PVQ with a memoryless VQ, where the memoryless VQ plays a role in reducing the error propagation due to the prediction structure in PVQ [11]. For a given MFCC vector, it is required to select one of either PVQ or the memoryless VQ in the safety-net PVQ. Hence, we use the Euclidean distance measure to select one of the VQs. In other words, PVQ is selected if the distance from PVQ is smaller than the from the memoryless VQ, and vice versa. In other words, an input MFCC vector of the nth frame is split into 2 subvectors as

$$\mathbf{C}[n] = \begin{bmatrix} \mathbf{C}_1[n] \\ \mathbf{C}_2[n] \end{bmatrix} = \begin{bmatrix} c_0 \\ c_1 \\ \vdots \\ c_{12} \end{bmatrix} \tag{1}$$

where $\mathbf{C}_1[n]$ and $\mathbf{C}_2[n]$ are an 1-dimensional subvector and a 12-dimensional subvector as described before. Then, each subvector is quantized by its corresponding safety-net PVQ, where a selector determines one of either PVQ or the memoryless VQ depending on the Euclidean distance measure. In PVQ, prediction is based on a past one quantized MFCC vector such as

$$\mathbf{C}_{ip}[n] = \alpha_i \hat{\mathbf{C}}_i[n-1], \tag{2}$$

where α_i is the prediction coefficient of the past one frame of the ith subvector in Eq. (1). Especially, we construct the memoryless VQ and PVQ for $\mathbf{C}_2$ with a

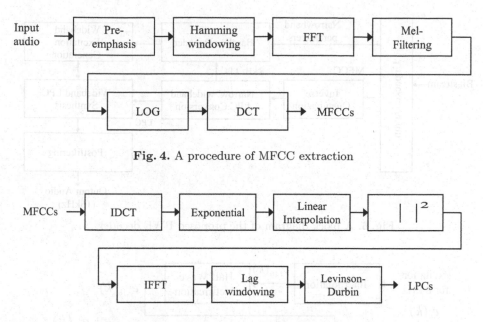

Fig. 4. A procedure of MFCC extraction

Fig. 5. A procedure of the MFCC to LPC conversion

multi-stage VQ because the multi-stage VQ is generally known to be efficient in search and training of VQ for high dimensional vectors [12]. Finally, we need to assign the number of bits to five quantization indices as described in Table 2.

As a result, we need more 5 bits/frame for the proposed BWE coder compared with the LSF quantization in the iLBC. In addition, the numbers of bits assigned for residual samples, codebook indices, and gains remain as in the iLBC as shown in Table 1.

3   Proposed BWE Decoder

Figure 6 shows a block diagram of the proposed BWE decoder. It mainly consists of two parts: the baseline decoder and the enhancement layer decoder. The enhancement layer decoder decodes 12 MFCCs from the transmitted bit-stream and converts these MFCCs into 10 narrowband LPCs and 16 wideband LPCs. And then, it generates wideband excitation signals from the excitation decoded by the baseline decoder. Finally, audio signals whose bandwidth is about 8 kHz are obtained by filtering the wideband excitation.

The 10 narrowband LPCs are obtained from MFCCs by using the identical procedure described in Section 3, whereas we can obtain 16 wideband LPCs from MFCCs by using the power density spectrum ranging from 0 to 8 kHz during the conversion procedure.

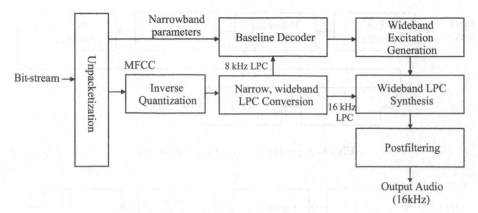

Fig. 6. A block diagram of the proposed BWE decoder

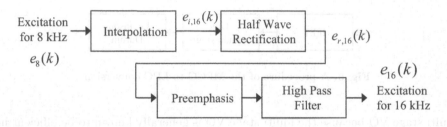

Fig. 7. A procedure of wideband excitation generation for the enhancement layer

In order to synthesize high frequency signal, the wideband excitation is needed. A detailed block diagram of wideband excitation generation is shown in Fig. 7. First, the narrowband excitation from the baseline decoder is interpolated as

$$e_{i,16}(k) = \begin{cases} e_8(k/2) & k = 0, 2, \cdots, 2N - 2 \\ 0 & k = 1, 3, \cdots, 2N - 1 \end{cases}, \tag{3}$$

where N is the number of samples per frame in the baseline decoder, and $e_8(k)$ and $e_{16}(k)$ are the k-th samples of the narrowband excitation and the interpolated excitation, respectively. Next, half wave rectification is performed to the interpolated excitation for generating high frequency components as

$$e_{r,16}(k) = \begin{cases} e_{i,16}(k) & if \ e_{i,16}(k) > 0 \\ 0 & otherwise \end{cases} \quad 0 \le k < 2N, \tag{4}$$

where $e_{r,16}(k)$ is the k-th sample of the half wave rectified excitation. In order to compensate for the reduced dynamic range due to the baseline coder, the half wave rectified excitation is emphasized through the pre-emphasis filter, $1 - 0.9z^{-1}$. A high pass filter whose cutoff frequency is 4 kHz is applied to the pre-emphasized excitation. Thus, the wideband excitation, $e_{16}(k)$ is finally obtained.

In order to obtain decoded audio signals, $e_8(k)$ and $e_{16}(k)$ are passed through the filters constructed by the narrowband LPCs and the wideband LPCs, respectively. Then, $s_8(k)$ filtered from $e_8(k)$ is interpolated by a factor of 2 and then pre-emphasized to increase the dynamic range of the highband spectrum. That is,

$$s_{p,8}(k) = s_{i,8}(k) - \beta\, s_{i,8}(k), \ \ 0 \le k < 2N, \tag{5}$$

where $s_{i,8}(k)$ and $s_{p,8}(k)$ are the k-th samples of the interpolated and pre-emphasized audio signals from the baseline decoder, and β is set to 0.2. Note that $s_{i,8}(k)$ and $s_{p,8}(k)$ are samples at a rate of 16 kHz.

Finally, we add two sets of audio signals, $s_{p,8}(k)$ and $s_{16}(k)$ to make the resultant audio signals. In this case, there are two factors we have to consider: one is a weighting factor between $s_{p,8}(k)$ and $s_{16}(k)$, and the other is a delay, D, occurred from the decimation at the proposed BWE encoder to make narrowband audio signals from the input audio signals. As a result, the decoded audio signals, $\hat{s}_{16}(k)$ are obtained by

$$\hat{s}_{16}(k) = w_{16} \cdot s_{16}(k) + w_8 \cdot s_{p,8}(k + D) \qquad 0 \le k < 2N, \tag{6}$$

where w_8 and w_{16} are the weighting factors for the decoded signals from the baseline decoder and those from the enhancement layer, respectively. From the exhaustive experiments, it was found that $w_8 = 1.2$ and $w_{16} = 0.5$ provided the best audio quality. Also, the delay was set at $D{=}48$.

4   Performance Evaluation

The performance comparison has been done in two ways: spectrum comparison and informal listening tests.

Figure 8 shows the performance comparison in the spectrum domain. Here, the 70th track in the sound quality assessment material (SQAM) was used as an original signal [13]. Because SQAM audio files were recorded at a sampling rate of 44.1 kHz, each file was down-sampled from 44.1 kHz to 16 kHz and we only took right channel signals. In the figure, the spectrum of the input audio signal is displayed in Fig. 8(a). During the process of the proposed BWE algorithm, the baseline coder requires audio signals sampled at 8 kHz, which is depicted in Fig. 8(b). Figures 8(c) and (d) show the spectrum of the decoded audio signal only by the baseline decoder and that by the proposed coder, respectively. It was shown that the spectrum of the audio signals by the proposed BWE coder was very close to that of the original audio.

To compare the quality in a subjective way, we performed an AB-comparison test with the baseline coder and the proposed BWE coder. For this end, we chose four genres such as pop, classical, jazz, and rock music. Each genre consisted of five audio files, which resulted in 20 audio files as a total. Here, audio files were also prepared from SQAM. Seven people without having any auditory disease participated in this test.

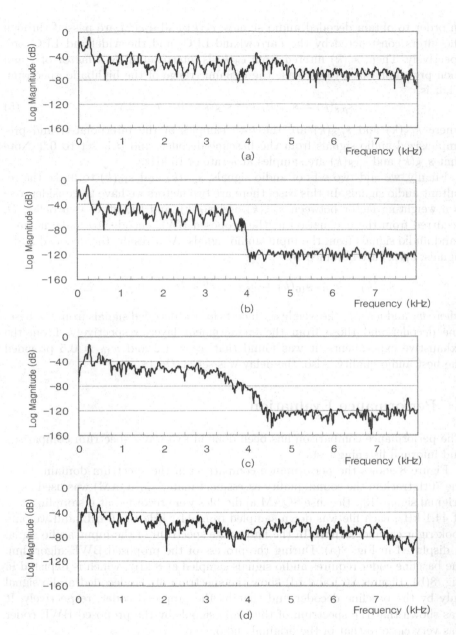

Fig. 8. Performance of the proposed BWE coder by comparing the spectra from (a) the original audio signals sampled at 16 kHz, (b) the signals used as an input to the baseline coder, (c) the signals decoded by the baseline decoder, and (d) the audio signal decoded by the proposed BWE coder

Table 3. Preference test results for the proposed BWE coder and iLBC

	Preference Score (%)	
Genre	Proposed BWE coder	iLBC
Pop	85.7	14.3
Classical	88.5	11.5
Jazz	80.0	20.0
Rock	97.2	2.8
Average	87.9	12.1

Table 3 shows the percentage of the relative preference of the proposed BWE coder compared to the baseline coder, iLBC. From the table, it was shown that the proposed BWE coder was significantly preferred with 87.9% to the iLBC for all the genres and all the audio files.

5 Conclusion

In this paper, we have proposed a bandwidth extension of the iLBC for music delivery over IP networks. The proposed BWE coder was on the basis of an embedded structure constructed by the iLBC as a baseline coder. The proposed coder was designed to increase the bandwidth with a minimal bit-rate increase. This was achieved by sharing spectral envelope parameters of the enhancement layer with those of the baseline coder. In addition, the excitation signals for the enhancement layer were generated by using the excitation decoded by the baseline coder. As a result, the increased bit-rate of the proposed coder was only 0.25 kbit/s. The performance of the proposed BWE coder was evaluated in terms of spectrum comparison and the preference listening test. It was shown that the proposed bandwidth extension coder gave significantly better quality than the iLBC.

Acknowledgments. This work was supported by the Ministry of Information and Communication (MIC) Korea, under the Information Technology Research Center (ITRC) support program supervised by the Institute for Information Technology Advancement (IITA-2006-C1090-0603-0017) and Immersive Contents Research Center (ICRC) at Gwangju Institute of Science and Technology (GIST).

References

1. ITU-T Recommendation G.729: Coding of speech at 8 kbit/s using conjugate-structure algebraic-code-excited linear-prediction (CS-ACELP) ITU (March 1996)
2. ITU-T Recommendation G.723.1: Dual rate speech coder for multimedia communications transmitting at 5.3 and 6.3 kbit/s ITU (March 1996)

3. ITU-T Recommendation G.728: Coding of speech at 16 kbit/s using low-delay code excited linear prediction, ITU (October 1992)
4. IETF RFC 3951, Internet Low Bit Rate Codec specification (December 2004)
5. Pan, D.Y.: Digital Audio Compression. Digital Technical Journal 5(2), 1–14 (1993)
6. Goode, B.: Voice Over Internet Protocol. Proc. IEEE 90, 1495–1517 (2002)
7. Andersen, S.V., Kleijn, W.B., Hagen, R., Linden, J., Murthi, M.N., Skoglund, J.: iLBC-A Linear Predictive Coder with Robustness to Packet Losses. In: Proc of IEEE 2002 Workshop on Speech Coding, Tsukuba, Japan, pp. 23–25 (2002)
8. Davis, S., Mermelstein, P.: Comparison of parametric representations for monosylable word recognition in continuously spoken sentences. IEEE Trans. Acoustic, Speech and Signal Processing 28, 357–366 (1980)
9. Kataoka, A., Kurihara, S., Sasaki, S., Hayashi, S.: A 16-kbit/s wideband speech codec scalable with G.729. In: Proc. Eurospeech, Rhodes, Greece, pp. 1491–1494 (1997)
10. Lee, G.H., Yoon, J.S., Kim, H.K.: A MFCC-based CELP speech coder for server-based speech recognition in network environments. In: Proc. of Eurospeech, Lisbon, Portugal, pp. 3169–3172 (2005)
11. Eriksson, T., Lindén, J., Skoglund, J.: Interframe LSF quantization for noisy channels. IEEE Trans. Speech Audio Process 7(5), 495–509 (1999)
12. Juang, B.H., Gray, A.H.: Multiple stage vector quantization for speech coding. In: Proc. of ICASSP, Paris, France, pp. 597–600 (May 1982)
13. EBU Tech Document 3253, Sound Quality Assessment Material (SQAM) (1988)

A User-Oriented GIS Search Service Using Ontology in Location-Based Services

Hyunsuk Hwang[1], Seonghyun Shin[2], and Changsoo Kim[2,*]

[1] Research Institute of Industrial Science and Technology,
PuKyong National University, Busan, Korea
hhs@pknu.ac.kr
[2] Div. of Electronic Computer and Telecommunication Engineering,
Pukyong National University, Busan, Korea
ccary001@hanmail.net, cskim@pknu.ac.kr

Abstract. Geographical Information Systems (GIS) technology plays an increasingly important role in Location-Based Services (LBS), which include applications such as car navigation, tour guide information, route-guide information, and tracking systems. Most GIS applications focus on showing the stored information on a map, not providing user-oriented map services to register user's preferred information. We propose an ontology-based approach to register and to search personal information on maps. We also implement a GIS search prototype considering user preferences using favorite food information and having a connection function to Web sites on the map using our digital map format. Our contribution is to provide a personalized service by connecting LBS/GIS services and ontology technology.

Keywords: Geographical Information Systems, Location-based Services, Ontology-based Search, Semantic Web, RDQL, RDF content, Protégé/OWL.

1 Introduction

Location-based Services (LBS) can be defined as services that integrate a mobile device's geographic position with other information to provide additional values for a user [1],[2],[3]. In this sense it is essential that LBS must include Geographical Information Systems (GIS) and Global Positioning Systems (GPS) to provide information geographically close to the user's location. The LBS applications are widely used in location tracking, tourist assistance, traffic monitoring, and security [4],[5].

As the Web continues to grow, we can attain a huge quantity of information. Users want to find abstracted and customized information and recognize information on the move in a geographic space. Thus, as stated in [6], it is an urgent need to provide geographical, rapid, stable, and semantic services based on the user's preferences and its history.

---

* Corresponding author.

M.S. Szczuka et al. (Eds.): ICHIT 2006, LNAI 4413, pp. 209–218, 2007.

The LBS applications require techniques of software agent and data integration as well as basic information technology such as GIS, GPS, database, and security to receive the knowledge tailored for mobile users from enormous contents. Ontology technology [7],[8] is considered as a solution to integrate and to analyze dispersed data sources from vender offerings and Web [9],[10]. The ontologies, which are an essential component of the semantic Web, which is the next generation of the Web to provide intelligent services, define the common words and concepts used to describe and represent an area of knowledge. The construction of ontologies in some domains such as travel, education, and medical data has developed to integrate different data structures on the Web and to provide semantic information with less redundancy, integrated terms, and inferred knowledge.

Research about semantic issues to adaptive information to users on the move in LBS has been processed within recent years. The CRUMPET project of the EU [11],[12],[13] is to provide a new delivery service in mobile tourism services, which focuses on location-based and personalized information. Yu et al. [14],[15] has proposed a multi-layered abstraction method to organize users' heterogeneous data. The research has been demonstrated as a method to satisfy users' needs on mobile services. However, there still remain several challenges to overcome limitations of data integration issues, agents for meditation, and static and dynamic information.

In this paper, we focus on providing a user-oriented map service in registering users' preferred information on a map and in searching the personalized information based on ontologies on an efficient GIS map model [16] we developed. We implement a prototype to provide results of a user's query, for instance, which is to find closest restaurant to a user's location when a user wants to search for his or her favorite meal for lunch. Our work is to propose an implemental method for connecting between GIS and homogenous data sources based on ontologies and to show possibility to provide semantic, associated, and personalized information in LBS.

This paper is organized as follows. In the next section, we describe related works of LBS with personalized and semantic function. In section 3, we design a search system based on ontologies in LBS and describe detailed implemental modules. We implement a prototype based on the ontology for describing BuildingType, Location, and Person objects in section 4. Finally, we summarize this research and describe future works.

2   Personalized and Semantic Location-Based Services

In order to provide semantic information to user's location in mobile environments, various studies have been executed on Mobile GIS and LBS. GIS information is needed to support development of more advanced LBS applications. In particular, on considering geographical data management of LBS, the user's personalized interaction and integration technology of data structure between resources of different domains is an important field.

Gandon et al. [17] has developed a context-aware application Semantic e-Wallet, which is a semantic Web architecture aimed at supporting automated discovery and access of personal resources for a given user. The e-Wallet acts as an agent for modifying and managing a user's personal resource and includes categorized functions to keep the user's preference information, which is static knowledge, dynamic knowledge, service invocation rules, and privacy preferences.

Zipf et al. [11][12][13] have executed the European project; CReation of User-friendly Mobile Services Personalized for Tourism (CRUMPET), which is to implement and test tourism-related value-added services for nomadic users across mobile and fixed networks and to evaluate agent technology in terms of user-acceptability, performance, and best-practice as a suitable approach for fast creation of robust, scalable, seamlessly accessible, and nomadic services.

Schmidt-Belz et al. [18] executed the user validation based on a field test about tourist's needs and user's perceived qualities of the system and got results that a high percentage of the test for users' added benefit of the system compared to other available information sources.

Yu et al. [14],[15] has proposed a multi-layered abstractions method to organize the users' heterogeneous data. The research has demonstrated a method to satisfy users' needs on mobile services. However, there still remain several challenges to overcome limitations of data integration issues, agents for meditation, and static information.

The research of ontologies and semantic Web in mobile environments are an ongoing field in recent years. The research aim is to provide semantic and personalized information that is tailored to mobile users. For the purpose, it is important to integrate data resources scattered on the Web and other diverse databases [19][20][21]. Although these systems provide semantic information in mobile environments, there still remain issues of technical implementation to be solved such as data integration, agents for meditation, and dynamic contents services.

3   System Architecture

In this section, we propose an architecture including primary components and implemental modules to provide services based on the ontologies in LBS.

3.1   System Components

The entire system structure for providing GIS search services based on ontologies with the information of user's preference is shown in Figure 1. Our system can provide dynamic information by considering user's preference from the user's current position on a map. For example, a user can search a closest restaurant that has his/her favorite food at the user's current position and can attain more information by connecting to the Web. To provide the information, our system includes the client component for map services, the Ontology Server component for search services based on ontologies, the GIS servers for updated geographical information services, and the contents providers for dynamic information services.

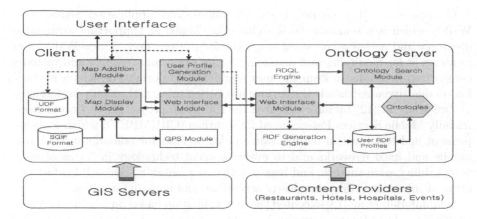

Fig. 1. Ontology-based Search Architecture of LBS

The main components of our system include as follows.

o The Client for map services needs a GPS module to determine user's location, the GIS map module to display a map on a screen and to add URL information on the map. For digital map services, we use a reduced digital map model developed in our previous work [16]. Also, a module to register a user's profile is needed.

o The Ontology Server needs to provide preferred information with users. To provide the information, this component needs the RDQL engine to generate user's query statement according to the defined ontology. Also, the ontology search module contacts the Ontologies and user's profiles to provide search services.

o The Content Providers can provide dynamic information by connecting with the Ontology Server because they include recent information in each domain. To do so, this component needs agent technology for integrating different resources.

o The GIS Servers include geographical spatial information and can provide recent map information with clients.

The detailed procedures of the proposed system are classified into two parts that one is an input process of the Web URL and users' preferences, and another is a search process of personalized information. The input process is denoted by arrows with solid lines. (1) A user registers the URL of a building location on a map through GPS to store a UDF file. (2) When the user wants to store related information of preferences, the information is stored in a RDF profile by an interface module and a RDF Generation Module after inputting the information of user's preferences if the user wants to store the information.

The search process is denoted by an arrow with dotted lines. (1) The user receives a current location of the user through GPS. The map display module works to display the map on a defined screen. (2) The user sends a query to the ontology server to receive information of a nearest restaurant in an area from the user's personalized profile. (3) The requested parameters by Web interface module and the user's location information of the Map Display on the client is sent to the Web interface module on the ontology server. (4) The parameters are used to generate a query in a RDQL Engine. (5) The Ontology Search Module extracts results from the user's profile using the generated query, and they are conveyed to the Map Display Module through the Web interface module. (6) The user can assure a location of favorite restaurants on a digital map. The closest restaurant on the user's location is automatically connected to the Web site.

3.2   Implementation Modules

We focus on the Client component and the Ontology Server to realize this system. Our system includes five modules to convey the connected results between geographical and ontological information.

Web Connection Module. Users can add URL information on a map to attain more detailed information. The URL Data Format includes the information for connecting to a Web site according to the location. The Map Module is connected with UDF files and Web interface modules for providing searches considering user's preference information.

Map Module. A Map Display Module, which shows a map on the screen, the Map Addition Module, which appends and edits new items, and the GPS Module, which sends current location coordinates of objects is included in the Map Module. The Map Display Module requires Simple Geographic Information File (SGIF), which is a digital map file with reduced size format and URL Data File (UDF), which has the information for connecting to a Web site of the location of the map. A Map Addition Module is connected with the UDF files and Web interface modules for providing searches considering the preference information of users.

Ontology Search and User Profile RDF Generation Module. The Ontology Search Module of the Ontology Server requires user's RDF contents, which include the user's preference information and location of related buildings. The Ontology Search Module provides search results from RDF files which have the individual preference information through the Web Interface Module. Ontology contains hierarchy and relation information between items, and the ontology-based search can provide the associated search information between resources. The RDF Generation Engine is to create and edit RDF files with preference information of users. The RDF contents are created on the defined ontology structure.

Web Interface Module. This module is responsible for connecting search services of the Ontology server with map services of the Client. If a user send preference and location information, the RDF file of the user is generated or modified on a server. The response results from the user's query are displayed on a map.

4 Prototype Implementation

We implemented a prototype to realize a search by connecting a GIS and an ontology server on a Tablet Personal Computer with a GPS module and a Wireless Internet function. A client system with Map services is developed in a .NET environment, and the Ontology server module of search services is developed in the Java Web Services Developer Pack 1.1(JWSDP) environment.

4.1 Web Connection Module

The Map Display Module requires UDF files for connecting to a requested Web site. Table 1 expresses UDF format structure including coordinates and URL information of buildings. We make 4040 area of coordinates of current location to display and handle location information on the screen. Also, the category code needs to display icons according to a kind of building.

Table 1. Structure of UDF Format

Field Name	Type	Comment
Category Code	Integer	Category Number
X, nX	Integer	X -20, X +20 on the Current Location X
Y, nY	Integer	Y -20, Y +20 on the Current Location Y
Location Name	String	Building Name on the Current Location
URL Name	String	URL on the Current Location

4.2 Map Module

SGIF Format. A DXF format provided by National Geographic Information Institute Spatial Data Warehouse in Korea needs a minimum of 12 bytes to mark a plane coordinate because it consists of an ASCII file. We express it as 8 bytes by converting a file into a binary file format. Therefore, our proposed file format can be easily read by extracting essential layers and can reduce time of search on a map.

Map Display Module. The Map Display Module works to receive a user's current location information from GPS and to show location on the map. Also, it is responsible for displaying the basic map from the SGIF digital map file and additional map inputted by users from the UDF file. Figure 2 shows a process of the digital map. The process is needed to express a related map on the screen by connecting information of a digital map and Web between records of SGIF and UDF on clients. The system shifts the map to the memory buffer instead of marking objects to be drawn on a screen immediately. When the map to be displayed was represented, the contents of the map are displayed in the memory buffer on the screen. Therefore, our map system can prevent a screen twinkling phenomenon, and a function of memory management is easy.

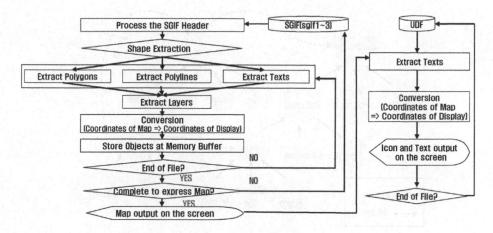

Fig. 2. Process of Digital Map Display

The Map Addition Module. This Module is executed when a user wants to register URL information of buildings on map. The information is stored in a UDF file. When the user moves an arrow on the map and movement occurs, the module converts a coordinator of the current arrow position into a coordinate of the map by extracting UDF files. Next, if the arrow position is in the ranges of the rectangle with values of X, Y, nX, nY in the UDF file, the arrow is changed into a handed display.

4.3   Ontology Search Module

The technology such as ontologies, RDF contents, and the RDQL query needs to support ontology-based search. The recent researches use OWL ontology language [22], which is widely accepted as a standard language for sharing semantic Web contents [9, 14] to express ontologies. Resource Description Framework (RDF) based on XML is used to express contents of the semantic Web and must be generated by ontologies which can define hierarchy and association of resources [5,9]. RDF Query Language (RDQL) is a query language for querying RDF contents.

Figure 3 is an ontology expressing hierarchy and relation of classes such as the BuildingType, the Location, and the Person for providing a GIS search as to individual preferences. The circles represent objects and lines with dots mean object properties, which is used to express relation between classes. We construct the BuildingType, which includes School, Hospital, and Restaurant class, and the Location including the CityDo, the GuGun, the DongRee, and the ZipCode. Also, the Korean class is specifically categorized into Bulgogi, Califonia Roll, and Noodles classes. In this paper, we use Protégé/OWL tool to create the ontology and RDF contents.

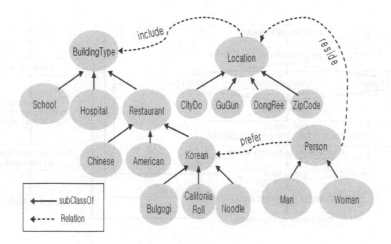

Fig. 3. Ontology for Location, BuildingType, and Person Object

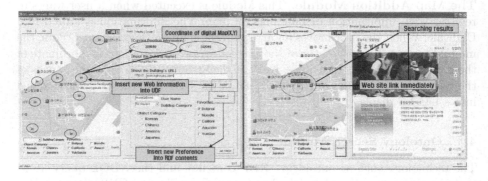

Fig. 4. Registration Interface of Preference Information and the Search Result

4.4   Web Interface Module

The Web Interface Module on a client is to send the user's location, query information, and preference information into the interface module on the ontology server and is to receive search results based on a defined ontology. We use HTTP protocol and Get method for transmitting input values of a client into the server. JSP and Servlet for interface programs are used.

4.5   An Example of Ontology-Based Search

We imagine that users want to register some restaurants with their preferred food on a GIS map, and to connect to its Web site to get more specific information. We implement a prototype using a Korean GIS map model.

The left-hand side of Figure 4 shows an interface screen shot on a tablet PC in which users can register and edit information such as name, URL, user's preference information for the building. The preference condition of a user is stored in the individual RDF content file of the Web server. The URL name of a related Web site and building's location is stored in an individual UDF file. The right-hand side of Figure 3 shows a search result which indicates coordinates of restaurants with the bulgogi in registered information of the user and connects automatically to a web site of the nearest restaurants. To search preference information, we use RDF Query Language (RDQL) which is a query language of RDF.

5   Conclusion

In this paper, we presented an ontology-based search method for supporting a user-oriented GIS search information in Location-Based Services. We proposed the architecture of the ontology-based GIS search system. The advantages of this system include that (1) we use our reduced GIS digital map model to display a map in rapid speed and (2) provide a personalized GIS search services using the ontology. Also, we described detailed implementation modules to realize LBS services and implemented a search prototype to find a favorite and closest restaurant in a range of a user's location.

The combined research of LBS/GIS services and ontology technology is still at an invoice stage. It is important to extract semantically dynamic data sources from huge Web contents delivered from content providers. To do this, a standard format based on ontologies for data integration must be constructed. From a system development point of view, we will improve our system by adding a knowledge-based search module of GIS, which can provide the associated and abstract information by inferring a user's registered information.

Acknowledgement. This research was supported by the Program for the Training Graduate Students in Regional Innovation which was conducted by the Ministry of Commerce, Industry and Energy of the Korean Government.

References

1. Spiekermann, S.: General Aspects of Location-Based Services, in Location-Based Services. Morgan Kaufmann, San Francisco (2004)
2. Tilson, D., Lyytinen, K., Baxter, R.: A Framework for Selecting a Location Based Service (LBS) Strategy and Service Portfolio, System Science, pp. 78–87 (2004)
3. Adusei, I.K., Kyamakya, K., Erbas, F.: Location-based Service: Advances and Challenges. Electronic and Computer Engineering 1, 1–7 (2004)
4. Li, Q., Huang, X., Wu, S.: Applications of Agent Technique on GIS. IEEE 1, 238–243 (2001)
5. Virrantaus, K., et al.: Developing GIS-Supported Location-Based Services. In: Proceedings of ENTER 2002, Innsbruck, Austria, Springer, Heidelberg (2002)
6. Bharat, R., Minakakis, L.: Evolution of Mobile Location-based Services. Communications of the ACM 46(12), 61–65 (2003)

7. Uschold, M., Gruning, M.: Ontologies Principles, Methods, and Application, Artificial Intelligence Applications Institute (AIAI), the University of Edinburgh, AIAI-TR-191 (1996)
8. Devedzic, V.: Understanding Ontological Engineering. Communications of the ACM 45(4), 136–144 (2002)
9. Berners-Lee, T., Hendler, J., Lassila, O.: The Semantic Web. Scientific American BOl. 284(5), 34–43 (2001)
10. Guarino, N., Masolo, C., Vetere, G.: OntoSeek: Content-based Access to the Web. IEE Intelligent Systems 14(3), 70–80 (1999)
11. Schmidt-Belz, B., Nick, A., Poslad, S., Zipf, A.: Personalized and Location-based Mobile Tourism Services. In: Workshop on Mobile Tourism Support Systems in conjunction with Mobile HCI 2002, Pisa (2002)
12. Zipf, A.: User-Adaptive Maps for Location-based Services for Tourism. In: Proc. of ENTER, Innsbruck, Austria, pp. 329–338 (2002)
13. Zipf, A.: Personalized mobile Maps for Telecommunication, Human Factors in Telecommunications Symposium. Sophia Antipolis, France (2006)
14. Yu, S., Spaccapietra, S., Cullot, N., Aufaure, M.: User Profiles in Location-based Services: Make Humans More Nomadic and Personalized. In: Proceedings of the IASTED I International Conference on Databases and Applications, Innsbruck, Austria, pp. 17–19 (2004)
15. Yu, S., Al-Jadir, L., Spaccapietra, S.: Matching User's Semantics with Data Semantics in Location-based Services, Semantics in Mobile Environments, Ayia Napa, Cyprus (2005)
16. Kim, J.W., Park, S.S., Kim, C.S., Lee, Y.G.: The Efficient Web-based Mobile GIS Service System through Reduction of Digital Map. In: Laganà, A., Gavrilova, M., Kumar, V., Mun, Y., Tan, C.J.K., Gervasi, O. (eds.) ICCSA 2004. LNCS, vol. 3043, pp. 410–417. Springer, Heidelberg (2004)
17. Gandon, F.L., Sadeh, N.M.: Semantic Web Technologies to Reconcile Privacy and Context Awarness. Journal of Web Semantics 1(3) (2004)
18. Schmidt-Belz, B., Poslad, S.: User Validation of a Mobile Tourism Service. In: Workshop HCI in Mobile Guides, Italy (2003)
19. Hessling, A., Kleemann, T., Sinner, A.: Semantic User Profiles and their Application in a Mobile Environment, Artificial Intelligence in Mobile Systems, Nottingham (2004)
20. Sonntag, D.: Toward Interaction Ontologies for Mobile Devices Accessing the Semantic Web, on HCI in Mobile Cuides and AIMS (2005)
21. Flury, T., Privat, G., Ramparany, F.: OWL-based Location Ontology for Context-aware Services, Artificial Intelligence in Mobile Systems, Nottingham (2004)
22. Knublauch, H., Fergerson, R.W., Noy, N.F., Musen, M.A.: The Protégé OWL Plugin: An Open Development Environment for Semantic Web Applications. In: McIlraith, S.A., Plexousakis, D., van Harmelen, F. (eds.) ISWC 2004. LNCS, vol. 3298, pp. 229–243. Springer, Heidelberg (2004)

A Filtered Retrieval Technique for Structural Information*

Young-Ho Park, Yong-Ik Yoon, and Jong-Woo Lee

Department of Multimedia Science
SookMyung Women's University, Korea
{yhpark,yiyoon,bigrain}@sookmyung.ac.kr

Abstract. We present a filtered retrieval technique for structural information on internet-scale knowledge such as large-scale XML data in the Web. The technique evaluates XML standard queries on heterogeneous XML documents using information retrieval technique based on the relational tables in relational database management systems. The XML standard queries, XPath queries, in their general form are partial match queries, and these queries are particularly useful for searching documents of heterogeneous schemas. Thus, our technique is geared for partial match queries expressed as the queries. This indexes the elements in label paths, which are sequences of node labels, like keywords in texts, and finds the label paths matching a given query.

1. Introduction

There have been significant research on processing queries against XML documents recently. However, most of them considered small documents such as tens or hundreds of documents with a fixed schema, and thus are not suitable for large-scale applications. A novel method is needed for such applications, and we address it in this paper.

A typical example of a large-scale application is an Internet search engine. The following shows example queries written as *path expressions* used in XPath[5].

- Q1: : document("*")/movie/actor/last
- Q2: : document("*")//animation//production//title/sub

The document("*") here denotes all XMLs, and the queries find all the XML documents whose schema contain the label paths, which are sequences of node labels formally defined in Definition 1) matching the path expressions as a query. Between these two queries, Q1 has only parent-child relationships '/' specified on the paths, whereas Q2 has ancestor-descendent relationships '//' as well. The '//' allows for retrieving all documents as long as they partially match the query. In the regard, Q2 is called a partial match query as opposed to Q1, which is called a full match query [10]. The Partial match facility such as in Q2 is particularly useful for finding matched XML documents of heterogeneous schemas [1].

We focus on queries like Q1 and Q2 as the target query type because this is a basic query form used very popularly in XML document retrieval. In the paper,

---

* This Research was supported by the Sookmyung Women's University Research Grants 2006.

M.S. Szczuka et al. (Eds.): ICHIT 2006, LNAI 4413, pp. 219–228, 2007.

our objective is to present an efficient filtered retrieval method to evaluate the partial match queries on web-scale XML documents of heterogeneous schemas.

Related Existing methods that can handle the partial match queries will appear in Section 2.2 in detail. As the basis of our method, we use the schema-level methods using relational tables and improve their efficiency significantly for web-scale heterogeneous documents. For this purpose, we apply the inverted index technique used traditionally in the information retrieval field for finding a very large number of documents. Our method builds an inverted index on the label paths occurring in XML documents and uses the index to process queries efficiently.

For the technique, we first describe the structures for storing the XML structure information. Then, we present the algorithm for the partial match query processing. For this purpose, we then present the rules for mapping a partial match query to a search expression on the inverted index and present an algorithm for finding the nodes matching the query. Then, we experiment the performance of the method using XMLs collected by crawlers from the current Internet. The results show that our method outperforms other works by many orders of magnitude and is scalable as the number of queried XML documents increases.

2   Background

2.1   XML Document and Query Model

In this section, we describe the XML document model and the query model used in the remainder of the paper.

Our XML document model is represented as a rooted, ordered, labeled tree. A *node* in the tree represents an element, an attribute, or a value; an *edge* in the tree represents an element-subelement relationship, element-attribute relationship, element-value relationship, or attribute-value relationship. Element and attribute nodes collectively define the document structure, and we assign labels (i.e., names) and unique identifiers to them. We modify this model so that a node represents either an element (*element node*) or an attribute (*attribute node*) but not a value. We also modify the model with the notion of label paths as defined in Definition 1. Figure 1 shows an example XML tree of a document.

Definition 1. A *label path* in an XML is defined as a sequence of node labels $l_1, l_2, ..., l_p$ $(p \geq 1)$ from the root to a node p in the tree, and is denoted as $l_1.l_2. \cdots .l_p$.   □

We say a label path *matches* a partial match query. For example, in Figure 1, `issue.editor.first` is a label path matching a path expression `//editor//first`. Note that there may be multiple nodes belonging to the same label path.

Our query model belongs to the tree pattern query class [11]. The query model supports two kinds of path expressions: 1) LPE: linear path expressions, and 2) BPE: branching path expressions. Among them, we define the LPE, the query type of our focus in the paper, as a sequence of labels connected with '/' or '//' as in Definition 2.

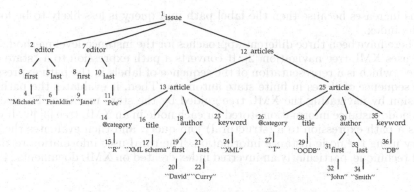

Fig. 1. An example XML tree of a document

Definition 2. A *LPE: linear path expression* is defined as $l_0 o_1 l_1 o_2 l_2 \cdots o_n l_n$, where l_i $(i = 0, 1, \cdots, n)$ is the i-th label in the path, and o_j $(j = 1, 2, \cdots, n)$ is either '/' or '//' which, respectively, denotes a parent-child relationship or an ancestor-descendant relationship between l_{j-1} and l_j. Here, l_0 is the root of the XML tree denoting the set of all XMLs (i.e., `document("*")`) and may be omitted. □

For example, the LPE `/movie//actor/last` is for searching all `last` elements that are children of the `actor` elements that are descendants of the `movie` elements.

2.2 Related Work

As mentioned in Introduction, there are two kinds of methods for evaluating path expressions: schema-level methods and instance-level methods. A schema-level method uses structural information like the label paths to find nodes matching a path expression, whereas an instance-level method uses only node identification information like the start and end positions of a node. In this section, we discuss representative methods in each category with a primary focus on the methods using relational tables, which provide a basis for our our method method.

XRel [12] and XParent [8] are considered the two representative methods in the methods using relational tables. In XRel, in order to evaluate LPEs, XRel needs only the tables `Path` and `Element`. It first finds the label paths matching the query's path expression from the `Path` table. The matching is done using the SQL string match operator `LIKE`. All label paths in the `Path` table must be scanned in this case because an index like the B+-tree cannot be used to search for a partially matching label path. Then, XRel joins the set of matching label paths with the table `Element` via the column `label_path_id` to obtain the result nodes. XParent [8] is similar to XRel, but uses a slightly different table schema for a different node identification mechanism, i.e., the node identifier (*node_id*) instead of the interval (*start_position, end_position*). In query processing, XParent evaluates an LPE in the same way as XRel, using the tables `Labelpath` and `Element`.

Index Fabric [6] is the ones in the methods using special-purpose indexes. These two methods achieve very efficient query execution for label paths registered in the index. However, this becomes less effective as the number of label paths increases because then the label path in a query is less likely to be found in the index.

There have been three different approaches for the instance-level method. The first uses XML tree navigation [2]. It converts a path expression to a "state machine", which is a representation of the sequence of labels in the path expression as a sequence of states in finite state automata. Then, it evaluates the path expression by navigating the XML tree guided by the state machine. The second uses node instance information stored for each node in an XML tree [3] [9]. It converts a path expression to a (structural) join query, and then evaluates the join query using the node instance information. The third uses information retrieval (IR) technique, particularly an inverted index created on XML documents [4] [7].

3   Storage Structures and Query Processing

In this section we present the storage structures and query processing of our method, with particular attention to the special index mechanism of the storage and the efficiency of query processing.

3.1   Storage Structures

Our method uses three tables and an inverted index to store information about XML document structure:

```
LabelPath(label_path_id, label_path)
Element(document_id, label_path_id, node_id, sibling_order)
Data(document_id, label_path_id, node_id, sibling_order, value)
Inverted index on label_path of the table LabelPath.
```

Figure 2 shows the LabelPath table and the inverted index for the example XML tree in Figure 1. The table LabelPath represents the *schema-level information* and stores all the distinct label paths occurring in XML documents and their path identifiers (label_path_pid). We add the labels prefixed with '$' and '&' to denote the first label and the last label of each label path. The first label is to match the root label of the document, and the last label is to match the last label in a partial match query. The details on their use in query processing will appear in Section 3.2.

The LabelPath inverted index is created on the label_path column of the LabelPath table. Here, we consider label paths as text documents and labels in these label paths as keywords. Like the traditional inverted index, the LabelPath inverted index is made of the pairs of a keyword (i.e., a label) and a posting list. Each posting in a posting list has the following fields: label_path_id, occurrence_count, offsets, label_path_length, where label_path_id is the identifier of the label path in which the label occurs, occurrence_count is the number of occurrences of the label within the label path, offsets is the set of the positions of the label from the beginning of the label path, and label_path_length is the number of labels in the label path. For instance, in the posting of the label section in a label path $chapter.chapter.section.section.section.paragraph.&paragraph, the occurrence_count of section is 3, the offsets of section is {3, 4, 5}, and the label_path_length is 7. The tables Element and Data represent the

(a) LabelPath table.

label_path_id	label_path
1	$issue.issue.&issue
2	$issue.issue.editor.&editor
3	$issue.issue.editor.first.&first
4	$issue.issue.editor.last.&last
5	$issue.issue.articles.&articles
6	$issue.issue.articles.article.&article
7	$issue.issue.articles.article.@category.&@category
8	$issue.issue.articles.article.title.&title
9	$issue.issue.articles.article.author.&author
10	$issue.issue.articles.article.author.first.&first
11	$issue.issue.articles.article.author.last.&last
12	$issue.issue.articles.article.keyword.&keyword

(b) LabelPath Inverted Index.

keyword	posting list
$issue	: <1, 1, {1}, 3> <2, 1, {1}, 4> <3, 1, {1}, 5> ...
issue	: <1, 1, {2}, 3> <2, 1, {2}, 4> <3, 1, {2}, 5> ...
&issue	: <1, 1, {3}, 3>
article	: <6, 1, {4}, 5> <7, 1, {4}, 6> <8, 1, {4}, 6> ...
&article	: <6, 1, {5}, 5>
articles	: <5, 1, {3}, 4> <6, 1, {3}, 5> <7, 1, {3}, 6> ...
&articles	: <5, 1, {4}, 4>
editor	: <2, 1, {3}, 4> <3, 1, {3}, 5> <4, 1, {3}, 5>
&editor	: <2, 1, {4}, 4>
author	: <9, 1, {5}, 6> <10, 1, {5}, 7> <11, 1, {5}, 7>
&author	: <9, 1, {6}, 6>
first	: <3, 1, {4}, 5> <10, 1, {6}, 7>
&first	: <10, 1, {7}, 7>
last	: <4, 1, {4}, 5> <11, 1, {6}, 7>
&last	: <11, 1, {7}, 7>
title	: <8, 1, {5}, 6>
&title	: <8, 1, {6}, 6>
keyword	: <12, 1, {5}, 6>
&keyword	: <12, 1, {6}, 6>
@category	: <7, 1, {5}, 6>
&@category	: <7, 1, {6}, 6>

Fig. 2. An example `LabelPath` table and inverted index

(a) Element table.

document_id	label_path_id	node_id	sibling_order
1	1	1	1
1	2	2	1
1	3	3	1
1	4	5	2
1	2	7	2
1	3	8	1
1	4	10	2
1	5	12	3
1	6	13	1
1	7	14	1
1	8	16	2
1	9	18	3
...	...	...	...
1	12	35	4

(b) Data table.

document_id	label_path_id	node_id	sibling_order	value
1	3	4	1	Michael
1	4	6	1	Flanklin
1	3	9	1	Jane
1	4	11	1	Poe
1	7	15	1	R
1	8	17	1	XML schema
1	10	20	1	David
1	11	22	1	Curry
1	12	24	1	XML
1	7	27	1	T
1	8	29	1	OODB
1	10	32	1	John
1	11	34	1	Smith
1	12	36	1	DB

Fig. 3. An example `Element` table and `Data` table for the XML document of Figure 1

instance-level information and are identical to those of XParent. They are used to identify the nodes in the XML documents that belong to the label path selected from `LabelPath` table representing the schema-level information, which is two tables of XParent correspond to three tables `Element`, `Text`, and `Attribute` of XRel. We prefer those of XParent due to the join efficiency for BPE [8] of using node identifiers (`node_id`) over using start_position and end_position in XRel. Figure 3 shows an example of the `Element` table and `Data` table for the XML tree in Figure 1.

3.2 Query Processing

The method for evaluating an LPE based on the our storage structures described in Section 3.1. It first finds matching label paths in the `LabelPath` table using the `LabelPath` inverted index, and then, performs an equi-join between the set of the label paths found and the `Element` table via the column `label_path_id`. It then returns the matching nodes as the query result. The table `Data` can be subsequently used to retrieve the values of the selected nodes if needed.

Example 1. Consider the LPE `//article//author/first`. Using the rule LPE-to-IRExp, our method converts the LPE to the IRExp `article near(∞)` `author near(1) first near(1) &first`. Then, finding the `LabelPath` inverted index in Figure 2 (b) returns the *pidSet* {10}, and joining this set with the `Element` returns the set of nodes { 19, 31 }. Figure 4 shows the SQL generated for evaluating the LPE. The `MATCH` in the `WHERE` clause makes use of the `LabelPath` inverted index. The symbol `MAXINT` is a system-defined maximum integer. □

```
SELECT   DISTINCT e1.document_id, e1.node_id
FROM     LabelPath p1, Element e1
WHERE    p1.pid = e1.pid
AND      MATCH(p1.label_path, `article' NEAR(MAXINT) `author'
                  NEAR(1) `first' NEAR(1) `&first');
```

Fig. 4. SQL statement for the LPE `//article/author/first`

Our performance advantage over XRel and XParent comes from the use of inverted index search as opposed to string match for finding label paths matching the LPE. That is, for our method, the label path match time is determined by the number of labels in the label path and the lengths of the posting lists associated with these labels in the index whereas, for XRel or XParent, the time is proportional to the number of label paths stored in the `Labelpath` (or `Path`) table. The query processing algorithms of XRel, XParent, and our method share the same outline, but have quite different implementations leading to their performance differences. The query processing algorithms of XRel, XParent, and XIR share the same outline, but have some different implementations leading to their performance differences. We show a summary of comparing the three methods in views of performance related features, in turns of XRel, XParent, and our method as follows.

First, the label path match method is a string match for XRel, XParent and an inverted index search for our method. Second, the element identification uses an interval between nodes for XRel, a node identifier of a node for XParent, and an identifier sequence from root to a node for our method. Third, factors affecting label path match time are the number of label paths in `Path` table for XRel, the number of label paths in `LabelPath` table for XParent, and the sum of lengths of the posting lists for the labels in the linear path expression.

4   Performance Evaluation

In this section we compare the performance of our method with those of XRel and XParent, with particular attention to the efficiency of query processing.

4.1 Experimental Setup

Databases. We have collected above 10000 XML documents from the Web using two web crawlers [14] [13]. Note that we have not used a synthetic data set because these data are confined within a particular domain and, consequently, do not have sufficiently heterogeneous structures. Using the collected XML documents, we have constructed five sets of data files of different sizes. Each set contains approximately 5000, 10000, 20000, 40000, and 80000 distinct label paths. The last set has 1460000 nodes. A larger set contains all label paths in a smaller set, i.e., is a superset of smaller sets. Each set has been loaded into three databases, each containing tables used by XRel, XParent, and our method methods. The total number of databases thus generated is fifteen.

Queries. Queries divides two groups organized as sets of XPath LPE queries: one is on `issue` documents; the other is on `movie` documents. The former has far more document instances than the latter. LPEs within each group have different numbers of labels and/or different combinations of '/' and '//'. The `selectivity` field in follows is the ratio between the number of nodes resulting from the LPE and the total number of nodes in the database. There are two groups in Queries. The first group is LPEs on issue documents. The second group is LPEs on movie documents. The first group linear path expressions are four queries for book issue information as follows.

- LPE1: `/issue/editor`, Selectivity: $10^{-7} \sim 10^{-6}$
- LPE2: `//issue//first`, Selectivity: $10^{-4} \sim 10^{-3}$
- LPE3: `//issue//author/first`, Selectivity: $10^{-5} \sim 10^{-4}$
- LPE4: `//issue//article//author/first`, Selectivity: $10^{-5} \sim 10^{-4}$

The second group linear path expressions are also four queries for movie information as follows.

- LPE5: `/movie/cast`, Selectivity: $10^{-7} \sim 10^{-6}$
- LPE6: `//movie//first`, Selectivity: $10^{-6} \sim 10^{-5}$
- LPE7: `//movie//actor//first`, Selectivity: $10^{-6} \sim 10^{-5}$
- LPE8: `//movie/cast//actor//first`, Selectivity: $10^{-6} \sim 10^{-5}$

Figure 5 shows label frequencies counted according to the number of distinct label paths, that is the number of postings for each label. The figure represents the characteristics of the crawled documents as the distribution of labels. Among labels, labels such as `article`, `issue`, `author`, `first` have a high frequency in turn. The label frequencies of `article`, `issue`, `author`, `first` appear 5288, 3690, 2624, 1286 times in 80000 label paths.

4.2 Experimental Results

New label paths are added as new documents are added by crawling since the crawlers collect *arbitrary* documents from the Internet. We have extracted the number of distinct label paths from the XML documents collected. Figure 6 shows the result. The database size is 301 Mbytes for our method, which is called XIR-Linear in the paper figures, 271 Mbytes for XRel, and 248 Mbytes for XParent when the number of XML documents is 10,000 (and the number of distinct label paths is 80,000). The database size for our method is slightly

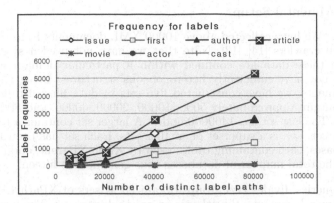

Fig. 5. Label frequencies according to the increment of the label paths

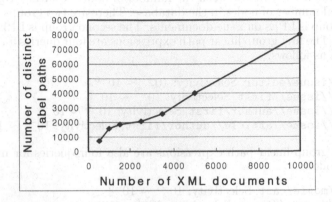

Fig. 6. The number of distinct label paths as the number of XML documents increases

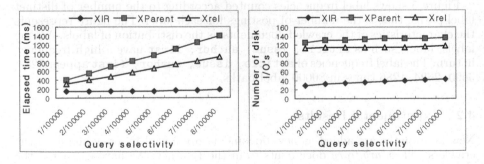

Fig. 7. The variations for the Elapsed Time and I/O counts as the query selectivities grow

LPE1	Our method	XParent	XRel
5000	76	170	170
10000	67	384	370
20000	99	763	759
40000	123	1406	1425
80000	119	2931	2865

LPE2	Our method	XParent	XRel
5000	119	229	236
10000	107	497	484
20000	235	1071	1007
40000	394	1833	1858
80000	779	3777	3799

LPE3	Our method	XParent	XRel
5000	125	225	236
10000	112	498	486
20000	239	1010	986
40000	404	1830	1850
80000	722	3766	3799

LPE4	Our method	XParent	XRel
5000	131	229	233
10000	131	493	475
20000	245	1010	968
40000	391	1811	1874
80000	745	3789	3800

LPE5	Our method	XParent	XRel
5000	46	192	190
10000	70	413	388
20000	80	794	759
40000	97	1454	1417
80000	90	2932	2865

LPE6	Our method	XParent	XRel
5000	65	220	221
10000	58	485	455
20000	85	928	922
40000	99	1748	1708
80000	138	3634	3568

LPE7	Our method	XParent	XRel
5000	52	212	213
10000	51	497	455
20000	97	952	915
40000	110	1750	1832
80000	135	3546	3534

LPE8	Our method	XParent	XRel
5000	76	216	212
10000	52	503	455
20000	74	946	904
40000	110	1751	1736
80000	122	3672	3546

Fig. 8. The Elapsed Time (ms) Comparison for the LPEs Queries

larger due to inclusion of the inverted index. Since the crawlers collect and classify arbitrary XML documents from the Web depends on the XML content types such as movies, animations, and articles, new label paths are added as new documents are added by crawling. We have extracted the number of distinct label paths from the XML documents collected. Figure 8 shows the results in elapsed micro-second time as the execution of the LPEs. We see that, for the database used, our method is more efficient than both XRel and Xparent by a factor of minimum 3 times to maximum 75 times in elapsed time and the number of disk I/O's in the experiment database size. Note that the performance gap widens as the database size grows since the costs increase linearly for XRel and XParent while sublinearly, almost constant, for our method. This confirms the significant performance advantage of our method in a large-scale database environment.

5   Conclusions

We have presented a novel approach for evaluating linear path expressions on a large number of heterogeneous XML documents. In this approach, the label paths occurring in XML documents are treated as texts, and an inverted index is created on them. This inverted index supports much faster partial match than XRel's or XParent's string match when evaluating a linear path expression. We have presented the storage structures of our method, including the inverted index as well as the tables storing all the label paths and nodes of the XML documents. Based on these structures, we have presented the query processing algorithm for a linear path expression. Then, we have compared the performance of our method with those of XRel and XParent through experiments using real XML documents collected from the Internet. The results show that our method outperforms both XRel and XParent by an order of magnitude with the performance gap widening as the database size grows.

References

1. Aboulnaga, A., Alameldeen, A.R., Naughton, J.: Estimating the Selectivity of XML Path Expressions for Internet Scale Applications. In: VLDB. Proc. the 27th Int'l Conf. on Very Large Data Bases, Rome, Italy, pp. 591–600 (2001)
2. Altinel, M., Franklin, M.J.: Efficient Filtering of XML Documents for Selective Dissemination of Information. In: VLDB. Proc. the 26th Int'l Conf. on Very Large Data Bases, Cairo, Egypt, pp. 53–64 (September 10-14, 2000)
3. Al-Khalifa, S., Jagadish, H.V., Koudas, N., Patel, J.M.: Structural Joins: A Primitive for Efficient XML Query Pattern Matching. In: ICDE. Proc. the 18th Int'l Conf. on Data Engineering, San Jose, California, pp. 141–152 (2002)
4. Bremer, J.-M., Gertz, M.: XQuery/IR: Integrating XML Document and Data Retrieval. In: WebDB 2002. Proc. the Fifth Int'l Workshop on the Web and Databases, Madison, Wisconsin, pp. 1–6 (2002)
5. Clark, J., DeRose, S.: XML Path Language (XPath), W3C Recommendation (November 1999), http://www.w3.org/TR/xpath
6. Cooper, B.F., Sample, N., Franklin, M.J., Hjaltason, G.R., Shadmon, M.: A Fast Index for Semistructured Data. In: VLDB. Proc. the 27th Int'l Conf. on Very Large Data Bases, Rome, Italy, pp. 341–350 (September 11-14, 2001)
7. Halverson, A., Burger, J., Galanis, L., Kini, A., Krishnamurthy, R., Rao, A.N., Tian, F., Viglas, S., Wang, Y., Naughton, J.F., DeWitt, D.J.: Mixed Mode XML Query Processing. In: VLDB. Proc. the 29th Int'l Conf. on Very Large Data Bases, Berlin, Germany, pp. 225–236 (September 9-12, 2003)
8. Jiang, H., Lu, H., Wang, W., Xu Yu, J.: An Efficient RDBMS-Based XML Database System. In: ICDE. Proc. the 18th Int'l Conf. on Data Engineering, San Jose, California, pp. 335–336 (February 26 - March 1, 2002)
9. Jiang, H., Wang, W., Lu, H., Yu, J.X.: Holistic Twig Joins on Indexed XML Documents. In: VLDB. Proc. the 29th Int'l Conf. on Very Large Data Bases, Berlin, Germany, pp. 273–284 (September 9-12, 2003)
10. Mandreoli, F., Martoglia, R., Tiberio, P.: Searching Similar (Sub)Sentences for Example-Based Machine Translation. In: Proc. SEBD 2002, Isola d'Elba, Italy (June 2002)
11. Ramanan, P.: Covering Indexes for XML Queries: Bisimulation - Simulation = Negation. In: VLDB. Proc. the 29th Int'l Conf. on Very Large Data Bases, Berlin, Germany, pp. 165–176 (September 9-12, 2003)
12. Yoshikawa, M., Amagasa, T., Shimura, T., Uemura, S.: XRel: A Path-based Approach to Storage and Retrieval of XML Documents using Relational Databases. ACM Transactions on Internet Technology(TOIT) 1(1), 110–141 (2001)
13. ReGet Deluxe 3.3 Beta (build 173), http://deluxe.reget.com/en/
14. Teleport Pro Version 1.29, http://www.tenmax.com/teleport/pro/home.htm

An Analysis of a Lymphoma/Leukaemia Dataset Using Rough Sets and Neural Networks

Kenneth Revett[1] and Marcin Szczuka[2]

[1] University of Westminster
Harrow School of Computer Science,
London, UK
revettk@westminster.ac.uk
[2] Institute of Mathematics, Warsaw University
Banacha 2, 02-097 Warsaw, Poland
szczuka@mimuw.edu.pl

Abstract. In this paper, we describe a rough sets approach to classification and attribute extraction of a lymphoma cancer dataset. We verify the classification accuracy of the results obtained from rough sets with a two artificial neural network based classifiers (ANNs). Our primary goal was to produce a classifier and a set of rules that could be used in a predictive manner. The dataset consisted of a number of relevant clinical variables obtained from patients that were suspected of having some form of blood based cancer (lymphoma or leukaemia). Of the 18 attributes that were collected for this patient cohort, seven were useful with respect to outcome prediction. In addition, this study was able to predict with a high degree of accuracy whether or not the disease would undergo metastases.

1 Introduction

Lymphoma is a general term for a group of cancers that originate in the lymphatic system. The lymphomas are divided into two major categories: Hodgkin lymphoma and all other lymphomas, called non-Hodgkin lymphomas. Hodgkin lymphoma was named for Thomas Hodgkin, an English physician who described several cases of the disease in 1832. Hodgkin lymphoma represented about 12.7 percent of all lymphomas diagnosed in 2004 [1]. About 62,250 Americans were be diagnosed with lymphoma in 2004. This figure includes approximately 7,880 new cases of Hodgkin lymphoma (4,330 males and 3,550 females), and 54,370 new cases of non-Hodgkin lymphoma (28,850 males and 25,520 females). The annual incidence of lymphoma has nearly doubled over the last 35 years. The cause of Hodgkin lymphoma is uncertain [1]. Many studies investigating the role of environmental, especially occupational, linkages have been conducted with ambiguous results. For example, woodworking exposure has been associated with the disease, but causality has not been established. The Epstein-Barr virus has been associated with about one-third of cases of the disease, although it has not been established conclusively as a cause of Hodgkin lymphoma. Persons infected

M.S. Szczuka et al. (Eds.): ICHIT 2006, LNAI 4413, pp. 229–239, 2007.

with HTLV and HIV also have an increased probability of developing Hodgkin lymphoma [1].

In this study, we investigate a dataset containing data on 148 cases with four decision classes (2 normal, 81 metastases, 61 with malignant lymphoma, and 4 with fibrosis) of patients that were hospitalised for suspected lymphoma. The dataset contains 19 attributes including the decision attribute (see section 2.1 for a listing of the attributes) with no missing values. We investigated this dataset with respect to the following: i) attribute pruning, ii) classification accuracy and iii) rule induction. Pruning (dimensionality reduction) removes variables that are not directly related to the classification process. This feature of rough sets makes the dataset much easier to work with and may help to highlight the relevant classification features contained within the data. Once the redundant features have been pruned from the dataset, rough sets is used in the classification process, mapping attributes and their values to decision classes. In many cases, rough sets is able to produce a classification accuracy that is superior to more 'traditional' classification algorithms. Lastly, rough sets provide a set of decision rules that are readily interpretable by a domain expert. These three inherent facilities available in the rough set paradigm provide a unique and consistent approach to extracting knowledge from data. In the next section, we present an overview of rough sets, followed by the use of rough sets to classify this particular dataset, followed by a results section and lastly a summary of this work.

2   Data Mining

Rough set theory is a relatively new data-mining technique used in the discovery of patterns within data first formally introduced by Pawlak in 1982 [9,10]. Since its inception, the rough sets approach has been successfully applied to deal with vague or imprecise concepts, extract knowledge from data, and to reason about knowledge derived from the data [7,8]. We demonstrate that rough sets has the capacity to evaluate the importance (information content) of attributes, discovers patterns within data, eliminates redundant attributes, and yields the minimum subset of attributes for the purpose of knowledge extraction.

The first step in the process of mining any dataset using rough sets is to transform the data into a decision table. In a decision table (DT), each row consists of an observation (also called an object) and each column is an attribute, one of which is the decision attribute for the observation.

Formally, a DT is a pair $\mathbf{A} = (U, A \cup \{d\})$ where $d \notin A$ is the decision attribute, U is a finite non-empty set of objects called the universe and A is a finite non-empty set of attributes such that $a : U \mapsto V_a$ for $a \in A$, where V_a is called the value set of a. Once the DT has been produced, the next stage entails cleansing the data.

There are several issues involved in small datasets - such as missing values, various types of data (categorical, nominal, continuous, and interval) and multiple decision classes. There are several techniques available for addressing all of these particular issues that arise in datasets, which generally enhance the information gain from a DT. Missing values is very often a problem in biomedical datasets and can arise in two different ways. It may be that an omission of a value for one or more subject was intentional - there was no reason to collect that measurement for this particular subject (i.e. 'not applicable' as opposed to 'not recorded'). In the second case, data was not available for a particular subject and therefore was omitted from the table. We have 2 available options: remove the incomplete records from the DT or try to estimate (impute) what the missing value(s) should be. The first method is obviously the simplest, but we may not be able to afford removing records if the DT is small to begin with. So we must derive some method for filling in missing data without biasing the DT. In many cases, an expert with the appropriate domain knowledge may provide assistance in determining what the missing value should be - or else is able to provide feedback on the estimation generated by the data collector. In this study, we employed a conditioned mean/mode fill method for data imputation. In each case, the mean or mode is used (in the event of a tie in the mode version, a random selection is used) to fill in the missing values, based on the particular attribute in question, conditioned on the particular decision class the attribute belongs to. There are many variations on this theme, and the interested reader is directed to [5,6] for an extended discussion on this critical issue.

Once missing values are handled, the next step is to discretise the dataset. Rarely is the data contained within a DT all of ordinal type - they generally are composed of a mixture of ordinal, continuous, and interval data. Discretisation refers to partitioning attributes into intervals - tantamount to searching for "cuts" in a decision tree. All values that lie within a given range are mapped onto the same value, transforming the data into categorical form. As an example of a discretisation technique, one can apply equal frequency binning, where a number of bins n is selected and after examining the histogram of each attribute, $n - 1$ cuts are generated so that there is approximately the same number of items in each bin. See the discussion in [6,11] for details on this and other methods of discretisation that have been successfully applied in rough sets.

Rough sets can be successfully applied to datasets with multiple decision classes. The critical issue is whether there are a sufficient number of objects for each class. There is no hard and fast rule for the number of required objects - this is dependent on the quality of the data in most cases. Once the DT has been preprocessed, the rough sets algorithm can be applied to the DT for the purposes of supervised classification. The basic philosophy of rough sets is to reduce the elements (attributes) in a DT based on the information content of each attribute or collection of attributes (objects) such that the there is a mapping between similar objects and a corresponding decision class.

In general, not all of the information contained in a DT is required: many of the attributes may be redundant in the sense that they do not directly influence which decision class a particular object belongs to. One of the primary goals of rough sets is to eliminate attributes that are redundant. Rough sets uses the notion of the lower and upper approximation of sets in order to generate decision boundaries that are employed to classify objects.

Consider a decision table $\mathbf{A} = (U, A \cup \{d\})$ and let and $X \subset U$. What we wish to do is to approximate X by the information contained in the subset of attributes $B \subset A$ by constructing the B-lower (BL) and B-upper (BU) approximation of X. The objects in BL (BLx) can be classified with certainty as members of X, while objects in BU are not guaranteed to be members of X. The difference between the two approximations: BU - BL, determines whether the set is rough or not: if it is empty, the set is crisp otherwise it is a rough set.

The object of the classification task is to partition the objects in the DT such that objects that are similar to one another (by virtue of their attribute values) are treated as a single entity (they are indiscernible). One potential difficulty arises in this regard if the DT contains inconsistent data. In this case, antecedents with the same values map to different decision outcomes (or the same decision class maps to two or more sets of antecedents). This is unfortunately not uncommon, and in the case of small biomedical datasets, such as the one used in this study, can have a substantial impact on the quality of the classification results. There are means of handling this and the interested reader should consult [8,12] for a detailed discussion of this interesting topic. The next step is to reduce the DT to a collection of attributes/values that maximises the information content of the decision table. This step is accomplished through the use of the indiscernibility relation $IND(B)$ and is defined for any subset $B \subset A$ as follows:

$$IND(B) = \{(x, y) \in U^2 : \forall_{a \in B} \quad a(x) = a(y)\}$$

The elements of $IND(B)$ correspond to the notion of an equivalence class. The advantage of this process is that the members of the equivalence class can be used to represent the entire class – thereby reducing the dimensionality of the objects in the DT. This leads directly to the concept of a reduct, which is the minimal set of attributes from a DT that preserves the equivalence relation between conditioned attributes and decision values. It is the minimal amount of information required to distinguish objects within U, the universe of objects. The collection of all reducts that together provide classification of all objects in the DT is called the $CORE(A)$. The $CORE$ specifies the minimal set of elements/values in the decision table which are required to correctly classify objects in the DT. Removing any element from this set reduces the classification accuracy.

It should be noted that searching for minimal reducts is an NP-hard problem, but fortunately there are good heuristics that can compute a sufficient amount of reducts in reasonable time. In the software system that we employ an order based genetic algorithm (o-GA) was used to search through the decision table for approximate reducts [11]. The reducts are approximate because we do not perform an exhaustive search via the o-GA which may miss one or more attributes that should be included as a reduct. Once we have our set of reducts, we are ready to produce a set of rules that will form the basis for object classification.

Rough sets generates a collection of "if..then.." decision rules that are used to classify the objects in the DT. These rules are generated from the application of reducts to the decision table, looking for instances where the conditionals match those contained in the set of reducts and reading off the values from the DT. If the data is consistent, then all objects with the same conditional values as those found in a particular reduct will always map to the same decision value. In many cases though, the DT is not consistent, and instead we must contend with some amount of indeterminism. In this case, a decision has to be made regarding which decision class should be used when there are more than 1 matching conditioned attribute values. Simple voting may work in many cases, where votes are cast in proportion to the support of the particular class of objects. With respect to the production of rules, one wishes to generate rule sets with low cardinality. If the rules are too detailed (i.e. they incorporate reducts that are maximal in length), they will tend to overfit the training set and classify weakly on test cases. This idea is analogous to the building block hypothesis used in genetics algorithms, where we wish to select for highly accurate and low defining length gene segments (building blocks hypothesis) [13]. There are many variations on rule generation, which are implemented through the formation of alternative types of reducts such as dynamic and approximate reducts. Discussion of these ideas is beyond the scope of this paper and the interested reader is directed towards [12] for a detailed discussion of these alternatives.

The rules that are generated are in the traditional conjunctive normal form and are easily applied to the objects in the DT. What we are interested in is the accuracy of the classification process – how accurately can the generated rule set acquired during the training process classify new objects? In addition, what sort of confidence do we have in the resulting classification of particular validation training set? These are standard issues that hold true for any machine learning application. In addition questions arise regarding methods for handling biomedical datasets that contain an unequal distribution of decision class objects. The quality of the results - measured by their statistical validity is a validation issue. In this work, we employed 10-fold cross validation, and it is these results that are reported in this paper. In the next section, we describe the dataset that was used in this study.

2.1 Dataset Description

The dataset used in this study was obtained from the internet and contained clinical data from a series of 148 patients that were hospitalised and/or diagnosed with suspected blood cancer (lymphoma/leukaemia). A total of 18 attributes were collected (see Table 1 for details) from typical routine medical diagnostics consistent with a diagnosis for lymphoma/leukaemia. The attributes contained both ordinal and continuous variables. There were four decision classes: normal (2), metastases (81), malignant (61), and fibrosis (4). The dataset was complete, so no data imputation was required. As required, any continuous attributes were discretised, using a minimum description length (MDL) conditioned on the decision attributes. Any other requirements for data processing are indicated in the methods section which is presented next.

3   Methods

The structure of the dataset consisted of 19 attributes, including the decision attribute (labelled 'result') which is displayed for convenience in table 1 below. We calculated the Pearson's Correlation Coefficient of each attribute with respect to the decision class. These correlation values can be used to determine if one or more attributes are strongly correlated with a decision class. As is indicated in Table 1, none of the attributes had a significantly correlated Correlation coefficient with respect to any of the decision classes. The data was subjected to a 10-fold cross validation scheme for classification results. After several experiments, we decided to use dynamic reducts based on the resulting classification accuracy. With the collection of dynamic reducts, decision rules were generated. The resultant rules were then applied to the testing datasets and the resulting values are used to create a confusion matrix for the classification task (this process of training/testing segregation is automatically performed using the cross-validation algorithm in the rough sets software system employed in this study). These are the primary results that are reported in this paper - which represent the classification accuracy (reported as the average of the cross-validation results) - along with the list of rules that are predictive of whether or not the lymphoma will metastases. In addition, we also examined the dataset using the Linear Transfer Function Classifier (LTF-C) (the implementation that is incorporated into RSES) and a multi-layer perceptron neural network trained with trained with the back-propagation learning algorithm. LTF-C (cf [2]) employs a novel neural network learning algorithm which is based on the architectures of the Radial Basis Function neural network (RBF) and Support Vector Machines (SVM). The network has a hidden layer with Gaussian neurons connected to an output layer of linear units. The number of inputs corresponds to the number of attributes while the number of linear neurons in output layers equals the number of decision classes. There are some additional restrictions on values of output weights that enables this classifier to use an entirely different training algorithm and to obtain very high accuracy in real-world problems [14]. The training algorithm of LTF-C comprises four types of modifications of the network, performed after every presentation of a training object. Namely the network can: change positions (means) of Gaussians in hidden layer, changing widths (deviations) of Gaussians separately for each hidden neuron and attribute, insert new hidden neurons, remove unnecessary or 'harmful' hidden neurons. As one can see, the LTF-C structure is extremely dynamic. The training process starts with an empty hidden layer, adding new hidden neurons when the accuracy is insufficient and removing the units which do not positively contribute to the calculation of correct network decisions. This feature of LTF-C enables automatic choice of the best network size, which is much easier than setting the number of hidden neurons manually. Moreover, this helps to avoid getting stuck in local minima during training, which is a serious problem in neural networks trained with gradient-descend. The results we obtained are described in the next section.

4   Results

The classification accuracy obtained in this study was significantly affected by the extent of the pre-processing procedure. Without any pre-processing at all, we obtained an average classification accuracy of approximately 60% (10 trials). As a first pre-processing step, we calculated the Pearson Correlation coefficients for all attributes in the decision table (excluding the decision attribute). The summary results for the correlation analysis are displayed in Table 1 below. From our experience, attributes with very low correlation coefficients (positive or negative) can be removed from the decision table without compromising classification accuracy [6,11].

Table 1. Attributes along with the Pearson Correlation coefficient for the elements in the decision table. Note that the entries marked with a '*' in the correlation coefficient column were the largest positively correlated attributes from the decision table.

Attribute Name	Correlation coefficient
Lymphatics	0.0176
Block of efferents d	0.093
Block of lymphatics c	0.176 *
Block of lymphatics s	0.169 *
By pass	0.157 *
Extravasates	0.197 *
Regeneration of nodes	-0.083
Early uptake	0.251 *
Lymph node diminishing	-0.031
Lymph nodes enlarging	-0.112
Change in lymph	0.101 *
Defect in nodes	0.077
Changes in nodes	-0.181
Changes in structure	-0.047
Special forms	-0.012
Dislocation	-0.060
Exclusion of lymph	0.101 *
Number of nodes	0.081

We used dynamic reducts with no reduct/rule filtering and 10-fold cross validation using a randomly selected set of classification tasks. Results are presented in Table 2 in the form of composed confusion matrix. The sensitivity is listed in the top right column and the specificity is directly underneath the sensitivity (both are highlighted). Please note - '0' corresponds to metastases, '1' to malignant and '2' to fibrosis. Note, we removed the two 'normal' objects from the decision table.

We next pooled together the malignant and metastases classes (a total of 65 objects) in order to examine directly whether or not the attributes could be used to determine if the tumour would undergo metastasis. Note that malignant

Table 2. Confusion matrices from a randomly selected set of classification tasks

Test 1	0	1	2	
0	19	2	1	0.864
1	3	13	1	0.813
2	0	0	5	1.0
	0.864	0.867	0.714	0.841
Test 2	0	1	2	
0	18	2	2	0.818
1	1	14	2	0.824
2	0	0	5	1.00
	0.900	0.875	0.50	0.841
Test 3	0	1	2	
0	20	1	1	0.909
1	2	14	1	0.824
2	0	1	4	0.80
	0.857	0.750	0.571	0.864

Table 3. Confusion matrices from a randomly selected set of classification tasks. The data was derived from dynamic reducts with no reduct/rule filtering, and 10-fold cross-validation. The sensitivity is listed in the top right column and the specificity is directly underneath the sensitivity (both are highlighted. Note that '0' corresponds to metastases and '1' to malignant and fibrosis combined.

Test 1	0	1	
0	20	1	0.953
1	2	14	0.875
	0.909	0.933	**0.919**
Test 2	0	1	
0	19	2	0.905
1	2	14	0.875
	0.950	0.882	**0.892**

Table 4. Results when the RSES 2.0 built-in LTF-C classification algorithm was employed on the complete dataset. Note the data used was the same as that sued for the rough sets analysis.

	Normal	Metastases	Malignant	Fibrosis
LTF-C	78.1%	91.4%	88.9%	83.7%

lymphomas have a very high probability of metastasing. The resulting confusion matrices are presented in Table 3.

The results from the LTF-C are consistent (though higher in accuracy generally) with those from the classification resulting from the rough sets paradigm - with an average classification accuracy for each class presented in Table 4.

Table 5. Results of a back-propagation classification using the full and the reduced dataset. This is the same pair of datasets utilised in Table 4. Note that 'Full' represents the complete dataset and 'Reduced' is as indicated in Table 3. The reduced dataset represents the classification results when malignant and metastases case are combined.

	Normal	Metastases	Malignant	Fibrosis
MLP - Full	79.5%	88.9%	88.3%	86.1%
Reduced	89.9%	85.4%	85.4%	89.9%

Table 6. This table presents examples of rules that indicate which attributes and their values were most correlated with a lymphoma undergoing metastasis. Note that a decision of '0' refers to metastasis occurs and '1' indicates metastasis does not occur.

If *block of efferents* = 'yes' **and** *extravasates* = 'yes' **0**
If *block of efferents* = 'yes' **and** *extravasates* = 'yes' **and** *lymph node enlargements* = '2' **then 0**
If *extravasates* = 'yes' **and** *lymph node enlargements* = '3' **then 0**
If *lymphatics* = 'normal' **and** *block of efferents* = 'no' **then 1**
If *extravasates* = 'yes' **and** *lymphatics* = 'normal' **and** *lymph node enlargement* = '2' **then 1**

In addition, we classified both the complete dataset (with all four decision classes) and the reduced dataset (malignant and fibrosis classes) using a multi-layer perceptron trained with the back-propagation algorithm. These results are presented in Table 5.

The resulting rule sets was instructive with regards to deciding whether or not the cancer would undergo metastasis. A sample rule set is presented in Table 6. Note that the rules generated were not filtered in any way.

5   Conclusion

In this paper, we present an analysis of data obtained through the clinical diagnosis of suspected lymphoma/leukemia patients. Although the dataset was heavily weighted towards the positive diagnosis for blood cancer, valuable information regarding whether or not the attributes were indicative of metastasis were conclusive with an accuracy of approximately 85%. These results are superior to other values reported in the literature [4,5]. In addition, the classification accuracy was slightly higher when examining metastasis versus non-metastasis cases (by pooling the malignant and metastases cases), compared to normal cases alone.

The results from the rough sets approach were consistent with those acquired from using two neural networks: the LTF-C and a standard back-propagation based MLP. One of the primary advantages of the rough sets approach is that it yields in addition to a very accurate classifier, decision rules that directly relate attributes and their values to their respective decision classes. The rules are readily interpreted by an individual/system with the appropriate domain knowledge. This feature provides an objective method for evaluating the relevance of the model produced by the classifier. The results from this study indicate that blockage of afferents, extravasates, lymph node enlargements, and dislocation were indicative of metastasis. This is a clinically relevant result that can be directly verified by a larger clinical study. It is hoped that studies such as this can provide the impetus for enhanced cooperation between the medical profession and the computational community. Only by examining the results obtained in this study on a larger scale will we be able to ascertain whether they are truly clinically relevant.

Acknowledgements

The dataset was donated by: Igor Kononenko, University E. Kardelj, Faculty for Electrical Engineering, Trzaska 25 61000 Ljubljana, Slovenia.

We also acknowledge use of the software Package Rosetta, available from the Internet at: http://www.idi.ntnu.no/aleks/rosetta/

Marcin Szczuka is partly supported by Polish grant 3T11C02226.

References

1. http://www.leukemia-lymphoma.org
2. Bazan, J., Szczuka, M.: The Rough Set Exploration System. In: Peters, J.F., Skowron, A. (eds.) Transactions on Rough Sets III. LNCS, vol. 3400, pp. 37–56. Springer, Heidelberg (2005), http://logic.mimuw.edu.pl/~rses
3. Bazan, J., Nguyen, H.S., Nguyen, S.H., Synak, P., Wróblewski, J.: Rough set algorithms in classification problems. Studies in Fuzziness and Soft Computing 56, 49–88 (2000)
4. Cestnik, G., Konenenko, I., Bratko, I.: Assistant-86: A Knowledge-Elicitation Tool for Sophisticated Users. In: Bratko, I., Lavrac, N. (eds.) Progress in Machine Learning, pp. 31–45. Sigma Press (1987)
5. Clark, P., Niblett, T.: Induction in Noisy Domains. In: Bratko, I., Lavrac, N. (eds.) Progress in Machine Learning, pp. 11–30. Sigma Press (1987)
6. Khan, A., Revett, K.: Data mining the PIMA Indian diabetes database using Rough Set theory with a special emphasis on rule reduction. In: INMIC2004, Lahore, Pakistan, pp. 334–339 (2004)
7. Komorowski, J., Pawlak, Z., Polkowski, L., Skowron, A.: Rough sets: A tutorial. In: Pal, S.K., Skowron, A. (eds.) Rough Fuzzy Hybridization - A New Trend in Decision Making, pp. 3–98. Springer, Singapore (1999)
8. Ohrn, A.: Discernibility and Rough Sets in Medicine - Tools and Applications. Department of Computer and Information Science, Norwegian University of Science and Technology, Trondheim, Norway (1999)
9. Pawlak, Z.: Rough Sets. International Journal of Computer and Information Sciences 11, 341–356 (1982)
10. Pawlak, Z.: Rough sets - Theoretical aspects of reasoning about data. Kluwer, Dordrecht (1991)

11. Revett, K., Khan, A.: A Rough Sets Based Cancer Classification System. In: METMBS. Proceedings of International Conference on Mathemtics and Engineering, pp. 315–320 (2005)
12. Slezak, D.: Approximate Entropy Reducts. Fundamenta Informaticae 53(3-4), 365–387 (2002)
13. Wroblewski, J.: Theoretical Foundations of Order-Based Genetic Algorithms. Fundamenta Informaticae 28(3-4), 423–430 (1996)
14. Wojnarski, M.: LTF-C: Architecture, training algorithm and applications of new neural classifier. Fundamenta Informaticae 54(1), 89–105 (2003)

A Frequency Adaptive Packet Wavelet Coder for Still Images Using CNN

N. Venkateswaran, J. Vignesh, S. Santhosh Kumar, S. Rahul,
and M. Bharadwaj

Department of Electronics and Communication Engineering
Sri Venkateswara College of Engineering, Anna University, India
nvenkat@svce.ac.in

Abstract. We present the packet wavelet coder implemented with Cellular Neural Network architecture as an example of the applications of cellular neural networks. This paper also demonstrates how the cellular neural universal machine (CNNUM) architecture can be extended to image compression. The packet wavelet coder performs the operation of image compression, aided by CNN architecture. It uses the highly parallel nature of the CNN structure and its speed outperforms traditional digital computers. In packet wavelet coder, an image signal can be analyzed by passing it through an analysis filter banks followed by a decimation process, according to the rules of packet wavelets. The Simulation results indicate that the quality of the reconstructed image is improvised by using packet wavelet coding scheme.

1 Introduction

The wavelet coding method has been recognized as an efficient coding technique for lossy compression. The wavelet transform decomposes a typical image data to a few coefficients with large magnitude and many coefficients with small magnitude. Since most of the energy of the image concentrates on these coefficients with large magnitude, lossy compression systems just using coefficients with large magnitude can realize both high compression ratio with good quality reconstructed image at the same time The wavelet transform does a good job of de-correlating image pixels in practice, especially when images have power spectra that decay approximately uniformly and exponentially. For images with non-exponential rates of spectral decay, and for images which have concentrated peaks in those spectra away from DC, one can do considerably better. The optimal sub-band decomposition for an image is one for which the spectrum in each sub-band is approximately flat. On the other hand Cellular neural networks (CNN) is a analog parallel-computing paradigm defined in space and characterized by locality of connections between processing elements such as cells or neurons. A CNN principal property is the ability to perform massive processing and such property fits perfect in image processing tasks where minimum processing time is demanded. In this paper we have presented packet wavelet coder using CNN architecture and compared the results to that of pyramidal wavelets

M.S. Szczuka et al. (Eds.): ICHIT 2006, LNAI 4413, pp. 240–248, 2007.

using CNN architecture. Packet wavelets allow the option to zoom in to high frequencies if we like, by analyzing at each level only the band with highest energy. The signal passes through filter banks, and it is sampled following the packet wavelet coder. In order to implement image coder using CNN, the kernels are considered the templates defined in the CNN mathematical representation. The image is passed through the CNN, which performs filtering operation. Image reconstruction is performed by inverse wavelet transform. The kernels for inverse wavelet transform are also treated as templates. In our work, we have used Quadrature mirror filters (QMF) coefficients for filtering operation. Contemporarily, Wavelets transform is considered a powerful tool, which provides a time-frequency representation. Wavelets were designed with such non-stationary data in mind, and with their generality and strong results have quickly become useful to a number of disciplines, such as image compression.

2 Cellular Neural Networks

Cellular Neural Networks (CNN) is a massive parallel computing paradigm defined in discrete N-dimensional spaces. Following the Chua-Yang [1] definition: A CNN is an N-dimensional regular array of elements (cells) The cell grid can be for example a planar array with rectangular, triangular or hexagonal geometry, a 2-D or 3-D torus, a 3-D finite array, or a 3-D sequence of 2-D arrays (layers) Cells are multiple input-single output processors; all described by one or just some few parametric functionals. A cell is characterized by an internal state variable, sometimes not directly observable from outside the cell itself More than one connection network can be present, with different neighborhood sizes A CNN dynamical system can operate both in continuous (CT-CNN) or discrete time (DT-CNN) CNN data and parameters are typically continuous values Cellular Neural Networks operate typically with more than one iteration, i.e. they are recurrent networks.

The definition of CNN is, a 2, 3 or N-dimensional array of mainly identical dynamical systems called cells which satisfies two properties namely:

1) Most interactions are local within a finite radius R.
2) All state variables are continuous valued signals.

The Cellular neural Network invention was based on the fact that many complex computational problems can be resolved as well defined tasks where signal values, coefficients in the case of images, are placed on a regular geometric 2-D or 3-D grid. The direct interactions between signal values are limited to a finite local neighborhood. Due to their local interconnectivity, cellular neural networks can be realized as VLSI chips, which operate at very high speeds and complexity. Since the range of dynamics is independent of the number of cells, the implementation is highly reliable and efficient. The design of cloning templates is based on the geometric aspects of the problem. The matrix also defines the interaction between each cell and all its neighboring cells in terms of their input state and output variables. In [2], it has been shown that the CNN is also a spatial approximation of a diffusion type partial differential equation. However, CNN is inherently local in nature. So, it cannot be expected to perform global operations of a coding scheme e.g., entropy coding scheme. However, most steps of the scheme can be implemented using CNN and its highly parallel nature makes its speed outperform traditional digital solutions. Defining equations of CNN [1] are:

$$\widehat{x}_{ij} = -x_{ij} + \sum_{C(k,l)\varepsilon S_r(i,j)} A(i,j;k,l) \times y_{kl} + \sum_{C(k,l)\varepsilon S_r(i,j)} B(i,j;k,l) \times u_{k,l} + z_{i,j} \quad (1)$$

$$y_{ij} = f(x_{ij}) = 0.5 \times |x_{ij} + 1| - 0.5 \times |x_{i,j} - 1| \quad (2)$$

where $x_{ij}\varepsilon\Re$, $u_{kl}\varepsilon\Re$, $z_{ij}\varepsilon\Re$ are called state output, input and threshold of cell C(i,j) respectively. A(i,j;k,l) and B(i,j;k,l) are called the feedback and input synaptic operators to be defined below. Figure (1) shows standard non-linearity $g(x^e)$. Hence Cellular neural networks are also called as cellular non-linear networks. Moreover it has also been demonstrated [1] that CNN paradigm is universal, being equivalent to the Turing Machine. A mathematical formal description of the discrete time case is contained in the following equations.

$$x(t+1) = g(x(t)) + I(t) + \sum(A(yk(t), PA(j))) + \sum(B(uk(t), PB(j))) \quad (3)$$

$$y(t) = f(x(t)) \quad (4)$$

where x is the internal state of a cell, y its output, u its external input and I a local value called bias. A and B are two generic parametric functionals, PA(j) and PB(j) are the parameters arrays (typically the inter-cell connection weights). The neighbor yk and uk values are collected from the cells present in

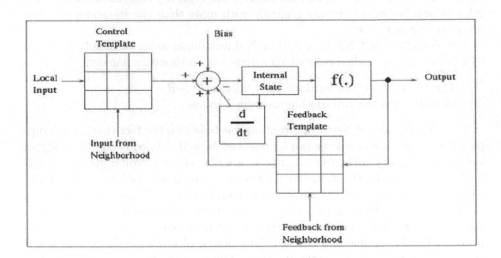

Fig. 1. Generic Cellular Neural Network

the two neighborhood Nr, for the feedback functional A, and Ns, for the control functional B. The two neighborhoods are potentially different. The functionals A and B are also called templates. The instantaneous local feedback function g expresses the possibility of an immediate feedback effect. This function is typically not used. f is the function that gives cell output from the internal state. Generally is used the chessboard distance convention, expressed by the equation

$$d(i,j) = max(|x(i) - x(j)|, |y(i) - y(j)| \tag{5}$$

In most of cases the system is nonMonrovian, i.e. the future internal state depends also from the past history of the system. In the special case of time-variant CNN all the above functions, neighborhoods and parameters can be also function of time. A block-schematic of a generic CNN is shown in Fig. 1.

3   Packet Wavelet Representation for Images

Images are 2-D signals. An image can be analyzed by passed it through four different combinations of low and high pass filter. At each level of resolutions, instead of splitting the signal in to two components, it can be split in to four. A schematic representation of such analysis is shown in Fig. 2. This is known as Packet wavelet representation. The tree we construct this way is called structure tree [2]. Zooming into a frequency band with a packet wavelet transform may be achieved by preferentially expanding a chosen band at each level of resolution. It is a usual practice that one chooses the band, which contains the maximum energy. The energy of a band is computed by adding the squares of the values of the individual pixels. In the method used, the band with maximum energy

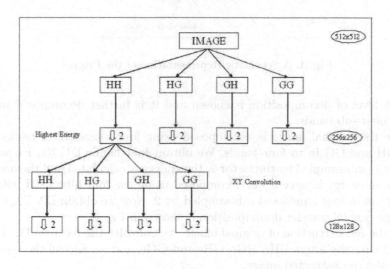

Fig. 2. Packet Wavelet Decomposition

D2 HH	E2 HG	E1
F2 GH	G2 GG	HG
F1 GH		G1 GG

Fig. 3. Wavelet Decomposition

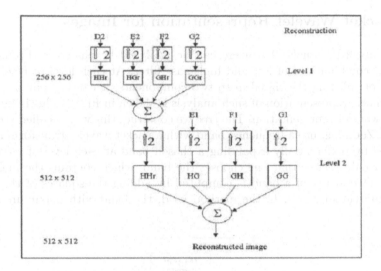

Fig. 4. A Schematic Representation of the Process

at each level of decomposition is chosen and it is further decomposed into its constituent sub-bands.

Here the original image is decomposed using four filters (filter banks) HH, HG, GH and GG in to four bands. We obtain four bands D1, E1, F1 and G1 Then it is sub-sampled by the factor 2, the energy is calculated and the one with maximum energy is once again decomposed using the four filters HH, HG, GH and GG in to four bands and sub-sampled by 2. Now we obtain D2, E2, F2 and G2. The general wavelet decomposition is shown in Fig. 3.

For the reconstruction of original image, we convolve the received D2, E2, F2 and G2 with the filters HHr, HGr, GHr and GGr, and we add all the results to obtain the reconstructed image.

A schematic representation is shown in Fig. 4.

4   Packet Wavelet Coder in CNN Computer

The Packet Wavelet scheme represented in the figure can be implemented in the CNN computer. While coding, for example, the 2D Haar Wavelet can be implemented using CNN universal machine, each convolution block is modeled according to [1].

In order to perform the Wavelet transform with the CNN chip, the relationship templates are:

$$A = \begin{pmatrix} 0\,0\,0 \\ 0\,0\,0 \\ 0\,0\,0 \end{pmatrix} \tag{6}$$

$$Bhh_f = \begin{pmatrix} 0 & 0 & 0 \\ 0 & 0.5 & 0.5 \\ 0 & 0.5 & 0.5 \end{pmatrix} Bhg_f = \begin{pmatrix} 0 & 0 & 0 \\ 0 & -0.5 & 0.5 \\ 0 & -0.5 & 0.5 \end{pmatrix} \tag{7}$$

$$Bgh_f = \begin{pmatrix} 0 & 0 & 0 \\ 0 & -0.5 & -0.5 \\ 0 & 0.5 & 0.5 \end{pmatrix} Bgg_f = \begin{pmatrix} 0 & 0 & 0 \\ 0 & 0.5 & -0.5 \\ 0 & -0.5 & 0.5 \end{pmatrix} \tag{8}$$

When reconstruction is performed, conditions stated for direct transform are preserved. We only change the B templates [5] which will be;

$$Bhh_i = \begin{pmatrix} 0.5 & 0.5 & 0 \\ 0.5 & 0.5 & 0 \\ 0 & 0 & 0 \end{pmatrix} Bhg_i = \begin{pmatrix} 0.5 & -0.5 & 0 \\ 0.5 & -0.5 & 0 \\ 0 & 0 & 0 \end{pmatrix} \tag{9}$$

$$Bgh_i = \begin{pmatrix} 0.5 & 0.5 & 0 \\ -0.5 & -0.5 & 0 \\ 0 & 0 & 0 \end{pmatrix} Bgg_i = \begin{pmatrix} 0.5 & -0.5 & 0 \\ -0.5 & 0.5 & 0 \\ 0 & 0 & 0 \end{pmatrix} \tag{10}$$

To implement the wavelet coder in a CNN computer we exploit the ability that CNN-UM has to perform in real time 2-D convolution operations. The 2-D convolutions are time consuming when they are carried out in a digital or programming environment; the circuitry implemented in CNN chip and its analog nature permits to overcome this handicap. In this way, the CNN computer can improve a wavelet coder, and almost any other processes, which involves analog operations.

5   Simulation Results

Extensive simulations with a number of images for the packet wavelet transform implementation in CNN were carried out. The results obtained for a 512*512 image is shown below. The values of the Peak Signal to Noise Ratio and the energy level of each sub-band during each level of decomposition are recorded for each image. The compression ratio achieved is equal to 1:4. The concept of packet wavelet transform in implemented in such a way that, at each iteration the energy of the four sub-bands are calculated. The sub-band with maximum energy

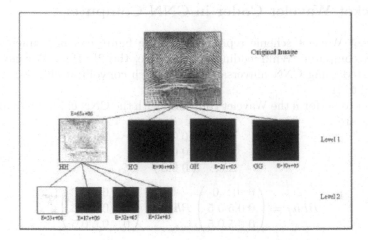

Fig. 5. Decompostion of Finger Print Image Based on Packet Wavelets

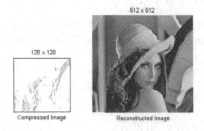

Fig. 6. The Compressed and Reconstructed Images

is now taken out and decomposition is performed on that particular sub-band again. For example the Barbara test image contains a narrow band component at high frequencies that comes from the tablecloth and striped clothing. Finger-print images contain similar narrow band high frequency components. To get the reconstructed image, convolution is performed at each compression level and the tempering matrices for the same are given same. Similarly, the matrices used for deconvolution are also shown. Fig. 5 shows the results obtained using fingerprint are done for the case of 2D-Haar wavelet transform image.

Simulations can be extended to other class of wavelets besides Haar wavelet. However, in real world, reported are limited by neighborhood ratio, it means tem-plates are limited. Such limitation is due to physical restrictions when the VLSI is constructed. Image size is not a limitation; we can also start with 1024 × 1024, although CNN-UM is limited to operate over an n × n cell array, image can be divided in n × n pixel size regions as necessary and boundary conditions deter-mined by adjacent blocks. There is no increase in error value. The results we have obtained for 512 × 512 holds good for 1024 × 1024, in terms of PSNR value. We have used traditional way of implementing wavelets, which uses four filter

Table 1. Energy at Four Bands for Different Test Images

TEST IMAGE for EXTRACTION OF PSNR VALUES	PSNR
Barbara	29.40
Lena	25.32
Desert	29.76
Boston City	29.40
Boat	25.95
Fingerprints	32.26

Table 2. Energy at Four Bands for Different Test Images

TEST IMAGE	ENERGY EHH	ENERGY EHG	ENERGY EGH	ENERGY EGG
Barbara	2.9e+09	3.4864e+07	5.8e+06	5.88e+06
Lena	4.62e+09	7384038	345094	1051472
Desert	7.6e+09	1.2e+06	4.04e+06	3.5e+05
Boston City	619963602	3974763	3974763	108336
Boat	4.6e+09	1049904	4734126	779104
Fingerprints	65e+06	98e+05	21e+05	10e+05

banks, incase we go for additional filter; there is no change in error value, yet it requires additional hardware.

6 Conclusion

We have presented the packet wavelet based image coding scheme implemented with the Cellular Neural Network architecture. The idea is to exploit CNN characteristics along with the frequency adaptive nature of the packet wavelet transform. From the simulations results and image reconstruction on various test images, we can say that high compression rates and high visual quality are possible for images that contain narrow band frequency components.

References

1. Chua, L.O., Yang, L.: Cellular Neural Networks: Theory. IEEE Transactions on Circuits and Systems 35, 1257–1272 (1988)
2. Petrou, M., Garcia, P.: Image processing, Dealing with textures. Wiley Edition 78 (2006)
3. Mallat, S.G.: Theory for Multi resolution signal decomposition: The wavelet representation. IEEE Trans. on PAMI 11(7), 674–693 (1989)
4. Cruz, J.M., Chua, L.O.: A 16 x 16 Cellular Neural Network Chip: the first complete single-chip Dynamic computer Array with distributed Memory and Grayscale I/O. Analog Integrated Circuits and Signal Processing 15(3), 227–228 (1998)
5. Liñan, G., Domínguez-Castro, R., Espejo, S., Rodríguez-Vázquez, A.: CNNUN3 User Guide. Inst.de Microelectrónica de Sevilla (1999)

6. Espejo, S., Domínguez-Castro, R., Liñan, G., Rodríguez-Vázquez, A.: 64x64 CNN Universal Chip with Analog and Digital I/O. In: IEEE 1998 Int.Conf on Electronic Circuits and Systems (1998)
7. Peter Venetianer, L., Roska, T.: Image Compression by CNN. Hungarian Academy of Sciences Computer and Automation Institute (MTA- SzTAKI) 13 (1995)
8. Peter Venetianer, L., Verblin, F., Roska, T., Chua, L.O.: Analogic CNN Algorithms for Some Image Compression and Restoration Tasks. IEEE Transactions on Circuits and Systems 78, 315–333 (1982)
9. Mrak, M., Grgic, S., Grgic, M.: Picture Quality Measurement in Image Compression Systems. In: Proc. EUROCON, Ljublijana, Slovenia (2003)

Reduced RBF Centers Based Multi-user Detection in DS-CDMA Systems[*]

Jungsik Lee[1], Ravi Sankar[2], and Jaejeong Hwang[1]

[1] School of Electronic and Information Engineering, Kunsan National University
Kunsan, Chonbuk 573-701, Korea
{leejs,hwang}@kunsan.ac.kr
[2] Department of Electrical Engineering,
University of South Florida Tampa, FL 33620, USA
sankar@eng.usf.edu

Abstract. The major goal of this paper is to develop a practically implemental radial basis function neural network based multi-user detector for direct sequence code division in multiple access systems. This work is expected to provide an efficient solution by quickly setting up the proper number of radial basis function centers and their locations required in training. The basic idea in this research is to select all the possible radial basis function centers by using supervised k-means clustering technique, select the only centers which locate near seemingly decision boundary, and reduce them further by grouping some of the centers adjacent to each other. Therefore, it reduces the computational burden for finding the proper number of radial basis function centers and their locations in the existing radial basis function based multi-user detector, and ultimately, make its implementation practical.

1 Introduction

A critical issue for future wireless communications systems is the selection of proper multiple access techniques for reliable and affordable communication, anywhere and anytime. One type of wireless technology which has become popular over the last few years is the direct-sequences code division multiple access (DS-CDMA). DS-CDMA uses spread spectrum modulation so that the narrow band signals of each user look like low power wideband noise to all other users. In a DS-CDMA system, the objective of the receiver is to detect the transmitted information bits of one (at mobile-end) or many (at base station) users.

1.1 Background

A variety of MUD has been proposed for DS-CDMA systems. Generally, the linear minimum mean square error (MMSE) MUD is widely used, as it is computationally very simple and can readily be implemented using standard adaptive

---

[*] This work was financially supported by the Kunsan National University's Long-term Overseas Research Program for Faculty Member in the year 2004.

M.S. Szczuka et al. (Eds.): ICHIT 2006, LNAI 4413, pp. 249–257, 2007.

filter techniques [1-3]. The conventional linear detectors, however, fail to achieve good performance when channel suffers from high levels of additive noise or highly nonlinear distortion, or when the signal-to-noise ratio is poor. The linear detector can only work when the underlying noise-free signal classes are linearly separable with the introduction of proper channel delays, where the channel is assumed to be stationary. In reality, the mobile channels are going to be non-stationary where it is hard to determine the proper channel delay that varies with time. If proper channel delay is not introduced in linear MUD, the signal classes from the channel output states will be nonlinearly separable.

In order to get around this problem, neural network technology has been considered in implementing MUD, because it has the capability of recovering the originally transmitted signals from nonlinear decision boundary cases [4-8]. In fact, neural networks have received much attention from a variety of fields, especially for telecommunication systems, because of its characteristics, such as inherent parallelism, noise immunity, knowledge storage, adaptability, and pattern classification capability [4-13].

Aazhang et al. [4] first reported a study of multi-layer perceptrons (MLP) in CDMA systems, and showed that its performance is close to that of the optimum receiver in both synchronous and asynchronous Gaussian channels. Although the simulation results proved that back-propagation learning rule outperforms the conventional detectors, it still leaves a lot of difficulties, such as long training time, performance sensitivity over network parameters including initial weights, and finding the proper number of hidden layer and hidden nodes. In [5], it is shown that the energy function of the recurrent neural network is identical to the likelihood function encountered in MUD. The dynamics of the network are geared toward minimization of a energy function. The performance of the receiver is near optimum; however, the receiver is non-adaptive.

For the last decade, radial basis functions (RBF) neural network have been the promising candidate for the application to various telecommunication fields, including channel equalization and detection [6],[7],[10-13]. Mitra and poor [6] applied a RBF network to the MUD problem. The simulation results show that the RBF based MUD is its intimate link with the optimal one-shot detector, and its training times are better and more predictable than the MLP. However, the RBF based MUD obviously requires more RBF centers, when both channel order and the number of users increase. Eventually, it leads to computational complexity. Thus, it becomes necessary to determine the proper number of RBF centers for the practically implemental RBF based MUD. In [15], Chen et al. employed support vector machine for MUD. It still has difficulty of finding proper number of support vectors or reducing them.

1.2   RBF Neural Network

Originally, RBF network was developed for data interpolation in multi-dimensional space [14]. The RBF is becoming an increasingly popular neural network with diverse application and is probably the main competitor to the MLP. Much of the

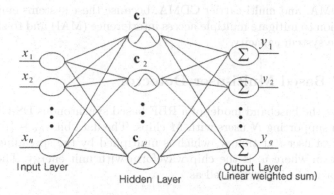

Fig. 1. The Structure of Radial Basis Function Neural Network

inspiration for RBF networks has come from the traditional statistical pattern classification techniques.

A RBF networks is a three-layer network whose output nodes form a linear combination of the basis (or Kernel) functions computed by the hidden layer nodes. The basis functions in the hidden layer produce a localized response to input stimuli. A diagram of a radial basis function network is shown in Figure 1.

1.3 Previous Neural Network Based MUD

As mentioned above, neural network based MUD has shown great promises over linear ones, such as adaptability for nonlinear separable problem and better error performance. However, MLP based MUD requires long training time and risks getting stuck to the local minima due to the initial training weights [4]. On the other hand, the RBF based MUD has some advantages over MLP based MUD, such as structural simplicity, fast training, and better error performance. It, however, still results in higher computational complexity, when both channel order and the number of users go high; more RBF centers are required as the channel order increases [6],[7]. This increase of centers will lead to the long training. Furthermore, the optimal number of RBF centers and the position of RBF centers should be determined before training.

1.4 The Proposed Techniques

The main purpose of this paper is to develop a technique for quickly determining the number of RBF centers and their location, and to apply the RBF neural network with the proposed techniques to the multi-user detection problem. The basic idea behind the proposed techniques is to select the centers which locate near the decision boundary. Furthermore, the selected RBF centers will be averaged especially when the distribution of RBF centers are dense.

This technique will also be applicable to various 3G/4G wireless cellular communication systems using multiple access schemes such as DS-CDMA,

OFDM-CDMA, and multi-carrier CDMA, because these systems employ multi-user detection to mitigate multiple access interference (MAI) and to significantly increase the system capacity.

2   RBF Based Multi-user Detector

Fig. 2 shows the baseband model of a RBF based synchronous DS-CDMA communication supporting N users with M chips. The data bit $s_{i,k} \in \{\pm 1\}$ denotes the symbol of user i at time k, which is multiplied by the spreading, or signature waveform where u_i is the chip waveform with unit energy. The signature sequence for user i is represented as

$$\mathbf{u}_i = [u_{i,1}, \cdots, u_{i,M}]^T \tag{1}$$

and the channel impulse response is

$$H(z) = h_0 + h_1 z^{-1} + \cdots + h_q z^{-q} \tag{2}$$

where q denotes the channel order. The baseband model for received signal sampled at chip rate is represented as [16]

$$\mathbf{r}_k = \begin{bmatrix} h_0\ h_1 \cdots h_q & & & \\ & h_0\ h_1 \cdots h_q & & \\ & & \cdots & \\ & & & \cdot \\ & & & h_0\ h_1\ h_q \end{bmatrix} \begin{bmatrix} \mathbf{UA} & 0 & \cdots & 0 \\ 0 & \mathbf{UA} & \cdot & \cdot \\ & & \mathbf{UA} & \\ \cdot & & \cdot & \\ 0 & \cdots & 0 & \mathbf{UA} \end{bmatrix} \begin{bmatrix} \mathbf{s}_k \\ \mathbf{s}_{k-1} \\ \vdots \\ \mathbf{s}_{k-P+1} \end{bmatrix} + \mathbf{n}_k$$

$$= \hat{\mathbf{r}}_k + \mathbf{n}_k \tag{3}$$

where the user symbol vector $\mathbf{S}_k = [s_{1,k}, s_{2,k}, \ldots, s_{N,k}]^T$, the white Gaussian noise vector $\mathbf{n}_k = [n_{1,k}, \ldots, n_{M,k}]^T$, $\hat{\mathbf{r}}_k$ denotes the noise free received signal. The first, second, and third part of $\mathbf{r}_k$ are $M \times PM$ channel impulse response matrix, $PM \times PN$, and $PM \times 1$, respectively. Thus, the $\mathbf{r}_k$ is the $M \times 1$ vector. $\mathbf{U} = [\mathbf{u}_1, \ldots, \mathbf{u}_N]$ denotes the normalized user code matrix, and the diagonal user signal amplitude matrix is given by $\mathbf{A} = diag\{A_1, \ldots, A_N\}$. The channel inter-symbol interference span P depends on the channel order q and the chip sequence length M [16]: $P = 1$ for $q = 0$, $P = 2$ for $0 < q \leq M - 1$, $P = 3$ for $M - 1 < q \leq 2M - 1$, and so on.

2.1   The Adaptive Training of RBF MUD

Considering the third part in (3), the user symbol vectors, $\mathbf{S}_k$, the number of user, N and the number of interference span, P, there are 2^{NP} combinations of the channel input sequence. Here, $\mathbf{S}_k$ is represented as

$$\mathbf{S}_k = [s_{1,k}, \ldots, s_{N,k}, s_{1,k-1}, \ldots, s_{N,k-1}, \ldots, s_{1,k-P+1}, s_{2,k-P+1}, \ldots, s_{N,k-P+1}]^T \tag{4}$$

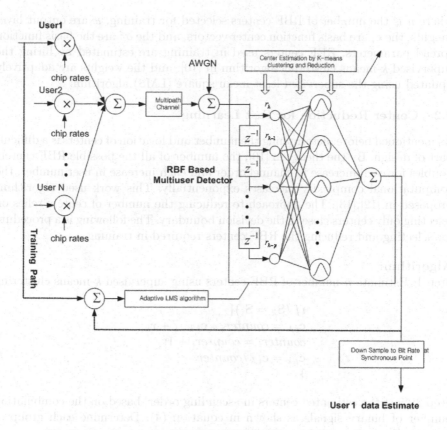

Fig. 2. The Structure of Radial Basis Function Multiuser Detector

This produces 2^{NP} values of the noise-free channel output vector

$$\mathbf{r}_k = [r_k, r_{k-1}, \ldots, r_{k-M+1}]^T \tag{5}$$

These vectors will be referred to as the desired channel states (used as RBF centers), and they can be partitioned into two classes according to the corresponding value in $\mathbf{S}_k$, depending on which user is considered in making decision (here no channel delay is assumed)

$$\begin{aligned} R_i^+ &= \{\hat{r}_k | s_{i,k} = 1\}, \\ R_i^- &= \{\hat{r}_k | s_{i,k} = -1\}. \end{aligned} \tag{6}$$

Once all the channel output states and corresponding desired state are determined, these values can be used as training pairs (input–output) for training RBF based multi-user detector. The output response of an RBF network is a mapping F

$$F(\mathbf{r}) = \sum_{i=1}^{n} w_i \exp\left(\|\mathbf{r} - \mathbf{c}_i\|^2 / 2\sigma^2\right) \tag{7}$$

where n is the number of RBF centers selected for training, w_i are output layer weights, the $\mathbf{c}_i$ are basis function center vectors, and the σ^2 are the basis function spread parameters. RBF centers used in training are estimated by hiring the supervised k-means clustering algorithm in [10], and the weights are adaptively updated using the supervised least mean square (LMS) algorithm.

2.2   Center Reduction for Fast Learning

As mentioned before, determining the number and location of centers is a difficult part of design. By the theory in [10], the number of all the possible RBF centers doubles for each increase in channel order and each increase in user number, the computational complexity increases exponentially. This work used algorithms proposed in [12],[13]. The approach to reducing the number of centers relies on selecting only centers close to the decision boundary. The following is a procedure for selecting and reducing the RBF centers required in training.

Algorithm:
Step 1: Estimate n number of RBF centers using supervised k-means clustering

$$if\ (\mathbf{S}_k = \mathbf{S}_i)\{$$
$$\mathbf{c}_{i,k} = counter_i * \mathbf{c}_{i,k-1} + \mathbf{r}_k;$$
$$counter_i = counter_i + 1;$$
$$\mathbf{c}_{i,k} = \mathbf{c}_{i,k}/counter_i$$
$$\}$$

Step 2: Sort the estimated centers in ascending order, based on the combination number of binary signals as shown in equation (4). Determine each group of centers, $Center^j, j = 1, 2, \ldots, 2^q$.

Step 3: Find the center category J, for which the maximum value of each center group, $Center^j$, is smaller than the values of other center groups.

$$J = \arg\left[\text{Min } C_{\max}^j\right], j = 1, 2, \ldots, 2^q$$

where $C_{\max}^j$ represents the maximum value of center among all the centers in $Center^j$. This step is for selecting only the centers which are close to decision boundary.

Step 4: Find all the centers in $Center^J$

Step 5: Sort the selected centers in ascending order based on the combination number of the associated binary signals, and set $\mathbf{c}_l^J, l = 1, 2, \ldots, L$, as a set of ordered centers, where $L = M/2^q = 2^{NP-q}$.

Step 6: Find all the average distance (AD) between the +1 centers and −1 centers through step below; here +1 centers stands for the center whose corresponding received symbol of ith user, $\mathbf{S}_{j,k}$ is equal to 1, while −1 centers stands for the center whose corresponding received symbol of ith user $\mathbf{S}_{j,k}$ is equal to −1.

$$for\ (l = 1;\ \ l \leq L/4;\ \ + + l)\{$$
$$AD = AD + fabs\left(c^l - c^{2^q-(l-1)}\right)$$
$$\}$$
$$AD = AD/(L/4)$$

Step 7: Check if $(AD \geq \gamma)$, where γ is a distance parameter in the range $0.5 \leq \gamma \leq 1.0$ used to keep the centers properly separated without overlapping (severe intersymbol interference causes the regions containing $+1$ and -1 centers to overlap). If not, stop (all the centers in category J will be used for network training).

Step 8: To deal with the channel, whose order is high enough to make the distribution of channel output states very dense, an additional step is taken. The method is to group 2^ρ numbers of consecutive centers, after sorting the centers in step 2 based on their combinations of user symbol signals as shown in equation (4), where ρ is arbitrary number that could increase or decrease due to the status of channel output distribution.

3 Simulation Studies

Simulation studies were performed to compare RBF based MUD with and without reduction in the number of RBF centers. For the purpose of showing that multiuser detection can be regarded as a classification problem, a very simple two user system with 2 chips per symbol was considered. The chip sequences of the two users were set as $(-1, -1)$ and $(-1, 1)$, respectively. The following are channel impulse responses used in simulation

$$H_1(z) = 1 + 0.4z^{-1}$$
$$H_2(z) = 0.8 + 0.5z^{-1} + 0.3z^{-2}. \tag{8}$$

The two users are assumed to have equal signal power. Simulation works consist of some procedures. The first is to estimate both the RBF centers and noise variances using supervised k-means clustering [10]. The next one is to select only the estimated centers which are close to decision boundary [13]. The center spread value was set to $2\sigma^2$ where σ^2 is the same as estimated noise variance. Once the RBF centers and center spreads are determined, weights updating begins using adaptive LMS algorithm. The learning rate used in training was between 0.01 and 0.05. The number of training samples was 20,000. The bit error rate (BER) performance was conducted with 100,000 inputs with Gaussian noise. The estimated centers shown in Fig. 3(a) are well separated, and the number of centers was reduced from 16 (only 12 points are shown in graph, because some of centers have same values) to 4 using a technique in [13]. As shown in Fig. 4(a), the error rate performance of an RBF based MUD with reduction in centers compared favorably with the RBF based MUD without reduction in centers, and better than a linear MUD. Fig. 3(b) illustrates both the selected centers and averaged centers. Fig. 4(b) shows that RBF based MUD with reduction in centers performed as well as the RBF based MUD without center reduction.

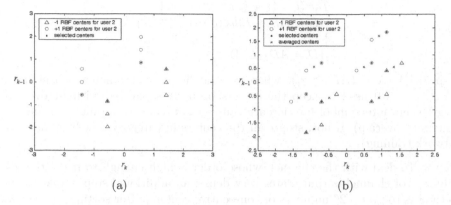

(a)   (b)

Fig. 3. The distribution of selected centers (a) $H_1(z) = 1 + 0.4z^{-1}$, (b) $H_2(z) = 0.8 + 0.5z^{-1} + 0.3z^{-2}$

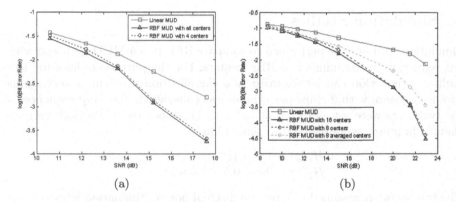

(a)   (b)

Fig. 4. Error rate performance comparing the RBF based MUD, with and without reduction in centers, to Linear MUD (a) $H_1(z) = 1 + 0.4z^{-1}$, (b) $H_2(z) = 0.8 + 0.5z^{-1} + 0.3z^{-2}$

4 Conclusions

The RBF based multiuser detector (MUD) described in this paper was developed primarily for overcoming the difficulties of using a large number of centers in RBF based MUD, especially when channel order and user number go to high.

RBF based MUD with and without reduction in the number of centers performed better than the linear multi-user detector, and error rate performance of the RBF based MUD with reduction in the number of centers was comparable, in most cases, to performance with the full number of centers. These improvements in RBF based MUD design could make its implementation more practical.

Research has been continuing into more complex cases with higher channel order, many users, and long chip sequences. It is expected that the results of this study can be extended to various 3G/4G wireless cellular communication systems using multiple access schemes such as DS-CDMA, OFDM-CDMA, and multi-carrier CDMA.

References

1. Madhow, U., Honig, M.L.: MMSE Interference Suppression for Direct-Sequence Spread –Spectrum CDMA. IEEE Transactions on Communications 42, 3178–3188 (1994)
2. Miller, S.L.: An Adaptive Direct-Sequence Code-Division Multiple-Access Receiver for Multiuser Interference Rejection. IEEE Transactions on Communications 43, 1746–1755 (1995)
3. Poor, H.V., Verdu, S.: Probability of Errors in MMSE Multiuser Detection. IEEE Transactions on Information Theory 43, 858–871 (1997)
4. Aazhang, B., Paris, B.P., Orsak, G.C.: Neural Networks for Multiuser Detection in Code-Division Multiple-Access Channels. IEEE Transactions on Communications 40, 1212–1222 (1992)
5. Miyajima, T., Hasegawa, T., Haneishi, M.: On the Multiuser Detection using a Neural Network in Code-Division Multiple-Access Communications. IEICE Transactions on Communications E76-B, 7–11 (1993)
6. Mitra, U., Poor, H.V.: Neural Network Techniques for Adaptive Multiuser Demodulation. IEEE Journal on Selected Areas in Communications 12, 1460–1470 (1994)
7. Cruickshank, D.G.M.: Radial Basis Function Receivers for DS-CDMA. IEE Electronic Letters 32, 188–190 (1996)
8. Matyjas, J.D., Psaromiligkos, I.N., Batalama, S.N., Medley, M.J.: Fast Converging Minimum Probability of Error Neural Network Receivers for DS-CDMA Communications. IEEE Transactions on Neural Networks 15, 445–454 (2004)
9. Gibson, G.J., Siu, S., Cowan, C.F.N.: Application of Multilayer Perceptrons as Adaptive Channel Equalizers. In: IEEE International Conference on Acoustics, Speech, Signal Processing, pp. 1183–1186 (1989)
10. Chen, S., Mulgrew, B., Grant, P.M.: A Clustering Technique for Digital Communication Channel Equalization using Radial Basis Function Networks. IEEE Transactions on Neural Networks 4, 570–579 (1993)
11. Lee, J., Beach, C.D., Tepedelenlioglu, N.: Channel Equalization using Radial Basis Function Network. IEEE International Conference on Acoustics, Speech, Signal Processing 3, 1719–1722 (1996)
12. Lee, J., Beach, C.D., Tepedelenlioglu, N.: Channel Equalization using Radial Basis Function Network. In: IEEE International Conference on Neural Networks, vol. 4, pp. 1924–1928 (1996)
13. Lee, J., Beach, C.D., Tepedelenlioglu, N.: A Practical Radial Basis Function Equalizer. IEEE Transactions on Neural Networks 10, 450–455 (1999)
14. Powell, M.J.D.: Radial Basis Functions for Multivariable Interpolation: a Review. Algorithm for Approximation, pp. 143–167 (1987)
15. Chen, S., Samingan, A.K., Hanzo, L.: Support Vector Machine Multiuser Receiver for DS-CDMA Signals in Multipath Channels. IEEE Transactions on Neural Networks 12, 604–610 (2001)
16. Chen, S., Samingan, A.K., Hanzo, L.: Adaptive Near Minimum Error Rate Training for Neural Networks with Application to Multiuser Detection in CDMA Communication Systems. Signal Processing 85, 1435–1448 (2005)

Approximate Life Cycle Assessment of Product Concepts Using a Hybrid Genetic Algorithm and Neural Network Approach

Kwang-Kyu Seo[1] and Won-Ki Kim[2]

[1] Department of Industrial Information and Systems Engineering, Sangmyung University, San 98-20, Anso-Dong, Chonan, Chungnam 330-720, Korea
[2] Department of Computer, Information and Telecommunication Engineering, Graduate School, Sangmyung University, 7 Hongji-Dong, Jongno-Gu, Seoul 110-743, Korea
{kwangkyu,wkkim}@smu.ac.kr

Abstract. Environmental impact assessment of products has been a key area of research and development for sustainable product development. Many companies copy these trends and they consider environmental criteria into the product design process. Life Cycle Assessment (LCA) is used to support the decision-making for product design and the best alternative can be selected by its estimated environmental impacts and benefits. The need for analytical LCA has resulted in the development of approximate LCA. This paper presents an optimization strategy for approximate LCA using a hybrid approach which incorporate genetic algorithms (GAs) and neural networks (NNs). In this study, GAs are employed to select feature subsets to eliminate irrelevant factors and determine the number of hidden nodes and processing elements. In addition, GAs will optimize the connection weights between layers of NN simultaneously. Experimental results show that a hybrid GA and NN approach outperforms the conventional backpropagation neural network and verify the effectiveness of the proposed approach.

1 Introduction

Environmental impact assessment of products is an important and widely studied issue since it can have significant impact on making a decision for environmentally conscious design and development. Traditionally, manufacturers mainly focus on how to reduce the cost that a company spends for materials acquisition, production, and logistics, but due to widespread awareness of global environment problems and environmental legislative measures, companies also should take environmental considerations into their decision-making process of product development. These trends are driving many companies to consider environmental impacts during product development. Product designers are challenged with questions of what and how to consider environmental issues in relation to the products they are developing. In particular, it is quite relevant to understand how design changes can affect the environmental performance of product concepts in the early design process.

M.S. Szczuka et al. (Eds.): ICHIT 2006, LNAI 4413, pp. 258–268, 2007.

Life Cycle Assessment (LCA) is now the most sophisticated tools to consider and quantify the consumption of resources and the environmental impacts associated with a product or process [1]. By considering the entire life cycle and the associated environmental burdens, LCA identifies opportunities to improve environmental performance. Conceptually, a detailed LCA is a very useful method, but it may be rather costly, time consuming and sometimes difficult to communicate with non-environmental experts. Further the use of LCA poses some barriers at the conceptual stage of product development, where ideas are diverse and numerous, details are very scarce, and the environmental data for the assessment is short. This is unfortunate because the early phases of the design process are widely believed to be the most influential in defining the LCA of products. Therefore, a new methodology to assess the environmental impacts of products is required in early design phase.

Statistical methods and data mining techniques have been used for developing more accurate prediction models. The statistical methods include regression, discriminant analysis, logistic models, factor analysis, etc. The data mining techniques include artificial intelligence (AI) approaches such as decision trees, neural networks (NNs), fuzzy logic, genetic algorithm (GA), etc. In these artificial intelligence (AI) approaches, NNs are powerful tools for prediction problems due to their nonlinear non-parametric adaptive-learning properties. Hybrid models also have advanced with various prediction models [2][3][4]. One of the popular hybrid models used GA. GA has been increasingly applied in conjunction with other AI techniques such as NN and CBR. However, few studies have dealt with integration of GA and NN although there is a great potential for useful applications in approximate LCA area.

This paper proposes a hybrid GA and NN model for an approximate LCA. In this study the GA is employed as an optimization method of relevant feature subsets selection, the determination of the number of hidden layer and processing elements. In addition, the GA globally searches and seeks optimal or near-optimal connection weights of NN to improve the prediction accuracy.

The rests of the paper are organized as follows. Section 2 shows the overview of the proposed method. Section 3 presents a description of the hybrid GA and NN model. Sections 4 describe the procedures of experiments and experimental results are shown in section 5. Finally, section 6 discusses the conclusions and future research issues.

2   Overview of the Proposed Method

In this section, we outline the proposed method. The reasonable environmental impact drivers ($EIDs$) and the meaningful product attributes are introduced and identified for the proposed model. The $EIDs$ stand for environmental impact categories and the product attributes are meaningful to designers during conceptual design. The proposed hybrid model predicts the LCA results of product concepts using the hybrid GA and NN approach with product attributes as inputs and $EIDs$ as outputs.

2.1 Environmental Impact Drivers ($EIDs$) and Product Attributes

In order to estimate the environmental impacts of products for the entire life cycle, $EIDs$ are introduced in this section. $EIDs$ represent the key environmental characteristics that determine the environmental impacts of products and are strongly correlated with the key factors in life cycle inventory. $EIDs$ eventually mean the environ-mental impact categories such as greenhouse effect, acidification, winter/summer smog, eutrophication, ozone depletion, solid material and energy and so on which capture the environmental performance of product concepts [5] and statistically tested for its ability to predict impact categories (see table 1) [6][7].

Table 1. The proposed environmental impact driver (EID) set

Impact category	Key factors	EID
greenhouse effect	carbon dioxide(CO_2)	EID_{green}
ozone layer depletion	chlorofluorocarbons (CFC)	EID_{ozone}
acidification	sulfur dioxide(SO_2), nitrous oxides(NO_x)	$EID_{acid.}$
eutrophication	carbon, nitrogen, phosphorus	$EID_{eutro.}$
summer smog	VOC, nitrous oxides(NO_x)	$EID_{s.smog}$
winter smog	sulfur oxide(SO_2), hydrocarbons, SPM	$EID_{w.smog}$
life-cycle energy	energy	EID_{energy}
solid material waste	material	$EID_{material}$
heavy metals	Cd, Pb	EID_{metal}
carcinogen	PAHs	$EID_{carcin.}$

Product attributes are envisioned to provide a learning interface between environmental experts and designers and can express the design changes. We identify product attributes as feature subsets and they are used as inputs in the hybrid GA and NN model. The product attributes need to be both logically and statistically linked to $EIDs$, and also be readily available during product concept design. The attributes must be sufficient to discriminate between different concepts and be compact so that the demands on the hybrid model are reasonable. Finally, designers must easily understand them. These criteria were used to guide the process of systematically developing a product attribute set.

To achieve these goals, a set of candidate product attributes, based upon the literature and the experience of experts, was investigated [8][9][10][11]. Experts in both product design and environment discussed about the effectiveness of candidate attributes. In practice, product attributes at the conceptual stage are defined as a few, simple, and expressed in a product-specific language. Also, different levels of information are available and used at the early stage of product design, depending on the purpose of the design. After candidate attributes identified, they were grouped for organizational purposes as shown in table 2 and reviewed for conceptual linkages to the $EIDs$ and potential coverage of the entire life cycle. They are specified, ranked, binary or not applicable according

Table 2. Product attribute set as feature subsets grouped for organizational properties

Group name	Associated product attributes
general design properties	durability, degradability
functional properties	mass (mass is represented as the component ratio of 9 materials), volume
manufacturing properties	assemblability, process
operational properties	lifetime, use time, energy source, mode of operation, power consumption, flexibility, upgradeability, serviceability, modularity, additional consumables
end-of-life properties	recyclability, reusability, disassemblability

to their properties such as an appropriate qualitative or quantitative sense or typically rank order concepts.

2.2 A Brief Description of the Proposed Method

Approximate LCA based on the hybrid GA and NN model is a different approach to other LCA methods [12][13][14]. The proposed hybrid model is trained to generalize on characteristics of product concepts using product attributes and corresponding *EIDs* from pre-existing LCA studies. The designer queries the hybrid GA and NN model for approximate LCA with product attributes to

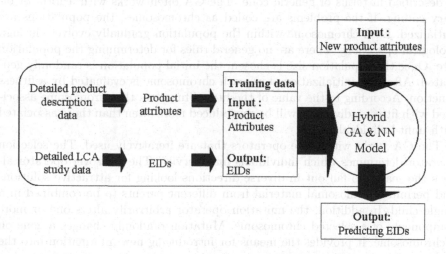

Fig. 1. The structure of the proposed method

quickly obtain an approximate LCA result for a new product concept as shown in Fig. 1.

Designers need to simply provide product attributes of new product concepts to gain LCA predictions. It has the flexibility to learn and grow as new information becomes available, but it does not require the creation of a new model to make as LCA prediction for a new product concept. Also, by supporting the extremely fast comparison of the environmental performance of product concepts, it does not delay product development.

This study proposes a hybrid GA and NN model to improve the prediction accuracy of approximate LCA. The GA part of the hybrid model as an optimization method is proposed to select relevant product attributes and determine the number of hidden nodes and processing elements. In addition, the hybrid model simultaneously searches the connection weights between layers in NN. In other words, the GA glob-ally searches and seeks an optimal or near-optimal NN topology and connection weights for approximate LCA.

3   A Hybrid GA and NN Approach

The GA is a general-purpose evolutionary algorithm that can be used for optimization. When compared to traditional optimization methods, the GA provides heuristic near-optimal solutions. The GA uses a parallel search approach for locating the optimal solution. In the GA, each population member is a potential solution. Recently, the GA has been investigated and shown to be effective in exploring a complex space in an adaptive way, guided by the biological evolution mechanisms of reproduction, crossover, and mutation [15][16].

The first step of the GA is problem representation. The problem must be represented in a suitable form to be handled by the GA. Thus, the problem is described in terms of genetic code. The GA often works with a form of binary coding. If the problems are coded as chromosomes, the populations are initialized. Each chromosome within the population gradually evolves through biological operations. There are no general rules for determining the population size. Once the population size is chosen, the initial population is randomly generated. After the initialization step, each chromosome is evaluated by a fitness function. According to the value of the fitness function, the chromosome associated with fittest individuals will be reproduced more often than those associated with unfit individuals [17].

The GA works with three operators that are iteratively used. The selection operator determines which individuals may survive. The crossover operator allows the search to fan out in diverse directions looking for attractive solutions and permits chromosomal material from different parents to be combined in a single child. In addition, the mutation operator arbitrarily alters one or more components of a selected chromosome. Mutation randomly changes a gene on a chromosome. It provides the means for introducing new information into the population. Finally, the GA tends to converge on a near optimal solution through these operators [18].

The GA is usually employed to improve the performance of AI techniques. For NN, the GA is popularly used to select neural network topology including optimizing relevant feature subsets, and determining the optimal number of hidden layers and processing elements. The feature subsets, the number of hidden layers, and the number of processing elements in hidden layers are the architectural factors of NN to be determined in advance for the modeling process of NN [3]. However, determining these factors is still part of the art. These factors were usually determined by the trial and error approach and the subjectivity of designer. This may lead a locally optimized solution because it cannot guarantee a global optimum.

In this paper, we propose the hybrid GA and NN model to resolve these problems of approximate LCA. In this study, the GA is used for the step of selecting relevant product attributes and optimizing the network topology of NN. And then, the GA search near optimal connection weights in NN. Eventually, the GA globally searches an optimal or near-optimal NN topology and connection weights in the hybrid model.

The overall framework of the hybrid model is shown in Fig. 2.

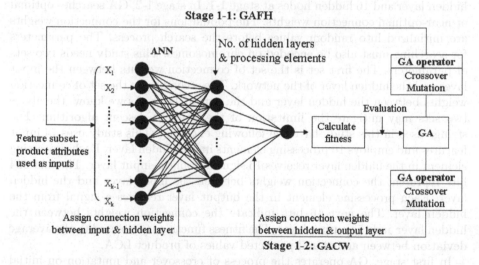

Fig. 2. The structure of the proposed hybrid model

The estimating process of the hybrid model consists of the two stages as follows: In the first stage, the GA searches optimal or near optimal feature subset and determines the number of hidden nodes and processing elements. The GA also fines optimal or near optimal connection weights of NNs. The first stage is divided into the following two sub-stages.

Stage 1-1 (GA for Feature subset selection and determination of the number of Hidden layers and processing elements: GAFH):
The populations, feature subset, the number of hidden nodes and the processing elements are initialized into random values before the search process. The feature subset, the number of hidden nodes and processing elements for searching must be encoded on chromosomes. The chromosomes for feature subset are encoded as binary strings stands for some subset of original feature set. Each bit of one chromosome represents whether the correspond-ing feature is selected or not. 1 in each bit means the corresponding feature is selected whereas 0 means it is not selected. In this study, the chromosomes for the feature subsets such as product attributes are encoded as 28-bit string. The number of hidden layer is encoded 2-bit string and that of hidden nodes is encoded as 8-bit string. The encoded chromosomes are searched to optimize a fitness function. In this study, the fitness function is the average deviation between actual and predicted values of product LCA. The parameters to be searched use only the information about the training data.

Stage 1-2 (GA for Connection Weights: GACW):
After determining the feature subset, the number of hidden nodes and processing elements, connection weights are optimized by GAs. In this study, NN has one hidden layer and 16 hidden nodes at stage 1-1. In stage 1-2, GA searches optimal or near-optimal connection weights. The populations for the connection weights are initialized into random values before the search process. The parameters for searching must also be encoded on chromosomes. This study needs two sets of parameters. The first set is the set of connection weights between the input layer and the hidden layer of the network. The second set is the set of connection weights between the hidden layer and the output layer. As we know, the above two sets may mitigate the limitation of the gradient descent algorithm. The strings used in this study have the following encoding. This study uses 14 input features and employs 16 processing elements in the hidden layer. Each processing element in the hidden layer receives 16 signals from the input layer. The first 224 bits represent the connection weights between the input layer and the hidden layer. Each processing element in the output layer receives a signal from the hidden layer. The next 16 bits indicate the connection weights between the hidden layer and the output layer. The fitness function is also used to the average deviation between actual and predicted values of product LCA.

In first stage, GA operates the process of crossover and mutation on initial chromosomes and iterates until the stopping conditions are satisfied. For the controlling parameters of the GA search, the population size is set to 100 organisms and the crossover and mutation rates are varied to prevent NN from falling into a local minimum. The range of the crossover rate is set between 0.7 and 0.9 while the mutation rate ranges from 0.05 to 0.1. As the stopping condition, only 5,000 trials are permitted.

The second stage, selected product attributes, and optimized or near-optimized NN topology and connection weights are applied to the holdout data. This stage is indispensable to validate the generalizability because NN has the eminent ability of learning the known data. If this stage is not carried out, the model may fall into the problem of over-fitting with the training data.

The proposed hybrid GA and NN algorithms are as follows:

In stage 1-1 (GAFH):

Step 1. Encode chromosomes of features and the number of hidden layers indent and processing elements.

Step 2. Define population size, probability of crossover and mutation.

Step 3. Generate binary coded initial population randomly.

Step 4. While stopping condition is false, do steps 5-13.

Step 5. Decode selected chromosome to obtain the corresponding feature subset and the number of hidden layer and processing elements.

Step 6. Apply feature subset and the number of hidden layer and processing elements to the NN model.

Step 7. Each processing element in the input layer receives an input signal and forwards this signal to all processing elements in the hidden layer.

Step 8. Each processing element in the hidden layer sums its weighted input signals and applies the sigmoid activation function to compute its output signal of the hidden processing element and forwards it to all processing elements in the output layer.

Step 9. Each processing element in the output layer sums its weighted signals from the hidden layer and applies the sigmoid activation function to compute its output signal of the output processing element and computes the difference between the output signal and the target value.

Step 10. Calculate fitness (Fitness function: average predictive accuracy).

Step 11. Select individuals to become parents of the next generation.

Step 12. Create a second generation from the parent pairs(Perform crossover and mutation).

Step 13. Stop the iterative step when the terminal condition is reached.

In stage 1-2 (GACW):

Step 14. Do step 1-13 for optimal or near-optimal connection weights.

4   Research Data and Experiments

With product attributes and *EIDs* defined, the hybrid model for approximate LCA was trained in an effort to validate the proposed model. In this study, the feasibility test was conducted focusing on EID_{energy}. Training data with product attributes and corresponding life cycle energy from true past studies were collected for 300 different products such as various electronic appliances and automobiles and so on.

This study compares GAFH and GACW to the backpropagation neural network (BPNN). BPNN uses the gradient descent algorithm to train the network. This is the conventional approach of previous studies. As mention earlier, GAFH employs the GA to select feature subsets and determine the number of hidden layers and process-ing element of NN. Using the selected feature subsets and the determined the number of hidden layers and processing element of NN, GACW employs the GA to determine the connection weights of NN. The number of hidden layer is one and the processing elements in the hidden layer are fixed at 16 as described in section 3.

About 20 percent (60) of the data is used for holdout and 80 percent (240) for training. The training data is used to search the optimal or near-optimal parameters and is employed to evaluate the fitness function. The holdout data is used to test the results with the data that is not utilized to develop the model.

5   Experimental Results

The GA selects 14 product attributes from all 28 product attributes. In addition, the GA recommends one hidden layer and 16 processing elements in NN. Table 3 presents mean absolute percent error (MAPE) by different models. In table 3, GACW has higher prediction accuracy than BPNN and GAFH for the holdout data. GAFH outperforms BPNN with the gradient descent algorithm. The experimental results show that the prediction accuracy performance of NN is sensitive not only to various feature subsets but also to different topology. Thus, this result shows that simultaneous optimization of the feature subsets and topology is need for the best prediction. In addition, we concluded that NN with the genetically evolved connection weights provides higher prediction results than those of GAFH.

Table 3. Prediction accuracy by conventional NN and hybrid model (MAPE)

BPNN	BPNN	GAFH	GAFH	GAFH+GACW	GAFH+GACW
Training	Holdout	Training	Holdout	Training	Holdout
18.75	26.32	11.82	17.25	10.17	15.43

The *McNemar* tests are used to examine whether the proposed model shows better performance than the conventional BPNN. This test is a nonparametric test for two related samples using the chi-square distribution. This test may be used with nominal data and is particularly useful with 'before-and-after' measurement of the same subjects [19]. The test results show that GAFH and GACW outperform the conventional BPNN at a 1 percent statistical significance level. Although GACW shows higher accuracy than that of GAFH, GACW does not outperforms GAFH at a 5 percent statistical significance level.

6   Conclusions

The lack of analytic LCA motivated the development of approximate LCA based on the hybrid GA and NN model. For approximate LCA, the product attributes and *EIDs* were identified to predict the environmental impacts of products.

In order to resolve the local optimum problem of NN, the hybrid GA and NN model has been proposed. In this study, the GA was used for the step of selecting relevant feature subsets and determining the number of hidden layers and processing elements. In addition, the GA genetically evolved connection weights of NN. The proposed model outperformed the conventional BPNN. The major advantage of the proposed model was simultaneous consideration of efficiency and effectiveness for approximate LCA.

For future work, we intend to apply the GA to other AI techniques to improve the accuracy and efficiency. We believe that there is great potential for further research with the optimization using the GA for other AI techniques. In addition, we also develop a classification method of products according to environmental and product characteristics to support the specialization of approximate LCA for different groups of products. We think that the different classes of products might further improve results without unduly restricting the data sets available for model training.

References

1. Curran, M.A.: Environmental Life-Cycle Assessment. McGraw-Hill, New York (1996)
2. Lee, K.C., Han, I., Kwon, Y.: Hybrid NN models for bankruptcy predictions. Decision Support Systems 18, 63–72 (1996)
3. Kim, K.-J., Han, I.: Genetic algorithms approach to feature discretization in artificial neural networks for the prediction of stock price index. Expert Systems with Applications 19(2), 125–132 (2000)
4. Liang, T., Chandler, J., Hart, I.: Integrating statistical and inductive learning methods for knowledge acquisition. Expert Systems with Applications 1, 391–401 (1990)
5. Park, J.-H, Seo, K.-K.: Approximate Life Cycle Assessment of Product Concepts using Multiple Regression Analysis and Artificial Neural Networks. KSME International 17(12), 1969–1976 (2003)
6. Goedkoop, M., et al.: The Eco-indicator 99: A Damage Oriented Method for Life Cycle Impact Assessment. Pre consultants B.V., Netherlands (1999)
7. SimaPro 4 User's Manual. The Netherlands: PRe Consultants BV (1999)
8. Clark, T., Charter, M.: Eco-design Checklists for Electronic Manufacturers, Systems Integrators, and Suppliers, of Components and Subassemblies (1999), http://www.cfsd.org.uk
9. Rombouts, J.P.: LEADS-. A Knowledge-based System for Ranking DfE-Options. In: Proceedings of the 1998 IEEE International Symposium on Electronics and the Environment, pp. 287–291 (1998)
10. Hubka, V., Eder, W.E.: Engineering Design: General Procedural Model of Engineering Design. Heurista, Zurich, Switzerland (1992)
11. Eisenhard, J.: Product Descriptors for Early Product Development: an Interface between Environmental Expert and Designers. MS thesis of Science in Mechanical Engineering, MIT (2000)
12. Sousa, I., Eisenhard, J., Wallace, D.R.: Approximate Life-Cycle Assessment of Product Concepts using Learning Systems. Journal of Industrial Ecology 4(4), 61–81 (2001)

13. Park, J.-H., Seo, K.-K., Wallace, D.R.: Approximate Life Cycle Assessment of Classified Products using Artificial Neural Network and Statistical Analysis in Conceptual Product Design. In: Second Intl. Symposium on Environmentally Conscious Design and In-verse Manufacturing, pp. 321–326 (2001)
14. Seo, K.-K, Min, S.-H., Yoo, H.-W.: Artificial Neural Network based Life Cycle As-sessment Model of Product Concepts using Product Classification Method. In: Gervasi, O., Gavrilova, M., Kumar, V., Laganà, A., Lee, H.P., Mun, Y., Taniar, D., Tan, C.J.K. (eds.) ICCSA 2005. LNCS, vol. 3483, pp. 458–466. Springer, Heidelberg (2005)
15. Goldberg, D.E.: Genetic algorithms in search, optimization, and machine learning. Addison-Wesley, Reading, MA (1989)
16. Adeli, H., Hung, S.: Machine learning: Neural networks, genetic algorithms, and fuzzy systems. Wiley, New York (1995)
17. Davis, L.: Genetic algorithms and financial applications. In: Deboeck, G.J. (ed.) Trading on the edge, pp. 133–147. Wiley, New York (1994)
18. Wong, F., Tan, C.: Hybrid neural, genetic, and fuzzy systems. In: Deboeck, G.J. (ed.) Trading on the edge, pp. 243–261. Wiley, New York (1994)
19. Cooper, D.R., Emory, C.W.: Business research methods. Irwin, Chicago, IL (1995)

A Solution for Bi-level Network Design Problem Through Nash Genetic Algorithm

Jong Ryul Kim, Jung Bok Jo, and Hwang Kyu Yang

Division of Computer and Information Engineering, Dongseo University,
San 69-1, Churye-2-dong, Sasang-ku, Busan, 617-716, Republic of Korea
xmaskjr@gdsu.dongseo.ac.kr, {jobok,hkyang88}@dongseo.ac.kr

Abstract. This paper presents a Nash genetic algorithm (Nash GA) as a solution for a network design problem, formulated as a bi-level programming model and designs a backbone topology in a hierarchical Link-State (LS) routing domain. Given that the sound backbone topology structure has a great impact on the overall routing performance in a hierarchical LS domain, the importance of this research is evident. The proposed decision model will find an optimal configuration that consists of backbone router for Backbone Provider (BP), router for Internet Service Provider (ISP), and connection link properly meeting two-pronged engineering goals: i.e., average message delay and connection costs. It is also presumed that there are decision makers for BP and the decision makers for ISP join in the decision making process in order to optimize the own objective function. The experiment results clearly indicates that it is essential to the effective operations of hierarchical LS routing domain to consider not only the engineering aspects but also specific benefits from systematical layout of backbone network, which presents the validity of the decision model and Nash GA.

Keywords: Network Topology Design, Bi-level Programming, Nash Genetic Algorithm (Nash GA), Link-State (LS) routing.

1   Introduction

There has been an explosive growth in internet systems since the 1990s. Network design problems are one of important issues in the building and expansion of computer networks and have attracted many related researchers' attentions. Especially, topology design problems for network systems have been taken attentions by many related researchers, as network size increases. The scalability issues like routing information overflow are emerging as one of the most critical network design and operation problems. To cope with scalability issues, an ISP hiring the LS routing protocols can hierarchically divide the logical configuration of its own network (so-called Autonomous System, AS), thereby partitioning the entire Topology DataBase (TDB) into two tiers: multiple local areas (bottom layer TDBs) and a single backbone (top layer TDB).

In LS routing protocol, routers exchange their piecemeal topological information and construct a map (TDB) representing the overall network topology. In

M.S. Szczuka et al. (Eds.): ICHIT 2006, LNAI 4413, pp. 269–280, 2007.
© Springer-Verlag Berlin Heidelberg 2007

order to maintain the entire connectivity (i.e., inter-area routing), a contiguous backbone connecting local areas should be constructed, i.e., all backbone routers should be connected to other backbone routers through backbone links and ensure direct connection of each local area ([1], [2], [3]). Therefore, the topology of backbone network is extremely important in hierarchically configured LS network. However, this gain comes with some side effects. Setting apart operational inconvenience, hierarchical configuration degrades the routing performance due to limitation of available routes on TDBs. Furthermore, the overall routing performance in a hierarchical LS AS becomes heavily dependent on the backbone performance.

Typically, in the Internet systems, some AS networks are connected with backbone network. Lastly, these network systems are well designed with fiber optic cable, because the requirements from users become increased, due to its potentially limitless capabilities [4]. Considering high cost of the fiber optic cable, it is more desirable that the initial structure of network generally consist of a spanning tree. Therefore, a suitable model of overall network architecture will be provide by connecting backbone routers by means of a backbone (transport) network with high-capacity multiplexed links, and connecting to it all the routers of AS by means of a distribution network.

One of important things designing network systems is finding out the best layout which optimizes some performance measures, such as connecting costs, average message delay, network reliability, and so forth. These performance measures are very important and largely affected by network topology. The optimal network topology design problems can be formulated as combinatorial optimization problems, generally which is difficult to solve with the classical method because that it is exponentially complex with network size, are characterized as a kind of NP-hard combinatorial optimization problem. Many ideas and methods for solving these networks design problems have been proposed and tested. Recently, there is an increasing interesting in applying genetic algorithms (GAs) to the problems related to computer networks. GA-based approaches have taken attention as the tool of solution method for the optimal design of network considering reliability and are often used to solve many real-world problems [5], [6].

Game theory has been improved since Von Neumann laid the mathematical foundation in the late 1920s and is a research field that how to determine the optimal strategy in order to maximize the benefit of the player in the game. Game theory is widely used in many academic fields, such as industrial engineering, economics, and so on. According to game theory, when solving an optimization problem with competing multiple objective functions, Nash game is defined as that during game process, players (decision makers) in the game have their own objective function respectively and each player processes decision making (game) considering given constraints until reaching the balance state (Nash equilibrium or non-cooperative strategy), which isn't able to optimize their own objective function any more.

Recently, game theory has taken attention as solution method for solving the bi-level programming network design problems [7]. Also, we can find the analysis that network optimization problems fundamentally have a game structure in the references [8], [9]. In this paper, we solve the network optimal design problem, formulated as bi-level programming problem, with Nash GA based on the idea of Nash game [10], [11], [12].

The structure of network design problems, considering in this paper, has two kinds of decision maker, i.e., a decision maker for backbone network and decision makers for distribution network, assuming that each decision maker non-cooperatively joins in decision making process to optimize their own objective function. These network design problems can be formulated as bi-level programming model and translated into Nash game. Therefore, in this paper, we employ the proposed Nash GA in order to solve network topology design problems, considering connection cost, average message delay network reliability, which are formulated as bi-level programming model and translated into Nash game. Also, we employ the encoding method that the connection between backbone routers is represented by Prüfer number and the connection between backbone routers and routers of AS is depicted by clustering string which describes distribution of routers into the backbone routers, because it needs only $n + m - 2$ memory for n backbone routers and m routers of AS. Finally from numerical examples, we confirm that the proposed Nash GA can be employed to solve the bi-level programming network design problems.

2   The Bi-level Network Design Problem

The Bi-level programming problem is a mathematical programming problem which is composed of upper-level problem and lower-level problem and can be translated into Nash game, when non-cooperatively optimizing upper-level problem and lower-level problem with each other. In this paper, we assume that bi-level programming problem is made up of the upper-level problem that the decision maker of backbone network minimizes connection cost and the lower-level problem that the decision makers of distribution network minimize average message delay.

In order to formulate the problem, we consider a network that connects n backbone routers and m routers of AS. The communication traffic demands between the routers are given by an $m \times m$ matrix U which is called the router traffic matrix. An element u_{ij} of matrix U represents the traffic from router i to router j. Note that traffic between a pair of routers can vary from heavy traffic to no traffic. Moreover, this traffic can be burst or constant. The traffic peak rate is probably an overestimation of the traffic requirement, since most of the time the actual traffic is far below the peak. On the other hand, taking the average traffic rate as the requirement may yield poor results when heavy traffic has to be forwarded. For our purposes, we shall assume that traffic characteristics are known and summarized in the router traffic matrix U.

Also, we assume the $n \times n$ backbone router topology matrix X which represents the connected appearance of backbone routers. An element x_{ij} is represented

$$x_{1ij} = \begin{cases} 1, & \text{if the backbone routers } i \text{ and } j \text{ are connected,} \\ 0, & \text{otherwise.} \end{cases} \tag{1}$$

We further assume that the network is partitioned into n segments (backbone routers or clusters). The routers are distributed over those n backbone routers. The $n \times m$ clustering matrix Y specifies which router connects with which backbone router. Thus,

$$y_{ij} = \begin{cases} 1, & \text{if the router } j \text{ connects with the backbone router } i, \\ 0, & \text{otherwise.} \end{cases} \tag{2}$$

A router can belong only to one backbone router. We define $n \times (n+m)$ matrix S called the spanning tree matrix ($[X\ Y]$) and define a $n \times n$ matrix T called the backbone router traffic matrix. An element t_{ij} of this matrix represents the traffic forwarded from backbone router i to backbone router j. Obviously, $T = Y^T U Y$. A backbone router is a segment with a known capacity. An M/M/1 model [13], [14] is used in this paper to describe a single cluster (segment) behavior. We define the following notations in order to formulate the bi-level programming network topology design problems:

$$a_{ij}^k(X) = \begin{cases} 1, & \text{if traffic from the backbone router } i \text{ to the backbone} \\ & \text{router } j \text{ through the backbone router } k \text{ exists} \\ 0, & \text{otherwise} \end{cases} \tag{3}$$

$$b_{ij}^{(k,l)}(X) = \begin{cases} 1, & \text{if traffic from the backbone router } i \text{ to the backbone} \\ & \text{router } j \text{ passes through an existing link} \\ & \text{connecting backbone routers } k \text{ and } l \text{ exists} \\ 0, & \text{otherwise} \end{cases} \tag{4}$$

Then we can formulate the network design problems represented by bi-level programming problems as follows:

$$\min_{Y} \sum_{i=1}^{n-1} \sum_{j=i+1}^{n} w_{1ij} \cdot x_{ij} + \sum_{i=1}^{n} \sum_{j=1}^{m} w_{2ij} \cdot y_{ij} \tag{5}$$

$$\min_{X} \frac{1}{\Gamma} \left[\sum_{i=1}^{n} \frac{F_i(S)}{C_i - F_i(S)} + \sum_{i=1}^{n} \sum_{j=1}^{n} \beta_{ij} \cdot f_{ij}(S) \right] \tag{6}$$

$$\text{s. t. } R(S) > R_{\min} \tag{7}$$

$$\sum_{i=1}^{n-1} \sum_{j=i+1}^{n} x_{ij} = n - 1 \tag{8}$$

$$\sum_{i=1}^{n} y_{ij} = 1, \; j = 1, 2, \cdots, m \tag{9}$$

$$F_i(S) < C_i, \; i = 1, 2, \cdots, n \tag{10}$$

where $R(S)$ is the network reliability, $R_{\min}$ means the requirement of network reliability, C_i is the traffic capacity of backbone router i, β_{ij} is the delay per bit due to the link between backbone routers i and j, w_{1ij} is the cost of the link between backbone routers i and j, and w_{2ij} is the cost of the link between the backbone router i and the router j of AS. Also, the total offered traffic Γ is represented as follows:

$$\Gamma = \sum_{i=1}^{n} \sum_{j=1}^{n} t_{ij}.$$

The total traffic at backbone router k, $F_k(\boldsymbol{X})$, is represented as follows:

$$F_k(\boldsymbol{X}) = \sum_{i=1}^{n} \sum_{j=1}^{n} t_{ij} \cdot a_{ij}^{k}(\boldsymbol{X}), \quad k = 1, \cdots, n.$$

The total traffic through link (k, l), $f_{kl}(\boldsymbol{X})$, is represented as follows:

$$f_{kl}(\boldsymbol{X}) = \sum_{i=1}^{n} \sum_{j=1}^{n} t_{ij} \cdot b_{ij}^{(k,l)}(\boldsymbol{X}), \qquad k = 1, \cdots, n \ \ l = 1, \cdots, n.$$

3   Nash Genetic Algorithm for Solving Bi-level Network Design

The idea of Nash GA is to bring together genetic algorithms and Nash game in order to cause the genetic algorithm to build the Nash equilibrium. In the Fig. 1, we present how such merging can be achieved with 2 players trying to optimize 2 different objectives.

Let $\boldsymbol{V} = \boldsymbol{XY}$ be the string representing the potential solution for a dual objective optimization problem. Then $\boldsymbol{X}$ denotes the subset of variables handled by Player 1 (P1) and optimized along criterion 1. Similarly $\boldsymbol{Y}$ denotes the subset of variables handled by Player 2 (P2) and optimized along criterion 2. According to Nash theory, P1 optimizes $\boldsymbol{X}$ with respect to the first criterion by modifying $\boldsymbol{X}$ while $\boldsymbol{Y}$ is fixed by P2. Symmetrically, P2 optimizes $\boldsymbol{V}$ with respect to the first criterion by modifying $\boldsymbol{Y}$ while $\boldsymbol{X}$ is fixed by P1. There are two different populations, i.e., P1's optimization task is performed by Population 1 (Pop1) whereas P2's optimization task is performed by Population 2 (Pop2). Let $\boldsymbol{X}_k$ be the best value found by P1 at generation k and $\boldsymbol{Y}_k$ be the best value found by P2 at generation k. At generation k, P1 optimizes $\boldsymbol{X}_k$ while using $\boldsymbol{Y}_{k-1}$ in order to evaluate $\boldsymbol{V}$. P2 also optimizes $\boldsymbol{Y}_k$ while using $\boldsymbol{X}_{k-1}$ in order to evaluate $\boldsymbol{V}$. After the optimization process, P1 sends the best value $\boldsymbol{X}_k$ to P2 who will use it at generation $k+1$. Similarly, P2 sends the best value $\boldsymbol{Y}_k$ to P1 who will use it at generation $k+1$. Nash equilibrium is reached when neither P1 nor P2 can further improve their criteria [11], [12].

3.1   Representation and Initialization

The genetic representation is a kind of data structure which represents the candidate solutions of the problem in coding space. Usually different problems have different data structures or genetic representations.

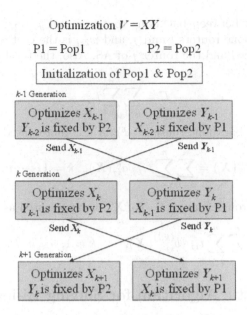

Fig. 1. The Process of Nash GA

We use the following representation model in order to represent active network configurations, i.e., spanning tree configurations: the model is that backbone routers are represented by a Prüfer number and routers in AS are depicted by a clustering string which describes distribution of routers into the backbone routers. We employ an encoding method used with Prüfer number and clustering string.

One of the classical theorems in graphical enumeration is Cayley's theorem there are $k^{(k-2)}$ distinct labeled trees on a complete graph with k nodes. Prüfer provided a constructive proof of Cayley's theorem by establishing an one-to-one correspondence between such trees and the set of all strings of $k-2$ digits [5], [6]. This means that we can use only $k-2$ digits permutation to uniquely represent a tree where each digit is an integer between 1 and k inclusive because that there are always at least two leaf nodes for any tree. This permutation is usually known as the Prüfer number.

We randomly generate the chromosome in the initialization process, i.e., backbone routers are composed of $n-2$ digits (Prüfer number) randomly generated in the range $[1, n]$ and terminals are make up of m digits (clustering string) randomly generated in the range $[1, n]$, which mean how to allocate the routers to backbone routers so that each router belongs to a specific service center. Note that our representation method has the $n+m-2$ size of chromosome.

3.2 Evaluation

In this paper, we calculate the evaluation value of a chromosome as their own objective function for each decision maker.

$$f_1(\boldsymbol{V}_k^i) = f_1(\boldsymbol{V}_{1k}^i, \boldsymbol{V}_{2k-1}^*) \tag{11}$$

$$f_2(\boldsymbol{V}_k^j) = f_1(\boldsymbol{V}_{1k-1}^*, \boldsymbol{V}_{2k}^i) \tag{12}$$

where $\boldsymbol{V}_{1k}^i$ and $\boldsymbol{V}_{2k}^i$ are i-th and j-th chromosome of P1 and P2, respectively. $\boldsymbol{V}_{1k-1}^*$ and $\boldsymbol{V}_{2k-1}^*$ are best chromosome selected by adversary at $k-1$ generation, respectively. $f_1(\cdot)$ and $f_2(\cdot)$ means the objective function of each decision makers.

3.3 Reliability Calculation

Because only spanning tree topologies can be used as active network configurations, the evaluation of the reliability of network can often be accomplished by breaking down the problem into several problems on trees, and a single portion [15].

We consider, as reliability measure, the probability of all operative nodes (backbone routers and routers) being connected. Now we want to calculate the reliability of a spanning tree network assuming the reliability of its elements, nodes and links, are known. Considering the tree to be a rooted tree, we associate a state vector with the root of each sub-tree. The state vector associated with a root node contains all information about that node relevant to our calculation. We then define a set of recursion relations which yield the state vector of a rooted tree given the state of its sub-trees. For sub-trees considering of single nodes the state is obvious. Then we join the rooted sub-trees into larger and larger rooted sub-trees using the recursion relations until the state of the entire network is obtained [15].

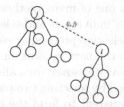

Fig. 2. The Recurrence Relation of Tree

Deriving the recurrence relations is somewhat mechanical. It comes simply from considering the situation depicted in the Fig. 2. We have two sub-trees, one with root i and the other having as its root j. We assume the state of node i and node j are known and we wish to compute the state of j relative to the tree obtained by joining node i into node j with the link (i, j), where node j is the father node of node i.

Then we assume the notations for the procedure tree-based reliability calculation as follows: we have associated with each node i a probability of node failure p_i^f and a probability $p_i^o (= 1 - p_i^f)$ of node being operative. Similarly, for the link (i, j) we have probabilities l_i^f and l_i^o of the link (i, j) failing and being operative respectively. We also define the following state vectors for each sub-trees:

e_i means the probability that all nodes in the sub-tree are failed, o_i means the probability that the set of operative nodes, including the root of the sub-tree, are connected, and r_i means the probability that the root of the sub-tree is failed and the set of operative nodes in the sub-tree is connected.

For the tree with root node 1 and n nodes, we can calculate the reliability of tree, i.e., the probability of all operating nodes communicating as follows:

Procedure: Reliability Calculation

Step 1: Set $r_i = 0$, $o_i = p_i^o$, $e_i = p_i^f$, $i = 1, 2, \cdots, n$. Set $i = n$. Go to step 2.

Step 2: If node j is the father node of node i, using the following recurrence relations, recalculate r_i, o_i, e_i:

$$r'_j = r_j \cdot e_i + r_i \cdot e_j + o_i \cdot e_j$$
$$o'_j = o_i \cdot o_j \cdot l_i^o + o_j \cdot e_i$$
$$e'_j = e_i \cdot e_j$$

Step 3: Set $i = i - 1$. If $i = 1$, go to step 4; otherwise go to step 2.

Step 4: Return $r_1 + o_1 + e_1$.

3.4   Genetic Operations: Selection, Crossover, and Mutation

In a GA, the selection plays a very important role. When we regard the genetic operation as the exploration for the search in solution space, the selection can be thought as the exploitation for the GA to guide the evolutionary process. The selection used here is the method combined with the *tournament* and *elitist* approach, in order to enforce the proposed GA to freely search solution space.

The tournament selection is one of many methods of selection in GAs which runs a tournament among a few individuals and selects the winner (the one with the best fitness). Using elitist selection process, we can keep the best chromosome from the current generation to the next generation.

We employed the multi-point crossover (or called uniform crossover). This type of crossover is accomplished by selecting two parent solutions and randomly taking a component from one parent to form the corresponding component of the offspring. Simple mutation is used in this paper, which simply selects one position at random and random perturbation within the permissive range from 1 to n (the number of backbone routers).

4   Experimental Results

The test problems are experimented to confirm that the proposed Nash GA can be used effectively for bi-level programming network design problems. We will have two problems to see the performance of our proposed method. The experimented device is PC (Windows XP SP2) with Pentium4 3.20GHz CPU and 1GB RAM. Data for problems are referred by reference [14] due to unavailability of real network systems information.

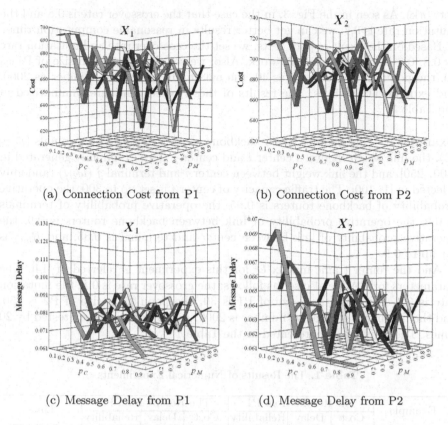

(a) Connection Cost from P1   (b) Connection Cost from P2

(c) Message Delay from P1   (d) Message Delay from P2

Fig. 3. The Results of Pilot Tests

Example 1. This example has 4 backbone routers ($n = 4$), 8 terminals ($m = 8$), the link weight between center i and center j (w_{1ij}) randomly generated in [100, 250], and the link weight between center i and terminal j (w_{2ij}) randomly selected in [1, 100]. The traffic capacity of center i (C_i) is set up to 50, the operative probability of backbone routers is 0.95, the operative probability of terminals is 0.9, the operative probability of link between backbone routers is 0.9, the operative probability of link between center and terminal is 0.85, and $R_{\min}$ is set up to 0.9. And the router traffic matrix U for this experiment is referred in [14]. The parameters for Nash GA are experimented in the following method. Let the population size of backbone network be 20, the population size of distribution networks be 40, and the maximum generation number be 500. The rates of crossover and mutation are changed between 0.1 0.9, respectively. With these setting, the pilot tests are experimented by 10 times. The results of pilot tests are depicted in the Fig. 3, which represent the average objective function value of 10 experiments. Fig. 3 (a) and (c) mean the connection cost and average message delay from P1 (for backbone network). Fig. 3 (b) and (d) also mean the connection cost and average message delay from P2 (for distribution network). As seen in the Fig. 3, in the case that the crossover rate is 0.8 and the mutation rate is 0.7, we can get better results in reasonable computation time.

Based on the result of pilot tests, we set the crossover rate and mutation rate as 0.8, 0.7 respectively, in this work. Also, we set the population size of P1 as 20, the population size of P2 as 40, the maximum generation number as 2000, and experiment 20 times. The results of these experiments are summarized in the Table 1.

Example 2. This example has 6 backbone routers ($n = 6$), 30 terminals ($m = 30$), the link weight between center i and center j (w_{1ij}) randomly generated in [100, 250], and the link weight between center i and terminal j (w_{2ij}) randomly selected in [1, 100]. The traffic capacity of center i is set up to 300, the operative probability of backbone routers is 0.95, the operative probability of terminals is 0.9, the operative probability of link between backbone routers is 0.9, the operative probability of link between center and terminal is 0.85, and R_{min} is set up to 0.9.

And the router traffic matrix U for this experiment is referred in [14]. The parameters for GA are also set as follows: the crossover rate is 0.8, the mutation rate is 0.7, the population size of P1 is 30, the population size of P2 is 150, and the maximum generation number is 2000. And experiments are tested by 20 times. The results are represented in the Table 1.

Table 1. The Results of Numerical Experiments

Example		The Results from P1			The Results from P2			Time
		Cost	Delay	Reliability	Cost	Delay	Reliability	
1	Avg.	551.00	0.06182	0.91441	553.15	0.06104	0.91314	168.4
	Var.	26.38	0.00239	0.00945	25.56	0	0.00897	1.8
	Best	538.00	0.06104	0.91199	538.00	0.06104	0.91199	165.0
	Worst	625.00	0.06104	0.90227	625.00	0.06104	0.90227	171.0
2	Avg.	1307.80	0.03754	0.99958	1320.15	0.03448	0.99950	9638.1
	Var.	33.01	0.01078	0.00059	28.47	0.01064	0.00100	133.3
	Best	1306.00	0.04569	0.99989	1306.00	0.04569	0.99989	9763.0
	Worst	1379.00	0.02446	0.99951	1394.00	0.02218	0.99989	9440.0

As known in the Table 1, we can find that Nash equilibrium from our proposed method converges on the near-optimal of objective function for each decision makers (connection cost for P1 and average message delay for P2), because two kinds of decision makers optimize their own objective function with non-cooperation.

5   Conclusion

In this paper, we propose a Nash GA with encoding method based on Prüfer number and clustering string, in order to solve a bi-level programming network design problems which have two kinds of objective functions and some constraints, and the decision makers non-cooperatively join in a decision making process. So far as we know that the proposed Nash GA is the first try on employing Nash theory to solving a network topology design problems for a hierarchical LS routing domain with multiple objective functions and some constraints, formulated as a bi-level programming problems. From numerical experiments, we confirm that the proposed method can get the satisfactory results in most cases.

In order to testify the potentiality of the proposed method, we took two kinds of numerical examples in this paper and experimented with these examples. From the Fig. 3 and Table 1, we can see that our proposed method is able to effectively get a solution for bi-level programming network design problems in reasonable computation time.

Recent researches for the genetic representation of spanning trees [16], [17] have reported the weak point of Prüfer number encoding method because it has poor locality and heritability. So, our future research will be tried to overcome this shortcoming of the Nash GA employed in this paper and also think up faster algorithm method with parallel computing or distributed computing structure, such as MPI, PVM, and so, on.

References

1. Moy, J.T.: OSPF - Anatomy of an Internet Routing Protocol. Addison Wesley, Reading (1998)
2. Parkhurst, W.R.: Cisco OSPF Command and Configuration Handbook. Cisco Press (2002)
3. Thomas, T., Thomas II, M.: OSPF Network Design Solutions. Cisco Press (2003)
4. Mukherjee, B.: Optical Communication Networks. McGraw-Hill, Inc., New York (1997)
5. Gen, M., Cheng, R.: Genetic Algorithms and Engineering Design. John Wiley & Sons, New York (1997)
6. Gen, M., Cheng, R.: Genetic Algorithms and Engineering Optimization. John Wiley & Sons, New York (2000)
7. Lim, Y.T., Lim, K.W.: Game Theory and Traffic Network Problems. The Journal of Environmental Studies 42, 107–121 (2004) (in Korean)
8. Kim, D.H., Hyun, C.H.: Game Structure of Info-telecommunication Policy and its Dynamics. The Journal of National Basic Information System 2(4), 36–48 (1995) (in Korean)
9. Yun, K.L.: A Game Theoretic Analysis of the Interconnection Pricing in Network Industries, KAIST Graduate School of Management, Dissertation (1997)
10. Wang, J.F., Periaux, J.: Multi-point Optimization using GAs and Nash/Stackelberg Games for High Lift Multi-airfoil Design in Aerodynamics. In: Proc. of the 2001 Congress on Evolutionary Computation, vol. 1, pp. 552–559 (2001)
11. Sim, K.B., Ki, J.Y., Lee, D.W.: Optimization of Multi-objective Function based on the Game Theory and Co-evolutionary Algorithm. Journal of Fuzzy Logic and Intelligent Systems 12(6), 491–496 (2002) (in Korean)
12. Sefrioui, M., Periaux, J.: Nash Genetic Algorithms: Examples and Applications. In: Proc. of the 2000 Congress on Evolutionary Computation, pp. 509–516 (2000)

13. Bertsekas, D., Gallager, R.: Data Networks, 2nd edn. Prentice-Hall, New Jersey (1992)
14. Elbaum, R., Sidi, M.: Topological Design of Local-area Networks using Genetic Algorithms. IEEE/ACM Transactions on Networking 4(5), 766–778 (1996)
15. Kershenbaum, A., Van Slyke, R.: Recursive Analysis of Network Reliability. Networks 3, 81–94 (1973)
16. Raidl, G.R., Julstrom, B.A.: Edge Sets: an Effective Evolutionary Coding of Spanning Trees. IEEE Transaction on Evolutionary Computation 7(3), 225–239 (2003)
17. Gen, M., Cheng, R.: Evolutionary Network Design: Hybrid Genetic Algorithms Approach. International Journal of Computational Intelligence and Applications 3(4), 357–380 (2003)

An Alternative Measure of Public Transport Accessibility Based on Space Syntax

Chulmin Jun*, Jay Hyoun Kwon, Yunsoo Choi, and Impyeong Lee

The University of Seoul, Jeonnong-dong, Dongdaemun-gu, Seoul, Korea
{cmjun,jkwon,choiys,iplee}@uos.ac.kr

Abstract. The local governments of major cities in Korea are giving focus on public transportation to reduce congestion and improve accessibility in city areas. In this regards, the proper measurement of accessibility is now a key policy requirement for reorganizing the public transport network. However, Public transport routing problems are considered to be highly complicated since a multi-mode travel generates different combinations of accessibility. While most of the previous research efforts on measuring transport accessibility are found at zone-levels, an alternative approach at a finer scale such as bus links and stops is presented in this study. We propose a method to compute the optimal route choice of origin-destination pairs and measure the accessibility of the chosen mode combination based on topological configuration. The genetic algorithm is used for the computation of the journey paths, whereas the space syntax theory is used for the accessibility. The resulting accessibilities of bus stops are calibrated by O-D survey data and the proposed process is tested on a CBD of Seoul using the city GIS network data.

1 Introduction

The City of Seoul recently reformed the entire bus system as part of an intensified effort to relieve traffic congestion by encouraging public transportation. Although the reformed system receives relatively positive reputation, it still has some room for improvements in route organization. Especially, the provision and proximity of public transport networks are important factors for leading the people toward public transportation. Therefore, accessibility planning is a key area of public-oriented transportation strategy.

The concept of accessibility based on recent studies is generally categorized into two areas; zone-based and individual-based (e.g., Hanson 2004, Miller 1999, and Kwan 1998). Here, zone-based measures capture the accessibility between zones whereas individual measures take into account individual daily travel pattern with more detailed characteristics such as travel time and means (Berglund 2001). Based on a simpler definition on accessibility such as "a measure of ease of access to opportunities" (Harris 2001), the concept of accessibility of these

---

* Corresponding author.

M.S. Szczuka et al. (Eds.): ICHIT 2006, LNAI 4413, pp. 281–291, 2007.
© Springer-Verlag Berlin Heidelberg 2007

two areas becomes identical and evaluated only at different scales and grains (Litman 2003). Typical accessibility measures take the general form as:

$$A_i = \sum_j O_j d_{ij}^{-1},$$ (1)

where A_i is the accessibility of person i (in case of individual measure) or zone i (in case of zone-based measure), O_j is the number of opportunities at distance j (individual-based) or zone j (zone-based), and d_{ij} is a measure of separation between i and j (Hanson 2004). Regardless of the type or scale of accessibility measure, two components are always considered in the formulation; opportunities at the distance and distance or, in a wide sense, travel cost (Berglund 2001, Geertman and Van Eck 1995). However, if the accessibility is measured at a finer level such as between two bus stops, counting the number of opportunities has less meaning because mostly there is only one stop at the destination in a strict sense. Unlike the movement of free-moving vehicles like private cars, the travel costs in public transport network should include the transfers between modes in addition to the distance or travel time.

In this paper, we present an alternative method to assess individual path-based accessibility of public transport network. The main idea is based on the fact that the more transfers take place, the more difficult or 'the deeper' the mobility becomes making that area evaluated as less advantageous as for accessibility. We employed this '*depth-based*' concept from a theory called space syntax. Space syntax is the technique that has been used to derive the connectivity of urban or architectural spaces (Hillier 1996). The primary principle of space syntax is to model a spatial structure as a set of axial lines and compute spatial indices of a space based on '*the depth*' to other spaces to derive the relationships between different parts of urban or indoor spaces. The resulting index is expressed as the integration of that space which is the degree to which that space is integrated and connected with other spaces in the defined area. Although transport network problems are not included in typical applications of space syntax, one can see the analogy between space syntax's spatial integration and public transport links since both are based on the hierarchical transitions between spaces. In other words, the transfer of vehicles can be considered as a connection node that links two different routes. Based on the analogy, we suggest an algorithm to show how topological accessibility based on depths of public transport routes, rather than their physical distance, can be computed.

In the following sections, we first take a look at the principle of space syntax theory briefly. Then, how the theory is applied to public transport problem is described. The proposed process is tested on a Central Business District (CBD) of Seoul. We compared the resulting depth-only accessibility measure with the weighted values by the O-D trips data. The public transport network data including different types of bus and subways routes are built using GIS data of Seoul and relational database. In the discussion section, we suggested how other travel costs than transfers can be incorporated into the depth-based accessibility measure.

2 Hierarchical Network Connectivity in Space Syntax

In space syntax, when converting the continuous space into a connected set of discrete units, it uses the concept of convex space partitioning or simply axial mapping. The procedure to generate the convex map involves taking a given spatial structure and partitioning it into a set of "fewest and fattest" convex spaces. The convex map, or the axial map can be drawn by laying down the longest strait lines that passes through theses convex spaces (Fig. 1-b). On the other hand, traditional way of abstracting street network follows different procedure. It generally uses center lines of streets. Whenever two center lines intersect each other, an intersection is created (Fig. 1-c). When representing the configured lines as a graph, space syntax represents each line by a node and each intersection as an edge, while in traditional method, the situation is vice versa, that is, an intersection becomes a node and a line connecting two nodes becomes an edge.

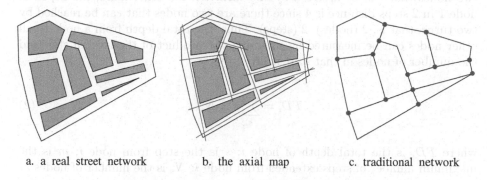

a. a real street network   b. the axial map   c. traditional network

Fig. 1. Comparing the network representation of streets

In the literature of space syntax theory (Bafna 2003), the transition of spaces is described thought a variable called depth, and only the number of turns along a path rather than actual journey length is counted. The cost such as distance or travel time along an edge is not regarded as significant factor in space syntax.

Fig. 2 shows how a meandering streets network is mapped to a graph capturing the hierarchical relationship of component units. Line 2 is accessible from line 1 by one turn, whereas line 4 is accessed by two turns. In other words, the relationship of 2 and 3 is called symmetrical with respect to 1 whereas the relationship of 4 to 1 is asymmetrical. If one were to represent each component with a node and a turn with a link connecting their respective nodes, one could then describe the hierarchy from each node as shown in Fig. 3.

Depth of one node from another can be directly measured by counting the number of steps (or turns) between two nodes. The greater the depth of two nodes, the greater the hierarchical difference between them. The depth of a node (or a street) in certain step distance is defined by the number of nodes distant from a given number of steps to that node. For example, the depth of

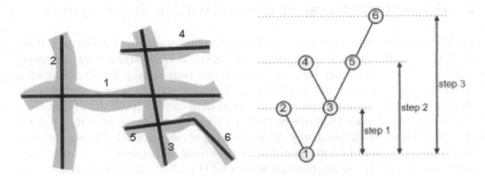

Fig. 2. Streets network in space syntax   **Fig. 3.** Hierarchical structure of *street* 1

node 1 for immediate neighbors (eg. step 1 nodes) in Fig. 3 is 2 since there are two nodes that can be accessed by one turn. On the other hand, the depth of node 1 in 2 steps distance is 4 since there are two nodes that can be reached by two turns, that is, 2 (nodes) 2 (steps). Thus, the total depth from a node to all other nodes can be measured by summing the product of the level of step and the number of nodes in that step as given by:

$$TD_i = \sum_{s}^{m} s \times N_s, \tag{2}$$

where TD_i is the total depth of node i; s is the step from node i; m is the maximum number of steps extended from node i; N_s is the number of nodes at step s.

The mean depth (MD) then is given by the total depth divided by $k-1$, where k is the total number of nodes in the graph (Hillier 1996). Fig. 4 shows extreme cases from node 1 in a network of same number of nodes. Case 'a' contains only neighboring nodes to a node while case 'b' contains nodes that extend to the maximum number of steps. We call the case 'a' as completely symmetrical structure having MD_i of 1 whereas case 'b' is completely asymmetrical structure with MD_i of $k/2$. Therefore, the range of MD is given as $1 \leq MD \leq k/2$, which can be normalized as follows:

$$0 \leq ND_i := \frac{2(MD_i - 1)}{k - 2} \leq 1. \tag{3}$$

Now, the depth from a node in a graph can be represented by normalized depth (ND) ranging from 0 to 1. In space syntax studies, the inverse of ND called *the Integration value* (I), is typically used since the Integration value helps more intuitive interpretation about the depth of a node than ND. That is, higher integration values of nodes indicate that the node is less deep on an average from all other nodes and more integrated in the spatial system.

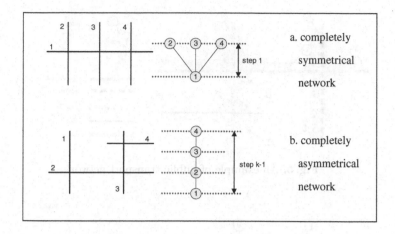

Fig. 4. Spatial depths of two extreme cases

3   Applying to Public Transport Problem

The hierarchical description in space syntax summarized in the previous section can be applied in the analysis of public transport accessibility. Noticing the public transport problem generally entails transfers between vehicles, one can map turns of spaces to transfers between transport modes. The cost in public transportation generally increases as the number of transfers between different modes increases. In this case, the cost can be either total fares or time taken in transfers, or it can even be seen as the mental burden that a traveler feels when he or she moves to or waits for the next vehicle in transfer areas. If we map the components using nodes and links in the previous section to a public transport network, a node (or a street in Fig. 2) can be seen as a stop, regardless of bus or subway, whereas a link between two nodes can be mapped to a transfer between two vehicles. This relationship is illustrated in Fig. 5, and the corresponding hierarchical representation in Fig. 6.

If a person moves from stop 1 to stop 2, 3 or 4, he or she does not need to transfer because these stops are on the same route, subway line 1. However, if the traveler wants to go to stop 5 or stop 6 from stop 1, he or she has to transfer in the zone A. Similarly, one transfer is needed in zone B from stop 1 to stop 7 or 8 and, if the destination is stop 9, he or she has to transfer two times, one in zone B and then in zone C. One transfer from a transportation mode to another is the 'spatial transfer' which becomes one depth between spaces. If this network were that of pedestrian streets, when a person were to move from point 6 to point 7, he or she needs two turns, which means these points are two depths away from each other. However, in case of using pre-laid transport routes, the existence of a route that connects two points is first taken into account in computing the depths. Therefore, no transfer is needed in case of moving from stop 7 to 6 because there is a direct line connecting these two points.

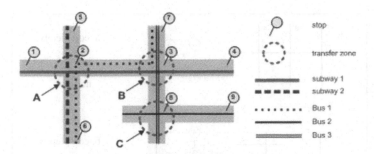

Fig. 5. An example of public transport network

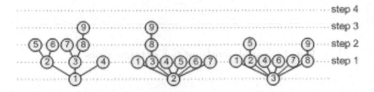

Fig. 6. Hierarchical representation of bus stops (stops 1, 2 and 3)

The procedure for generating the graph like Fig. 6 is iterative, starting with a stop and then progressively identifying the next neighboring stops until the entire stops are covered. We can then compute TD, MD and ND as shown in Table 1. Note that *the Integration*, the reciprocal of ND is also calculated for better intuitive interpretation. As one may easily expect, stop 3 shows the highest accessibility, 14.0, followed by 9.33 of stop 2. Note that stop 7 ranks in the third because stop 7 has the routes that pass transfer areas.

Table 1. Depths from each stop in Fig. 5

Stop No.	TD_i	MD_i	ND_i	$I(ND_i^{-1})$
1	14	1.750	0.214	4.67
2	11	1.375	0.107	9.33
3	10	1.250	0.071	14.00
4	14	1.750	0.214	4.67
5	17	2.125	0.321	3.11
6	13	1.625	0.179	5.60
7	12	1.500	0.143	7.00
8	14	1.750	0.214	4.67
9	21	2.625	0.464	2.15

4   The Case Study

In this section, we show how the proposed method is applied to a site in Seoul. It should be mentioned that computing the depth of a stop or a station in a target

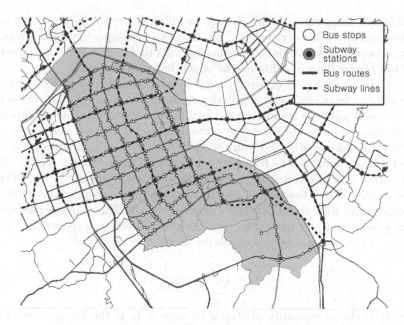

Fig. 7. The Test Area (Kangnam District of Seoul)

area entails finding paths from that stop to all others in the area, each of which being the optimal cost path. In this study, an optimal cost path is defined as the one that has the minimum number of transfers among minimum distance (or, cost) paths between O-D. Finding an optimal path in a public transport network is more complex than in the network of privately-owned cars. It is because the network is composed of multi-modes of vehicles and also has time-constraints at transfer zones. If a vehicle has a list of pre-specified departure times and transfers to other modes take place, the departure time is constrained for each available mode and its departure schedule, and comparison among these different departure times needs to be performed in order to explore the minimum time path (Desrochers et al. 1988). Genetic algorithms aim at such complex problems. Since the description about GA is well described in many publications, it is not described in this paper. For more description about the application of GA to the transportation problem, see Jun (2005).

The test site of this study is a CBD called 'Kangnam District' of Seoul which has highly complex multimodal network structure (Fig. 7). The bus stops located within walking distance are grouped into a transfer zone, and the Integration value (ND-1) was computed for each transfer zone. Fig. 8 shows the Integration values of stops depicted with gradient colors, with higher values meaning higher degree of overall depth-based accessibility.

The computed accessibility of a stop or a transfer zone implies aggregate easiness of movement from that spot in the aspect of network structure. In other words, it only measures the supply of transport routes and does not take into account the demand side, which is impractical in real situations. For instance, two identical Integration values do not mean that people in these two zones feel the same level of 'serviceability' as for public transport system. Thus, the transport service level needs to be balanced with real demand. In this paper, we used O-D survey data to calibrate the depth-based accessibility.

2002 O-D survey report of Seoul (Seoul Development Institute 2003) contains the number of trips by modes originating from survey zones. Among these trips data, we used public transport modes such as town bus, city metro, limousine and subway. The average size of 26 zones is about 1.17 km^2 and the average population is around 20,000. From the O-D matrix, we can compute the total number of trips using public transport emanating from a zone by adding all the trips to other zones from that zone. Then, the weighted accessibility by O-D trips can be measured by 'discounting' the Integration of the stop by the number of originating trips from the zone where the stop belongs. This makes sense because the Integration (current level of service) should be inversely proportional to trip demands. Thus, the weighted accessibility of stops is given by:

$$A_i^s = \frac{I_s}{\sum_j T_{ij}} = \frac{I_s}{O_i}, \tag{4}$$

where A_i^s is the accessibility of stop s in zone i; I_s is the Integration of stop s; T_{ij} is the number of trips between origin i and destination j; O_i is the total number of trips originating in zone i.

Fig. 9 shows the weighted accessibility calculated using Eq. (4) and then normalized to the range of 0 and 1.

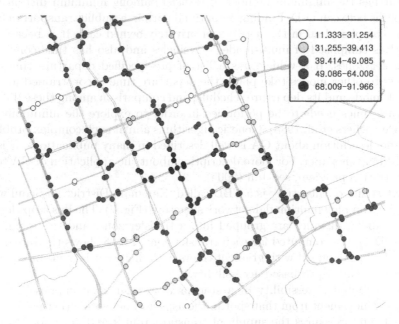

○	11.333–31.254
◔	31.255–39.413
●	39.414–49.085
●	49.086–64.008
●	68.009–91.966

Fig. 8. The Integration values (ND^{-1})

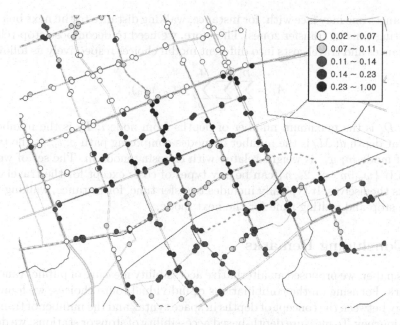

○ 0.02 ~ 0.07
◐ 0.07 ~ 0.11
● 0.11 ~ 0.14
● 0.14 ~ 0.23
● 0.23 ~ 1.00

Fig. 9. The weighted accessibility

5   Discussion: Modifying the Depth-Based Accessibility

In computing the impedance, we considered '*depth-penalty*' only, which is translated into the number of transfers in the paper. The space syntax-based measure, however, has a major drawback in not taking into account traditional travel costs such as distance or travel time. To overcome the drawback, one can combine the depth-based accessibility measure and the traditional one. Usually traditional accessibility measure balances the benefit of having places to visit j with the costs of moving to those places from a location i as described in Eq. (1). By considering both transfers and travel costs, accessibility of node i to all other nodes can be described as:

$$A_i = \sum_d^{D_i} f(w_d) \left(\sum_n^{N_d} f(t_n) \right), \tag{5}$$

where D_i is the maximum number of depths from node i; N_d is the number of destination nodes at depth d; t_n is travel cost between i and n; w_d is weight at depth d. Here we introduced weight value w_d that increases as the depth gets deepened. Then some function of this depth-weight (eg. w_d-1) is multiplied with the computed accessibility at each depth. However, in reality, travel costs should vary according to modes respectively. Also, assigning the same weight to all the paths at a certain depth may be impractical. In space syntax, all kinds of turns or visual transitions are considered to have the same amount of depths. But in multimodal travel choice, taking or changing to a vehicle may become diverse depending on the circumstances one may face with, for instance, walking distance to the next bus stop or waiting time at transfer zones. Therefore, we need to decompose stop-related impedances and travel costs into different modal choices respectively as follows:

$$A_i = \sum_d^{D_i} \sum_p^{P_d} \sum_m^{M_p} f(t_m, w_m), \qquad (6)$$

where D_i is the maximum number of depths from node i; P_d is the number of paths at depth d; M_d is the number of modes composing path p; t_m is the travel cost of mode m; w_m is weight related with choosing mode m. The set of weight values $W\{w_m|m = 1, 2, , n\}$ can be any types of costs except for the travel costs of links themselves. It typically includes transfer time, for instance, walking time to the stop and waiting time for the next mode.

6   Concluding Remarks

In this paper, we proposed an alternative accessibility measure of public transport network. Focusing on the combinations of individual route choices, we found an analogy between the concept of depths in space syntax and the number of transfers in the journey. To measure depth-based accessibility of stops or stations, we developed an algorithm to automate the computing process using GIS data. In order to generate optimal O-D paths during the process, we employed genetic algorithm. However, the proposed measure only assesses the level of supply of network connection from a stop without looking at demand side. Therefore, when we applied the proposed method to a CBD of Seoul, we compared the depth-only measure and weighted measure by O-D trips. Here, the O-D trips are interpreted as travel demand from a zone. The limitation of the proposed method is clearly pointed out in the last section. That is, it does not consider the travel cost generated from in-vehicle travel time or distance. To overcome this limitation, we also suggested a general form of equation which accommodates the depth as well as the cost.

The accessibility measure at individual routes level may especially be useful for the policy makers who are interested in assessing and tuning current network configuration for improving inter-modal connection. However, simulating alternative paths in real time shows more or less slow performance in current system. The reason comes from the GA process requiring the generation of multiple candidate solutions and multiple repetitions with 'the more the better' basis. GA, however, is a good example where parallel processing can be applied. Since processes for creating chromosomes are independent each other, the speed can be significantly increased by using parallel processing.

References

1. Bafna, S.: Space syntax: a brief introduction to its logic and analytical techniques. Environment and Behavior 35(1), 17–29 (2003)
2. Berglund, S.: Path-based accessibility. Journal of Transportation and Statistics 4(2/3), 79–91 (2001)

3. Desrochers, M., Soumis, F.: A Reoptimization Algorithm for the Shortest Path Problem with Time Windows. European Journal of Operational Research 35, 242–254 (1988)
4. Geertman, S.C.M., Van Eck, J.R.R.: GIS and models of accessibility potential: an application in planning. International Journal of Geographical Information Systems 9, 67–80 (1995)
5. Hanson, S.: The context of urban travel. In: Hanson, S., Giuliano, G. (eds.) The Geography of Urban Transportation, pp. 3–29. The Guilford Press, New York, NY (2004)
6. Harris, B.: Accessibility: concepts and applications. Journal of Transportation and Statistics 4(2/3), 15–30 (2001)
7. Hillier, B.: Space is the Machine. Cambridge University Press, Cambridge (1996)
8. Jun, C.: Route selection in public transport network using GA. In: Proc. 2005 ESRI International User Conference, San Diego, California (July 25-29, 2005)
9. Kwan, M.-P.: Space-time and integral measures of individual accessibility. Geographical Analysis 30(3), 191–216 (1998)
10. Litman, T.: Measuring transportation: traffic, mobility and accessibility. ITE Journal 73(10), 28–32 (2003)
11. Miller, H.J.: Measuring space-time accessibility benefits within transportation networks: basic theory and computational procedures. Geographical Analysis 31, 187–212 (1999)
12. Seoul Development Institute, 2003. Seoul O-D Survey (2002)

Adaptive Routing Algorithm Using Evolution Program for Multiple Shortest Paths in DRGS

Sung-Soo Kim[1] and Seung B. Ahn[2]

[1] Department of Industrial Engineering
Kangwon National University, 192-1, Hyoja-2-Dong
Chunchon, Kangwon, 200-701, Korea
kimss@kangwon.ac.kr
[2] Graduate School of Logistics
University of Incheon, 7-46, Yeonsu-Gu
Incheon, 406-840, Korea
sbahn@incheon.ac.kr

Abstract. There are several search algorithms for the shortest path problem: the Dijkstra algorithm and Bellman-Ford algorithm, to name a few. These algorithms are not effective for dynamic traffic network involving rapidly changing travel time. The evolution program is useful for practical purposes to obtain approximate solutions for dynamic route guidance systems (DRGS). The objective of this paper is to propose an adaptive routing algorithm using evolution program (ARAEP) that is to find the multiple shortest paths within limited time when the complexity of traffic network including turn-restrictions, U-turns, and P-turns exceeds a predefined threshold.

Keywords: Shortest path, Evolution program, Adaptive routing algorithm using evolution program (ARAEP), Dynamic route guidance system.

1   Introduction

The aims of Intelligent Transport System (ITS) are to promote safe driving, to lower the pollution and to provide shorter travel time for the drivers. The DRGS of ITS makes the use of information for promoting the safe, smooth flow of traffic for users. The objective of the DRGS is also to promote the efficiency of vehicle operation [1,2]. The overall system design for DRGS is shown in Figure 1. This system consists of vehicles with an on-board unit (OBU), an operation center, traffic information sources, and users. In this system, the two-way communication enables cars to work as sensors of information. Cars measure their own travel times and locations. They send their locations, travel time, information about vehicles, and inquiries to the operation center on the uplink. This allows better traffic modeling, prediction, and route calculation. Based on information gathered from street networks, this system can help drivers find the optimal path for driving. On the downlink the vehicle equipment receives new traffic information (e.g., the new shortest path). There are several search algorithms

M.S. Szczuka et al. (Eds.): ICHIT 2006, LNAI 4413, pp. 292–301, 2007.

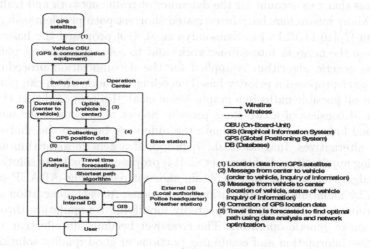

Fig. 1. Overall System Design for DRGS

for the shortest path problem: the breadth-first search algorithm, the Dijkstra algorithm, Floyd-Warshall algorithm and the Bellman-Ford algorithm, to name a few [3,4]. These algorithms can solve the shortest path problems in polynomial time. But, they exhibit unacceptably high computational complexity when there are dynamic changes of travel time in large size of traffic network including many cross-roads, turn-restrictions, and U-turns. Modified existing shortest path algorithms for turn-restriction, U-turn, and P-turn in Traffic Network were proposed [5,6]. Theses methods search only for the shortest route, but they can not determine any other short routes in one search. Therefore, these methods need re-routing to search for another route if the shortest route conditions are not satisfied [7]. For small spaces, classical shortest path methods usually suffice; for larger spaces special artificial intelligence techniques must be employed. The ARAEP that is proposed method in this paper is among such techniques. The ARAEP not only selects the shortest route but also determines multiple semi-shortest routes by memorizing the possible solutions produced in the search process. The purpose of this paper is to study ARAEP for finding the multiple shortest paths within limited computation time when the complexity of traffic network exceeds some modest limits and the traffic network also includes turn-restrictions, U-turns, etc.

2 Development of ARAEP

Yang et al. [8] proposed an algorithm for determining optimal routing decisions where arc travel times vary over time in stochastic and time varying networks. Fu [9] also proposed an adaptive routing algorithm for in-vehicle route guidance systems with real-time information. The DRGS require fast shortest-path algorithms that can account for the dynamics of traffic network and traffic regulations. Many researchers have investigated shortest path problem using genetic algorithm [7],[10,11,12,13,14]. Sannomiya et al. [10] proposed the basic idea to decompose the network into sub-networks and to solve the shortest path problem. The genetic algorithm is applied for the decomposition procedure. Gen et al. citex11 proposed a priority-based encoding method, which can potentially represent all possible paths in a graph. Voicu et al. [13] proposed that the shortest method consists of a two-stage process. Seo et al. [12] used the number of overlapped links or nodes to evaluate the difference between the shortest path and the alternatives. Inagaki et al. [7] proposed a genetic algorithm approach for routing applications. Uchimura et al. [14] propose approximate solution using genetic algorithm in real time to obtain shortest paths. The ARAEP performs a search by maintaining a population of solutions. At each generation good solutions reproduce. It works by repeatedly modifying a population through the application of genetic operators. The crossover becomes an effective means of exchanging information and combining portions of good quality solutions. The mutation is simply an occasional random change of a genotype position based on probability of mutation. The mutation operator helps in avoiding the possibility of obtaining a local minimum. The proposed ARAEP procedure can be shown in Figure 2. The step 1 is to create an initial population of potential shortest path solutions. The genetic representation for potential solutions to the shortest path problem in this paper is to start from both source and destination nodes and successively add new randomly selected nodes to both sides until the two path segments become connected. Once completed, the path is saved. For example, each path starts with starting node 1 and destination node 12 in traffic network of Figure 3. Each branch from starting node and destination node will connect randomly selected nodes 2 and 11 by successive cloning operations until the two branches become connected by node 7 as shown in Figure 3. Step 2, 6, and 8 are the elimination of unnecessary circulation (U-turn and etc) after random generation of path (chromosome) in step 1, crossover in step 5, and mutation in step 7. We make chromosomes (paths) by eliminating the unnecessary circulations considering turn-restrictions, U-turns, and P-turns. Step 3 and 9 use an evaluation function that plays the role of the environment, rating solutions in terms of their fitness. The chromosomes are evaluated using some measure of fitness. In each generation, we evaluate each chromosome and select new population based on fitness values. For the selection process, a roulette wheel with slots sized according to fitness is used. Obviously, some chromosomes would be selected more than once. The best chromosomes get more copies, the average stay even, and the worst die off. The modified elitist pool method is used to preserve the best chromosome for the next generation. With the elitist selection, if the best individual in the current generation is not reproduced into the new generation, remove one individual randomly from the new population and add the best one to the new population. An undirected network comprises a set of nodes V_i and a set of edges connecting nodes. Corresponding to each edge, there is a non-negative value w_{ij} representing the cost (distance, traveling time,

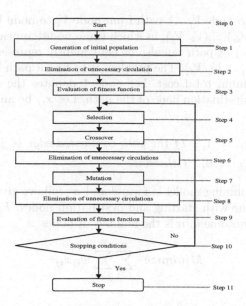

Start	Step 0
Generation of initial population	Step 1
Elimination of unnecessary circulation	Step 2
Evaluation of fitness function	Step 3
Selection	Step 4
Crossover	Step 5
Elimination of unnecessary circulations	Step 6
Mutation	Step 7
Elimination of unnecessary circulations	Step 8
Evaluation of fitness function	Step 9
Stopping conditions	Step 10
Stop	Step 11

Fig. 2. Flowchart of Proposed ARAEP Procedure

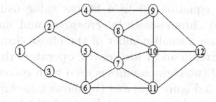

Fig. 3. Example of traffic network

Sequence 1:   *node 1* starting node
                  node 12 destination node
Sequence 2:   *node 1*   adjacent nodes of node 1 (2, 3)
all nodes generated from destination node (12)
Sequence 3:   *node 1, 2*   adjacent nodes of node 2 (4, 5)
all nodes generated from destination node (12)
Sequence 4:   *node 11, 12*   adjacent nodes of node 11 (6,   7, 10)
all nodes generated from starting node (1, 2)
Sequence 5:   *node 1, 2, 4*   adjacent nodes of node 4 (8, 9)
all nodes generated from destination node (11, 12)
Sequence 6:   *node 7, 11, 12*   adjacent nodes of node 7 (5, 6, 8, 10)
all nodes generated from starting node (1, 2, 4)
Sequence 7:   *node 1, 2, 4, 8*   adjacent nodes of node 8 (4, 7, 10)
all nodes generated from destination node (7, 11,12)
Sequence 8:   Chromosome (*node 1, 2, 4, 8, 7, 11, 12*) is generated by connecting node 7.

Fig. 4. Example for generation of a chromosome

etc) from node V_i to node V_j. A path from node V_i to node V_j is a sequence of edges (V_i, V_l), (V_l, V_m),...,(V_k, V_j), in which some nodes appears more than once (U-turn, P-turn, etc). A path can also be equivalently represented as a sequence of nodes $(V_i, V_l, V_m, ..., V_k, V_j)$. The problem is to find a path between two given nodes having minimum total cost. Let node 1 denotes the starting node and node n denote the destination node of the path. Let x_{ij} be an indicator variable defined as follows:

$$x_{ij} = 1, \quad If \ the \ path \ contained \ edge \ (i,j)$$
$$x_{ij} = 0, \quad Otherwise \tag{1}$$

The integer programming model is formulated as follows given in equation (2) for each chromosome with starting and destination nodes. F_k is the evaluation value of each chromosome k. n is the number of nodes.

$$Minimize \ \sum_{i=1}^{n} \sum_{j=1}^{n} w_{ij} x_{ij} \tag{2}$$

$$F_k = \frac{1}{\sum_{i=1}^{n} \sum_{j=1}^{n} w_{ij} x_{ij}} \tag{3}$$

Since this problem is a minimization problem, we have to convert the original objective value using equation (2) to a fitness value using equation (3). We evaluate only changed chromosomes by crossover and mutation operators in each generation. We use a small number of the elitist chromosomes to increase population diversity. Step 5 and 7 are genetic operators that alter the solution. Two-point crossover of two chromosomes is used with two common nodes (6 and 10) is shown in Figure 5. Figure 5 (a) and (b) shows two chromosomes before and after crossover. The first chromosome of Figure 5 (b) has a U-turn circulation $(9- > 10- > 9)$. If there is no turn-restriction in $8- > 9- > 12$, the circulation is not needed and should be eliminated in step 6. Otherwise, it should be not eliminated.

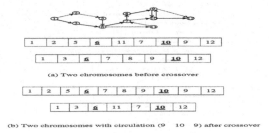

(a) Two chromosomes before crossover

(b) Two chromosomes with circulation (9   10   9) after crossover

Fig. 5. Two-point (6 and 10) Crossover with Common Nodes

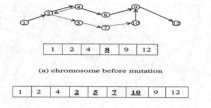

(a) chromosome before mutation

(b) A chromosome with circulation (2->4->2) after mutation

Fig. 6. Mutation

The randomly selected node 8 for mutation would be replaced by a partial path segment $(2- > 5- > 7- > 10)$ from node 4 to node 10 as shown in Figure 6 (b). The partial path segment is generated from randomly selected nodes 2, 5, 7, 10 from former node 4 and latter node 9 of node 8 by successive cloning operations until a chromosome is completed. A chromosome after mutation has a U-turn circulation $(2- > 4- > 2)$ in Figure 6 (b). If there is no turn-restriction in $1- > 2- > 5$, the circulation should be eliminated in step 8. Otherwise, it should be not eliminated For a more focused search in later generations we adapt the crossover rate and the mutation rate over evolution. Mutation rate and crossover rate decreases with generation, all according to the equations given below. Fc is the crossover rate and Fm is the mutation rate in generation g. In generation 0, c is the crossover rate and m is the mutation rate. α, β is the weighting factors, $0 < \alpha < 1$, $0 < \beta < 1$. g is the generation 0, 1, 2,3, ..., G. The values we use are equal for α and β and are 0.999.

$$F_c = c * \beta^g \tag{4}$$

$$F_m = m * \alpha^g \tag{5}$$

Step 10 is to stop the algorithm when no further improvement is observed. It is better if the algorithm terminates, when the chance for an improvement is slim. Whenever the best candidate in the population does not violate the constraints, the search may terminate. Also, the search can stop at user defined maximum number of generations.

3   Evolution Procedure for Case Study

The evolution procedure for the traffic network using ARAEP is shown in this section. Figure 7 is the traffic network of Seoul with restrictions. In this network, node 1 is the starting node and node 46 is the destination node. We have considered 11 turn-restrictions and 5 U-turn restrictions as shown in Figure 7.

Genetic algorithm to this traffic network to obtain the optimal path is done in the following manner. An initial population of 10 solutions (paths) is randomly generated. In these initial solutions the unnecessary circulations are removed. Then the cost of each path is computed using the objective function and the

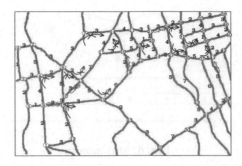

Fig. 7. Traffic Network of Seoul with Restrictions

Table 1. Path, Cost & Evaluation Value for Each Chromosome in the First Generation

Chromosome	Path	Cost	Evaluation Value
1*C*1	1 8 5 3 9 2 4 6 7 6 10 13 11 14 16 20 22 24 27 35 32 36 40 46	75	0.0133
1*C*2	1 8 4 3 9 2 6 13 10 15 19 23 29 30 44 45 39 41 42 41 37 35 32 31 28 33 38 46	77	0.0130
1*C*3	1 8 5 3 4 6 13 19 23 29 30 44 43 35 37 45 39 41 36 40 46	70	0.0143
1*C*4	12 15 10 13 11 14 17 21 26 28 31 34 33 38 46	47	0.0213
1*C*5	1 8 4 3 9 2 6 13 10 15 19 18 16 20 17 21 26 28 31 32 35 43 44 45 39 41 42 41 36 34 33 38 46	97	0.0103
1*C*6	1 8 5 3 4 6 10 13 11 16 20 17 21 26 28 31 34 46	51	0.0196
1*C*7	1 12 15 10 13 19 23 29 30 44 45 37 35 32 36 40 46	58	0.0172
1*C*8	1 12 15 8 4 3 9 2 6 10 13 19 23 29 30 44 45 39 41	67	0.0149
1*C*9	1 12 15 10 6 7 6 2 6 4 6 13 19 23 19 18 22 24 27 43 35 37 45 39 41 36 32 31 34 46	99	0.0101
1*C*10	1 12 15 10 13 11 16 18 16 20 17 21 26 28 31 34 46	48	0.0208
Total		689	0.1549

fitness is assigned to these paths. Equations (2) and (3) are used for this. A roulette wheel selection method is used to select solutions from the initial population for genetic operations. Genetic operations (crossover and mutation) are performed on selected solutions for the given crossover rate and mutation rate, for that generation. In this manner, new solutions are produced for this problem. The best 10 solutions are the initial population for next generation. This process of evolution continues generation after generation until the program terminates or maximum number of generations are reached. In Table 1 the initial population of 10 solutions, their cost and fitness are shown. The solutions obtained after removing the unnecessary circulations, are selected based on roulette wheel selection and is given in Table 2. Two solutions are selected from Table 2 for crossover operation. Let the solutions selected be 2C4 and 2C5. In this solutions

Table 2. Path, Cost & Evaluation Value for Each Chromosome in the Second Generation

Chromosome	Path
2C1	1 8 5 3 4 6 13 19 23 29 30 44 43 35 37 45 39 41 36 40 46(1C1)
2C2	12 15 10 13 11 14 17 21 26 28 31 34 33 38 46 (1C4)
2C3	1 8 5 3 4 6 10 13 11 16 20 17 21 26 28 31 34 46 (1C6)
2C4	1 8 4 3 9 2 6 13 10 15 19 23 29 30 44 45 39 41 42 41 37 35 32 31 28 33 38 46 (1C2)
2C5	1 8 4 3 9 2 6 13 10 15 19 18 16 20 17 21 26 28 31 32 35 43 44 45 39 41 42 41 36 34 33 38 46 (1C5)
2C6	1 8 5 3 9 2 4 6 7 6 10 13 11 14 16 20 22 24 27 35 32 36 40 46 (1C1)
2C7	1 12 15 10 13 11 16 18 16 20 17 21 26 28 31 34 46 (1C10)
2C8	1 8 5 3 9 2 4 6 7 6 10 13 11 14 16 20 22 24 27 35 32 36 40 46(1C1)
2C9	1 12 15 10 13 11 16 18 16 20 17 21 26 28 31 34 46 (1C10)
2C10	1 8 5 3 4 6 13 19 23 29 30 44 43 35 37 45 39 41 36 40 46 (1C3)

Table 3. Path, Cost & Evaluation Value for Each Chromosome in the Last Generation

Chromosome	Path	Cost	Evaluation Value
C1	1 8 15 19 23 19 18 22 20 16 20 25 21 26 28 33 34 46	51	0.0196
C2	1 8 15 19 23 19 18 16 14 17 21 26 28 33 34 46	42	0.0238
C3	1 8 15 10 13 11 16 18 22 24 25 21 26 28 33 34 46	45	0.0222
C4	1 8 15 19 13 11 16 20 17 21 26 28 31 34 46	45	0.0222
C5	1 8 15 19 23 19 18 16 14 17 21 26 28 33 34 46	42	0.0238
C6	1 8 15 19 23 19 18 22 20 16 20 25 21 26 28 33 34 46	51	0.0196
C7	1 8 15 19 18 22 24 25 21 26 28 31 34 46	34	0.0294
C8	1 8 15 19 23 19 18 16 14 17 21 26 28 33 34 46	42	0.0238
C9	1 8 5 3 4 6 13 19 23 19 18 22 24 25 21 26 28 33 38 46	48	0.0208
C10	1 8 15 19 18 22 24 25 21 26 28 33 34 46	36	0.0278
Total		362	0.2771

(2C4) and (2C5), two common points (nodes) are randomly selected. Let the nodes selected be 15 and 41. Using the crossover operation explained in Figure 5, we obtain two new solutions 3C4, and 3C5. In these two new solutions 3C4 and 3C5 the unnecessary circulation are eliminated. After elimination the 3C4 and 3C5 obtained are: 3C4: 1 8 4 3 9 2 6 13 10 15 19 18 16 20 17 21 26 28 33 38 46 3C5:184 3 9 2 6 10 15 19 23 29 30 44 45 39 41 36 34 33 38 46.

In mutation operation only one solution is selected from Table 2. Let the solution selected be 2C2. One node from 2C2 is selected randomly. Let the node selected be 15. A new solution 3C2 is obtained after mutation using the operation explained in Figure 6. In 3C2 also unnecessary circulation is eliminated. After elimination the 3C2 obtained is: 3C2: 1 8 5 3 4 6 10 13 11 14 17 21 26 28 31 34 33 38 46.

After genetic operations are done, we will have more solutions. The best 10 solutions (population size) are taken as the initial population for the next generation. This process continues till the termination criteria are satisfied. At the termination, the paths obtained and their cost are given in Table 3. From Table 3, the shortest path is 34 with path 1 8 15 19 18 22 24 25 21 26 28 31 34 46.

4   Discussions and Conclusions

Discussions: Genetic algorithm for our problem starts with an initial population of solutions. The solutions in the population are modified by using genetic operators and a new population of solutions is obtained. The new population of solutions is the initial population of solutions for the next generation. The solutions obtained in the next generation will be better solutions than the solutions in the previous generation. In this way, the better paths that are missing in the initial solutions are generated in genetic algorithm. Since, genetic algorithm uses the survival of the fittest strategy; better solutions are obtained from generation to generation. The conventional methods like Dijkstra and Floyd-Warhall algorithms consider all the paths from starting node to the end node are possible paths. But in our problem, this condition is not true, because of turn restrictions. In the conventional methods there are no turn restrictions. Hence, we have used a genetic algorithm which can handle the turn restrictions.

Conclusions: This paper has addressed a combinatorial optimization problem that has growing importance in DRGS. The proposed ARAEP finds a collection of paths between source and destination considering turn-restrictions and U-turns that are genetically evolved until an acceptable solution to the shortest path problem is reached. The series of specific node number is used as a representation of chromosome to save computation time and memory for the progress of improved ARAEP. Only the changed chromosomes by crossover and mutation are evaluated in each generation. This method uses high population diversity and low selective pressure in the initial generations and low population diversity and high selective pressure in the last generations. This paper also shows the procedure to find the multiple shortest paths in real traffic network using ARAEP as an efficient shortest path algorithm.

Acknowledgement. This paper is sponsored by University of Incheon (2004 Research Fund).

References

1. Ben-Akiva, M., Koutsopoulos, H.N., Mukandan, A.: A dynamic traffic model system for ATMS/ATIS operations. IVHS Journal 1(4), 1–19 (1994)
2. Shinsaku, Y.: The strategy and deployment plan for VICS. IEEE Communications Magazine, 94–97 (1996)
3. Ahuja, R.K, Magnamti, T.L., Orlin, J.B.: Network Flows. Prentice-Hal, Englewood Cliffs (1993)

4. Neapolitan, R., Namipour, K.: Foundations of Algorithms using C++ Pseudo Code, 2nd, Jones and Bartlett (1998)
5. Namkoong, S., et al.: Development of the tree-based link labeling algorithm for optimal path-finding in urban transportation networks. Math. Compute. Modeling 27(9-11), 51–65 (1999)
6. Kim, S., Lee, J.: Modified Dijkstra and Floyd-Warshall Algorithms for Turn-restriction, U-turn, and P-turn in Traffic Network. In: Proceeding, INFORMS Seoul 2000 Conference (2000)
7. Inagaki, J., Haseyama, M., Kitajima, H.: A Genetic Algorithm for Determining Multiple Routes and Its Applications. In: Proc. of the 1999 IEEE Int. Symposium on Circuits and Systems (1999)
8. Yang, B., Miller-Hooks, E.: Adaptive routing considering delays to signal operations. Transportation Research Part B 38, 385–413 (2004)
9. Fu, L.: An adaptive routing algorithm for in-vehicle route guidance systems with real-time information. Transportation Research Part B 35, 749–765 (2001)
10. Sannomiya, N., Tatemura, K.: Application of genetic algorithm to a parallel path selection problem. Int. Journal of Systems Science 27(2), 269–274 (1996)
11. Gen, M., Cheng, R., Wang, D.: Genetic Algorithms for Solving Shortest Path Problems. In: Proceedings of the 1997 IEEE International Magnetics Conferences (1997)
12. Seo, K., Choi, G.: The Genetic Algorithm Based Route Finding Method for Alternative Paths. In: Proc. of the 1998 IEEE Int. Conference on Systems, Man, and Cybernetics (1998)
13. Voicu, L., Myler, H.: Cloning operator and its applications. In: Proceedings of the Applications and Science of Computational Intelligence (1998)
14. Uchimura, K., Takahashi, H., Saitoh, T.: Demand responsive services in hierarchical public transportation system. IEEE Trans. on Veh. Techn. 51(4), 760–766 (2002)

Particle Swarm Optimization for a Multi-UCAV Cooperative Task Scheduling

Xiaohua Huo, Lincheng Shen, and Tao Long

Mechatronics and Automation School, National University of Defense Technology,
Changsha 410073, China
cindy.huo@gmail.com

Abstract. Task scheduling is one of the core steps to effectively exploit the capabilities of cooperative control of multiple uninhabited combat aerial vehicles(UCAVs) team. The main function of multi-UCAV cooperative task scheduling is to allocate tasks which should be implemented by vehicles, and arrange the sequence of these tasks to be carried out for each vehicle simultaneously, while optimizing the team objective and satisfying various constrains of vehicles and tasks. By analyzing the characters of tasks and UCAVs, we presented a general mathematical model based on a combinatorial optimization. By defining a suitable particle structure, the Particle Swarm Optimization (PSO) algorithm was applied to solve this problem. Adaptive weight values and stochastic turbulence strategies were added to the algorithm. Simulation results indicate that the PSO algorithm proposed in this paper is a feasible and efficient approach for task scheduling in multi-UCAV cooperative control.

1 Introduction

Using multiple uninhabited combat aerial vehicles (UCAVs) to complete "the dull, the dirty and the dangerous" missions, including suppression of enemy air defenses (SEAD), intelligence, surveillance and reconnaissance (ISR), destruction of enemy air defense (DEAD), electronic attack (EA), etc., will be an important combat modality in the future [1, 2]. Recent interest in using collaborating UCAVs team to complete these complex missions with coordinate coupled tasks missions have led to numerous investigations into cooperative control technology, which is suggested as a possible solution to automatically control and coordinate UCAVs team [2, 3, 4, 5, 6]. Hierarchical decomposition control structure [4], cooperative task allocation [5, 6], cooperative path planning [1], cooperative searching an unknown area [10], multi-point attack [9] problems in multiple UCAVs cooperative control have been studied in recent decade.

In this paper, we address the task scheduling problem in multiple UCAV cooperative control, which the main function is to allocate tasks for every UCAV, and arrange the sequence or schedule of these tasks to be carried for each vehicle at the same time. In other words, task scheduling consists two sub-problems: task allocating and task ordering. Maximizing the whole efficiency and minimizing the

M.S. Szczuka et al. (Eds.): ICHIT 2006, LNAI 4413, pp. 302–311, 2007.
© Springer-Verlag Berlin Heidelberg 2007

attrition of UCAVs team are the main objectives of the problem, while satisfying various constrains of every UCAV and task.

Since tasks locate different positions, a UCAV must visit its tasks in a certain sequence or a route. Then cooperative tasks scheduling for multi-UCAV can be viewed as a problem of finding a set of tasks sequence with optimal routings for a fleet of aerial vehicles, in order to finish a set of tasks requires and be sure that constrains of tasks and UCAVs be satisfied. Analogizing tasks as customers, the task scheduling problem is similar to the vehicle routing problem(VRP), a famous NP-hard combinatorial optimizing problem. Similarly the tasks may be carried out in a certain time window, or need multiple vehicles to serve them. But the multi-UCAV cooperative task scheduling is more complex to be modeled, and it is difficulty to be solved since it has more complex constraints. By analyzing the characteristics and constraints about UCAVs and tasks, a general mathematics model is designed in section 2.

Recent decades, a new algorithm called Particle Swarm Optimization (PSO) has proved very efficient in solving many practical problems, which is motivated from the simulation of social behavior. In this paper, we propose a PSO based heuristic method for solving task scheduling problem by redefining the particle's structure, which will be discussed in section 3. Then experiments on multi-UCAV cooperative task scheduling problem are given in section 4. Simulation results indicate that algorithm is a feasible and effective for task scheduling problem in multi-UCAV cooperative control. At last a discussion conclusions are discussed in section 5.

2   Multi-UCAV Cooperative Task Scheduling Formulation

Consider a scenario where a group of n UCAVs are required to finish m tasks. UCAVs team is denoted by $U = \{U_1, U_2, ..., U_n\}, n \in R$. The set of tasks is $M = \{M_1, M_2, ..., M_m\}, m \in R$. Tasks that will be assigned to UCAVs locate different positions. The location of each task is defined as: $\{MP_j = (x_j, y_j), j \in M\}$. in the 2-dimensional space. The location of each UCAV is defined as $\{UP_i = (x_i, y_i), i \in U\}$.

We assume each UCAV can execute entire or portion tasks in tasks set, but one UCAV just can do only one task in one time, and the tasks that need UCAV to completed are predetermined, the cost and award of each task executed by each UCAV can be calculated before planning.

The aerial vehicles team may be heterogeneous since weapons are depleted and their mission capabilities are not consistent across the team. So the cost of different UCAV finished same task may be different too, the cost may includes the depletion of weapons and fuel, vehicles attrition, et al. Let c_{ij} denotes the cost for task j completed by UCAV i. Then the cost vector of task j can be denoted by vector $c_j = (c_{1j}, c_{2j}, ..., c_{nj})$.

Tasks are each only available for a special window of time, which is important for a planner to consider when ordering of these tasks. The permitted started time and expected completed time of tasks are denoted by $\{t_{s1}, t_{s2}, ..., t_{sm}\}$ and

$\{t_{e1}, t_{e2}, ..., t_{em}\}$ separately. Then the available time window of task j is D_j : $[t_{sj}, t_{ej}]$, which means task j should be executed in this time window. Since UCAVs are heterogeneous, the time different UCAVs need for finishing same task may be different too. Let T_{ij} denotes the time UCAV i need to complete task j. Note that UCAV's carrying capacity is limited. Let $L_i, i \in U$ denotes the load limit of UCAV i.

The main objective of the multi-UCAV cooperative task scheduling is to find a set of tasks sequence with optimal routings for a fleet of vehicles, in order to finish all of the tasks while minimize the cost of UCAVs team for finishing all tasks, the tasks finished time and the number of delayed task separately. Set binary variable x_{ij} and continuous variable t_j as the decision variables. It is 1 if task j is assigned to UCAV i otherwise it is 0. Time variable t_j is the time when task $j \in M$ to be visited. Then the multi-UCAV cooperative task scheduling problem can be are formulated as:

$$min J = \beta \sum_{i \in U} \sum_{j \in M} c_{ij} x_{ij} + \gamma \{max\{(t_j + T_{ij}) x_{ij}\} + \delta \sum_{j \in M} \omega_j b_j \tag{1}$$

$$c_{ij} = a_1 * f_{ij} + a_2 * l_{ij} \tag{2}$$

$$\sum_{j \in M} l_j^{iw} x_{ij} \leq L_j^{iw} \tag{3}$$

$$t_j = max\{t_{sj}, t_k + T_{ik} + t_{kj}^{ifly}\}, \forall k, j \in S_i, i \in U \tag{4}$$

$$t_{sj} \leq t_j \leq t_{ej} \tag{5}$$

$$b_j = \begin{cases} 0 & : (t_j + T_j) * x_{ij} \leq t_{ej} \\ 1 & : t_{ej} < (t_j + T_j) * x_{ij} \end{cases} \tag{6}$$

$$\forall x_{ij} \in \{1, 0\}; \sum_{i \in U} x_{ij} = 1, \forall j \in M \tag{7}$$

Equations (1) is the objective function of the problem, β, γ, δ convey the weight of each component. The fuel cost and threat cost are primary cost of a UCAV to execute tasks, which are both relative to the route of a UCAV, so we incorporate them as one, marked as f_{ij}(refer to [1]). l_{ij} denotes the weapon cost of UCAV i to execute task j. Fuel cost, threat cost and weapon cost compose the cost of a UCAV together, and c_{ij} can be calculated by equation (2).

Function (3) means the total number of each type of weapons UCAV used is no more than the number of that it carried. Equation (4,5,6) are time constrains which ensure that the task will be carried out in the valid time window and be logical. U_j is a binary variable, it is 1 if task j was delayed, otherwise it is 0. ω_j is the weight of each task, reflecting the importance degree of each task. Equation (7) ensures that one task be executed once. It is obvious that multi-UCAV cooperative task scheduling problem is a combination optimization problem with complex constrains.

3 Particle Swarm Optimization for Task Scheduling

3.1 General Particle Swarm Optimization

Particle swarm optimization (PSO) was first proposed by Kenedy and Elberhart in 1995 [11], which was inspired by the choreography of a bird flock. PSO is a stochastic search algorithm that conducts searches using a population of particles. Particles change their position by flies in a problem space looking for the optimal position. The status of a particle can be described by its position and velocity. The new velocity and position of a particle for the next are calculated using the following equations:

$$V_i^{k+1} = \omega * V_i^k + c_1 * rand1 * (pBest_i^k - P_i^k) + c_2 * rand2 * (gBest^k - P_i^k) \quad (8)$$

$$P_i^{k+1} = P_i^k + V_i^{k+1} \quad (9)$$

where V_i^k and P_i^k indicate for velocity and position of particle i at k. $pBest_i^k$ represents the best position of particle i in its past experience, and $gBest_i^k$ is the best position of the population. $\omega, c_1, c_2 \geq 0$. ω is the inertia weight. c_1 and c_2 convey the weight information of cognition model and social model. $rand1$ and $rand2$ are two random values in the range $[0, 1]$.

The formulation includes three components. First is a particle's present velocity, dedicating a particle's current status, it can balance global and local searching capability. Second is cognition model, expressing a particle's cognition ability, it can make a particle has enough global searching capability avoiding local maximum. Last is social model, representing communication among particles. Particles land the best position effectively with three components acting together. Otherwise, as a particle adjusting its position according to velocity, it must keep the velocity in the range of min-velocity and max-velocity.

3.2 PSO for Multi-UCAV Cooperative Tasks Scheduling

State representation of particle is the key of applying PSO to solve real-life problems successful, which is finding a suitable mapping between real-life problems solution and PSO status of particles. As we addressed above, we should to make decisions: which task assigned to which UCAV and the time that each to be carried. The start time of one task can be calculated if the executed order of the task has been known. So we set up a search space of $2*m$ dimension. Let the first m dimension indicates the tasks which will be assigned. Each dimension has a discrete set of possible assignment limited to n, which is the number of UCAVs. The second m dimension indicates the sequence of each task to be carried out. The mapping between one possible schedule instance and a particle position is showed in figure 1.

The position value of a particle is continuous variable, but the task's index is integer, so does the sequence of a task to be carried out. So we should transform the continuous particle's position to discrete task's index and executing sequence(task plan of a vehicle) before evaluate its fitness, which is the key of

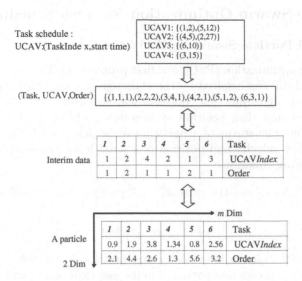

Fig. 1. Task schedule maps to a PSO particle

using PSO algorithm solving discrete optimizing problems. First, we rounded the first m dimension of particles up to an approximate integer. Second, we transform the second m dimension to the executing sequence of tasks to be carried out for each UCAV, by sorting the corresponding bits of the tasks that have been allocated to the same vehicle.

For example, the value of the first m dimension of a particle is [0.9 1.9 3.8 1.34 0.8 2.56], the second m dimension is [2.1,4.4,2.6,1.3,5.6,3.2], as shown in figure 1. By modified the particle, the first m dimension integer vector [1 2 4 2 1 3] is got. In this instance, both task 2 and 4 are allocated to UCAV 2. The corresponding values of corresponding bits in the second m dimension are 4.4 and 1.3. The corresponding bit's value of task 4 is smaller than that of task 2, then task 4 will be executed before task 2. Using this modification method the PSO algorithm will be fit for discrete optimization problems.

The performance of each particle is evaluated according to a predefined fitness function, which is usually proportional to the cost function associated with the problem to be solved. In this paper, the particle's fitness is determined by (1).

As shown in equation (8), the inertia factor ω is employed to control the impact of the previous history of velocity on the current velocity. In some sense, it balance the global and local exploration ability of the swarm. A large inertia weight facilitates global exploration, while with a small one the particle is more intended to do local exploration [12]. A a linearly decreasing inertia weight is usually set according to the following equation:

$$\omega = \omega_{max} - \frac{\omega_{max} - \omega_{min}}{iter_{max}} * iter \qquad (10)$$

where ω_{max} is the initial weight, ω_{min} is the final weight, $iter_{max}$ is maximum iteration number, $iter$ is current iteration number. Using above equation, particles search in wide range at prophase, and in nearby neighborhood at anaphase. In other words, diversification characteristics is gradually decreased.

Similarly, the learn factors c_1, c_2 are set as:

$$c^{1,2} = c_{max}^{1,2} - \frac{c_{max}^{1,2} - c_{min}^{1,2}}{iter_{max}} * iter \tag{11}$$

where $c_{max}^{1,2}$ is the initial value of learn factors, $c_{min}^{1,2}$ is the final value.

We have implemented the turbulence factor as below:

$$P_i^{k+1} = P_i^k + V_i^{k+1} + Rand_T * P_i^k \tag{12}$$

where $Rand_T$ is a random value in $[0,1]$, which is added to the updated position of each particle with a probability.

4 Experiments

First, a simple computational experiment was carried out on the following instance: 3 tasks were carried by 3 UCAVs. Then a more complex test was carried out on: 3 UCAVs should finish 14 tasks. All tests were done with $\omega_{max} = 1.2, \omega_{min} = 0.8, c_{max}^{1,2} = 0.8, c_{min}^{1,2} = 0.4$. The test experiment is Pentium 4, CPU 2.40 GHz, EMS memory 512M.

The parameters of tasks and UCAVs in first scenario are given in table 1.

Table 1. Parameters of tasks and UCAVs in 3 to 3 scenario

Task j	$MP_j : x_j$	$MP_j : y_j$	$D_j : t_{sj}$	$D_j : t_{ej}$	T_{1j}	T_{2j}	T_{3j}
1	22	5	30	92	60	96	45
2	72	55	265	338	99	12	78
3	30	50	128	273	134	78	89

	UCAV 1	UCAV 2	UCAV 3	End Rendezvous
$UP_i : y_i$	5	95	65	55
$UP_i : y_i$	5	5	95	55

Tests are done for 100 particles, 20 generations with different model parameters. The results are shown in figure 2. In test (a), parameters are set as : $\beta = 1/100, \gamma = 1/300, \delta = 1/3$. In test (b), parameters are set as: $\beta = 1/300, \gamma = 1/100, \delta = 1/3$.

In this scenario, 3 UCAV started from different locations, gathered at a certain location while they finished their tasks. In Fig.2(a), the total length of UCAV

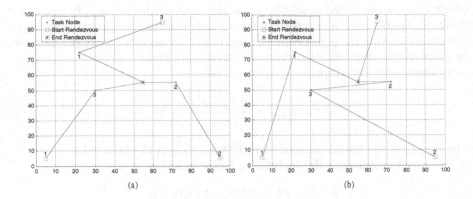

Fig. 2. The test results in scenario 3 to 3

team's route is 235.0208, the time that last UCAV arrived at the rendezvous is 363, namely the total finish time. In Fig.2(b), total route length is 290.2068, the total finish time is 279. In this plan, one UCAV has been allocated two tasks, while another one has no task. The state representation of particle proposed in this paper is flexible to deal with this special instance.

The parameters of tasks in second scenario are given in table 2. The start rendezvous and end rendezvous of UCAVs locate (5,5) and (95,5) respectively.

Set $\beta = 1/500, \gamma = 1/100, \delta = 10/19$. Tests are done for 1000 particles, 100 generations. One result is shown in figure 3.

The task plan of UCAVs team is :UCAV1: $1 \to 2 \to 12 \to 3 \to 10$; UCAV2: $4 \to 7 \to 5 \to 14$; UCAV3: $6 \to 13 \to 11 \to 8 \to 9$. The corresponding start time

Table 2. Parameters of tasks in 3 to 14 scenario

Task j	$MP_j : x_j$	$MP_j : y_j$	$D_j : t_{sj}$	$D_j : t_{ej}$	T_{1j}	T_{2j}	T_{3j}
1	17.0	30.0	0	34	6	6	6
2	16.0	42.0	0	34	6	6	6
3	30.5	43.2	0	44	8	8	8
4	29.5	27.0	0	34	6	6	6
5	50.4	45.3	16	56	8	8	8
6	45.0	62.1	0	54	6	6	6
7	40.2	42.1	0	54	6	6	6
8	76.1	47.0	0	54	6	6	6
9	82.0	30.0	0	54	8	8	8
10	68.3	28.5	0	54	6	6	6
11	62.1	48.5	0	54	6	6	6
12	62.1	48.5	16	80	6	6	6
13	52.4	49.1	16	80	6	6	6
14	78	33.5	16	80	6	6	6

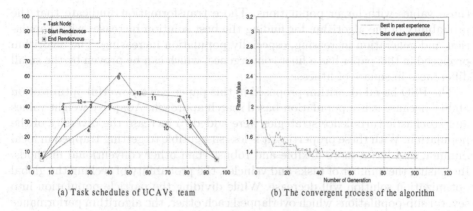

(a) Task schedules of UCAVs team   (b) The convergent process of the algorithm

Fig. 3. One result of the tests in scenario 3 to 14

of tasks are $3 \to 11 \to 19 \to 26 \to 39, 4 \to 12 \to 20 \to 32 \to 42, 7 \to 16 \to 23 \to 31 \to 39$. The total finish time is 51, the total route length is 410.4375. No task was delayed.

From the Fig. 3 (b), we can see that the PSO based algorithm presented in this paper can find the optimum resolution in less generations with small population, or in another word, a very short computing time.

More tests have been done with different populations and generations, and each group of tests have been done 50 times. Statistical data of these tests have been listed in table 3, including population size, generation number, difference in square of fitness value, computing time and the min, max, mean value in the results's fitness of these tests. The difference in square of fitness value is defined as: $D = \sum (Fit._{result} - Fit._{Opt.})^2 / N$, where $Fit._{result}$ is the best solution's fitness value of each test, $Fit._{Opt.}$ is the best solution's fitness value in all these tests, N is the test times. In this paper $Fit._{Opt.} = 1.3309, N = 50$.

As the population of particle swarm increasing, the mean fitness and the difference in square of fitness was decreased obviously, even with 100 hundred generations only. And they both were decreased too as the generation increasing with a small population. But the min, max, mean fitness values still higher than

Table 3. Statistical data of different populations and generations run for 50 times

Pop.Size	Gen.Num	Min.Fit.	Max.Fit.	Mean Fit.	D	Com. Time
100	100	1.3708	1.5589	1.4390	0.0131	2.6228
100	500	1.3538	1.4969	1.4267	0.0106	13.1397
100	1000	1.3342	1.5173	1.4328	0.0123	27.1528
200	100	1.3344	1.4835	1.4041	0.0072	5.3756
500	100	1.3190	1.4616	1.3700	0.0025	16.3303
800	100	1.3342	1.4311	1.3633	0.0017	32.3600
1000	100	1.3309	1.4067	1.3583	0.0010	48.3941

those values with larger populations. Those transformations indicate that the probability of the algorithm for finding the best solution increases with population and generation increasing, especially population increasing. The algorithm proposed in this paper has a finer convergence, which can be proven by the small differences in square of fitness.

The PSO based algorithm proposed in this paper has a powerful optimization ability because it is a population-based search approach. Adaptive weight values and stochastic turbulence strategies were added to increase the diversity of the population. With these properties, it is less susceptible to getting trapped in local minima, and has a more flexible and robust than other conventional methods. Increasing the number of tasks and vehicles, the probability of seeking the global optimization solution will decrease. While dividing the particle population into several sub-populations which overlapped each other, the algorithm performance in seeking optimization will be enhanced obviously.

Unlike genetic algorithm and other heuristic algorithms, PSO has the flexibility to control the balance between the global and local exploration of the search space. This unique feature of PSO overcomes premature convergence and enhances the search capability. And the velocity-position update of particles needs only primary computations, which consume very little computing time. So the PSO algorithm has a high computing effectiveness. This character is very important for PSO solving task scheduling problem in multi-UCAV cooperative control.

5   Conclusion and Future Work

In this paper, we formulated task scheduling problem in multi-UCAV cooperative control. The model presented in this paper is an integration of task allocating and task ordering. Using this model, the order or execute time for tasks to be executed will be planned while task allocating for vehicles, whereas traditional single task allocating can't satisfy this need. By mapping the solution of the problem to a particle, Particle Swarm Optimization algorithm is applied to solve the problem. The tests show that this is a feasible and efficient way for solving multi-UCAV cooperative scheduling problem.

In the future, we would like to incorporate some local searching methods to improve the performance of the algorithm for solving more complex instances, such as simulating anneal, new population topologies.

References

1. Li, S.M., Boskovic, J.D., Seereeram, S., Beard, R.W., et al.: Autonomous Herichical Control Multiple Unmanned Combat Air Vehicles (UCAVs). In: ACC. Proc. American Control Conferemer, Anchorage AK, pp. 274–279 (2002)
2. Darrah, M.A., Niland, W.M., Stolarik, B.M.: Multiple UAV dynamic task allocation using mixed integer linear programming in a SEAD mission. American Institute of Aeronautics and Astronautics (2005)

3. Li, M., Jovan, D., et al.: Autonomous hierachical control of multiple unmanned combat air vehicles (UCAVs). In: Proc. American Control Conference, Anchorage, AK, pp. 274–279 (2002)
4. Chandler, P.R., Pachter, M.: Hierarchical Control for Autonomous Teams. In: AIAA Guidance, Navigation, and Control Conference and Exhibit, Montreal, Canada, AIAA-2001-4149 (2001)
5. Schumacher, C., Chandler, P.R., Pacher, M., Pachter, L.: Constrained Optimization for UAV Task Assignment. In: AIAA Guidance, Navigation, and Control Conference and Exhibit, Providence, Rhode Island, AIAA-2004-5352 (2004)
6. Nygard, K.E., Chandler, P.R., Pachter, M.: Dynamic network flow Optimization Models for Air Vehicle Resource Allocation. In: Proc. American Control Conference, Argington, VA, pp. 1853–1858 (2001)
7. Huo, X.H., Zhu, H.Y, Shen, L.C.: Particle Swarm Optimization for Task Allocation in multi-UAV Cooperative Control. In: ICCES 2005. Proc. International Conference on Computational & Experimental Engineering and Sciences, India, pp. 610–615 (2005)
8. Chen, G., Curz, J.B.: Genetic Algorithm for Task Allocation in UAV Cooperative Control. In: AIAA Guidance, Navigation, and Control Conference and Exhibit, Austin, Texas, AIAA-2003-5582 (2003)
9. Lua, C.A., Altenburg, K., Nygard, K.E.: Synchronized multi-Point Attack by Autonomous Reactive Vehicles with Simple Local Communication, 95–102 (2003) IEEE 0-7803-7914-4/03
10. Maza, I., Ollero, A.: Multiple UAV Cooperative Searching Operation using Polygon Area Decomposition and Efficient Coverage Algorithms. In: Proc. 7th International Symposium on Distributed Autonomous Robotic Systems, Toulouse, France, pp. 211–220 (2004)
11. Kennedy, J., Eberhart, R.C.: Particle swarm optimization. In: Proc. 4th IEEE International Conference on Neural Network, pp. 1942–1948 (1995)
12. Shi, Y.H., Eberhart, R.C.: A Modified Particle Swarm Optimizer. In: Proc. IEEE International Conference on Evolutionary Computation, Piscataway, pp. 69–73 (1998)

Expert System Using Fuzzy Petri Nets in Computer Forensics

Hyun-Uk Hwang[1], Min-Soo Kim[2], and Bong-Nam Noh[3,*]

[1] National Security Research Institute, Korea
hhu@etri.re.kr
[2] Div. of Information Engineering, Mokpo Nat'l Univ.
phoenix@mokpo.ac.kr
[3] Div. of Electr-Comput. & Inform-Engine., Chonnam Nat'l Univ.
bbong@chonnam.ac.kr

Abstract. In the past, computer forensics was only used by means of investigation. However, nowadays, due to the sharp increase of awareness of computer security, computer forensics becomes very significant even to the nonprofessionals, and it needs inference as well as the integrity and reliability of the procedure. In this paper, we describe the inference rules using Fuzzy Petri Nets and adapt the collected data in a compromised system to a proposition for inference of the intrusion information. The inferred results are expressed as formalized 5W1H format. The COM-FEX(COMputer Forensic EXpert system) is inferable, even if the data is damaged in certain section, and the inference function of uncertainty is improved. This is useful to a system administrator who has weak analyzing ability of hacking, and it has improved capacity of managing the system security.

Keywords: Computer forensics, fuzzy Petri nets, inference rule, hacking, expert system.

1 Introduction

Security and hacking will be improved in the complementary relation, and computer forensics is more important to treat an event after an attack. Computer forensics deals with preservation, detection, analysis, and documentation of data. Its fields are classified into law enforcement, information warfare, and industrial security infrastructure [1]. The traditional computer forensics is focused on the law enforcement. It lays emphasis on legal issue in order that it should guarantee to protect a raw image and to verify the integrity, which means the evidence is not changed [2]. On the other hand, the computer forensics on military or commercial issue stress that the occurrence of an attack should be quickly detected and the threat of attack is quickly removed [1]. Because the existing tools for computer forensics show only simple results, the administrators have difficulty in analyzing the state of the damaged system without expert knowledge.

---

* Corresponding author.

M.S. Szczuka et al. (Eds.): ICHIT 2006, LNAI 4413, pp. 312–322, 2007.

Actually, a forensic expert uses the simple forensic tools to collect the evidence and to investigate the site of the event. Because system administrators are not able to analyze the hacking, they need some aid of the expert for analyzing hacking. Therefore, an advanced forensic tool is needed so that a non-expert administrator can utilize it with ease.

The crafty hackers remove their traces, install a tool for re-accessing anytime, and hide the tool. This prevents normal inferring the intrusion situation. The forensic experts are able to infer from the damaged data, but the non-experts can't get the intrusion information be-cause they depend only on the forensic tool. For this reason, even when the data is lost or modified, the inferring function of the hacking information is also needed.

In this paper, we present the collected information with the same result of 5W1H, in order to prove as an evidence of intrusion by analyzing it. This is called the abstraction process, and those who have no knowledge for computer forensics can use as an expert system through the process. We use FPN model to represent the abstraction process. The FPN is formed by making rules from the logical relationship of the digital evidences which are obtained from experiential analysis and defining them as the propositions. This FPN shows the inferring process from damaged evidence as well as whole evidence, we are sure that this method is more effective.

2   Computer Forensics

Forensics' is generally defined as a scientific and technical method of investigation or evidence proof at a trial of civil or criminal cases [3]. Similarly, a Computer Forensics,' which investigates a computer crime with a scientific and technical way and proves some facts, deals with maintaining, detecting, documenting, and analyzing computer data [4]. The basic principle of computer forensics is as follows. First, evidence must be obtained without any interference or modification to the original data. Second, the obtained evidence must be identified with the original data [5]. Last, the original data must be analyzed without any modification.

There are many tools for computer forensics according to its procedure and method. Representative forensic tools are Encase [6], TCT, TCTutils, and Sleuth kit [7]. Also, to support forensics, there are the other tools for detecting rootkits, informing file system and network (lsof), and for copying file blocks.

A basic difference can be found between Encase and TCT, which are the most famous computer forensics tools. Encase analyzes a duplicated image to preserve the integrity of a compromised system, on the other hand, TCT analyzes directly on the compromised system. Encase guarantees the system integrity from legal view, but it takes no thought of real-time information. TCT collects and analyzes the real-time information, but it doesn't have any guarantee of integrity. After all, the advanced computer forensics system has to collect the real-time information and guarantee the integrity as well.

Both Encase and TCT have the only function of showing the adaptable evidences. They do not support collecting evidences of specific attack, nor detect the kernel rootkits. Moreover, they don't have any function of inferring the intrusion situation from raw information. Therefore, it is difficult for a general administrator to understand the intrusion situation with these tools.

3   The Inference in Computer Forensics

In this section, we propose COMFEX(Computer Forensics Expert System) [8] that supports the function of collecting various evidences and inferring intrusion situation from the evidences. COMFEX has inference function as well as the searching function, and it helps easy understanding of the analyzed results. The information of existing forensic tools makes it difficult for a system administrator with no expertise to have a clear grasp of intrusion situation. Moreover, a thorough analyzing can't be expected with damaged information by attacks. Therefore, we present a method which is inferable from the damaged information.

3.1   Abstraction Process

Braian Carrier proposed the method of inferring significant information from raw data through abstraction layer in computer forensics system [9]. The abstraction layer generates the new information with the raw data and translation rule set. These data and translation rule set can be interpreted by using of the proposition and inferring rule set.

   The inference in computer forensics can be automated, and it gives analyzer convenience by quick showing formalized results. The inference, a reasoning process of a new fact from old ones, has deduction and induction [10]. Also, it is classified into forward inference for inferring the results from an actual fact and backward inference for inducting the results into the fact [11]. As the computer forensics is inferring the intrusion situation from the actual information, forward inference is more suitable.

   The information which is generated through the abstraction process is presented by 5W1H. Table 1 classifies the forensic evidence and inferential result into 5W1H. Who' which means who is intruder, is the intruder's ID and whois information of the intruder's IP address in computer system. When' means the time when the intruder invades the system, acts something, and sneaks out. Where' means where the intruder comes from and goes away. How' means how the intruder can invade. What' means what the intruder acts after invasion. Why' means the intruder's intention.

   However, the adaptation of general inference rules to computer forensics has a problem. Because the collected data in the actual compromised system may be modified or deleted, the inference can't work normally. So, the inference process must present how exact the intrusion situation is explained with the damaged data.

3.2   Computer Forensics Using FPN

We propose the inference process using FPN model for the purpose of inferring the intrusion situation from the data in compromised system. Generally, Petri

Table 1. Forensic evidence and inferential result with 5W1H

	Evidence		Inferential Result
	Command or COMFEX Analysis module	System Information	
Who	Connection situation (w, netstat), whois information	Behavior log of intruder	Intruder information
When	Login information, last, last-log	Acting time	Invading time
Where	Network connection, netstat, w, lsof, ps_detect	Connection log	Intruder's connection information
How	log_detect	Action in log files	Intrusion method
What	checkmd5, lkms, detect_rootkit, sniffer_check, rpm_check	Integrity check of binary files, Installation of rootkits	System damage, Installation of hacking tools
Why	data_check, lsmod, index_check, detect_rootkit	xferlog for ftp, Modification of index page, Search the attack tools and back-doors, Search the hacking tools	Information leakage, Homepage modification, Stepping stone, Agent for DDoS attack

Nets is tool for modeling the information processing system [12, 13]. The concept of FPN is derived from Petri nets [13]. A generalized FPN model can be defined as a 9-tuple [14]:

$$FPN = \{P, T, D, I, O, \mu, \lambda, \alpha, \beta\},$$

where

$P = \{p_1, p_2, \ldots, p_n\}$ is a finite set of places,

$T = \{t_1, t_2, \ldots, t_m\}$ is a finite set of transitions,

$D = \{d_1, d_2, \ldots, d_n\}$ is a finite set of propositions,

$P \cap T \cap D = \emptyset, |P| = |D|,$

I: $T \rightarrow P^\infty$ is the input function, representing a mapping from transitions to bags of places,

O: $T \rightarrow P^\infty$ is the output function, representing a mapping from transitions to bags of places,

μ: $T \rightarrow [0,1]$ is an association function, a mapping from transitions to real values between zero and one,

λ: $T \rightarrow [0,1]$ is an association function, a mapping from transitions to real values between zero and one,

α: $P \rightarrow [0,1]$ is an association function, a mapping from places to real values between zero and one,

β: $P \rightarrow D$ is an association function, a bijective mapping from places to propositions.

In realistic terminology, α_i represents the fuzzy belief (FB) of place p_i, i.e., $\alpha_i = \alpha(p_i)$; $\mu_j = \mu(t_j)$, and $\lambda_j = \lambda(t_j)$, represent the certainty factor (CF) and threshold of transition t_j, respectively. Further $d_i = \beta(p_i)$. A transition t_j is enabled if $\alpha_i : p_i \in I(t_j) > \lambda_j$. An enabled transition fires, resulting in a fuzzy

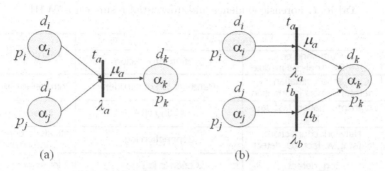

Fig. 1. The computation of fuzzy belief

truth token (FTT) at all its output arcs. The value of the FTT is a function of the CF of the transition and fuzzy belief of its input places.

Definition 1. If an inference model is inferable from the damaged data, it is called *the loss tolerance inference model*.

When the FPN is used in computer forensics, it should be explained how the damaged data is processed. A generalized FPN uses the minimum function (AND operation) or maximum function (OR operation) to compute a fuzzy belief. However, inferring from the damaged data is not possible because a transition is not fired in the state. Therefore, we use the average function for tolerating the loss information and make the transition fired.

Definition 2. (AND operation) If $I(t_a) = \{p_i, p_j\}$, $O(t_a) = \{p_k\}$, and average $\{\alpha_i, \alpha_j\} > \lambda_a$, a new fuzzy belief α_k is average(α_i, α_j) * μ_a (figure 1(a)).

Definition 3. (OR operation) If $O(t_a) = O(t_b) = \{p_k\}$, $I(t_a) = \{p_i\}$, $I(t_b) = \{p_j\}$, $\alpha_i > \lambda_a$, and $\alpha_j > \lambda_b$, a new fuzzy belief α_k is $(\alpha_i * \mu_a + \alpha_j * \mu_b)$ - $(\alpha_i * \mu_a)$* $(\alpha_j * \mu_b)$ (figure 1(b)).

Definition 4. A place with no input arcs is called *a starting place*. The initial fuzzy belief value of all places is 0. If the proposition of the starting place is true, the fuzzy belief value of the starting place varies as the experiential importance of a forensic expert.

Definition 5. A place with no output arcs is called *a final place*. The inferred results in the final places are expressed as 5W1H format.

4   The Inference Experiment on Computer Forensics

4.1   A Test Intrusion Scenario

In the purpose of inferring the intrusion situation, it is necessary to know where the data is and how it can be gained. The data is divided into two categories according to its decidability. The very important data, if it is not changed, is the system log file without doubt. The unchangeable data by an intruder is the

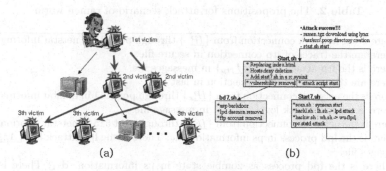

Fig. 2. The attack scenario of ramen worm

process and network information in system memory, rootkit information, the integrity check, etc [8]. The various attack scenarios are expressed as the FPN based on these data.

Now, we are about to express the intruding procedure of the system exploiting ramen worm [15] as the FPN. Figure 2 shows the attack scenario of ramen worm. The ramen worm is a self-propagating worm for Redhat Linux. The structure of ramen worm is consisted of start.sh(for attempting to alter the compromised server, bd7.sh(for patching the vulnerability and installing a backdoor), and start7.sh(for attacking another server).

4.2   The Inference Rule of the Scenario

The propositions for attack scenario of ramen worm are as follows(Table 2).

The ramen worm attack scenario is expressed as FPN (Figure 3). The undetermined value on FPN is the initial fuzzy belief in the starting place and the certainty factor in the transition. Determining the values needs experimental knowledge of an expert and adequate experimentation. This paper uses the following method to determine the values. The value of fuzzy belief corresponds to the trust value about the proposition of the starting place. If the proposition is true, the value of fuzzy belief is 1.0. In case of false, it is 0. However, if the proposition has uncertain value, the value of fuzzy belief will be between 0 and 1. The value of certainty factor for reasoning the different proposition from the given proposition means the certainty of inferring. Actually, on computer forensics, the evidence may be sufficient or insufficient information for legal. Thus if the proposition shows certain information, the value will be close to 1.0. If it shows uncertain information, the value will be lower.

The value of certainty factor is divided as the following definitions according to the relative importance as the legal evidence. There are certain evidences (CE) which are important inference rules and supportable evidences (SE) which are relatively less important inference rules. For instance, in case of CE, the proof of an attack appears clearly on file, *messages*, when ramen worm attacks the buffer overflow vulnerability of the compromised system. Simultaneously, the *ftp* authentication information on file, *secure*, appears not only from the

Table 2. The propositions for attack scenario of ramen worm

d_1: There is the network connection from $\{\underline{IP_x}\}$ through port 21 in netstat information.
d_2: There are the traces of ftp connection in secure file.
d_3: There is the ftp attack log $\{\underline{AM_{ftp}}\}$ in messages file.
d_4: There are the traces of ftp connection in last log
d_5: There is the network connection from $\{\underline{IP_x}\}$ through port 111 in netstat information.
d_6: There is the rpc attack log $\{\underline{AM_{ftp}}\}$ in message file.
d_7: There is the network connection from $\{\underline{IP_x}\}$ through port 515 in netstat information.
d_8: There is the lpd process in ps information. d_9: There is the lpd attack log $\{\underline{AM_{ftp}}\}$ in messages file.
d_{10}: There is the lpd process as zombie state in ps information. d_{11}: There is the /usr/src/.poop directory. d_{12}: There is the /usr/src/.poop/start.sh directory.
d_{13}: There is the ramen.tgz file in /tmp directory. d_{14}: There is the asp program in /usr/sbin directory d_{15}: There is the asp service in /etc/xinetd.d file. d_{16}: There are the .w file and the .l file
d_{17}: The port 27374 is opened. d_{18}: The binary file /bin/netstat is changed.
d_{19}: The binary file /bin/ps is changed. d_{20}: The binary file /bin/login is changed.
d_{21}: There is the index.html file of 373 bytes.
d_{22}: The E-mail is sent from the victim system to an attacker.
d_{23}: There is the $\{\underline{IP_x}\}$ address in /usr/src/.poop/myip file.
d_{24}: There is a ftp account in /etc/ftpusers file. d_{25}: There is an anonymous account in /etc/ftpusers file.
d_{26}: The size of /sbin/rps.statd, /usr/sbin/rpc.statd, and /usr/sbin/lpd becomes zero.
d_{27}: There is the /usr/src/.poop/start7.sh in /etc/rc.d/rc.sysinit.

activity of an attack also from the normal behavior. This paper gave $0.9(\mu_{ce})$ to CE and $0.5(\mu_{se})$ to SE. 0.9 means value of certain evidences and 0.5 is value of supportable evidences in extent of value between 1 and 0. The output places of next transition have valid propositions so that the certainty factor of the other transition is set $0.9(\mu_{other})$. The propositions of ramen worm can be divided into the following definitions according to the type of evidence.

CE: d_3, d_6, d_9, d_{11}, d_{12}, d_{13}, d_{14}, d_{15}, d_{18}, d_{19}, d_{20}, d_{23}, and d_{27}
SE: d_1, d_2, d_4, d_5, d_7, d_8, d_{102}, d_{16}, d_{17}, d_{21}, d_{22}, d_{24}, d_{25}, d_{26}, d_{27}, and d_{28}

The threshold value is the factor that determines the transition to fire. Since an error in fixing the critical value can derive the wrong inference, it must be the accurate value as far as possible. Thus, in this paper, the value can be decided through many experiments. The experimental result of the threshold value will be presented in the next section.

4.3   The Result Analysis of Inferring

Figure 4 is FPN for inferring 'how' of the ramen worm attack. The propositions are d_1, d_2, d_3, d_4, d_5, d_6, d_7, d_8, d_9, and d_{10}, and fuzzy belief and certainty factor is configured like figure 4. The final place is p_{how}. The expression is as follows.

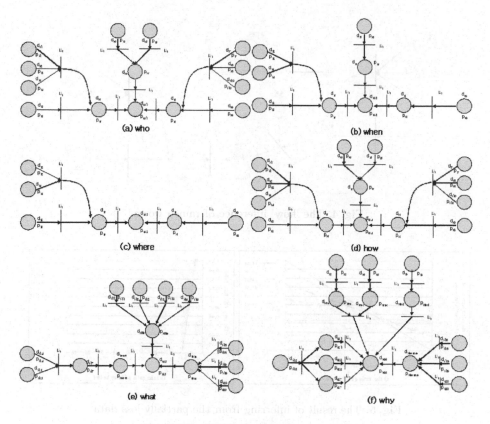

Fig. 3. The FPN of ramen worm attack

$I(t1) = (1.0 +1.0 +1.0)/3 * 0.5 = 0.5$ $I(t2) = 1.0* 0.9 = 0.9$
$\alpha(p_{af}) = 0.5 + 0.9 - (0.5* 0.9) = 0.95$ $\alpha(p_{ar}) = 0.5 + 0.9 - (0.5 * 0.9) = 0.95$
$\alpha(p_{al}) = 0.5 + 0.9 - (0.5 * 0.9) = 0.95$ $O(t7) = (0.95 + 0.95)/2 * 0.9 = 0.86$
$O(t8) = 0.95 * 0.9 = 0.86$ $\alpha(p_{how})= O(t7) + O(t8) - O(t7) * O(t8) =\mathbf{0.98}$

When no proposition loss exists, the final fuzzy belief value of target node p_{how} becomes 0.98.

This paper tested the inferring process. This assumes that the basic information of the system is lost. Figure 5 presents the result of inferring from the partially lost data out of 10 propositions. In figure 5(a), starting from 1 data loss to 6 data losses, the inclinations represent the ratio of successful inference. If there is 1 loss, which has the same effect as none, the graph shows the result of constant proportion. However, in case there are 5-6 data losses the inclination decreases, which implies less successful inference result. In figure 5(b), the graph shows the distribution about inferring result of the data loss from 1 to 6. We conclude with the fact that fuzzy belief value beyond 0.7 is in general many and also that proper threshold value used to FPN is 0.7.

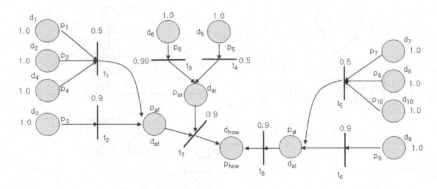

Fig. 4. The 'how' inference of ramen worm

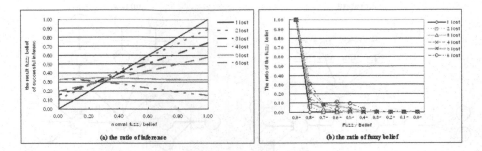

Fig. 5. The result of inferring from the partially loss data

On the basis of the result above, the cases that the log-file was deleted and the vola-tile information was lost are stimulated. In case that attacker deleted the messages file, the fuzzy belief of propositions of d3, d6, d9 set 0. This is shown as follows.

$I(t1) = 0.5,\ I(t2)= 0.0,\ \alpha(p_{a1}) = 0.5,\ \alpha(p_{a2}) = 0.5,\ \alpha(p_{a3}) = 0.5$
$O(t_7) = 0.45,\ O(t_8) = 0.45$
$\alpha(p_{how}) = O(t_7) + O(t_8) - O(t_7) * O(t_8) = 0.698$

Final fuzzy belief value of target node phow becomes 0.698. The result is less than threshold value 0.7 and the significant log-files are deleted so that the credibility of the result is obscured. Thus, additional process recovering log-files and discovering hidden information should be required to complete the evidence.

In case that there is no volatile information collected, the fuzzy belief of the propositions d1, d5, d7, d8, d9 and d10 set 0. This is shown as follows.

$I(t1) = 0.335,\ I(t2)= 0.9,\ \alpha(p_{a1}) = 0.934,\ \alpha(p_{a2}) = 0.9,\ \alpha(p_{a3}) = 0.9$
$O(t_7) = 0.825,\ O(t_8) = 0.81$
$\alpha(p_{how}) = O(t_7) + O(t_8) - O(t_7) * O(t_8) = 0.967$

At the target node, p_{how}, the final value of fuzzy belief becomes 0.967. Since there are log information and supportable information that is decisive evidence about the attack, fuzzy belief of the ultimate node has reliably high value. The certain value, 0.967 can enhance the rate of trust about the reasoning. Thus, the proposed method in this paper has the following merits. First, it reports the result with a fixed form according to 5W1H. Second, it can enhance the rate of trust about the reasoning from the value of fuzzy belief. Third, the administrator can deeply investigate the insufficient portion on the investigation procedure. Forth, the administrator with no expert knowledge can guarantee the level of trust of forensic operations.

5   Conclusion

When a system is invaded, an investigator usually uses the computer forensics techniques or tools. However, the existing computer forensics has some difficulty in inferring intrusion information, as the hacking is getting craftier and doesn't leave any traces. So, we proposed the inferring function with even uncertain data.

We presented the intrusion situation with the same result of 5W1H. This helps non-experts to know the full account of the event. Fuzzy Petri net was used to extract data from compromised system and convert this data to easily understood information. Inference function was also changed for suitability to computer forensics.

This paper presented the ramen worm attack scenario and the result of inferring. The inferring function was experimented using damaged data. That is, partly lost data was input into inference rule and the result was successful. This proved that the inference from dam-aged data is possible. In the future, data association of inference and finding the accurate value for the variables of fuzzy Petri net should be studied.

References

1. Palmer, G.: A road map for digital forensic research. In: Digital Forensics Research Workshop (2001, 2002, 2003)
2. Carrier, B.: Open source digital forensics tools: the legal argument, @stake (October 2003)
3. Kruse, W.G., Heiser, J.G.: Computer Forensics: Incident Response Essentials. Addison Wesley, Reading (2001)
4. Vacca, J.R.: Computer Forensics: Computer Crime Scene Investigation. Charles River Media (2002)
5. Marcella, A.J., Greenfield, R.S.: Cyber Forensics: A field Manual for Collecting, Examining, and Preserving Evidence of Computer Crimes, Auerbach (2002)
6. Guidance Software, EnCase Product Description, whitepaper (April 2005)
7. Carrier, B.: Sleuthkit v2.04 (May 2006), available at: http://sleuthkit.org
8. Hwang, H., Kim, M., Noh, B.: COMFEX: Linux forensic system supporting original informa-tion collection and hacking analysis. In: WISA 2003 (August 2003)

9. Carrier, B.: 'Defining digital forensic examination and analysis tools using abstraction layers. International Journal of Digital Evidence 1(4) (2003)
10. Russell, S.J., Norvig, P.: Artificial Intelligence: A modern Approach. Addison-Wesley, Reading (1995)
11. Yager, R.R.: Approximate reasoning as a basis for rule-based expert systems. IEEE Trans. Syst. SMC-14(4), 636–643 (1984)
12. Murata, T.: Petri nets: properties, analysis and applications. Proceedings of the IEEE 77(4), 541–580 (1989)
13. Peterson, J.L.: Petri Net Theory and the Modeling of Systems. Prentice-hall, Englewood Cliffs (1981)
14. Chen, S., Ke, J., Chang, J.: Knowledge representation using fuzzy Petri nets. IEEE Transac-tion on Knowledge and Data Engineering 2(3), 311–667 (1990)
15. SANS Institute, Ramen worm (2001), Available at: http://www.sans.org/y2k/ramen.htm

MMORPG Map Evaluation
Using Pedestrian Agents

Christian Anthony L. Go, Tristan Basa, and Won-Hyung Lee

Chung-Ang University,
Graduate School of Advanced Imaging Science, Multimedia and Film
221 Hukseok-Dong, Dongjak-Gu, Seoul, Korea
chipgo@gmail.com, trisb@hotmail.com, whlee@cau.ac.kr

Abstract. Massive Multiplayer Online Role-Playing Games (MMO-RPG) increasingly become places of social engagement, by providing spaces for social interaction and relationship beyond home and work-place. The center of this is the virtual environment, in which players interact. Today's savvy players are demanding progressive, more playable and navigable game environments. Traditional methods of the game map evaluation, though reliable, do not offer a quantitative measure of players discomfort and efficiency. This paper introduces a complementary method of the MMORPG map evaluation by utilizing pedestrian agents. The agents employ calculable measures of walking efficiency and discomfort, providing objective criteria, against which the game maps are evaluated. Thus, they promote social interaction by improving the space wherein the players interact.

Keywords: MMORPG, Online Game, Multi-Agent Systems, Game AI.

1 Introduction

The explosive growth of the Internet, rapid advances in computer and network technologies, along with the widely deployed access and mobile connectivity to Internet infrastructure and services, motivated the designers of computer games to create new, interactive, and distributed multiplayer games, or even to add online multiplayer capabilities to already existing ones. Leveraging these technologies and experiencing phenomenal growth is the Massively Multiplayer Online Game (MMOG). Majority of the MMOG market is dominated by the MMORPG sub genre.

MMORPG stands for massively multiplayer online role-playing game. In an MMORPG, thousands of players exist in the same game world at the same time. This creates an incredibly rich and active environment in which interesting things are constantly taking place. MMORPGs provide thousands of hours of game play, with a nearly infinite variety of goals to achieve across a vast world covering miles of land and sea.

Of vital importance to the MMORPG experience is the realism and immersiveness of the virtual world. These virtual worlds are experienced more like places and the users feel to be part of something they perceive of as a place [5].

M.S. Szczuka et al. (Eds.): ICHIT 2006, LNAI 4413, pp. 323–332, 2007.

For virtual worlds to be immersive, they must not only be realistic, they must likewise be navigable. When it comes to the virtual realm, today's savvy gamers not only demand realism, they demand ease of use and intuitive navigation.

Traditional methods of map evaluation involve human operators/testers exploring the virtual environment and, based on certain criteria, judging whether it is navigable and intuitive. Other methods of game map evaluation include the use of bots-a type of computer program to do automated tasks, to simulate real players in a game. Yet another method is open and closed beta testing. This involves actual users testing the beta version MMOORPG game environment online. All the above methods, while effective, each have their own limitations.

This paper introduces an alternative, quantifiable method of evaluating the navigability and intuitiveness of an MMORPG Map by utilizing pedestrian agents and their feedback.

2   Related Work

Multi-agent systems (MAS) have been extensively applied to virtual worlds. Dickerson and Kosko [6] proposed the use of Fuzzy Cognitive Maps (FCM) for modeling virtual worlds. The agents in this previous research, however, did not focus on goal-oriented behaviors. Converseley, Catelfranchi [7] discussed the reasons behind the exigency for goal-autonomy. He describes an "executive autonomy", which is that goal-oriented behavior which chooses among alternative courses of action.

Shared virtual worlds also have numerous applications in computer supported collaborative work (SCCW) and computer games. Leong and Chunyan [8] proposed a new type of intelligent virtual agents within an MMOG. These fuzzy cognitive goal net (FCGN) agents exhibited goal-oriented behavior and non-deterministic behavior. They also demonstrated that FCGN agents are able to fit the constraints for game AI in a real-time MMOG. Massive numbers of instantiations of intelligent virtual agents within the MMOG was performed [8].

Sharifi, Mousavian and Aavani [10] were able to predict the state of the environment in an MAS by predicting the future behavior of each agent and its relationship behavior with other agents and its environment. This prediction agent was implemented both heuristically and using a neural network. The game involved the 2002 Robocup simulation league as test data. Their experiments showed that an acceptable level of prediction under both heuristic and neural network was attainable. Tveit et al. [9] implemented a scalable and flexible agent-based simulation of players in an MMOG through their parallel mobile agent platform-Zereal. Players and NPCs (Non-Player Characters) on this platform have a sense-reaction-act behavior whose reasoning is founded on Markov Chains and reasoning plans. This implementation has been shown to be close to linearly scalable in terms of players simulated.

Go and Lee [12] likewise proposed the idea of using pedestrian agents to navigate virtual terrain. These agents employed the social forces model and we will implement their proposal in this paper.

3   MMORPG

The aim in an MMORPG is to allow a set of player's avatars to interact in a virtual world, by following specific game rules. The typical characteristics of an MMORPG are:

- Thousands of simultaneous players.
- A very big 3D environment with several cities and vast areas between them.
- Character classes of varied complexity.
- A set of skills for the player to choose from and develop for the character during the game by usage and by assigning experience points.
- Combat system, in game mostly used for fighting NPC (Non-player characters) foes like monsters, but optionally to combat other players.
- Magic system that ties into combat system and skill system.
- Items in game world that can be used by players as equipment or modified and used by using acquired skills.
- In game trading between players, which often extends to out of game trading with real money.
- Homes, areas in the game that a single or several players have ownership over and can modify by placing and storing items in them.
- Quests for players to perform, either in the form of items or NPCs in the game world leading or motivating players to perform a certain series of actions, or events initiated by a game master or implemented by a live team.
- Evolving story line, i.e. the history of the game world.
- Social systems allowing players to form permanent or temporary groupings.
- NPCs of several types, usually including monsters, humanoid NPCs that trade items and humanoid NPCs that deliver quests [1].

One of the major trends in MMORPGs involves virtual worlds. Game developers are striving to create worlds that provide an alternative to the real world but with similar perceived levels of complexity and realism. This in-game complexity is born when developers are able to create systems that are so dynamic that a massive number of players can use in-game features to create systemic complexity by interacting according to different frameworks for social structure, politics and economics. The idea and praxis of a virtual life parallel to real life is promoted to the users [1].

4   The Social Forces Model

The social forces model states that the motion of pedestrians can be described as if they would be subject to 'social forces'. These forces are not directly exerted by their environment, but rather, are a measure of the internal motivations of a pedestrian to perform certain movements. In this case, the motivation of a pedestrian evokes the physical production of an acceleration or deceleration force as a reaction to the perceived information that is obtained about the environment [3]. The following effects determine the motion of a pedestrian:

1. He/She wants to reach a certain destination as comfortable as possible. Therefore, a pedestrian normally walks the shortest possible way without detours. This shortest path to the destination will therefore have the shape of a polygon. The pedestrian will try to avoid taking detours moving opposite the desired direction even if the direct way is crowded.
2. The motion of a pedestrian is influenced by other pedestrians and borders. The pedestrian will keep a certain distance from other pedestrians and borders of walls in order to avoid being hurt, this is dependent on pedestrian density and desired speed. This distance is smaller the more a pedestrian hurries and it also decreases as pedestrian density increases.
3. Pedestrians are sometimes attracted by other persons or objects. These can be modeled by time-dependent attractive forces in a way similar to the repulsive effects but with an opposite sign. This attractiveness, while responsible for the formation of pedestrian groups, is normally decreasing with time since interest is declining. These effects are responsible for the formation of pedestrian crowds which demonstrate the tendency of pedestrians to exhibit joining behavior [3].

Combining these factors, the social force model is summarized by the equation:

$$\vec{F}_\alpha(t) = \vec{F}^0_\alpha(\vec{v}_\alpha, v^0_\alpha \vec{e}_\alpha) + \sum_\beta \vec{F}_{\alpha\beta}(\vec{e}_\alpha, \vec{r}_\alpha - \vec{r}_\beta)$$

$$+ \sum_\beta \vec{F}_{\alpha\beta}(\vec{e}_\alpha, \vec{r}_\alpha - \vec{r}^\alpha_\beta) + \sum_\iota \vec{F}_{\alpha\iota}(\vec{e}_\alpha, \vec{r}_\alpha - \vec{r}_\iota, t) \qquad (1)$$

wherein the first term of the equation is the acceleration term. The second term represents the distance to other pedestrians. The distance to borders of walls is the fourth term and the last term represents attraction to other pedestrians and street artists.

Given this mathematical representation of a pedestrian's movement we can derive mathematical measures of performance such as efficiency and discomfort. The Efficiency Measure is given by:

$$E := \frac{1}{N} \sum_\alpha \frac{\overline{\vec{v}_\alpha \cdot \vec{e}_\alpha}}{v^0_\alpha} \qquad (0 \le E \le 1) \qquad (2)$$

where N is the number of pedestrians α and the bar denotes a time average. This measure calculates the mean value of the velocity component into the desired direction of motion in relation to the desired walking speed [4].

Discomfort/uncomfortableness is given by:

$$\bigcup := \frac{1}{N} \sum_\alpha \frac{\overline{(\vec{v}_\alpha - \overline{\vec{v}_\alpha})^2}}{(\overline{v_\alpha})^2} := \frac{1}{N} \sum_\alpha (1 - \frac{\overline{\vec{v}_\alpha}^2}{(\overline{v_\alpha})^2}) \qquad (0 \le \bigcup \le 1) \qquad (3)$$

This reflects the degree and frequency of sudden velocity changes, the level of discontinuity of walking due to necessary avoidance measures [4]. Hence, optimal pedestrian facilities would be ones with the highest values of efficiency and comfortableness. Optimization of pedestrian facilities using these measures could be applied not only to new facilities but to existing ones with the goal of reducing bottlenecks by introducing subtle modifications.

The social force model of pedestrian dynamics, despite its simplicity, very realistically describes a lot of observed phenomena. Computer simulations of pedestrian groups demonstrated 1. lane development consisting of pedestrians who walk in the same direction, 2. oscillating changes of walking direction that occurs at narrow passages and 3. the spontaneous formation of roundabout traffic at crossings [4]. These self-organized spatio-temporal patterns arise due to the nonlinear interactions of pedestrians. Since variables of pedestrian motion are easily measurable, this model is easily comparable with empirical data [3]. These include self-organization under certain conditions and collective behavioral patterns.

5 Present Methods of Map Evaluation

Most people engaged in the construction of theories for the design of Virtual worlds come from a myriad of backgrounds. These range from a background in traditional architecture, media and fine arts and design, to computing science and systems development. These three groups tend to look at the virtual world in very different ways. Architects see them as new places to fill with architectural constructs, artists see a new medium for expression and computer people look at them as an advanced distributed information and communication system [2]. Critical to development of virtual worlds is finding a balance between all three which also appeals to the end user.

Traditional methods of map evaluation involving human testers and operators, are dependent upon the subjective taste of the evaluators, making justification for the selected terrain a matter of taste, experience and gut feel. Furthermore, it is likewise time-consuming and potentially wasteful of valuable human resources. The use of bots for testing, while able to identify possible areas of congestion and simulate user loads, also has its limitations. Bots can only simulate player movements but cannot capture the individual players' satisfaction/dissatisfaction with the virtual environment. Open server testing, can accurately reflect end-user sentiment and acceptance of the game map, however, this is considered the final step before full launch and any major design modifications would severely affect game production and release timelines. Moreover, it would be extremely difficult to obtain quantifiable, objective feedback from participants since individual player preference is likewise subjective.

6 Experimental Setup

Due to the proprietary nature of most MMORPGs and game companies' reluctance to divulge their source code, we sought an alternative that had all the elements of an MMORPG yet allowed scripting and code modifications. With this in

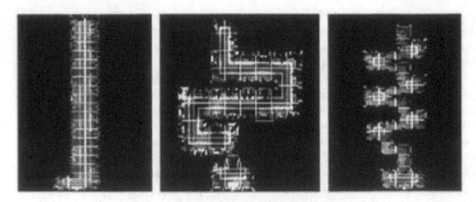

Fig. 1. Game Maps

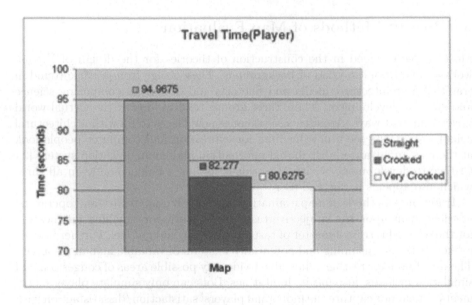

Fig. 2. Player Travel Times

mind, experiments were conducted on the Aurora Game Engine of BioWare-the game engine used for the massively popular Neverwinter Nights RPG franchise, which has been expanded into an MMORPG [11]. We likewise selected this game engine due to the large number of players that regularly interact on it online.

For the experiments, three game maps of identical lengths, albeit differing designs were specially created. (Figure 1) The first game map was a simple, straight map with no turns, the second was a moderately twisting one with a number of turns and the final map was a convoluted serpentine one. Initial testing was done with human players to set a satisfaction/dissatisfaction benchmark and for future comparison. Thirty (30) volunteers were instructed to navigate each game map thrice and traveling times were computed for each iteration and averaged. Volunteers were comprised of computer game players and all were adept at navigating virtual terrain. The results are illustrated by figure 2. Pedestrian agents were implemented on the game engine and instantiated to run on each of the game maps. Due to processing power constraints and the need to minimize system overhead, a maximum of eight (8) agents at a time per map were implemented. The results of their leaving times are shown by figure 2. All travel times were recorded and averaged. After the exercise, volunteers were instructed to rate each map's discomfort on a scale of 1-40, with 40 being the most uncomfortable.

7 Results

Human testers demonstrated, surprisingly, that while games maps with more turns are tedious to navigate, these can be traversed in less time given certain conditions. This is attributed to the fact that human testers unconsciously seek the shortest distance between any two points in the game map, "cutting corners", thereby minimizing travel time. Pedestrian agents likewise displayed this to a degree as is evident in figure 3. When the number of pedestrian agents was increased however, leaving times increased in direct proportion due to necessary collision avoidance measures. It is interesting to note that as maps become more difficult to navigate, each additional pedestrian agent increases overall leaving time by a larger margin (Figure 4).

A slight disparity was observed between agent discomfort and player discomfort, while pedestrian agents deemed the straight map to have the least discomfort,

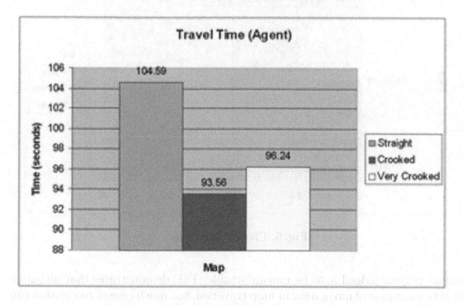

Fig. 3. Agent Travel Times

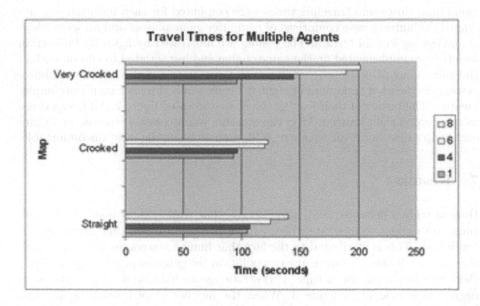

Fig. 4. Multiple Agent Travel Times

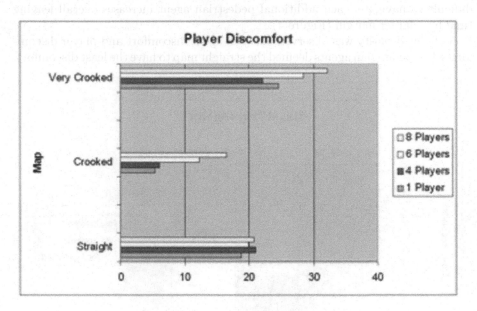

Fig. 5. Player Discomfort

human players judged it to be uncomfortable. This demonstrates that although players want ease of navigation in map traversal, too much ease of navigation can also be uncomfortable because it leads to boredom (Figure 5 and 6).

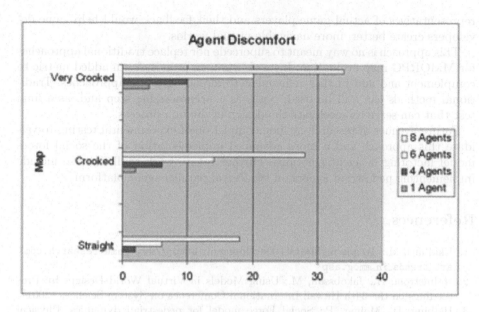

Fig. 6. Agent Discomfort

Moreover, a positive correlation was observed between leaving time and discomfort. Due to the complexity of implementing pedestrian agents and monitoring agent discomfort on other game engines, travel time can substitute as a reasonable measure of agent discomfort. This, however, must be precisely compared against a clearly set benchmark.

8 Conclusion

Using pedestrian agents, we can represent actual player avatars and simulate player movements across the map, thereby encapsulating the actual player experience as the virtual world is being navigated. At the same time, we collect data from the agent with regards to its satisfaction or discomfort with the terrain. With precise measures, both efficiency and discomfort can be accurately obtained from each individual pedestrian agent based on its independent experience of the virtual terrain map. The goal is to maximize efficiency and minimize discomfort, thereby reducing areas where potential bottlenecks may occur, while, at the same time creating an easily traversed game map. An optimal map would be one with the highest value if efficiency and the lowest value of uncomfortableness.

The advantage of this method over more traditional ones is that each pedestrian agent will be able to precisely report whether the map is an efficient or an uncomfortable one based on its individual explorations. This removes the element of subjectivity when employing human testers, by introducing quantitative measures of efficiency and discomfort. The pedestrian agents act as representatives of actual game players and their feedback would help game developers create better, more navigable virtual worlds.

This approach is no way meant to supercede nor replace traditional approaches for MMORPG map evaluation. Instead, what we propose is an added metric to complement and add further refinement to already-existing approaches. Traditional methods can still be used, both, as a preprocessing step and, as a final test that can serve to accelerate the design evaluation process.

Future avenues of research include completion of experimental testing to validate this approach and a more advanced implementation of the social forces model factoring in agent emotions. Further future work shall likewise include implementing pedestrian agents on the Zereal parallel agent platform.

References

1. Eladhari, M.: Trends in MMOG Development, http://www.game-research.com/art_trends_in_mmog.asp
2. Holmstrom, H., Jakobsson, M.: Using Models in Virtual World Design. In: Proceedings on the 34th Hawaii International Conference on System Sciences (2001)
3. Helbing, D., Molnar, P.: Social Force model for pedestrian dynamics. Physical review E 51, 4282–4286 (1995)
4. Helbing, D., Molnar, P.: Self-Organization Phenomena in Pedestrian Crowds, Self-Organization of Complex Structures: From Individual to Collective Dynamics, Gordon and Breach, pp. 569–577 (1997)
5. Croon Fors, A., Jakobsson, M.: Beyond Use and Design. The dialectics of being in virtual worlds, presented at Internet Research 1.0, Lawrence, Web Publicaton, 28 (September 2000)
6. Dickerson, J.A., Kosko, B.: Virtual Worlds as Fuzzy Cognitive Maps. Presence 3(2), 173–179 (1994)
7. Castelfranchi, C.: Guarantees for Autonomy in Cognitive Agent Architecture. In: ECAI Workshop on Agent Theories, Architectures and Languages, pp. 56–70 (1994)
8. Leong, P., Chunyan, M.: Fuzzy Cognitive Agents in Shared Virtual Worlds. In: Proceedings of the 2005 International Conference on Cyberworlds (2005)
9. Tveit, A., et al.: Scalable Agent-based Simulation of Players in Massively Multiplayer Online Games. In: Proceedings of the 8th Scandinavian Conference on Artificial Intelligence (2003)
10. Sharifi, M., et al.: Predicting the Future State of the Robocup Simulation Environment: Heuristic and Neural Networks Approacher. In: 2003 IEEE Internatinal Conference on Systems, Man and Cybernetics (2003)
11. Neverwinter Nights (NWN), produced by BioWare and published by Infogrames (now Atari)
12. Go, C.A., Lee, W.H.: Virtual Terrain Evaluation Using Pedestrian Agents. In: Proceedings of the Second International Conference on Design Computing and Cognition (2003)

The Analysis of Game Playing Experiences: Focusing on Massively Multiplayer Online Role-Playing Game

Seungkeun Song[1], Joohyeon Lee[1], and Jun Jo[2]

[1] Cognitive Science Program, The Graduate School, Yonsei University, 134 Shinchondong, Seodaemungu, Seoul 120-749, Korea
[2] Robotics and Games Laboratory, PMB50 Gold Coast Mail Centre, Griffith University, Gold Coast Campus, Queensland, Australia
{songsk,ljhyeon}@yonsei.ac.kr, j.jo@griffith.edu.au

Abstract. The purpose of this study is to (1) develop an analytic framework to systematically code the players' cognitive process during Massively Multiplayer Online Role-Playing Game (MMORPG) gameplay, and (2) to empirically explore the players' cognitive process. To construct the analytical framework of MMORPG gameplay, previous studies regarding gameplay and problem solving theory are reviewed. The specific gameplay actions and contents are derived by using a concurrent protocol analysis method. Consequently, gameplay actions are categorized into kinematics, perception, function, representation, methodology, and simulation. A new framework suitable for MMORPG gameplay is built. In order to study the players' cognitive process, the empirical experiment is executed during MMORPG gameplay. As a result of this study, we find a new problem space in the methodological scheme of gameplay. This study concludes with a number of key implications for game design in order to improve the quality of future gaming products.

Keywords: MMORPG, Game Design, Protocol Analysis, Problem Solving Theory, Play Theory, Analytical Framework, Affordance.

1 Introduction

Computer games are a high-tech, composite art including graphics, interfaces, gameplay, and stories. According to the game developer Crawford [1], the elusive trait of gameplay is derived from the combination of pace and cognitive effort required by the game. The gameplay is defined as the problems and challenges a player must overcome to win a game from the HCI designer's viewpoint [2]. Ludologists focus on game mechanics rather than analyzing games as an extension of the narrative, while narratologists argue that games are closely connected to their stories. They claim that gameplay is a configurative practice, rather than a more passive behavior, such as reading a novel or watching a movie [3]. These statements address uniformly the common point that gameplay is a human cognitive activity. However, most scholars and developers focus their efforts on

M.S. Szczuka et al. (Eds.): ICHIT 2006, LNAI 4413, pp. 333–343, 2007.

gameplay characteristics that reflect their academic backgrounds or the games they study.

One of the major problems for current game designers is that there is no commonly accepted theoretical basis to their craft. If there were, communication of ideas between game developers would be simpler and common mistakes could be avoided [4]. 'Level Design' is one of the most important design elements and directly affects gameplay. Unfortunately, it is carried out based on the past experiences of game designers often without sufficient theoretical background. Consequently, it is difficult to systematically manage their knowledge with current industry vocabulary due to the lack of a theoretical framework for computer games. This is because little empirical research has been conducted specifically on gameplay as it relates to a game players experience.

The primary goals of this research are (1) to develop an analytic framework to systematically code the game players cognitive processes in MMORPG gameplay and (2) to explore the game players cognitive process empirically.

2   The Development of Gameplay Framework

In order to systematically structure the player's playing experience we have devised the current gameplay framework using four sources of information.

Table 1. Gameplay Framework for MMORPGs

This Study			Related Studies				
Category		Action & Content	Kavakli, Thorne (2002)	Kim et. al. (1995)	John, Vera (1992)	Problem Solving (1972)	Play (1944, 1958)
Motor	Kinematical	<K-action> Move, View, Attack, Chat, Acquire	Physical Action		Keystroke-Operator		
Percep -tion	Perceptual	<P-action> Identify, Read	Perceptual Action				
Cogni -tion	Functional	<F-action> Search, Battle, Gather, Avoid, Communication	Functional Action		Functional-Operator		
	Represent-ational	<R-content> Goal: Entity, Relation <R-action> Understand, Construct	Conceptual Action	Represent -ation Space	Goal	Initial State Goal State Constraint	
	Methodol-ogical	<M-content> Template Method: (1) Find-No-Constraint, (2) First-Remove-Danger Sequential Method, Other Methods: Ambient, Pulling, Move-close <M-action> Evoke, Derive, Infer		Rule Space	Method, Selection Rule		Property: (1)Rule, (2)Free Type: (1)Competi -tion, (2)Chance, (3)Simula -tion, (4)Vertigo
	Simulative	<S-action> Generate, Refine, Evaluate		Instance Space			

First, Kavakli and Thorne's study [5] proposed the basis for the content and form of the present framework. For example, the concept of Kavakli and Thorne's information categories (i.e. physical, perceptual, functional, and conceptual categories) is the basis for action categories in our framework, such as kinematical, perceptual, functional, and representational schemes. Second, Kim et al. [6] and John's studies [7] built the basis for the process-oriented approach in the present framework. Kim's study [6] shows that the problem spaces of representation, rule, and instance are closely connected, and therefore the Problem Solving Theory (PST) is useful for the development of information systems, such as 'Object-Oriented Programming (OOP)'. Since John and Vera's study [7] presented the high possibility of applying GOMS methodology to computer game studies, game designers could use this to identify goals, operators, methods and selection rules of gameplay. The two studies constructed the basis which made it possible to define dynamic information processes, such as switching between problem spaces for the development of OOP programming [6] and selecting methods in order to overcome and win a video game (i.e., Nintendo's Super Mario Bros. 3®) [7]. Third, PST was classified in more detail to allow for the inclusion of the initial state, the goal state, and the constraint in the representational category [8]. According to Caillois [9], 'Ludus' is defined as an "activity organized under a system of rules that defines a victory or a defeat, a gain or a loss". 'Paidea' is defined as "prodigality of physical or mental activity which has no immediate useful objective, nor defined objective, and whose only reason to be is based in the pleasure experimented by the player". The concepts of 'Ludus' (rule) and 'Paidea' (free) within Play Theory cast an important insights into how to learn the criteria for selecting one of several alternative methods for the same subgoal in our GOMS analysis[1]. In other words, the player's intent and rule are classified into four categories - 'Agon' (competition), 'Alea' (chance), 'Mimicry' (simulation), and 'Ilinx' (vertigo) - which have an important influence on the investigation of other criteria for selection between the 'Template-Method' and the 'Sequential-Method' for collecting quest items as a subgoal. For example, the 'Template-Method' means that the player applies the game rules to his or her gameplay based on a sufficient understanding of the game while a holding template prepared on the rules in a game from the beginning. The notion of 'Agon' and 'Alea' based on 'Ludus' led our GOMS analyses to elucidate the 'Template-Method' from alternative methods. The 'Sequential-Method' indicates that the player holds the strategy to sequentially learn the rules in a game through trial and error without prior knowledge in the beginning of the gameplay. The idea of 'Mimicry' and 'Ilinx' based on 'Paidea' caused them to clarify the 'Sequential-Method' from several other potential methods. Consequently, the criterion for selection between them was the player's abstraction level of the game. A description of the full set of criteria is beyond the scope of this paper; we will refer the reader to a subsequent article for a more detailed account. Fourth, the intensive

---

[1] We conducted a GOMS analysis of World of Warcraft (WOW) as a small study. The task used in this analysis is the quest to collect Milly's harvest which is related to picking up quest items while avoiding dangerous monsters in the early stages of the game. Although a game manual provides the overall goals of game, operators, and some methods, it does not describe the selection rules necessary to decide between alternative methods. We produced selection rules for our analysis of verbal comments and in-game actions of three experts (Expert 1, Expert 2, and Expert 8).

analysis of verbal and action protocols of the player has provided an enormous number of concrete examples for generalized categorization and has developed an analytical framework for studying gameplay [10]. It was through a repeated process of adjusting a set of categories in a top-down approach, and then testing it on the examples, that we have finalized the framework into the present form (see Table 1).

3   Method

Twenty seven undergraduate students from Kyonggi University, Seoul, Korea participated in the experiment. The participants were classified into three groups according to their level of expertise with MMORPGs. The first level represented experts who were sufficiently experienced with MMORPGs and other console games as well as the World of Warcraft (WOW) for more than a year. The second level represented intermediate players that were experienced with MMORPGs but new to the WOW game. The third level represented novices who were new to all MMORPGs. The sample population contained 19 males and 8 females, including 9 experts, 9 intermediate players, and 9 novices. Ages ranged from 18–26, with a mean of 21.66 and a standard deviation of 2.336. The participants chosen for the study represent the general demographic distribution of computer players [11].

This study chose MMORPG as the target genre as it matched well with viewing gameplay as a problem solving process. The WOW game was selected from among the various MMORPG titles currently available. The selection was made based on its commercial success [11] and its excellence as commented on by many game critics. Specifically, for WOW, the game progresses through a series of quests. Quests can be conceptualized as a smaller game or task within the larger context of the WOW game. Moreover, due to the excellent design of WOW, which allows for proper constraints per game content, players are able to easily perceive the problems and challenges that they must face to win the game. These design factors in WOW are vital to the nature of gameplay. WOW, therefore, is an excellent platform for studying the problem-solving aspects of MMORPG gameplay.

The experimental sessions were divided into five sections. First, the participants were provided with instructions about the general nature of the experiment and the fact that verbal data would be collected. The participants were trained to "think-aloud" using two traditional tasks [10]. All participants were asked to select only the warrior class from the human race of the Alliance[2], one of the player characters in the game, in order to guarantee a homogeneous experimental environment for the participants. The actual playing session ranged from 1 to 2 hours. The participants' utterances and actions were recorded during gameplay, and were later asked to comment on their gameplay activities while reviewing

---

[2] Two character types, either alliance or horde members, exist in the game of WOW. Alliance characters represent good, and horde characters represent evil. The two character types face off in conflict.

the videotape. Finally, the participants were debriefed by the researcher about the primary purpose of this experiment.

In the early stages of the game, the participant would encounter 12 quests. The task used in this experiment is to choose the 'Milly's Harvest' quest among them. According to results reviewed by several game experts and designers, care was taken for an objective selection of representative quests. As a result, 'Milly's Harvest' quest was selected because it included all of the game-related characteristics of 12 quests more explicitly than the others. Moreover, it is the most difficult among them. 'Milly's Harvest' quest indicates that the player brings eight quest items to Milly (i.e., Non-Player Character) after player must pick up eight quest items to attack against dangerous monsters that guard around a vineyard.

A protocol analysis was conducted, which included collecting data on players' verbal comments and in-game actions, so that a more complete trace of their problem-solving behavior could be established. Two important preconditions of protocol analysis were set to identify an appropriate unit of analysis and to develop an objective coding system for each unit. We elected to use episodes - small self-contained phases of highly organized activities [8] - as the unit of analysis, considering the volume of protocol data from the experiment. While a motor-level action was defined as a simple code, a functional-level action was defined as each episode (i.e., the unit of analysis) based on the framework for gameplay because the process of gamplay would be explicitly identified by functional-level actions accomplished through motor-level actions, which correspond to clicking or pressing buttons on the computer input device [7]. According to what players played during each episode, the actions were classified as one of the following coding schema: 'Search', 'Avoid', 'Battle', 'Communication', 'Gather Type A' (Find-No-Constraint-Method as Template-Method), 'Gather Type B' (First-Remove-Danger-Method as Template-Method), and 'Gather Type C' (Sequential-Method). For each experiment, the data analysis was conducted step by step across all the players in order to increase the reliability of the protocol results gathered from the twenty seven players. The steps were: transcription, verification of transcription, initial segmentation, segmentation test, episode coding, test for episode coding, and construction of 'Gameplay Behavior Graphs'. For each of the protocols, a certain step was finished before the next step was started for any subject. The gameplay behavior was summarized into a 'Gameplay Behavior Graph' (GBG), which is similar to the 'Problem Behavior Graph' (PBG) proposed by Newell and Simon. A PBG portrays problem solving activities as a series of searches in the problem space [8].

4   Result

4.1   The Description of GBG

In the GBG, the ordinate included five functional actions of seven columns: 'Gather Type A', 'Gather Type B', 'Gather Type C', 'Communication', 'Battle', 'Avoid', and 'Search'. The first and the second columns describe the type of the 'Template-Method'. The former means the 'Find-No-Constraint-Method' which corresponds to the strategy to search only the place where there are no constraints, such as monsters and opponents with an expectation of a chance (Alea). The latter signifies the 'First-Remove-Danger-Method' which is related to the strategy to remove dangerous monsters that guard the quest item first, and then to obtain the item with competition (Agon). According to Juul [12], there are rules in a game. These "rules specify limitations and affordances". "They prohibit players from performing actions, but they also add meaning to the allowed actions and this affords players meaningful actions that were not otherwise available; rules give games structure". In this study, therefore, affordance was defined as the features to afford player's action based on the rules. Consequently, the 'Find-No-Constraint-Method' is equivalent to the method to fully understand limitations and affordances in a game, to search bugs, and to transcend rules in a game. We call this a 'Transcend Affordance'. Moreover, the 'First-Remove-Danger-Method' coincides with the method to apply the rules in a game, which the game designer embedded in the game, to his or her gameplay to win the game. We call this a 'Follow Affordance'. The third column means the 'Sequential-Method' which is related to a way to try first without prior knowledge in a game and then learn the rules in it as mentioned previously. The fourth column indicates the functional action to communicate with other player characters. The fifth column represents the functional action used to battle against the opponents. The sixth column signifies the functional action applied to avoid dangerous opponents. The seventh includes the functional action to search the place where the player conducts a quest. The horizontal axis contained episodes, start times, and elapsed times for each of the players, indicating the progression of gameplay. Time flows from left to right to show the dynamics of the gameplay process.

4.2 Experts' Gameplay Pattern

The GBG of Expert 2 (Fig. 1) shows the application of the 'Find-No-Constraints-Method' among the 'Template-Methods'. Expert 2 started the search process twice, first by avoiding the monster's approach to the utmost. He applied the 'Find-No-Constraints-Method' to his gameplay five of the eight times during the process of gathering quest items. The other three times resulted from making mistakes; Expert 2 did not yet identify the monsters present during gameplay. Moreover, he never utilized the 'First-Remove-Danger-Method' as a 'Template-Method'. In his early gameplay phase, he concentrated on the gathering action rather than battling against monsters. In particular, use of the 'Ambient-Method' was captured by action protocol analysis. The 'Ambient-Method' is one of the methodological schemes. It indicates that the players hold the tactic to identify the surroundings before moving closely to a quest item. This presents the critical clue to broadly securing the player's field of vision in terms of his game environment, to fully understand in and around information, and to determine the spot where to move. Consequently, it was a vital method to gather quest items effectively. In this study, the gameplay pattern of Expert 2 is defined as 'Gather Type A', which indicated the strategy to transcend the affordance of the player presented in section 4.1. This study revealed that 'Gather Type A' was discovered as a new problem space in methodological domain. In the traditional study of problem solving, most scholars believed that there existed the three problem spaces of rule, instance, and representation [6, 13]. However, the results of this study found that a new problem space, 'Transcend Affordance', was present in the methodological domain. Consequently, we argue that rule space includes two problem spaces: the new problem space, 'Transcend Affordance', and the traditional rule space, 'Follow Affordance'.

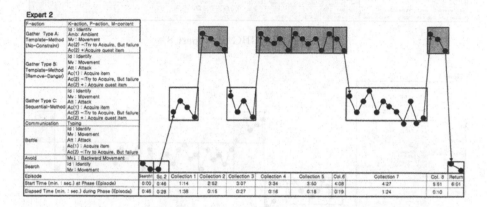

Fig. 1. The GBG of Expert 2

The GBG of Expert 8 (Fig. 2) significantly differed from the GBG of Expert 2 (Fig. 1). The GBG of Expert 8 shows the application of the 'First-Remove-Danger-Method' among the 'Template-Method'. Expert 8 started the battle process twice, after which he gathered a quest item. He applied the 'First-Remove-Danger-Method' to his gameplay five times, the 'Find-No-Constraints-Method' one time, and the 'Sequential-Method' twice. In the third gathering process, he did not take care of the enemy's re-generation while obtaining a quest item because he delayed equipping items to effectively attack monsters. Thus he applied the 'Sequential-Method' to his gameplay. In the case of the final gathering process, he used the 'Sequential-Method' since a conversation with the experimenter resulted from an impediment to his gameplay. The gameplay pattern of Expert 8 was mainly considered to use the 'First-Remove-Danger-Method' except for this three time gathering process as mentioned above. In this study, the gameplay pattern of Expert 8 is defined as 'Gather Type B', which indicated the strategy to follow the affordance of the player. The result of this study revealed that 'Gather Type B' was regarded as a traditional rule space and we refer the reader to Kim et al. [6] for a more detailed account.

The GBG of Expert 1 (Fig. 3) significantly differed from the GBG of other eight experts. He shows the application of the 'Sequential-Method' even though he was an expert. He started the search process first, and next had several battles

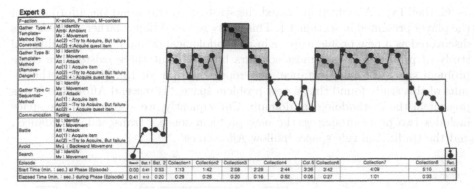

Fig. 2. The GBG of Expert 8

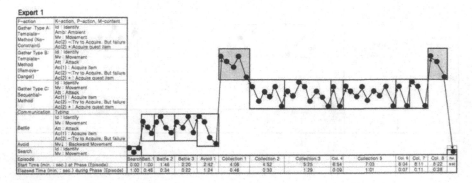

Fig. 3. The GBG of Expert 1

against enemies, and then had avoidance, and finally gathered quest items. He applied the 'Sequential-Method' to his gameplay six times, the 'First-Remove-Danger-Method' to the first gathering process and the final twice. The gameplay pattern was elucidated by the longer time period and the number of steps taken by Expert 1 compared with the other eight experts during the game. It is assumed that he did not use any rule template. In this study, the gameplay pattern of Expert 1 is defined as 'Gather Type C', which indicated the strategy to learn the affordance of the player through trial and error. Although comprehensive analysis of the verbal and action protocols most accurately shows the dynamic nature of gamers' cognitive process, it is impossible to present all the data in detail for all the subjects due to the space limit. Therefore we refer the reader to Table 2 that contains the aggregate protocol results of all subjects for a more detailed account.

4.3   Intermediate Players' Gameplay Pattern

The GBGs of all intermediate players show incremental learning curves from the 'Sequential-Method' (Gather Type C) to the 'First-Remove-Danger as

Template-Method' (Gather Type B) to the 'Find-No-Constraints as Template-Method' (Gather Type A). In particular, the GBG of Intermediate 15 (Fig. 4) shows a more obvious learning curve than the other eight intermediate players. Intermediate 15 started the search and battle process first. And then he applied the 'Sequential-Method' to his gameplay three times, the 'First-Remove-Danger-Method' twice, and the 'Find-No-Constraints-Method' three times. The GBGs of the other intermediates show similar incremental learning curves to that of Intermediate 15. When considering their whole gameplay pattern, they conduct the change from the 'Sequential-Method' to the 'Template-Method' first. Next, the 'Template-Method' is divided into two method parts: the 'First-Remove-Danger-Method' and the 'Find-No-Constraints-Method'. Following this, the transition occurs in two method parts. In other words, they show the transition from the 'First-Remove-Danger-Method' (Follow Affordance) to the 'Find-No-Constraints-Method' (Transcend Affordance).

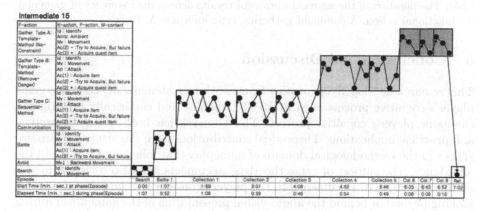

Fig. 4. The GBG of Intermediate 15

4.4 Novices' Gameplay Pattern

The GBGs of novices show similar incremental learning curves to that of intermediates: the change from the 'Sequential-Method' which is related to learning the rules of a game to the 'Template-Method' which is concerned with following and transcending them. However, all novices took a longer time period and a greater number of steps to accomplish the quest than the intermediate players. Most novices applied the 'Find-No-Constraints-Method' to their final gameplay 2~3 times. The gameplay patterns of novices revealed in the result was that a new problem space existed to transcend the affordance as in the case of most other players. The result of this study demonstrated that there was a new problem space rather than the traditional rule space [6, 13] in MMORPG gameplay.

In summary, the aggregate protocol results of all subjects for gameplay are summarized in Table2. 'Gather Type B' (51%) occurs superiorly in the expert group, and 'Gather Type B' (39%) appears inferiorly in the intermediate player group. In addition, 'Gather Type C' (48%) exists predominately in the novice group. Moreover, 'Gather Type A' (a new problem space) occurs uniformly in the expert group (32%), the intermediate group (33%), and the novice group (29%). Consequently, the result of the study reveals that there is a new problem space in most subjects' gameplay.

Table 2. The aggregate protocol results of all subjects for gameplay

Subjects	Expert group 1	2	3	4	5	6	7	8	9	Intermediate game player group 10	11	12	13	14	15	16	17	18	Novice group 19	20	21	22	23	24	25	26	27	Expert group Avg.		Intermediate Avg.		Novice group Avg.	
Gather Type A	5	5	5	2	1		1		4	7		2	1		3	5	1	5	1	3	3	1		1	6	4	2	2.6	32%	2.7	33%	2.3	29%
Gather Type B	2		3	3	6	6	8	5	4	1	8	2	7		2	2	6		1	3	1	2	1	4	1	3	1	4.1	51%	3.1	39%	1.9	23%
Gather Type C	6	3			1		3			4			8	3	1	1	3		6	2	4	5	7	3	1	1	5	1.4	17%	2.2	26%	3.8	48%
Dominant Type	C	A	A	A	B	B	B	B	A-B	A	B	C	B	C	A-C	A	B	C	C	A-B	C	C	C	B	A	A	C	B	51%	B	39%	C	48%
Battle	4	3	2	1	2	5	2	1		1	4		1	2	1	8	1		6	6	1	5	13	6	5	5	8	2.2		2		6.1	
Avoid	1		3	2	3		1			1	1		2	1		6	2		5	12			5	2	3		3	1.1		1.4		3.3	
Search	1	2	2	1	1	3	1	1	1	1	3	1	2	1	1	2	1	1	3	7	9	4	5	2	3	7	1	1.4		1.4		4.6	
Dead					1								1						1		2	1		1				0.1		0.1		0.5	

Note: The numbers of the aggregate protocol results denote the frequency of gameplay for functional action. A dominant gathering type indicates A, B, or C.

5   Conclusion and Discussion

This research developed a gameplay framework to systematically code the game player's cognitive process in MMORPG gameplay and empirically investigated the game player's cognitive process. This research has both strong theoretical and practical implications. Theoretical contributions from this study explore two spaces in the methodological domain of gameplay that follow and transcend the affordance. The subject of game theorists' arguments was also reviewed in this study. According to Friedman [14], the players consider the 'heart' of the game, looking beyond or behind the audio-visual presentation of the gameworld during gameplay. Looking behind the game indicates the action of the following affordance, and looking beyond the game represents the behavior of the transcending affordance during gameplay. Moreover, this demonstrated that a new problem space exists in problem space, and is another dimension beyond the traditional aspects Problem Solving Theory: the problem solver conducts the task through the interaction within representation space, rule space, and instance space on a problem domain. This study argues that two rule spaces exist in problem spaces rather than only the one rule space that is considered to be embedded in them, as the result of the traditional research related to the problem solving process [6, 13]. According to the difficulty of the task, a problem solver was found to switch between the two problem spaces approximately every minute.

Practical contributions of this research will provide MMORPG designers with design guidelines for players' behavior patterns and a new framework for gameplay. First, the six action categories of the gameplay derived from this study holds for a theoretical base of teaching and designing in the computer game. When the designers create a plan for the game, it is possible to manage their knowledge in a systematic way using the vocabulary of the scheme. Second, communication would be easier and common mistakes could be avoided as a new framework for the gameplay can provide a theoretical basis for teaching and development in computer games. The results of this study have clear implications for overall game development, and should serve to improve game quality at any stage of the design process, and the game design process is crucial for guaranteeing continuous success of the product.

This study examined participants playing WOW at a specified game entry point with a specified character class for approximately 1~2 hours. Consequently, the study results cannot be extended to the explanation of 'Deep Play' or gameplay with other races or classes. Future studies will include an assessment of playing experience for late-game design based on the gameplay in the latter stages of the game.

References

1. Crawford, C.: The Art of Computer Game Design (1982)
2. Clanton, C.: An Interpreted Demonstration of Computer Game Design. In: Proceedings of the conference on CHI 98 summary: human factors in computing systems, pp. 1–2 (1998)
3. Eskeline, M.: The Gaming Situation. Game Studies 1(1) (2001)
4. Heaton, T.: A Circular Model of Gameplay. Gamasutra.com (2006)
5. Kavakli, M., Thorne, J.: A Cognitive Modeling Approach to Capturing the context of complex Behavior in Gameplay. In: Proceedings of ICITA 2002, IEEE, Australia (2002)
6. Kim, J., Lerch, F., Simon, H.: Internal Representation and Rule Development in Object-Oriented Design. ACM Transactions on Computer-Human Interaction 2(4), 357–390 (1995)
7. John, B.E., Vera, A.: A GOMS Analysis of a Graphic, Machine-Paced, Highly Interactive Task. In: Proceedings of the conference on CHI 1992, pp. 251–258 (1992)
8. Newell, A., Simon, H.: Human problem solving. Prentice Hall, Englewood Cliffs (1972)
9. Caillois, R.: Man, Play and Games. University of Illiois Press, Urbana, Chicago (2001)
10. Ericcson, K.A., Simon, H.: Protocol Analysis. MIT Press, Cambridge (1993)
11. Entertainment Software Association: 2005 Essential Facts About the Computer and Video Game at E3 (2005)
12. Juul, J.: Half-Real: Video Games Between Real Rules and Fictional Worlds. The MIT Press, Cambridge (2006)
13. Schunn, C.D., Klahr, D.: A 4Space Model of Scientific Discovery. In: Proceedings of the 17 Annual Conference of the Cognitive Science Society, pp. 106–111 (1995)
14. Friedman, T.: Making sense of software: computer games and interactive textuality. In: Jones, S.G. (ed.) Cybersociety, Sage Publications, London (1995)

How to Overcome Main Obstacles to Building a Virtual Telematics Center

Bong Gyou Lee

Graduate School of Information
Yonsei University, 134 Shinchondong, Seodaemungu, Seoul, 120-749, Korea
bglee@yonsei.ac.kr

Abstract. This paper describes the limiting factors for building a virtual Telematics center which integrates ITS (Intelligent Transportation Systems) information collected by the system of each of the three government agencies. It also provides some details into how some of the obstacles and barriers can be solved such as data standardization, system mechanisms and regulations. There are very few papers and case studies regarding real problems and solutions for integrating systems of different organizations in general, and Telematics, in particular. The processes and outcomes of implementing systems in this study can be useful to develop other ITS and Telematics systems.

Keywords: Telematics, virtual center, ITS, data standardization, integration.

1 Introduction

The advances of ITS (Intelligent Transportation Systems), multimedia and communications technologies have generated various traffic information as well as rapidly increasing demands for real-time ITS information. However, existing ITS centers in Korea are faced with diverse limiting factors for gathering and providing nationwide real-time traffic information in terms of scarce funding and restricted coverage areas.

Although the physical establishment of a fully integrated or extended ITS center can be a straightforward and uncomplicated way to collect and supply nationwide ITS information, it can be a redundant investment or can require huge amounts of costs and time. Build an ITS/Telematics virtual center can be regarded as an alternative choice for constructing additional physical ITS centers. Obviously the circumstances for establishing the virtual center are quite different from those of the physical ITS center. Nevertheless, until recently, it has not been easy to find appropriate paradigms for building the virtual center in ITS and Telematics.

The process and experience of implementing the virtual systems in this study which has been funded approximately US$ 20 million by the MIC (the Ministry of Information and Communication) can be useful to develop other ITS and

M.S. Szczuka et al. (Eds.): ICHIT 2006, LNAI 4413, pp. 344–351, 2007.
© Springer-Verlag Berlin Heidelberg 2007

Telematics systems. In this study, some issues of technical, operational, and regal and policy areas have become critical tribulations to overcome. For example, the different standards to exchange traffic and road information between the existing ITS centers are serious problems for building the virtual Telematics center which integrates ITS information collected by the system of each of the three government agencies.

The purpose of this paper is to analyze the obstacles and barriers for building the virtual Telematics center and to present the problem-solving process of providing nationwide real-time traffic information. The proposed virtual center enables current ITS centers to collect and deliver real-time traffic information more accurately and economically.

This paper is divided into four parts. After the Introduction, Section two describes the conceptual framework and architectures of the virtual Telematics center. Section three explains critical problems and solutions for integration of virtual center. It provides some details into how some of the real problems can be solved such as data standardization and system mechanism. Final section summarizes the study and draws conclusions.

2   Conceptual Framework and Architecture of the Virtual Center

The conceptual framework, logical and physical architectures in Figure 1, 2 and 3, respectively, presents the intangible virtual center clearly. Building a virtual center is a part of the ongoing projects. In the project, approximately 40% of the total project cost goes towards this study.

In order for the virtual center to operate through linked servers within the system of each of the three agencies, as shown in Figure 1 and 2, various traffic data from government agencies should be integrated. The main government agencies includes the MOCT (the Ministry of Construction and Transportation), the KNPA (the Korea National Police Agency), and the MIC. The manipulating traffic information can be delivered directly to general users and to other agencies and private TSPs (Telematics Service Providers).

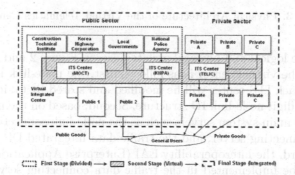

Fig. 1. Conceptual Framework of the Virtual Telematics Center

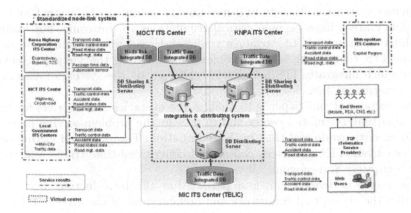

Fig. 2. Logical Architecture of the Virtual Telematics Center

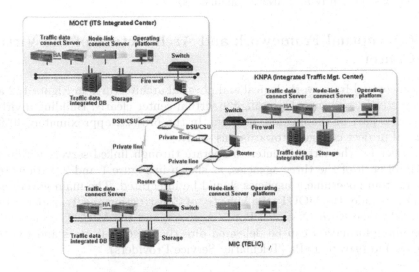

Fig. 3. Physical Architecture of the Virtual Telematics Center

Based on the logical and physical architectures in Figure 2 and 3, the process of developing the virtual center is as follows; first, the node-link integrated DB (database) of each ITS center will be modified and stored by taking the MOCT standard of 10-digit node-link ID structure. Second, based on the node-link integrated DB, the main system servers such as the node-link connecting server, the traffic data connecting server, and the traffic data integrated DB server will be developed. Third, the interoperability EAI (Enterprise Application Integration) programs will be implemented in the traffic data connecting servers. Last, the operating platform for operators in each ITS center will be developed.

3   Problems and Solutions for Integration of Virtual Center

The processes of integrating systems have led to unexpected troubles including legal, technical and operational problems. Because the existing ITS centers by government agencies have their own polices, standard data formats, and node-link systems for managing and controlling their specific transportation facilities like the traffic signals.

There have been several legal and policy issues such as the responsibility of inaccuracy, the ownership of information and the additional operating costs. There has been close correlation between these problems. For example, the ownership of information has closely related to the additional operating costs.

This section primarily describes operational and technical issues as the main problems and solutions for integration of virtual center. Most problems in this study have been solved by concrete system mechanisms, data standardization and policy revisions. Some solutions have been found by the process of trial and error.

3.1   Operational Issues

A serious problem in the operation stage is the difficulties of operation and management. It can be caused by the unclear responsibilities, connecting processes and regulations. Furthermore, unskilled operators, who are not familiar with connecting systems, tend to follow their own regulations.

In order to solve the problem, standard processes with diverse scenarios that include the detailed procedures have been developed. Figure 4 shows the connecting scope within the virtual center and Figure 5 presents the scenario for receiving information from the MOCT.

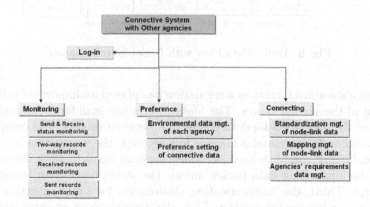

Fig. 4. Example of the Scenario for Receiving Information

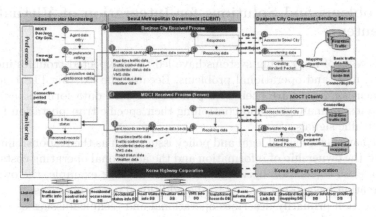

Fig. 5. Example of the Scenario for Receiving Information

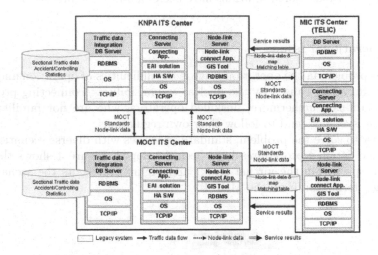

Fig. 6. Traffic Data Flow with NodeLinkInformation

Making a standard format as a regulation has played an important role in the operation of the virtual center. The 'data interfacing standard' can be a good example of the regulation for data connection between centers. The components of the 'data interfacing standard' are as follows; first, the 'organizing basic information and items' can be defined as connecting data items by basic information. Second, the 'organizing data packet' means the structure of data transmission and receipt. Third, the 'authenticating information for interconnection' is the regulation for releasing information. Last, the 'regulation on granting center ID' can be expressed as the rule for granting unique ID within each center, i.e., zip code +'00'.

On the other hand, the releasing data regulations in this study are composed of three parts; first, the agency only provides the request information by the connecting agency based on the standard node-link ID system. Second, the agency only provides the request information to the authorized agencies through the process of user authorization including login authentication. Last, connecting agencies should provide the connecting information to other agencies.

Also, to share and distribute traffic data efficiently, the virtual center has adopted the formal standards including data exchange standard in terms of the ASN.1 (Abstract Syntax Notation One) input standard and the standard of node-link ID system. Figure 6 presents the data flow within the three agencies of the virtual center that take the standards for data dictionary, communication protocol, and message sets for semantics and syntax. Figure 7 shows the communication protocol for the virtual center. Thereby, each agency of the virtual center can make and share the real-time traffic information including accident, vehicle speed, volume and occupancy information.

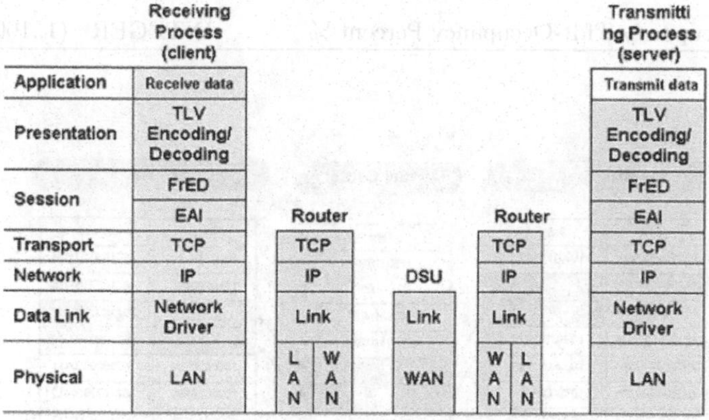

	Receiving Process (client)		Router	DSU	Router		Transmitting Process (server)
Application	Receive data						Transmit data
Presentation	TLV Encoding/ Decoding						TLV Encoding/ Decoding
Session	FrED						FrED
	EAI						EAI
Transport	TCP		TCP		TCP		TCP
Network	IP		IP	DSU	IP		IP
Data Link	Network Driver		Link	Link	Link		Network Driver
Physical	LAN		LAN | WAN	WAN	WAN | LAN		LAN

Fig. 7. Communication Protocol for the Virtual Center

3.2 Technical Issues

There have been a couple of technical key issues in terms of different data types and node-link ID systems. Since the traffic data based on traffic detections can have several data types such as text, audio and video formats, each agency has used different application software, middleware and DBMS.

In order to solve the problems of different data type, the standard data formats with regard to data dictionary, data registry, data type, data structure, and data element have been developed. Table 1 describes some standard data format for connecting traffic information between agencies.

The problem of the node-link ID system is that the MOCT uses the 10-digit node-link ID structure, but the MIC follows 8 digits for its node ID and 9 digits for its link ID, while the KNPA operates 8 digits for the node ID, and 13 digits for the link.

To solve the troubles of different node-link systems, the node-link matching table has been developed and utilized to initialize the node-link data of other agencies. Figure 8 presents the conceptual framework of the node-link matching table for making compatible node-link ID systems. An initialized data refers to a matching table that converts link data provided by the KNPA into the standard node format of the MOCT, and the standard link format of the MIC.

Table 1. Examples of Standard Data Format for Interconnecting Traffic Information

	ASN.1 Data Name	Unit	Data Type	Effective Value
Link ID No.	link-LinkIdNumber	Number	UTF8String	(1..40)
Link Speed	link-SpeedRate	Km/h	INTEGER	(1..300)
Link Traffic Volume	link-VoulmRate	Vehicle/h	INTEGER	(1..100000)
Travel Time	link-DensityRate	second	INTEGER	(1..10800)
Detector Occupancy	Tfdt-Occupancy Percent %		INTEGER	(1..100)

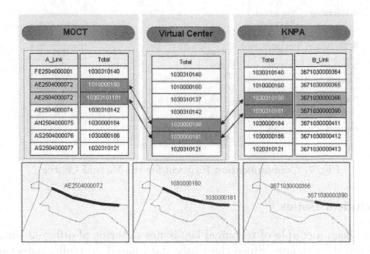

Fig. 8. Conceptual Framework of Node-Link Mapping Table

4   Conclusion

This paper reviews main issues of constructing and implementing the virtual Telematics center in Korea. There are very few papers and case studies regarding real problems and solutions for integrating ITS and Telematics systems.

The process of integrating heterogeneous systems has led to some unexpected troubles including legal, technical and operational problems. Since the existing

ITS centers have their own polices, standard data formats, and node-link systems for managing and controlling their own specific facilities. Fortunately, most serious problems in this study have been solved by concrete system mechanisms, data standardization, and policy revisions.

Due to its status as a technologically advanced case, this study can set an example for other ITS and Telematics centers.

References

1. Ministry of Construction and Transportation.: White Paper in Ministry of Construction and Transportation. MOCT. Seoul (2004)
2. Lee, B.G., Kim, K.Y., Lee, T.H., Song, J.Y.: u-Business Strategies for Telematics based on Demand Analysis of Real-time Traffic Information and Devices. In: Proceedings of the International Conference on Korean Management Information System, pp. 367–370 (2005)
3. Lee, B.G., Hong, I.G., Ryu, S.K., Moon, H.Y.: A Study on Integrating Wire & Wireless Communication Networks for Reducing Communication Costs in the National ITS Physical Architecture. Journal of the Korea open GIS Association 6(2), 77–84 (2004)
4. Kim, Y.K.: GIS/LBS/Traffic Information Technology. TTA Journal 89, 99–104 (2003)
5. Korea Research Institute for Human Settlements.: ITS Node-link System. Standardization and Operation Guideline. KRIHS. Seoul (2005)
6. National Computerization Agency.: A Study on Reinforcing ITS Communication Architecture. NCA. Seoul (2001)
7. U.S. Department of Transportation.: National ITS Architecture: version 5.1 (December 14, 2005), http://www.its.dot.gov/arch
8. ASN. 1 Information Site (May 3, 2005), http://asn1.elibel.tm.fr
9. Transport Canada.: ITS Architecture for Canada (May 13, 2005), http://www.its-sti.gc.ca

Real-Time Travel Time Estimation Using Automatic Vehicle Identification Data in Hong Kong

Mei Lam Tam and William H.K. Lam

Department of Civil and Structural Engineering, The Hong Kong Polytechnic University,
Hung Hom, Kowloon, Hong Kong, China

Abstract. This paper proposes a real-time traveler information system (RTIS) for estimating current travel times using automatic vehicle identification (AVI) data in Hong Kong. The current travel times, in RTIS, are estimated by real-time AVI data, the off-line travel time estimates and the related variance-covariance relationships between road links. The real-time AVI data adopted for RTIS are Autotoll tag data in Hong Kong; whereas the off-line link travel time estimates and their variance-covariance matrices are obtained from a traffic flow simulator. On the basis of integration of these real-time and off-line traffic data, the current traffic conditions on Hong Kong major roads can be estimated at five-minute intervals. A case study is carried out in Kowloon Central urban area to collect observed data for validation of the results of the proposed RTIS.

Keywords: Real-time traveler information system, automatic vehicle identification data, Hong Kong.

1 Introduction

The rapid development of Intelligent Transportation Systems (ITS) and electronic information and communication technologies has alerted researchers and practitioners to the potential of the application of advanced technologies to alleviate traffic congestion, particularly in large Asian cities like Hong Kong. The Advanced Traveler Information System (ATIS) is one of the options adopted in many Western countries. In ATIS, detectors are located at some sections of the road network to collect historical and real-time data such as traffic flow and speed. Based on these partial detected data, link flows and travel times within road networks can be estimated and delivered to drivers by various means, such as variable message signs, mobile phones and the internet. Drivers are then able to make their route choice decisions based on a combination of the received estimated travel times and their own driving experience.

An ATIS prototype, Journey Time Indication System (JTIS), was introduced in Hong Kong in mid-2003. This system provides current traffic conditions in terms of travel times via displays on gantry signs near major roads along Hong Kong harbour side. The displays are updated at five-minute intervals. This JTIS is being planned for extension to other urban areas in Hong Kong. Moreover, real-time closed-circuit television (CCTV) images and videos at major roads in Hong Kong are available on the Hong Kong Transport Department website (http://traffic.td.gov.hk/SwitchCenter.do). The CCTV images are updated at two-minute intervals. From these updated CCTV

M.S. Szczuka et al. (Eds.): ICHIT 2006, LNAI 4413, pp. 352–361, 2007.
© Springer-Verlag Berlin Heidelberg 2007

images, travelers can know the current traffic conditions on major roads in Hong Kong, in terms of traffic density. However, the number of CCTV cameras is limited rendering this method incapable of capturing detailed traffic conditions in the whole territory. As responses to pictures are to a degree subjective, digital information may be more useful and valuable for most drivers. Therefore, it is a value to collect and make use of additional real-time traffic data for development of a territory-wide real-time traveler information system in Hong Kong.

In view of this, an off-line short-term traffic forecasting platform with Geographic Information System (GIS) functions has been developed for Hong Kong ATIS applications [1]. In this system, a Traffic Flow Simulator (TFS) [2] has been calibrated for short-term forecasting of travel times by making use of the Hong Kong Annual Traffic Census (ATC) data. The models for short-term prediction of the hourly traffic flows at ATC detector locations have also been investigated by Lam *et al.* [3]. Based on these short-term traffic forecasting results at ATC detector locations, the TFS can be used to estimate the off-line short-term travel time forecasts and the variance-covariance relationships between road links for the whole territory of Hong Kong. Using these off-line travel time estimates, a website called SpeedOnRoad, shown in Fig. 1, has been developed by The Hong Kong Polytechnic University to provide Hong Kong traffic information via the internet. Website users can access various types of traffic information, such as traffic speed, the shortest path and corresponding travel time of a selected Origin-Destination (O-D) pair and zone-to-zone travel time information at selected areas during different peak hour periods. The off-line system is useful to travelers for pre-trip planning. However, the current traffic conditions have not yet been incorporated in the off-line system and hence travelers are unable to access road condition information whilst on route and on a real-time basis. In order for information to be given to drivers on route, a real-time traveler information system (RTIS), with such a capacity, is required.

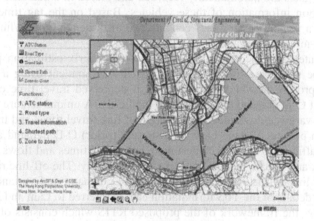

Fig. 1. SpeedOnRoad

In this study, an on-line travel time forecasting system is developed specifically for Hong Kong. This system is based on the off-line results obtained by TFS and real-time traffic data. One of the major products of this study is a website portal which delivers real-time traffic information to users via the internet. A case study for a major road network in Kowloon Central urban area is presented in this paper to demonstrate the RTIS performance. The travel time is estimated on the basis of the journey time of those vehicles equipped with automatic vehicle identification (AVI) system traveling from an origin to a destination. Real-time Autotoll tag data, which are a kind of AVI data in Hong Kong, are adopted in the proposed RTIS.

The remainder of the paper is structured as follows. The framework of the RTIS is given in the next section. The data filtering algorithm and on-line updating procedures are then described in Section 3. A case study set in the Kowloon Central urban area is presented in Section 4 together with validation results. Finally, conclusions are given and recommendations for further study suggested.

2 Framework of RTIS

An RTIS for Hong Kong has been recently developed by The Hong Kong Polytechnic University in collaboration with Autotoll Limited. Current travel times are easily updated in RTIS once every five minutes, by making use of real-time traffic data and results of the off-line travel time forecasting system. These real-time traffic data can be Autotoll tag data, Global Positioning System (GPS) data, CCTV data and image processing data captured by sensors/cameras [4]-[5]. In this study, current travel times are estimated on the basis of the real-time Autotoll tag data and the off-line estimates obtained by TFS.

There are about 550,000 licensed vehicles in Hong Kong since September 2006. Over 210,000 of these cars have been installed with Autotoll tags to enable toll charge payments at ten road tunnels or links in Hong Kong. Autotoll tag data are collected at the toll gates of these ten tunnels/links. The times of vehicles passing through the tunnels/links toll gates are automatically recorded and stored in a database together with the identification information of these vehicles. Based on the tag time records, the travel time of a vehicle passing between any two of these ten tunnels/links can be extracted at five-minute intervals.

The TFS integrates the origin-destination (O-D) matrix estimation problem with the probit-type stochastic user equilibrium (SUE) assignment using a bi-level programming approach, in which the variation of perceived travel time errors and the fluctuations of O-D demand are considered explicitly. A unique feature of the TFS is the use of the variance-covariance information of link travel times and traffic flows in the estimation process. On the basis of historical (prior) O-D demand and the Hong Kong ATC traffic data, the TFS can estimate link travel times and flows for the whole study network and update the O-D matrix simultaneously. The off-line results include the estimated link traffic flows, travel times, O-D demand and variance-covariance matrices. The details of the TFS formulation can be referred to Lam and Xu [6].

Fig. 2 shows the framework of the proposed RTIS which consists of both off-line and on-line travel time estimation components. The off-line results and the real-time traffic data are thus integrated for on-line travel time updating at five-minute intervals. The deliverables of RTIS are speed maps, the shortest path and corresponding travel time of a selected O-D pair, which can be displayed on website and/or 3-G mobile phones.

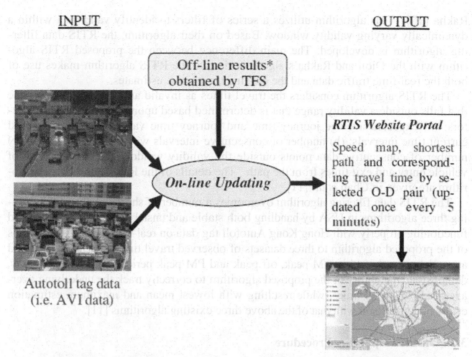

INPUT

Off-line results*
obtained by TFS

On-line Updating

OUTPUT

RTIS Website Portal

Speed map, shortest
path and correspond-
ing travel time by se-
lected O-D pair (up-
dated once every 5
minutes)

Autotoll tag data
(i.e. AVI data)

Note: Off-line results consist of the mean travel time by link and the link travel
time variance-covariance matrices

Fig. 2. Framework of the proposed RTIS

3   On-Line Updating

3.1   Data Filtering Algorithm

AVI tag readers provide time stamps at which vehicles pass successive monitoring
stations or checkpoints. Data that are gathered by these readers require some form of
filtering in order to remove outlier observations. The outlier observations may be due
to stops or detour takings by vehicles. Since these vehicles would experience a travel
time that is atypical, these observations should therefore be removed from the data set
of valid observations to avoid producing erroneous travel time estimates.

There are three existing algorithms in USA, including the TranStar system in
Houston [7], the TransGuide system in San Antonio [8] and the Transmit system in
the New York/New Jersey metropolitan area [9]. However, these algorithms all have
some shortcomings. TranStar algorithm is necessary to have high numbers of AVI
data for estimating the travel times, TransGuide algorithm cannot estimate the travel
times accurately under non-recurrent traffic congestion, whereas Transmit algorithm
cannot truly reflect the current traffic conditions as its weighting process combines the
current observed travel times with historical data. To address the shortcomings of the
existing state-of-the-art algorithms, a data filtering algorithm is proposed by Dion and

Rakha [10]. This algorithm utilizes a series of filters to identify valid data within a dynamically varying validity window. Based on their algorithm, the RTIS data filtering algorithm is developed. The main difference between the proposed RTIS algorithm with the Dion and Rakha's algorithm is that the RTIS algorithm makes use of both the real-time traffic data and the off-line travel time estimates.

The RTIS algorithm considers the travel times as invalid any observed travel time that falls outside a validity range that is determined based upon the following four factors: (a) expected average journey time and journey time variation in previous and current time intervals; (b) number of consecutive intervals without any valid data; (c) number of consecutive data points outside the validity window and (d) sequence of vehicle entry and exit times from the path. The details of the RTIS data filtering algorithm are given in Chan *et al.* [11].

The RTIS data filtering algorithm overcomes a number of shortcomings of the existing three algorithms in USA by handling both stable and unstable traffic conditions and functioning properly with Hong Kong Autotoll tag data on real-time basis. Applications of the proposed algorithm to three datasets of observed travel times, that were collected at a selected path during AM peak, off peak and PM peak periods on a typical Friday, demonstrate the ability of the proposed algorithm to correctly track the underlying average roadway travel times, while resulting with lowest mean and maximum estimation errors in comparison with that of the above three existing algorithms [11].

3.2 On-Line Updating Procedure

The real-time traffic information banked in the proposed RTIS is updated on the basis of off-line results and real-time traffic data. The on-line updating procedures of RTIS are outlined as follows:

Step 1: Filter out the invalid real-time traffic data by the proposed RTIS data filtering algorithm.

Step 2: Obtain the following from valid real-time traffic data, (a) the mean travel time on the links with real-time data $\hat{t}_d$; and (b) the updated link travel time variance $\hat{K}_{11}$ and covariance $\hat{K}_{12}$ and $\hat{K}_{21}$, which are used to replace the off-line variance-covariance matrix of link travel times $K = \begin{bmatrix} K_{11} & K_{12} \\ K_{21} & K_{22} \end{bmatrix}$ with the updated variance-covariance matrix of link travel times $\hat{K} = \begin{bmatrix} \hat{K}_{11} & \hat{K}_{12} \\ \hat{K}_{21} & K_{22} \end{bmatrix}$. It should be noted that, in off-line forecasting system, $t = \begin{bmatrix} t_d \\ t_e \end{bmatrix} \sim MVN \left(\begin{bmatrix} \overline{t}_d \\ \overline{t}_e \end{bmatrix}, \begin{bmatrix} K_{11} & K_{12} \\ K_{21} & K_{22} \end{bmatrix} \right)$ where t_d and t_e are the link travel times with respect to the links with and without detector data which follow a multivariate normal (MVN) distribution, K is the link travel time variance-covariance matrix that is part of the off-line results of the TFS. Note that K_{11} and K_{22} are the variance of the travel times corresponding to the links with and without detector data respectively, K_{12} is the travel time covariance of the links with and without detector data and $\mathbf{K}_{21}$ is the travel time covariance of the links without and with detector data.

Step 3: Estimate the mean travel times on the links without real-time data $\tilde{t}_e$ using Equation (1) based on partial real-time data and the updated variance-covariance matrices of link travel times [2].

$$\tilde{t}_e = \bar{t}_e + \hat{\mathbf{K}}_{21}\hat{\mathbf{K}}_{11}^{-1}\left(\hat{t}_d - \bar{t}_d\right) \tag{1}$$

where $\bar{t}_e$ and $\bar{t}_d$ are the mean link travel times obtained in the off-line system with respect to the locations without and with real-time data.

Step 4: Regenerate the link flows based on the mean travel times $\hat{t}_d$, $\tilde{t}_e$ and the updated variance-covariance matrix of link travel times $\hat{\mathbf{K}} = \begin{bmatrix} \hat{\mathbf{K}}_{11} & \hat{\mathbf{K}}_{12} \\ \hat{\mathbf{K}}_{21} & \mathbf{K}_{22} \end{bmatrix}$.

4 A Case Study

Kowloon Central road network is used as a test network in this case study to demonstrate the performance of the proposed RTIS in practice. The path from the Lion Rock Tunnel (LRT) to the Cross Harbour Tunnel (CHT) is selected for journey time estimation. This route is chosen as most of the major roads in Kowloon Central urban area take this path, whereas CHT is the most congested of the three road tunnels connecting Kowloon Peninsula with Hong Kong Island. The location of the selected path between LRT and CHT is illustrated in Fig. 3. The distance of this selected path from LRT to CHT is 6.23 km with average speed limit of 60 km per hour. Travel times of the selected path from LRT to CHT were estimated at five-minute intervals during the morning peak (08:00-10:00), off-peak (14:00-1600) and evening peak (17:30-19:30) periods of a typical weekday on 26 May 2006 (Friday) for the case study.

In order to validate the RTIS travel time estimates, observed journey times of all vehicles traveling on the selected path were collected by a manual license plate recognition survey carried out in the same time periods on the same day. There are two types of toll-gates before entering the road tunnels or bridges in Hong Kong. One type of these toll-gates is mainly for vehicles installed with Autotoll tags and hence the observations from Autotoll tag are obtained. Another type of toll-gates is used for other vehicles without Autotoll tags, at which there is no tolltag observation and the license plate readings are recorded by video cameras during the survey periods. Therefore, the combination of tolltag observations and license plate readings are the total observations at all the toll-gates for a particular road tunnel or bridge. In this survey study, video recording equipments were set at the toll plazas of LRT and CHT to record the license plate numbers of vehicles used manual toll-gates. The license plate numbers of vehicles recorded at LRT were then matched with those recorded at CHT using a computer program. The journey times of the matched vehicles traveling from LRT to CHT were then computed for RTIS validation together with the journey times obtained from Autotoll tag records.

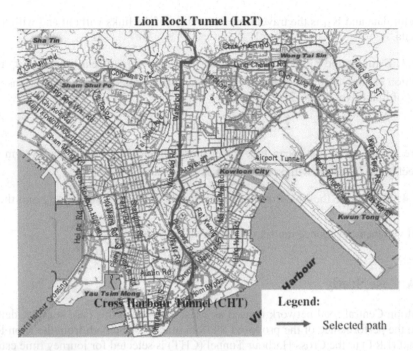

Fig. 3. Location of the selected path between Lion Rock Tunnel and Cross Harbour Tunnel in Hong Kong

The RTIS data filtering algorithm was applied to the observations collected from the manual license plate survey. The appropriate travel time window for travel between LRT and CHT at five-minute intervals was then determined. Table 1 shows the number of valid observations traveling on the selected path. In total, there are 1,922 valid observations obtained during these three survey periods, and about 40% of the valid observations are installed with tags. The observed travel times from LRT to CHT were in the range of 17.03 min to 28.01 min during the AM peak period, 17.77 min to 22.03 min during the off-peak period and 20.49 min to 28.83 min during the PM peak period, respectively. The journey times on the selected path were much varied during AM and PM peak periods due to heavy traffic demand from Kowloon Peninsula and the New Territories to Hong Kong Island using CHT.

Fig. 4 shows the profiles of the estimated travel times on the selected path during the three survey periods of 26 May 2006 (Friday). During the AM period, the travel time increases gradually until 08:45 and then has a smooth decrease. The travel times are well estimated from 08:00 to 08:45 although there are little underestimated after 08:45. For the off-peak period, the travel time is steadier and the estimation results are very satisfying. Pattern of the evening peak is less sharp than the morning peak and with some oscillations. However, the estimates follow the trend of the observed travel times quite well. In general, the estimation results well fit the observed travel times for all the three survey periods.

Table 1. Number of valid observations obtained during the survey periods[*]

	Valid observations from Autotoll tags	Valid observations from the manual license plate survey	Total
LRT to CHT	761 (39.59%)	1,161 (60.41%)	1,922 (100%)

[*] Survey periods are 08:00-10:00, 14:00-16:00 and 17:30-19:30 on 26 May 2006 (Friday).

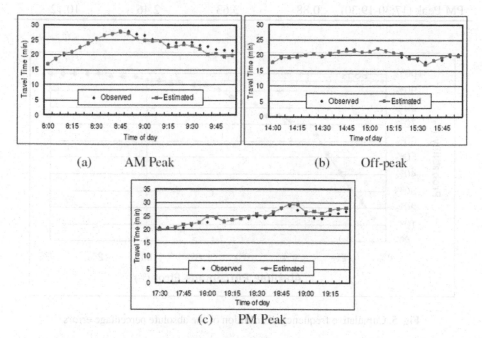

(a)　　　AM Peak　　　　　　　　　　(b)　　　Off-peak

(c)　　　PM Peak

Fig. 4. Profiles of the observed and estimated journey times of the selected path during AM, off and PM peak periods of 26 May 2006 (Friday)

The journey time estimates from LRT to CHT are compared with the observed travel times obtained from the manual license plate recognition survey. The estimation results are evaluated in terms of two measures of performance; namely, the absolute error (AE) and the absolute percentage error (APE) so as to determine the accuracy of the RTIS. Table 2 shows the average and the maximum AEs and APEs of the travel time estimates for the three survey periods. It can be seen in Table 2 that the average AEs are less than one minute for all the three periods, respectively. In terms of percentage, less than 5% of absolute errors are found in average. The cumulative frequency distribution of the APEs for LRT-CHT is also depicted in Fig. 5. It can be seen that over 95% of the travel time estimates are with the APEs of less than 10%. The results of the validation are satisfactory.

Table 2. Errors of the travel time estimates of the selected path during AM, off and PM peak periods of 26 May 2006 (Friday)

	Average		Maximum	
	AE (min)	APE (%)	AE (min)	APE (%)
AM Peak (08:00-10:00)	0.85	3.67	2.62	11.55
Off-peak (14:00-16:00)	0.41	2.10	1.22	6.30
PM Peak (17:30-19:30)	0.88	3.63	2.46	10.12

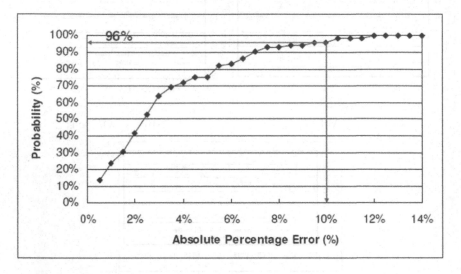

Fig. 5. Cumulative frequency distribution of the absolute percentage errors

5 Conclusions

In this paper, a real-time traveler information system (RTIS) for Hong Kong has been presented. The RTIS travel time estimates are based on the off-line results of the Traffic Flow Simulator (TFS) and the real-time Autotoll tag data in Hong Kong. A case study of validating RTIS results in the Kowloon Central urban area has been carried out. The travel time estimates on the selected path have been validated against the observed journey times collected from the manual license plate recognition survey during the morning peak, off-peak and evening peak periods of a typical Friday on 26 May 2006. The validation results are satisfactory with the average absolute errors of less than one minute, and over 95% of the travel time estimates are with less than 10% absolute percentage errors.

Algorithm development is continued to enhance the accuracy and the reliability of the travel time estimates by the developed RTIS. Different types of real-time traffic data, such as Global Positioning System (GPS) data and image processing data, will be incorporated in the RTIS. Supplementary car journey time surveys will also be conducted to measure travel times on alternative major paths in different urban areas of Hong Kong during different time periods. Hence, the current journey time estimates will be validated by road type for the whole strategic road network of Hong Kong. With satisfactory validation results, the travel times estimated by five-minute intervals for the whole territory are then fed into an on-line platform with Geographic Information System (GIS) functions for ATIS applications in Hong Kong.

Acknowledgements. The work described in this paper was collaborated with Autotoll Limited and jointly supported by research grants from the Research Committee of The Hong Kong Polytechnic University (Project Nos. 1-ZE10 and 1-BBZG).

References

1. Lam, W.H.K., Chan, K.S., Tam, M.L., Shi, J.W.Z.: Short-term Travel Time Forecasts for Transport Information System in Hong Kong. Journal of Advanced Transportation 39(3), 289–305 (2005)
2. Lam, W.H.K., Chan, K.S., Shi, J.W.Z.: A Traffic Flow Simulator for Short-term Forecasting of Travel Times. Journal of Advanced Transportation 36(3), 265–291 (2002)
3. Lam, W.H.K., Tang, Y.F., Chan, K.S., Tam, M.L.: Short-term Hourly Traffic Forecasts using Hong Kong Annual Traffic Census. Transportation 33(3), 291–310 (2006)
4. Tam, M.L., Lam, W.H.K.: Application of Information Technologies for Development of Real-time Traveler Information System in Hong Kong. In: Proceedings of the Joint International Conference on Computing and Decision Making in Civil and Building Engineering, June 14-16, 2006, Montreal, Canada, pp. 3784–3793 (2006)
5. Tam, M.L., Lam, W.H.K.: Development of Real-Time Traveler Information System for Hong Kong. In: Proceedings of the 8th Asia-Pacific Intelligent Transport Systems Forum and Exhibition 2006 [CD-ROM], Hong Kong (July 10–13, 2006)
6. Lam, W.H.K., Xu, G.: A Traffic Flow Simulator for Network Reliability Assessment. Journal of Advanced Transportation 33(2), 159–182 (1999)
7. Houston TranStar: TranStar description (accessed on 9 June, 2006), http://traffic. houstontranstar.org
8. Southwest Research Institute: Automatic Vehicle Identification Model Deployment Initiative - System Design Document. Report prepared for Texas Department of Transportation, TransGuide, San Antonio, TX (1998)
9. Mouskos, K.C., Niver, E., Pignataro, L.J., Lee, S.: Transmit System Evaluation. Final Report, Institute for Transportation, New Jersey Institute of Technology, Newark, NJ (1998)
10. Dion, F., Rakha, H.: Estimating Dynamic Roadway Travel Times Using Automatic Vehicle Identification Data for Low Sampling Rates. Transportation Research B 40(9), 745–766 (2006)
11. Chan, K.S., Tam, M.L., Tam, M.L., Lam, W.H.K.: Using Automatic Vehicle Identification Data for Estimating Current Travel Times in Hong Kong. In: Proceedings of the 86th Transportation Research Board Annual Meeting [CD-ROM], Washington, D.C. (January 21–25, 2007)

A Context-Aware Elevator Scheduling System for Smart Apartment Buildings*

Ohhoon Kwon[1], Hyokyung Bahn[2,**], and Kern Koh[1]

[1] School of Computer Science and Engineering, Seoul National University
{ohkwon,kernkoh}@oslab.snu.ac.kr
[2] Department of Computer Science and Engineering, Ewha University
bahn@ewha.ac.kr

Abstract. Ubiquitous computing technologies are becoming increasingly a part of our daily lives. For example, in the ubiquitous home environments, plenty of context information can be obtained from various home sensors, and this information can be exploited for smart home activities. This paper presents a new elevator scheduling system for smart apartment buildings that exploits the ubiquitous sensor technologies. In the proposed elevator scheduling system, floor sensors are located at each floor of the apartment building and detect the candidate elevator passengers' behavior before they come to the elevator door and push the elevator call button. The detected information is then delivered to the elevator scheduling system through the building network and the elevator scheduling system utilizes this information in the efficient control of the elevator. Through extensive simulations with various passengers' traffic conditions, we show that the proposed system performs better than the conventional elevator scheduling system in terms of the passengers' average waiting time, the maximum waiting time, and the energy consumption of the elevator significantly.

1 Introduction

In recent years, ubiquitous computing technologies increasingly become a part of our daily activities. In ubiquitous environments, we can obtain plenty of information from various sensors and exploit the information for smart home activities [2]. This paper presents a new elevator scheduling system, namely the Context-aware Elevator Scheduling System (CESS) for smart apartment buildings that exploits the ubiquitous sensor technologies. In the proposed elevator scheduling system, floor sensors are located at each floor of the apartment building and detect the candidate passengers' behavior before they come to the elevator door and push the elevator call button. The detected information is then delivered to the elevator scheduling system through the building network and the elevator scheduling system utilizes this information in the efficient control of the elevator.

---

* This work has been supported by KESRI (R-2005-7-059), which is funded by MOCIE (Ministry of commerce, industry and energy).
** Corresponding author.

M.S. Szczuka et al. (Eds.): ICHIT 2006, LNAI 4413, pp. 362–372, 2007.

There are several performance criteria of elevator scheduling systems such as minimizing passenger's waiting time, minimizing the riding time, and reducing the energy consumption. A plenty of studies have already been performed for elevator scheduling systems [5-7], [11-12], [13-19], [21-22]. Most of previous studies have focused on minimizing passenger's average waiting time since passenger's dissatisfaction grows rapidly as the waiting time increases [9]. Our elevator scheduling system focuses on reducing the energy consumption as well as minimizing the average waiting time. To minimize the average waiting time, previous studies exploited various optimization techniques such as genetic algorithms and fuzzy systems to predict passenger's traffic patterns [8]. However, their prediction is limited because the elevator scheduling system knows the passenger information only after he/she pushes the elevator call button. Unlike previous studies, our system detects the passenger information by floor sensors before passengers arrive at the elevator, and uses this detected information in smart elevator scheduling.

In the smart apartment buildings, we may also trace passengers' location using a variety of sensors, and we guess passengers' destination-floor using their ID card sensor and their history of elevator usage patterns. Therefore, we could obtain complete information of passengers and use them for elevator scheduling to minimize the passengers' waiting time and energy consumption of the elevator system.

To evaluate the performance of our elevator scheduling system, we conducted extensive simulations to compare the performance of our system with the conventional elevator scheduling system. Trace-driven simulations show that the proposed elevator scheduling system performs better than the conventional elevator scheduling system in terms of the passengers' average waiting time, the maximum waiting time, and the energy consumption of the elevator significantly.

2   Related Works

There have been a variety of studies on the elevator scheduling systems to reduce the passengers' waiting time. Igarashi et al. exploited fuzzy systems for the scheduling of group elevator systems [12]. In the group elevator systems, multiple elevators are scheduled together by a single control unit. In their studies, when a passenger pushes the elevator call button, the control unit of the system evaluates each elevator using a fuzzy function and assigns the elevator with the largest evaluation value to that request. Similar studies are also performed by Kaneko et al. [22]. Unlike the elevator control system proposed by Igarashi et al. that only considers the passengers' waiting time, Kaneko et al. considered the riding time of the elevator as well as the waiting time. Fujino et al. used genetic algorithms to optimize the elevator scheduling systems [15].

If an elevator scheduling system knows the destination floor of the passengers before they ride the elevator, the system could schedule the elevator more efficiently. A conventional elevator system, however, has only two buttons (upside and downside) at a floor, so it cannot predict passengers' destination in advance. To address this problem, Amano et al. proposed an elevator system that has the

destination-floor buttons at each floor. When a passenger wants to get into the
elevator, he/she pushes a specific destination-floor button at the floor. As a result, the destination-floor buttons at each floor provide more information to the
elevator control system than just up and down buttons [18].

In most elevator environments, traffic patterns occur periodically in a day,
and a number of studies have presented traffic pattern based elevator scheduling systems. Pepyne et al. analyzed up-peak traffic patterns of elevator systems
and exploited the pattern in an efficient elevator scheduling [14], [16]. Brand
et al. considered both up-peak and down-peak traffics. Their system dispatches
an empty elevator to a desired parking location even when there is no request
from users. The location is determined by the analysis results of the up-peak and
down-peak traffic patterns [6]. These parking strategies could minimize passengers' waiting time in common cases. The traffic of an elevator system changes
continuously as time progresses. This traffic change can be detected and modeled. Dewen et al. proposed a learning paradigm for elevator scheduling systems.
Their system detects the current flow and chooses a scheduling policy among a
set of ready-prepared policies [13]. Pepyne et al. proposed an elevator control
strategy that delays the elevator moving until the number of passengers inside
it reaches a certain threshold value [14].

3   The Proposed System

3.1   Limitations of Conventional Elevator Scheduling Systems

In the elevator scheduling systems, there are two types of calls. One is the hall
call and the other is the car call [8], [20]. A hall call is issued when a passenger
pushes one of the up or down buttons at a floor. A car call is issued when a
passenger pushes a destination-floor button inside the elevator car. In conventional elevator scheduling systems, a car moves when these calls are issued. If
the elevator scheduling system knows the time of a hall call or the destination
of the car call before it is actually issued, the system could schedule the car
more efficiently. This could eventually reduce the passengers' waiting time and
the energy consumption of the car. Conventional elevator scheduling systems,

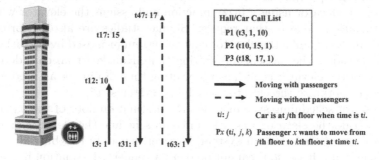

Fig. 1. An example of the conventional elevator scheduling system

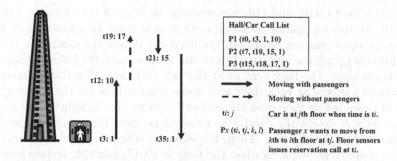

Fig. 2. An example of an efficient elevator scheduling system with the reservation call that is issued by the passenger information obtained from floor sensors

however, could not obtain this information until passenger pushes the hall call or the car call buttons.

Fig. 1 shows an example of this limitation of conventional elevator scheduling systems. In this example, there are three passengers, namely passenger 1, passenger 2, and passenger 3, waiting for the elevator. Passenger 1 is at the 1st floor and wants to move to the 10th floor. Passenger 2 is at the 15th floor and wants to move to the 1st floor. Passenger 3 is at the 17th floor and wants to move to the 1st floor. Each passenger pushes the hall call button at time t3, t10, and t18, respectively. We use the logical time that increases by one whenever the car moves up or down a floor and ignore all other times such as the boarding time in this simple example. The car first moves from the 1st floor to the 10th floor with passenger 1, and then moves up to the 15th floor to pick up passenger 2. When the time becomes t17, the car leaves the 15th floor for the 1st floor. Unfortunately, passenger 3 pushes a hall call button at the 17th floor immediately after the car moves to the 1st floor. If the elevator scheduling system knows the hall call at the 17th floor before the car moves to the 1st floor, the car may move to the 17th floor first instead of the 1st floor. In this case, the car picks up passenger 3 at t19 and passenger 3 waits only for one time unit. This could eventually reduce the average waiting time of the three passengers significantly. However, conventional elevator systems could not know the time of passengers' requests before they actually push the hall call buttons.

In recent years, ubiquitous computing technologies increasingly become a part of our daily lives. We can obtain plenty of information from various sensors and exploit the information for smart home activities [2]. In smart home environments, we can trace human behaviors using a large variety of sensors and equipments such as audio sensor, video sensor, camera, and optical sensor [1-4]. As a result, we can predict future human behaviors to a certain extent. In the elevator scheduling systems, if we know the passengers' boarding information such as the starting location, the target location, and the request time before they push the hall call and/or car call buttons, we can control the elevator systems more efficiently.

Fig. 2 shows a more efficient elevator scheduling with this information for the same example introduced in Fig. 1. The three passengers push the hall call buttons at time t3, t10, and t18, respectively, in front of the elevator, and the scheduling system recognizes these requests at that time. In a smart apartment building, however, various sensors in the floor can detect the candidate passengers before they push the buttons. As a result, we can make the following scenario with this example. The floor sensors at the 1st, 15th, and 17th floors detect the candidate passengers at each floor and make reservations for the passengers a certain of times before they push the hall call buttons. For simplicity, we assume that the sensors detect the candidate passengers three time units before they arrive in front of the elevator. Then, the elevator scheduling system recognizes the three passengers' requests when the time is t0, t7, and t15, respectively.

With this complete information, the car moves from the 1st floor to the 10th floor with passenger 1, and then goes to the 15th floor to pick up passenger 2. While going to the 15th floor, the scheduling system knows that passenger 3 wants to ride on the car. Hence, instead of stopping at the 15th floor, the elevator system passes by the 15th floor, and then goes to the 17th floor first to pick up passenger 3. Passenger 2 takes the car at time t21 after the elevator system goes down from the 17th floor. As compared with conventional elevator scheduling systems, passenger 2's waiting time increases by 4 time units. However, passenger 3's waiting time decreases by 18 time units and the total running time of the car also decreases by 28 time units. Due to the significant decrease in the running time of the car, energy consumption is also reduced. In a smart home environment, therefore, we can minimize passengers' average waiting time and energy consumption using the information from floor sensors.

3.2   The Context-Aware Elevator Scheduling System (CESS)

Conventional elevator scheduling systems could not obtain passengers' information before they push the hall call or car call buttons. Due to such limited scheduling information, passengers' waiting time and energy consumption increases. To address such problems, we propose the Context-aware Elevator Scheduling System (CESS). Our system obtains the passenger's information completely from the floor sensors and exploits the information in an efficient scheduling of the elevator systems.

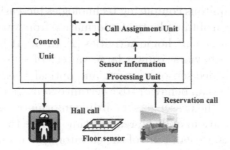

Fig. 3. Structure of the Context-aware Elevator Scheduling System

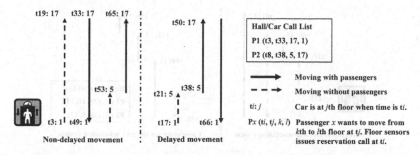

Fig. 4. An example of non-delayed movement and delayed movement

Our elevator scheduling system consists of three components, namely the Control Unit (CU), the Sensor Information Processing Unit (SIPU), and the Call Assignment Unit (CAU). The first component, CU controls the elevator system to accelerate or stop. The second component, SIPU processes the passenger information detected by the floor sensors. When a passenger comes to the elevator, the floor sensors embedded on the aisle detect passengers' location, and then send this information to the elevator control system through the building network. If SIPU receives a passenger's location, it calculates the time when the passenger arrives at the elevator. The third component, CAU evaluates the information received from SIPU, and then makes two decision values; Delay Time (DT) and Movement Direction (MD). DT is the time interval that should be delayed before the car starts movement for the reserved request, and MD is the movement direction, i.e., upside or downside, of the car. In CAU, the DT value is very important to minimize energy consumption. The DT value is computed as

$$DT = HC_i - (|CF - HF_i| * FH)/SC \qquad (1)$$

where HC_i is passenger i's hall call time, CF is the location of the car, HF_i is the location of passenger i, FH is the height of a floor, and SC is the speed of the car. Using expression (1), we estimate the time when the car starts to move, and then delay the car's movement for DT. Fig. 4 shows the profits of the delayed movement. There are two passengers in a smart apartment building. Passenger 1 is at the 17th floor and wants to move to 1st floor. Passenger 2 is at the 5th floor and wants to move to the 17th floor.

Each passenger makes reservations for the car at time t3 and t8, respectively, and they will be in front of the elevator at time t33 and t38, respectively. In the case of non-delayed movement, the car moves from the 1st floor to the 17th floor to pick up passenger 1 at time t3. Although passenger 2 makes a reservation for the car at time t8, the car passed by the 5th floor. As a result, passenger 2 waits for a long time, and also the system spends additional energy due to the long distance of movement. In the case of delayed movement, however, the car is delayed for DT although passenger 1 makes a reservation for the car at time t3. The car moves after the time interval of DT, and as a result, passenger 2 can take the car at this turn.

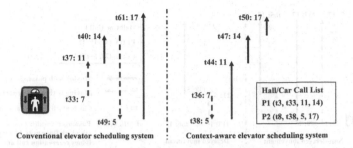

Fig. 5. An example of the Movement Direction (MD) value used in the Context-aware Elevator Scheduling System

```
1.    if a reservation call happens
2.    then
3.         delay_time = calculateDelayTime (reservation call information)
4.         the car waits for delay_time
5.         moving_direciton = decideDiretion()
6.         the car moves toward moving_direction
7.    fi
8.    ------------------------------------------------------------------------------------------
9.    calculateDelayTime (reservation call time, actual hall call time, hall call floor, destination floor)
10.        moving_time = the moving time of the car to the floor a passenger waits
11.        hallcall_time = actual hall call time
12.        delay_time = hallcall_time – moving_time
13.        return delay_time
14.   End calculateDelayTime          /* For Delay Time (DT) */
15.   ------------------------------------------------------------------------------------------
16.   decideDirection()                         /* For Movement Direction (MD) */
17.        if a new reservation call (NRC) is issued
18.        then
19.             if there are no reservation call
20.             then
21.                  decide the moving_direction of the car
22.                  /* a car's direction is set to upside or downside*/
23.                  target_floor = NRC's floor
24.             fi
25.             if the direction of a new reservation call is not same as moving_direction
26.             then
27.                  if (|NRC's floor – car's location| < |target_floor – car's location|)
28.                  then
29.                       target_floor = NRC's floor
30.                       decide the moving_direction of the car to the target_floor
31.                  fi
32.             fi
33.        fi
34.        return target_floor
35.   End decideDirection
```

Fig. 6. Algorithm of the Context-aware Elevator Scheduling System

MD is also important to minimize the passengers' average waiting time. Fig. 5 depicts an example of the profit of MD. There are two passengers and the elevator is at the 7th floor. Passenger 1 wants to move from the 11th floor to the 14th floor. Passenger 2 wants to move from the 5th floor to the 17th floor. In a conventional elevator scheduling system, the car first moves to the 11th floor to pick up passenger 1. At time t37, passenger 1 rides in the car and gets off at time t40. Then, to pick up passenger 2, the car moves from the 14th floor to the 5th floor, and arrives at the 5th floor when the time is t49. As a result, passenger 2 waits for 11 time units. On the contrary, in the Context-aware Elevator Scheduling System, the car first moves from the 7th floor to the 5th floor to pick up passenger 2, and then moves to the 11th floor to pick up passenger 1. As a result, the average passengers' waiting time and the elevator's moving distance are reduced significantly. Fig. 6 shows the algorithm of the proposed Context-aware Elevator Scheduling System.

4   Performance Evaluation

In this section, we present the performance evaluation results for the Context-aware Elevator Scheduling System (CESS). Table 1 shows the experimental conditions for performance evaluation. We use similar conditions with previous studies [5], [7]. We simulate the case of non-homogeneous arrival rates to adapt the real elevator system similar to the previous study [5]. To compare our elevator scheduling system with the conventional elevator scheduling system, we use the following three criteria.

- Average waiting time: a waiting time is the interval between a passenger's hall call time and a car's arrival time. Average waiting time is the average of all passengers' waiting times.
- Maximum waiting time: the maximum waiting time is the largest passenger's waiting time during the experiment period [11].
- Energy consumption: the energy consumption is the total amounts of energy consumed during the operation of the elevator system [5].

We conducted extensive simulations to compare the performance of our elevator scheduling system with the conventional elevator scheduling system. For a variety of experiments, CESS splits into three groups; CESS-30, CESS-60, and CESS-120. In CESS-30, the reservation call is issued 30 seconds before passengers make the actual hall call. Similarly, CESS-60 and CESS-120 issue the reservation calls 60 and 120 seconds before actual hall calls occur, respectively.

Fig. 7(a) shows the passengers' average waiting time for CS (Conventional elevator Scheduling system), CESS-30, CESS-60, and CESS-120. Since CESS

Table 1. Experimental Conditions

Items	setting
Number of floors	30
Elevator capacity	20
Moving speed	3 m/sec.
Floor height	3 m
Open/ close time	2 sec.
Boarding time	2 sec.

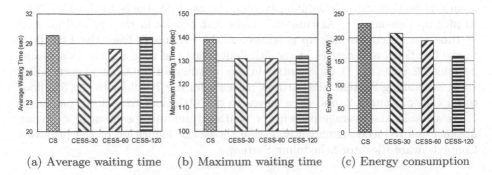

(a) Average waiting time   (b) Maximum waiting time   (c) Energy consumption

Fig. 7. Performance Results for the Conventional Scheduling system (CS) and the Context-aware Elevator Scheduling System (CESS)

groups (CESS-30, CESS-60 and CESS-120) exploit additional information, they perform better than CS. Specifically, the performance improvement of CESS-30 against CS is as much as 14.5%. The performance gain of CESS-120 is relatively small because the car picks up many passengers at a time. Fig. 7(b) shows the maximum waiting time. Similarly, CESS groups perform better than CS. In this result, the performance improvement of CESS against CS is as much as 6.1-6.8%. Fig. 7(c) shows the total amounts of energy consumption during the operation of the elevator system. CESS groups again perform better than CS. Specifically, CESS-120 shows the best performance for energy consumption because the car picks up more people at one time than other elevator scheduling systems. The performance improvement of CESS-120 against CS is as much as 30.6%.

5   Conclusion

We have presented a new elevator scheduling system for smart apartment buildings that considers both passengers' average waiting time and the energy consumption of the elevator system. The proposed elevator scheduling system knows the passengers' information before they push the elevator call buttons using floor sensors in the smart home environments. Our elevator scheduling system exploits complete passengers' information to control the elevator system efficiently. Experimental results show that the proposed elevator scheduling system significantly reduce passengers' average waiting time and the energy consumption.

References

1. Kidd, C.D., Orr, R.J., Abowd, G.D., Atkeson, C.G., Essa, I.A., MacIntyre, B., Mynatt, E., Starner, T.E., Newstetter, W.: The Aware Home: A Living Laboratory for Ubiquitous Computing Research. In: Proceedings of the Second International Workshop on Cooperative Buildings (1999)

2. Essa, I.: Ubiquitous Sensing for Smart and Aware Environments: Technologies towards the building of an Aware Home, the DARPA/NSF/NIST Workshop on Smart Environments (1999)
3. Orr, R.J., Abowd, G.D.: The Smart Floor: A Mechanism for Natural User Identification and Tracking. In: The 2000 Conference on Human Factors in Computing Systems (CHI 2000), The Hague, Netherlands (2000)
4. Addlesee, M., Jones, A., Livesey, F., Samaria, F.: The ORL Active Floor. IEEE Personal Communications (1997)
5. Lee, S., Bahn, H.: An Energy-aware Elevator Group Control System. In: 3rd IEEE International Conference on Industrial Informatics (2005)
6. Brand, M.E., Nikovski, D.N.: Optimal Parking in Group Elevator Control. IEEE International Conference on Robotics and Automation 1, 1002–1008 (2004)
7. Kim, J.H., Moon, B.R.: Adaptive Elevator Group Control with Camera. IEEE Transactions on Industrial Electronics 48(2), 377–382 (2001)
8. Kim, C., Seong, K.A., Lee-Kwang, H., Kim, J.O.: Design and Implementation of a Fuzzy Elevator Group Control System. IEEE Transactions on System, Man, and Cybernetics 28(3), 277–287 (1998)
9. Brand, M.E., Nikovski, D.N.: Risk-Averse Group Elevator scheduling, Elevcon World Congress on Vertical Transportation Technology (2004)
10. Kulkarni, A.B.: Energy consumption analysis for geared elevator modernization: upgrade from DC Ward Leonard system to AC vector controlled drive. In: The IEEE Industry Applications Conference, Rome, Italy, vol. 4, pp. 2066–2070 (2000)
11. Eguchi, T., Hirasawa, K., Hu, J., Markon, S.: Elevator Group Supervisory Control Systems Using Genetic Network Programming. Congress on Evolutionary Computation 2, 1661–1667 (2004)
12. Igarashi, K., Take, S., Ishikawa, T.: Supervisory control for elevator group with fuzzy expert system. In: Proc. IEEE Int. Conference Industrial Technology, pp. 133–137 (1994)
13. Dewen, Z., Li, J., Yuwen, Z., Guanghui, S., Kai, H.: Modern elevator group supervisory control systems and neural networks technique. In: IEEE Int. Conf. Intelligent Processing Systems, pp. 528–532 (1997)
14. Pepyne, D., Cassandras, C.: Optimal dispatching control for elevator systems during uppeak traffic. IEEE Transactions on Control Systems Technology, 629–643 (1997)
15. Fujino, A., Tobita, T., Segawa, K., Yoneda, K., Togawa, A.: An elevator group control system with floor attribute control method and system optimization using genetic algorithms. In: 21st Int. Conf. Industrial Electronics, Control, and Instrumentation (1995)
16. Pepyne, D., Cassandras, C.: Design and Implementation of an Adaptive Dispatching Controller for Elevator Systems During Uppeak Traffic. IEEE Transactions on Control Systems Technology, 635–650 (1998)
17. Chung, D.-W., Ryu, H.-M., Lee, Y.-M., Kang, L.-W., Sul, S.-K., Kang, S.-J., Song, J.-H., Yoon, J.-S., Lee, K.-H., Suh, J.-H.: Drive systems for high-speed gearless elevators. IEEE Industry Applications Magazine (2001)
18. Amano, M., Yamazaki, M., Ikejima, H.: The latest elevator group supervisory control system. Mitsubishi Elect. ADVANCE 67, 88–95 (1996)
19. Maeda, T., Komaya, K.: Revised design calculations of elevator systems. In: IEEE International Conference on Systems, Man, and Cybernetics, vol. 2, pp. 1041–1046 (2000)

20. Fujino, A., Tobita, T., Yoneda, K.: An On-line Tuning Method for Multi-Objective Control of Elevator Group. Proceedings of IEEE Industrial Electronics, Control, Instrumentation, and Automation 2, 795–800 (1992)
21. Fujino, A., Tobita, T., Segawa, K., Yoneda, K., Togawa, A.: An elevator group control system with floor-attribute system optimization using genetic algorithms. IEEE Transactions on Industrial Electronics 44(4), 546–552 (1997)
22. Kaneko, M., Ishikawa, T., Sogawa, Y.: Supervisory control for elevator group by using fuzzy expert system. In: 23rd Int. Conf Industrial Electronics, Control, and Instrumentation (1997)

A MOM-Based Home Automation Platform

Chun-Yuan Chen[1], Chi-Huang Chiu[1], and Shyan-Ming Yuan[1,2]

[1] Department of Computer Science,
National Chiao Tung University, Taiwan
[2] Department of Computer Science and Information Engineering,
Asia University, Taiwan
cychen@cs.nctu.edu.tw, chchiu@cis.nctu.edu.tw, smyuan@cs.nctu.edu.tw

Abstract. While there have been many home networking technologies such as UPnP and INSTEON, appliances supporting different home networking technologies cannot collaborate to finish Home Automation (HA). Although many studies of interoperability among home networking technologies have been done, researches on further HA in heterogeneous environments are still lacking. This paper proposes the MOM-based Home Automation Platform (MHAP), which accomplishes event-driven HA in incompatible home networks. MHAP is independent of any home networking technology and integrates home networking technologies in the home gateway. For users, MHAP provides the easy-to-use and standardized way to configure complex HA scenarios by rules. Through introducing Message Oriented Middleware (MOM) and Open Service Gateway Initiative (OSGi), MHAP offers reliable automatic operations, fault tolerant and reconfigurable HA, high extensibility and large scalable collaboration among appliances, other MHAP gateways and Internet services.

1   Introduction

In the last two decades, the science and the technology for accomplishing Home Automation (HA) have progressed tremendously. Nowadays, many digital appliances can join the HA system at home. Home Automation makes human life convenient through connection and collaboration between appliances and sub systems at home such as security system. Smart appliances and sub systems use home networking technology to communicate with each other. There have been more then twenty different kinds of home networking technologies on market such as X10, INSTEON, UPnP, jini and LonWorks. Traditionally, building a HA system should use exactly one home networking technology. A home networking middleware doesn't understand the messages of the other home network. Even one home network couldn't physically connect to the other home network. However, in Digital Home environments, the home gateway such as a PC or a set-top-box can physically connect to each home network and Internet. Recently, there have been substantial studies performed on one-to-one and one-to-many interoperability of heterogeneous home networking technologies. While some studies on device control in heterogeneous environments have been done, researches on HA in such situation are still critically lacking. Therefore, this

M.S. Szczuka et al. (Eds.): ICHIT 2006, LNAI 4413, pp. 373–384, 2007.

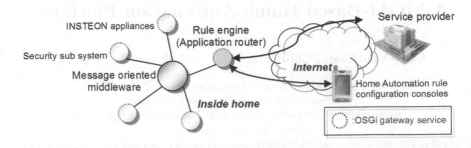

Fig. 1. The overview of the MOM-based Home Automation Platform (MHAP)

paper proposes the MOM-based Home Automation Platform (MHAP), which integrates all home networks and offers Home Automation in heterogeneous environments.

1.1 MHAP Design Goals

MHAP has six design goals such as the following:

- **Event-driven Home Automation:** In HA systems, events represent state changes of appliances and sub systems. Many HA operations need events to trigger them. A complete HA platform should support the event-based Home Automation. MHAP is proposed to provide the events across different home networks.
- **Device and Protocol Independent:** MHAP offers a device and protocol independent environment. HA programs can be designed by the consistent way and won't be limited by certain kinds of home networking technologies.
- **Reliability and Fault Tolerance:** Some HA operations, especially the operations about security, must be reliable. Beside, when some appliances or certain parts of the networks fail, MHAP should offer the fault tolerant function to ensure that the entire Home Automation will be still working.
- **Scalability:** Increasing the number of appliances at home should not apparently lower the performance of the MHAP. Besides, MHAP is designed to scale from home domain to building, campus and Internet. The appliances at home can cooperate with Internet services and other appliances at the building.
- **Extensibility:** There may be appliances only supporting new home networking technology on market. The extensibility of MHAP makes new smart appliances with additional home networking technologies join MHAP dynamically.
- **Easy to Use:** MHAP is designed for providing reconfigurable HA and the convenient way to configure HA scenarios. The configuration needs no complicated process and is easy to use.

1.2 MHAP Solution

Fig. 1 represents the MHAP solution. Inside home, the MHAP gateway connects each home network and the rule engine service performs as an application router. MHAP introduces the Message Oriented Middleware (MOM) into HA systems and accomplish Home Automation through platform and protocol independent MHAP message flow. The application router handles the message flow among appliances according to HA rules. MHAP also introduces the standard deployment and service management infrastructure to manage services and software bundles dynamically. The infrastructure in present MHAP implementation is based on Open Services Gateway initiative (OSGi) and has the following features:

- **Message Oriented Middleware:** MOM is a middleware decoupling message passing between the client and the server. When passing messages, the client and server don't have to be online simultaneously. MOM introduces event-driven model into the MHAP since an event is a special message. MHAP use. The publish/subscribe MOM, which scales well when the number of receivers increases. Because different appliance subscribes different topics in MOM, plugging in new appliance will not affect the HA execution. MHAP offers the reliable operation through durable subscription and persistent messages of MOM. MHAP uses Java Message Service (JMS), the standard MOM interface for Java, to access the function of MOM. Different MOM providers can be chosen to gain different degrees of performance.
- **Open Services Gateway initiative:** Through OSGi, software bundles on MHAP can be transferred from service providers through Internet and software providers and hardware providers can develop their own products without affecting each other. Such infrastructure makes MHAP extensible. The MHAP services such as protocol adaptors can be installed independently and dynamically without suspending the entire system and current automation operation.
- **Rule-based Home Automation:** MHAP provides the rule service instead of the traditional API or program. The HA scenario is indicated by high-level rules, which are the basic units representing automatic HA scenario assigned by users. All device and appliances are abstracted by XML descriptors. The user can use the consistent way to configure the rules in which the appliances supporting different home networking technologies participate.

2 System Architecture

2.1 Layered Architecture

For adapting heterogeneous HA environment and application, MHAP has a four-layered architecture showed by Fig. 2:

- **Physical Device and Network Layer:** The bottom layer consists of any home network and physical device supporting any home networking technology. MHAP aims at working over any device and home network including the future one.

- **Infrastructure Layer:** The layer introduces infrastructure such as OSGi to provide service management and deployment functions for MHAP services. For further extension, MHAP needs to install new service bundles and update the existing bundles. The layer provides standardized methods to dynamically control life cycle of service bundles including downloading, installing, activating and deactivating services through Internet with dependency check.
- **MHAP Service Layer:** The layer consists of MHAP services and provides functionalities constructing HA, which includes event notification, appliance control, HA rule configuration and device management. The layer is also an adaptation layer, which lets application operate appliances supporting different HA technologies through the consistent method. Internally, the layer uses MOM to support event-driven HA in heterogeneous environment. Furthermore, the message flow in this layer makes the Home Automation re-configurable.
- **Application Layer:** Facilitating Home Automation needs many different kinds of applications, for example, administration tools, rule configuration tools, appliance monitors and universal controllers. With assistance of the first three MHAP layers, developing all of these applications is easier and the developers needn't consider the complex problems in heterogeneous environment.

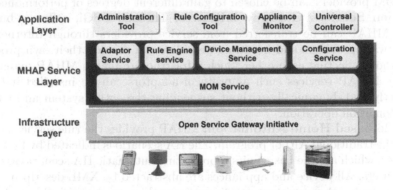

Fig. 2. The Four-layered architecture of MHAP

2.2   MHAP Message

MHAP services use messages to communicate with each other. The four main MHAP message types are as follows:

- **Command Message:** The MHAP controls appliances by command messages. For example, the rule engine can sends a "power on" command message to power on an appliance.
- **Query Message:** The function of query messages is to obtain the states of appliances. MHAP services can publish a query message if it needs the

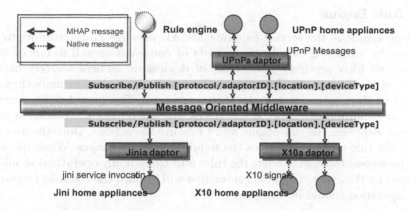

Fig. 3. Topic naming and message flow between independent channels in MHAP

state of an appliance. The MOM provider will push the query message to the adaptor handling the appliance.

- **State Message:** The message contains the state information of an appliance. When the adaptor of an appliance receives query messages, it responds state messages.
- **Event Message:** When some appliance state changes, the adaptor or service publishes event messages representing the change to notifying the other services.

2.3 Message Flow and Topic Naming

For introducing the publish/subscribe model of MOM, MHAP accomplishes high extensible Home Automation through message flow. The message follow in MOM is the main method to finish rule-based Home Automation. In MHAP, every component subscribes and publishes different topics. As Fig. 3 shows, a topic is an independent channel from a component to the MOM provider. Therefore, the joining of new appliance does not affect the existing Home Automation. The rule engine dispatches the messages among the components according to HA rules. An adaptor service is a proxy of certain kind of physical home appliances, which creates virtual appliance objects with MHAP descriptors. The handling of MHAP message flow could be simplified through topic naming. The naming rule of topics is *[protocol/adaptorID].[location].[deviceType]*. When a user set a rule, the rule engine will subscribe the event indicated in the rule. For example, a rule engine is handling a rule which will turn on the X10 light in room1 when the UPnP TV in room2 is powered off. The rule engine subscribes the topic *"upnp.room2.tv"* to receive the event. When receiving the event, the rule engine will publish a command message to the topic *"x10.room1.light"* and then the MOM will push the message to the adaptor handling the command.

2.4   Rule Engine

The rule engine is the service handling MHAP message flow according to the rules set by users (Fig. 4). A rule consists of one event as well as one or more actions and filter services. Two events of devices or services scarcely happen exactly at the same time. However, a rule may be applied when more than one services or devices are match the states assigned. The filter service provides these pre-condition of a rule. If the states are not on the pre-condition when the event happens, the rule engine won't execute the actions. Once the user sets a rule, the rule engine subscribes the required event messages. When receiving event messages, the engine refers the rules and check if any operations should be triggered by the event. Then the rule engine will publish the command messages if the operation should be executed.

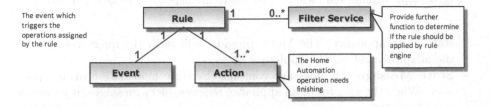

Fig. 4. The components of MHAP rules

2.5   Adaptor

The adaptor is responsible for bridging the MHAP message and native messages such as UPnP messages and subscribing/publishing MHAP messages as the proxy of appliances. The appliances belonging to a protocol or designed by a manufacturer may use the same adaptor. When new appliances join the MHAP, the adaptor needn't be re-implemented through the MHAP descriptors. For the low interoperability of present home network technologies, the appliances can't understand the native messages from another kind of home network. However, adaptors bridge the native message to universal and platform independent MHAP messages which are used to communicate each appliances and services. Therefore, the messages can flow among different home networks.

2.6   Other MHAP Services

The configuration service and the device management service are the other two important MHAP services. The configuration service offers the function to configure the Home Automation scenario. The functions include setting event, action and filter part of a rule, rule management and rule re-configuration. Besides, the configuration service also provides the interface to assign the MHAP descriptor. The adaptor service can automatically set the URL or content of descriptors or the user can assign them manually with a configuration terminal. The device management service records and maintains all the states of the device. It provides

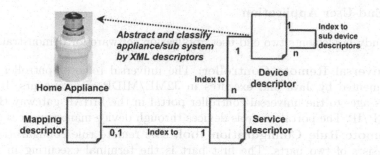

Fig. 5. The XML-based MHAP descriptors

universal and consistent method to get the information of device such as the associated adaptor and to manage devices.

2.7 MHAP Descriptors

MHAP descriptors provide profiles of the devices (Fig. 5). Through the descriptors, the adaptor acquires the required information to construct virtual appliances and drive actual appliances. Descriptors make MHAP services more flexible. When there is new appliances joining to the HA system, programs of the original adaptor needs no re-implementing. The descriptor is XML-based and independent of any platform. The MHAP services can easily parse the descriptors and attends the information.

3 Implementation

3.1 Environment

MHAP is deployed on any home gateway device with any OSGi compliant environment. The demonstration gateway in the paper is an IBM X 31 notebook installing Oscar, which is an open source OSGi framework implementation. The present MOM provider software is Active MQ, which offers JMS interface, persistent messages and durable subscription. The Active MQ can be replaced by any JMS compliant MOM. In the demonstration, the home gateway connects to the UPnP appliance such as the DSM 320 media player through IEEE 802.11b wireless LAN. The gateway also connects to X10 and INSTEON networks through X10 and INSTEON computer interfaces respectively, which send and receive signals in power line network. A Sony Ericsson K700i mobile phone which supports J2ME and connects to Internet through GPRS can control the appliances as a universal remote controller. Another HP iPAQ h6365 PocketPC connecting to the MHAP gateway through wireless LAN is running the rule configuration terminal.

380   C.-Y. Chen, C.-H. Chiu, and S.-M. Yuan

3.2   End-User Application

This study implements two end-user application software for demonstration:

- **Universal Remote Controller:** The universal remote controller is implemented by Java and executes in J2ME/MIDP environments. It sends messages to the universal controller portal in the MHAP gateway through TCP/IP. The portal controls devices through device management service.
- **Remote Rule Configuration tool:** The remote rule configuration tool consists of two parts. The first part is the terminal executing in mobile devices and the second part is the portal in the MHAP gateway. The terminal is a .NET program for Windows Mobile environments and implements the user friendly GUI to edit and manage rules. The terminal sends and receives configuration messages to and form the configuration portal, which uses the configuration service to manage and configure rules.

4   Demonstration and Discussion

4.1   Deployment Architecture

Fig. 6 shows the deployment architecture of MHAP. Inside home, the smart appliances connect to the MHAP home gateway through several home networks. Outside home, the MHAP gateway connects to each other and Internet services though Internet. JMS Internet services directly subscribe and publish MHAP messages and the other Internet services such as Web services do that through adaptors. Therefore, the legacy service outside home can join to the Home Automation without any change. MHAP can push message outside home to the service of a service provider such as a home security company and subscribe the messages form the services.

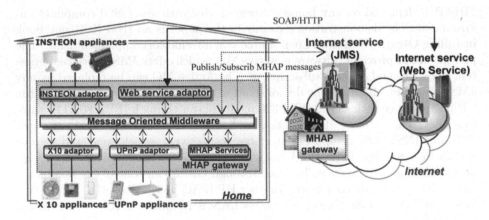

Fig. 6. Deployment architecture of MHAP

4.2   Demonstration

In the demonstration, the user controlled appliances with the MHAP universal controller. He also configured and managed HA rules by click through rule configuration tool. The user set a scenario which consists of three rules. The first one indicates the X10 electronic fan should be powered on when the state of the UPnP media player changes to *"playing"*. The second one indicates the INSTEON dimmable light should be turned on when the state of the UPnP media player changes to *"pause"*. The third one indicates the UPnP media player should resume playing when the state of the INSTEON dimmable light changes to *"off"* After configuration, the MHAP performed Home Automation according to the rules. When the UPnP media player had started to play, the X10 electronic fan was powered on automatically. When the user had paused the media player for walking out of the room, the state change triggered the turning on of the INSTEON dimmable light. When the user had been back to the room and turned off the light, the media player resumed playing again. Fig. 7 shows the message flow for rule 2. While the user had added the rule to MHAP with configuration tool, the rule engine subscribed the topic *"upnp.room1.avrenderer"* to receive the event message. After the UPnP media player is paused by the user, it sent a native UPnP message indicating the state change. The UPnP adaptor receiving the event message then published a MHAP event message to the topic *"upnp.room1.avrenderer"*. After the MOM provider had pushed the message to rule engine, the rule engine checked if any operation should be executed according to rules. To finish the operation assigned in rule 2, the rule engine published a command message to the topic *"insteon.room1.light"* to turn on the light. The INSTEON adaptor received the message and bridged it to power line network. Therefore, the assigned INSTEON dimmable light was turned on finally.

4.3   Scaling to Other Network and Internet

The services outside the home can also join to Home Automation through MHAP. Any legacy service which is implemented by any distributed technology such as MOM and Web service can cooperate with the Home. For example, the system of a home security company can subscribe the event of security sub system at home. When the security sub system at home detects a fire or a home burglar, MHAP will publish the message to the company and guarantee the message delivery. The MHAP can also scales to Internet. An Internet service can subscribe the event of media players at home to determine the most popular song playing at each home. The playing event messages will be pushed to the Internet service. Beside, the event of remote Internet service can also trigger HA operations. The UPnP media player can automatically play the most popular song every time when the Internet service publishes the event notifying that the hit parade is updated.

4.4   Benefit

– **Event-driven Home Automation in Heterogeneous Environment:**
  MHAP makes every smart appliance using incompatible home networking

technologies cooperate together. It can not only schedule operations of appliances but also let the state change of one appliance trigger operations of other appliances.

- **Device and Protocol Independent Platform:** MHAP imports each appliance as a corresponding abstract MHAP appliance through XML-based descriptors. In message flow of MOM, appliances communicate with each other through universal MHAP messages. Therefore, the applications and services needn't change when using different appliance.
- **Reliable Fault Tolerant Home Automation:** Through MOM, MHAP offers reliable operations and events which meet the requirements for HA related to security. Besides, MHAP divides the network into regions according to topic naming, which provides fault tolerant HA. Since each region subscribes and publishes the independent topic, the fail in a region doesn't affect the message flow of the other regions.
- **High Scalability:** The MOM core makes MHAP scales well when the numbers of appliance increasing through one-to-many publish/subscribe messaging approach. Beside, MHAP system can scale from home domain to Internet domain because the MOMs are able to cross subnet and connect to each other. The messages in one MHAP gateway can pass to other gateway outside home and push to Internet services. What participates the automation can not only an appliance or sub system at home but also an Internet service.
- **Reconfigurable Home Automation:** Publishing and subscribing messages through independent channel design offers complete reconfigurable HA. When user can dynamically change a certain part of the HA scenario, the other parts are still effective.
- **High Extensibility:** When a new appliance is plugged in, the MHAP will check if there is appropriate adaptor for it. If there is no appropriate adaptor in MHAP gateway, MHAP will automatically obtain it through Internet. New protocol, device and Internet services can dynamically join to the current Home Automation without any affection and suspending of operations.
- **Easy to use:** MHAP offers the easy-to-use and consistent way to configure Home Automation even if the appliances supporting different home networking technologies. MHAP separates HA scenario into rules and centralizes the settings and configurations. The user configures HA through editing rules without multiple processes loading each macro program to each smart appliance.

4.5   Related Work and Comparison

The researches about heterogeneous home network focused on interoperability among different home networking technologies. However, the one-to-one bridge middlewares such as "Experimental Bridge LONWORKS / UPnP" and "Jini Meets UPnP: An Architecture for Jini/UPnP Interoperability" are not suitable for HA in heterogeneous environments. When the number of home networking technologies is increasing, the $(n*(n-1))/2$ conversion complexity is too high.

The one-to-any approach is more suitable for HA in heterogeneous environment, for example, "A Framework for Connecting Home Computing Middleware" proposed by Eiji Tokunaga et al. and Universal Middleware Bridge (UMB) proposed by Kyeong-Deok Moon et al. With the former, all the converting interfaces should be regenerated via javassist tool when a user adds an appliance supporting new protocol. It is not convenient for HA. On the other hand, UMB integrates home networks dynamically like MHAP but doesn't meet the other HA requirements.

5 Conclusion and Future Work

5.1 Conclusion

In conclusion, MHAP provides a platform integrating every home network technology and Internet service. MHAP introduces MOM to supports event-driven, reliable, fault tolerant and reconfigurable HA. MHAP can scale to Internet, and cooperate with legacy Internet services and security systems of home security companies. Through introducing OSGi in infrastructure layer, MHAP has high extensibility and all components can be automatically updated and extended from Internet according to the service dependency. For user, MHAP also provides the easy-to-use way to manage and edit rules to configure the complex HA scenario.

5.2 Future Work

MHAP will offer the adaptive function for application developing, which will makes a HA application automatically generate suitable user interface for different device. On the other hand, the security mechanism will be discussed more because MHAP has scaled to Internet. MHAP will provide the filtering function protecting the HA system against the dangerous messages from other MHAP gateways or Internet. MHAP will also introduce Multimedia Home Platform (MHP) into infrastructure layer. Such services can be updated through terrestrial broadcast and the interactive Digital TV services can join HA at home.

Acknowledgments. The Ministry of Education of the Republic of China partially supported this work under grant nos: NSC94-2725-E009-006-PAE and NSC94-2213-E009-026.

References

1. Ryan, J.L.: Home Automation. Electronics and Communication Engineering Journal 1(4), 185–192 (1989)
2. Turnbull, J.G.: Introducing home area networks. BT Technology Journal 20(2) (April 2002)
3. Echelon Co., LonTalk Protocol Specification ver 3.0 (1994)
4. Valtchev, D., Frankov, I.: Service Gateway Architecture for a Smart Home. IEEE Communications Magazine 40(4), 126–132 (2002)

5. Chiu, C.-H., Lin, H.-T., Yeh, P.-J., Yuan, S.-M.: An OSGi Platform Connecting Heterogeneous Smart Home Appliances from Mobile Devices. WSEAS Transactions on Computers 5(7), 1549–1555 (2006)
6. Guttman, E., Kempf, J.: Automatic Discovery of Thin Servers: SLP, Jini and the SLP-Jini bridge, IECON 1999 (1999)
7. Ueno, D., Nakajima, T., Satoh, I., Soejima, K.: Web-Based Middleware for Home Entertainment. In: Jean-Marie, A. (ed.) ASIAN 2002. LNCS, vol. 2550, Springer, Heidelberg (2002)
8. LaPlant, B., Trewin, S., Zimmermann, G., Vanderheiden, G.: The Universal Remote Console: A Universal Access Bus for Pervasive Computing. IEEE Pervasive Computing 3(1), 76–80 (2004)
9. Latvakoski, J., Pakkala, D., Paakkonen, P.: A Communication Architecture for Spontaneous Systems. IEEE Wireless Communications Magazine (June 2004)
10. Banavar, G., Chandra, T., Strom, R., Sturman, D.: A Case for Message Oriented Middleware. In: Jayanti, P. (ed.) DISC 1999. LNCS, vol. 1693, Springer, Heidelberg (1999)
11. Hsiao, T.-Y., Cheng, M.-C., Chiao, H.-T., Yuan, S.-M.: FJM: a high performance Java message library. In: IEEE International Conference on Cluster Computing 2003, Hong Kong, pp. 460–463 (December 1-4, 2003)
12. Sun Microsystems, Java Message Service Specification Version 1.1 (April 2002)
13. Chemishkian, S., Lund, J.: Experimental Bridge LONWORKS / UPnP, CCNC (2004)
14. Allard, J., Chinta, V., Gundala, S., Richard III, G.G.: Jini Meets UPnP: An Architecture for Jini/UPnP Interoperability. In: SAINT 2003 (2003)
15. Tokunaga, E., Ishikawa, H., Kurahashi, M., Morimoto, Y., Nakajima, T., Framework, A.: A Framework for Connecting Home Computing Middleware, ICDCSW 2002 (2002)
16. Moon, K.-D., Lee, Y.-H., Lee, C.-E., Son, Y.-S.: Design of a universal middleware bridge for device interoperability in heterogeneous home network middleware. IEEE Transactions on Consumer Electronics 51 (2005)
17. An Adaptive Architecture for Secure Message Oriented Middleware WSEAS Transactions on Information Science and Applications. 3(7), 1239–1246 (July 2006)
18. Cheng, M.-C., Yuan, S.-M.: An Adaptive Mobile Application Development Framework. In: Yang, L.T., Amamiya, M., Liu, Z., Guo, M., Rammig, F.J. (eds.) EUC 2005. LNCS, vol. 3824, pp. 765–774. Springer, Heidelberg (2005)

An Error Sharing Agent for Multimedia Collaboration Environment Running on Pervasive Networks

Eung Nam Ko

Department of Information and Communication, Baekseok University,
115, Anseo-Dong, Cheonan, Chungnam, 330-704, Korea
ssken@bu.ac.kr

Abstract. This paper presents the design of an error sharing agent for multimedia collaboration environment which is running on RCSM. RCSM means Reconfigurable Context-Sensitive Middleware. RCSM provides an object-based framework for supporting context-sensitive applications. It has other services in optional components. A good example of other services in RCSM is a distance education system for multimedia collaboration environment. We propose an adaptive error hooking agent based on a hybrid software architecture CARV(Centralized Abstraction and Replicated View). This system can be automatically enforced according to different situations such as wired or wireless network environment.

Keywords: RCSM, object-based framework, context-sensitive applications, error hooking agent, wired or wireless network environment.

1 Introduction

In a ubiquitous computing environment, computing anytime, anywhere, any devices, the concept of situation-aware middleware has played very important roles in matching user needs with available computing resources in transparent manner in dynamic environments [1]. Distance Education system must be able to support real-time interaction and also support user synchronization including not only temporal synchronization and spatial synchronization but floor control for smooth interaction [2]. It is difficult to avoid a problem of the seam in the pervasive computing environment for seamless services. Thus, there is a great need for synchronization control algorithm in situation-aware middleware to provide dependable services in ubiquitous computing. RCSM facilitates the development and runtime operations of context-sensitive pervasive computing software [9]. We propose an adaptive error hooking agent based on a hybrid software architecture CARV which is adopting the advantage of CACV(Centralized-Abstraction and Centralized-View) and RARV(Replicated-Abstraction and Replicated-View) for situation-aware services.

M.S. Szczuka et al. (Eds.): ICHIT 2006, LNAI 4413, pp. 385–394, 2007.
© Springer-Verlag Berlin Heidelberg 2007

2 Related Works

The Context Toolkit was built based on this conceptual framework. There were five applications that were built to assess the actual benefits of the Context Toolkit. Seminal work has been done by Anind Dey, et al. [3] in defining context-aware computing, identifying what kind of support was required for building context-aware applications and developing a toolkit that enabled rapid prototyping of context-aware applications. They have laid out foundations for the design and development of context-aware applications by proposing a conceptual framework. The proposed conceptual framework separates concerns between context acquisition and the use of context in applications, to provide abstractions that help acquire, collect and manage context in an application independent fashion and identify corresponding software components. In the Context Toolkit, a predefined context is acquired and processed in context widgets and then reported to the application through application-initiated queries and callback functions.

In this Reconfigurable Context-Sensitive Middleware(RCSM), Stephen S. Yau et al. [4, 6] proposed a new approach in designing their middleware to directly trigger the appropriate actions in an application rather than have the application itself decide which method(or action) to activate based on context. Their motivation was to extend existing context-sensitive applications by adding new context sources and to easily let multiple concurrent contexts trigger a specific action. They already build a Smart Classroom to validate this RCSM model. RCSM provides an Object-based framework for supporting context-sensitive applications. Anand Ranganathan et al. [5] have built a middleware for developing context-aware applications.

This middleware is integrated into their infrastructure for Smart Spaces named GAIA. The middleware is based on a predicate model of context. This model enables agents to be developed that either use rules-based or machine learning approaches to decide their behavior in different contexts.

3 Our Approach

3.1 Pervasive Networks

A principal goal of pervasive computing is to make the actual computing part of it and its enabling technologies essentially transparent[4,7,8]. This transparency is partially possible because a pervasive computing environment is a collection of embedded, wearable, and handheld devices wirelessly connected, possibly to fixed network infrastructures such as the Internet.

RCSM facilitates the development and runtime operations of context-sensitive pervasive computing software [9]. A good example of other services in RCSM is a multimedia collaboration environment. The Object Request Broker of RCSM (R-ORB) assumes the availability of reliable transport protocols; one R-ORB per device is sufficient. The number of ADaptive object Containers (ADC)s depends on the number of context-sensitive objects in the device. ADCs periodically collect the necessary raw context data through the R-ORB, which in turn collects

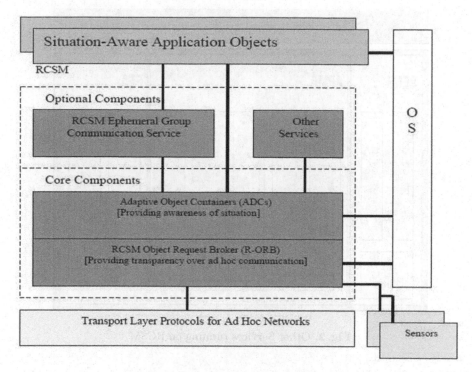

Fig. 1. RCSMs integrated components

the data from sensors and the operating system. Initially, each ADC registers with the R-ORB to express its needs for contexts and to publish the corresponding context-sensitive interface. RCSM is called reconfigurable because it allows addition or deletion of individual ADCs during runtime (to manage new or existing context-sensitive application objects) without affecting other runtime operations inside RCSM. Ubiquitous applications require use of various contexts to adaptively communicate with each other across multiple network environments, such as mobile ad hoc networks, Internet, and mobile phone networks. Figure 1 shows how all of RCSM's components are layered inside a device.

3.2 Multimedia Collaboration Environment

A Multimedia Distance Education System(MDES) is a good example of multimedia collaboration environment running on situation-aware applications. An example of smart classroom illustrated in [10]. However, it did not include other services support in the architecture. Our proposed model aims at supporting transparency requirements by using a mechanism in order to provide ubiquitous, seamless services. The development of multimedia computers and communication techniques has made it possible for a mind to be transmitted from a teacher to a student in distance system running on pervasive computing environment.

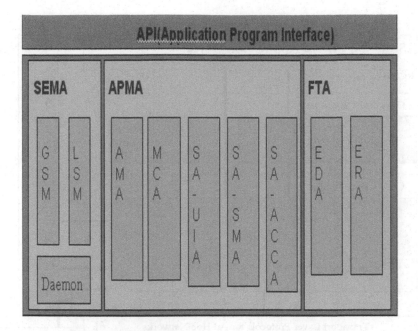

Fig. 2. Other Services running on RCSM

RCSM facilitates the development and runtime operations of context-sensitive pervasive computing software [9]. Figure 2 shows an example of other services of RCSM such as MDES which have many agents for multimedia distance education system running on pervasive computing environment such as smart classroom.

The functional modules, so called agents, consists of SEMA(Session Management Agent), APMA(Application Management Agent), and FTA(fault Tolerant Agent). SEMA supervises the beginning of each session and manages it during its lifetime. APMA keeps track of events occurred from collaborative sessions or applications, so that to process it according to management policy and requirement. APMA consists of several agents. They consist of AMA(Application Management Agent), MCA(Media Control Agent), FTA(Fault Tolerance Agent), SA-UIA(Situation-Aware User Interface Agent), SA-SMA(Situation-Aware Session Management Agent), and SA-ACCA(Situation-Aware Access and Concurrency Control Agent), as shown in Figure 3.

AMA consists of various subclass modules. It includes creation/deletion of shared video window and creation/deletion of shared window. MCA supports convenient applications using situation-aware ubiquitous computing. Supplied services are the creation and deletion of the service object for media use, and media share between the remote users. This agent limits the services by hardware constraint. FTA is an agent that plays a role in detecting an error and recovering it in situation-aware ubiquitous environment. SA-UIA is a user interface agent to adapt user interfaces based on situations. SA-SMA is an agent which plays a role in connection of SA-UIA and FTA as situation-aware

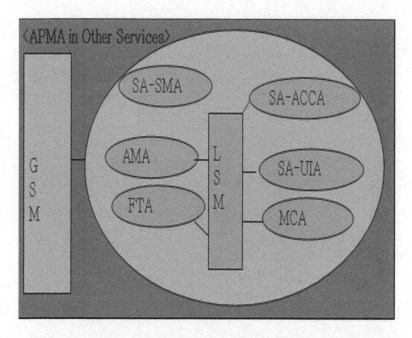

Fig. 3. The relationship between SEMA, APMA, and FTA

management for the whole information. SA-ACCA controls the person who can talk, and the one who can change the information for access and concurrency. SMA monitors the access to the session and controls the session. It has an object with a various information for each session and it also supports multitasking with this information. SMA consists of Global Session Manager (GSM), Daemon, Local Session Manager (LSM), Participant Session Manager (PSM), Session Monitor, and Traffic Monitor. SMA consists of GSM(Global Session Manager), Daemon, LSM(Local Session Manager) and PSM(Participant Session Manager). GSM has the function of controlling whole session when a number of sessions are open simultaneously. LSM manages only own session. For example, LSM is a lecture class in distributed multimedia environment. GSM can manage multiple LSM. Daemon is an object with services to create session. Daemon is an object with services to create session.

3.3 Fault Tolerance Agent

To ensure required reliability of multimedia communication systems based on situation-awareness middleware, FTA(fault Tolerant Agent) consists of 3 steps that are an error detection, an error classification, and an error recovery. FTA consists of EDA(Error Detection Agent) and ERA(Error Recovery Agent). EDA consists of ED(Error Detector) and ES(Error Sharing).

EDA is an agent which plays a role in detecting, and classifying errors. ED is an agent which plays a role as an interface to interact among an application.

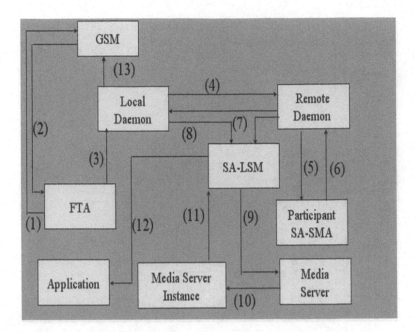

Fig. 4. Error Sharing Process

ED has functions which detect an error. This method detects an error by polling periodically the process with relation to FTE session. Windows 95/98/XP creates a process database to represent the process. Process database include a list of threads, a list of loaded modules, the heap handle of the default process heap, a pointer to the process handle table, and a pointer to the memory context that the process runs in. A process handle is essentially the same thing as a file handle. GetExitCodeProcess() function retrieves the termination status of the process specified by the Process handle passed in. While a process is still actively running, its exit code is 0x103(0x: hexadecimal code).

ES deals with learning in reactive multi-agent systems. Generally, learning rules may be classified as supervised or unsupervised. KB has a registration information of creation of service handle and session manager handle by Daemon and GSM. ES can decide whether it is hardware error or software error based on learning rules. In case of hardware error, it cannot be recoverable. In case of software error, it can be recoverable. This approach is based on the idea of comparing the expected error type which is generated by an ES with the actual error occurred from sites. This approach is based on the idea of comparing the expected error type which is generated with the actual error occurred from sites.

The sequence of recovery message flow can be shown in Figure 4. If an error is to be recovered, you can create sequences below. It creates a session with initial configuration information. It requests port ids for audio/video servers to build-up a Local Session Manager. It assigns port ids for audio/video servers of an application. It invites to the session and build-up a session instance monitor.

It sends invited messages to start build-up of session instance monitor. It builds up Session Instance Monitor using the configuration information from LSM. It sends joint message to the Local Session Manager. It sends session information to GSM(Global Session Manager) for set-up of GSM table. It begins a session. It exchanges message or command between LSM(Local Session Manager) and PSM(Participant Session Manager) and media data between media server based on interpretation of message handler.

4 Simulation Results

With the proposed conceptual framework for building context sensitive applications and the Context Toolkit laid a foundation for context ware middleware to develop. However, while the Context Toolkit does provide a starting point for applications to make use of contexts, its middleware paradigm still forces programmers to think about how to deal with the context widget for the appropriate contextual information. In general these projects do not provide much help on how to reason about contexts. They provide reusable sensing mechanisms but lack of reusable reasoning mechanisms. They do not provide any generic mechanism for writing rules about contexts, inferring higher-level context or organizing the wide range of possible contexts in a structured format [3].

RCSM has laid a further step towards middleware models that fully support for context awareness. RCSM deals with contexts dynamically and let context-sensitive application developers focus on implementing the actions in their favorite language without worrying about details of getting contextual information from different sources. While RCSM uses contextual information dynamically, it does not provide reasoning and/or learning mechanism to help agents reason about context appropriately [4]. The middleware paradigm for context awareness agents implemented in GAIA infrastructure provides a more generic way of specifying the behavior of context-aware-ness applications using different reasoning and learning mechanism. Since all the terms used in the environment are defined in the ontology, it is easy to frame rules for inferring contexts based on these terms. The developers do not have to worry about not using inappropriate terms or concepts, since they can refer to the definitions I the ontology when in doubt [5].

To evaluate the performance of the proposed system, an error detection method was used to compare the performance of the proposed model against the conventional model by using DEVS formalism. In DEVS, a system has a time base, inputs, states, outputs based on the current states and inputs. DEVS(Discrete Event System Specification) is a formalism of being developed by Bernard P. Zeigler. The structure of atomic model is as follows [11 - 17].

(Simulation 1)
The atomic models are EF, RA1, UA1, and ED1. The combination of atomic models makes a new coupled model. First, it receives input event, i.e., polling interval. The value is an input value in RA1 and UA1 respectively. An output

value is determined by the time related simulation process RA1 and UA1 respectively. The output value can be an input value in ED1. An output value is determined by the time related simulation process ED1. We can observe the result value through transducer.

(Simulation 2)
The atomic models are EF, RA2, and ED2. The combination of atomic models makes a new coupled model. First, it receives input event, i.e., polling interval. The value is an input value in RA2. An output value is determined by the time related simulation process RA2. The output value can be an input value in ED2. An output value is determined by the time related simulation process ED2. We can observe the result value through transducer. We can observe the following. The application or error detected time interval is as follows.

Conventional method:

$$(Poll - int)((App - cnt) + (App - cnt2)) \tag{1}$$

Proposed method:

$$(Poll - int)((App - cnt) + (Sm - t - a)) \tag{2}$$

Therefore, in case of

$$(App - cnt2) > (App - cnt), \tag{3}$$

$$((Poll-int)((App-cnt)+(App-cnt2))) > ((Poll-int)(App-cnt)+(Sm-t-a)) \tag{4}$$

That is, proposed method is more efficient than conventional method in error detected method in case of

$$(App - cnt2) > (App - cnt). \tag{5}$$

We have compared the performance of the proposed method with conventional method.

5   Conclusions

Distance Education system must be able to support real-time interaction and also support user synchronization including not only temporal synchronization and spatial synchronization but floor control for smooth interaction. It is difficult to avoid a problem of the seam in the pervasive computing environment for seamless services. Thus, there is a great need for synchronization control algorithm in situation-aware middleware to provide dependable services in ubiquitous computing. In a ubiquitous computing environment, the concept of situation-aware middleware has played very important roles in matching user needs with available computing resources in transparent manner in dynamic environments.

This proposed structure is distributed architecture but for error and application program sharing, centralization architecture is used. We proposed an adaptive synchronization control agent based on a hybrid software architecture which is adopting the advantage of CACV and RARV for situation-aware middleware. It described a hybrid software architecture that is running on situation-aware ubiquitous computing for a web based distance education system which has an object with an various information for each session and it also supports multicasting with this information. To evaluate the performance of the proposed system, an error detection method was used to compare the performance of the proposed model against the conventional model by using DEVS formalism. In DEVS, a system has a time base, inputs, states, outputs based on the current states and inputs. DEVS(Discrete Event System Specification) is a formalism of being developed by Bernard P. Zeigler. The merit of system is to make the actual computing part of it and its enabling technologies essentially transparent. This approach has no consideration of domino effect between processes.

And we remain simulations for fault-tolerance QOS based on RCSM and QoS resolution strategies for fault-tolerance QOS based on GAIA projects as future work.

References

1. Yau, S.S., Karim, F.: Contention-Sensitive Middleware for Real-time Software in Ubiquitous Computing Environments. In: Proc. 4th IEEE Int Symp. on Object-Oriented Real-time Distributed Computing (ISORC 2001), pp. 163–170 (May 2001)
2. Park, G.C., Hwang, D.J.: Design of a multimedia distance learning system: MIDAS. In: Proceedings of the IASTED international conference, Pittsburgh, USA (April 1995)
3. Dey, A.K., et al.: A Conceptual Framework and a Toolkit for Supporting the Rapid Prototyping of Context-Aware Applications, anchor article of a special issue on Context-Awareness Computing. Human-Computer Interaction HCI Journal 16 (2001), http://www.cc.gatch.edu/fce/contexttoolki
4. Yau, S., Karim, F., Wang, Y., Wang, B., Gupta, S.: Reconfigurable Context-Sensitive Middleware for Pervasive Computing. IEEE Pervasive Computing 1(3), 33–40 (2002)
5. Ranganathan, A., et al.: A Middleware for Context-Aware agents in Ubiquitous Computing Environments, GAIA Project (2003), http://choices.cs.uiuc.edu/gaia/papers/
6. Gupta, S.K.S., et al.: An Overview of Pervasive Computing. IEEE Pervasive Comm. 8(4), 8–9 (2001)
7. Abowed, G., Mynatt, E.D.: Charting Past, Present, and Future Research in Ubiquitous Computing, ACM Trans. Computer Human Interaction 7(1), 29–58 (2000)
8. Yau, S.S., Wang, Y., Huang, D., In, H.: A Middleware Situation-Aware Contract Specification Language for Ubiquitous Computing, FTDCS (2003)
9. Weiser, M.: The Computer for the Twenty First Century. Scientific Am. 265(3), 66–75 (1991)
10. Saha, D., Mukherjee, A.: Pervasive computing: a paradigm for the 21st century. IEEE Computer 36(3), 25–31 (2003)

11. Zeigler, B.P., Cho, T.H., Rozenblit, J.W.: A Knowledge-Based Simulation Environment for Hierarchical Flexible Manufacturing. IEEE Transaction on Systems, Man, and Cybernetics-Part A: System and Humans 26(1), 81–90 (1996)
12. Cho, T.H., Zeigler, B.P.: Simulation of Intelligent Hierarchical Flexible Manufacturing: Batch Job Routing in Operation Overlapping. IEEE Transaction on Systems, Man, and Cybernetics-Part A: System and Humans 27(1), 116–126 (1997)
13. Zeigler, B.P.: Object-Oriented Simulation with hierarchical, Modular Models. Academic Press, San Diego (1990)
14. Zeigler, B.P.: Multifacetted Modeling and Discrete Event Simulation. Academic, Orlando, FL (1984)
15. Zeigler, B.P.: Theory of Modeling and Simulation. John Wiley, NY, USA (1976) reissued by Krieger, Malabar, FL, USA (1985)
16. Conception, A.I., Zeigler, B.P.: The DEVS formalism: Hierarchical model development. IEEE Trans. Software Eng. 14(2), 228–241 (1988)
17. Lauwers, J.C., Lantz, K.A.: Collaboration Awareness in Support of Collaboration Transparency: Requirements for the Next generation of Shared Window Systems. In: Proc. of ACM CHI 1990, pp. 302–312 (April 1990)

A Hybrid Intelligent Multimedia Service Framework in Next Generation Home Network Environment*

Jong Hyuk Park[1], Jungsuk Song[1], Byoung-Soo Koh[2],
Deok-Gyu Lee[3], and Byoung-Ha Park[4]

[1] R&D Institute, Hanwha S&C Co., Ltd., Jangyo-dong, Jung-gu, Seoul, Korea
parkjonghyuk@gmail.com, songjs@hanwha.co.kr
[2] DigiCaps Co., Ltd., Jinjoo Bldg. 938-26 Bangbae-dong, Seocho-gu, Seoul, Korea
bskoh@digicaps.com
[3] ETRI, Kajung-dong, Yooseong-gu, Daejeon-si, Korea
hbrhcdbr@paran.com
[4] KETI, Yatap-Dong 68, Boondang-gu, Seongnam-si, Kyunggi-do, Korea
bhpark@keti.re.kr

Abstract. In Next Generation Home network Environment (NGHE), multimedia service will be a key concept of advanced intelligent and secure services which are different from the existing ones. In this paper, we propose a Hybrid Intelligent Multimedia Service Framework (HIMSF) which is mixed with application technologies like intelligent home Infrastructure or multimedia protection management through the ubiquitous sensor network based technology to provide a proper multimedia service suitable for NGHE. The proposed framework provides an interoperability among heterogeneous equipments regarding home network appliances like wireless devices, electronic appliances, PC. In addition, it provides adaptive application services.

1 Introduction

Recently an age of ubiquitous has come through the rapid progress of home network environment based on eager demands about mobile network. Ubiquitous computing environment is a smart environment where computers provide computing environment that users want by themselves, any time, any where. In addition, invisible computers, which are connected with one another, are embedded through networks so that users could use computers any time, any where.

The ubiquitous computing has following features. First, there are various dispersed computing equipments for some users who have a specific purpose and secondly, computing equipments, which are connected through networks without a disconnect, are existing. The third, it is invisible to users and users could feel only a humanizing interface [1, 2].

---

* This research was supported by the MOCIE (Ministry of Commerce, Industry and Energy) - HIMS of the HISP Project (Project Number: 10016508).

M.S. Szczuka et al. (Eds.): ICHIT 2006, LNAI 4413, pp. 395–403, 2007.

Because of those reasons, ubiquitous computing has lots of shortcomings especially in security. Therefore, it is possible for users or devices to attack masquerade in servers because there are various dispersed computing devices. In addition, although users authorize the installment at the devices regarding authenticated programs, codes which has a malicious intention might transmit it to the devices which do not have a computing capability [3].

Furthermore, in terms of multimedia services, it should be considered such as followings: user based on services, contents protection / management or adaptive application, inter-operability among devices, and so on [4, 5].

In this paper, we propose Hybrid Intelligent Multimedia Service Framework (HIMSF) mixed with application technologies like provided home infrastructure and multimedia protection and management in NGHE. The proposed framework provides multimedia inter-operability among different devices for a multimedia adaptive application under NGHE. Finally, it supports protection and management of multimedia contents in the 2nd devices.

This paper is organized as follows. In Section 2, related works including Ubiquitous Sensor Network, MPEG-21 IPMP or DIA are discussed. In Section 3, the proposed HIMSF including design, protocol, prototype implementation is discussed. A conclusion and discussions for future research direction are presented in Section 4.

2   Related Works

In this Section, we discuss Ubiquitous Sensor Network and MPEG-21 IPMP/DIA technologies which are core elements of the HIMSF.

Ubiquitous Sensor Network (USN) is not only a computing based network consisting of an ultra light, low-power sensor but also a wireless network. Lots of sensors, which are related with one network, sensed environmental changes of field, and then transmit the information to base station. After the information is delivered to users through network servers, the information gathering is completed.

Things recognize other things like a human being through the USN any time, any where, it becomes available for them to confirm the information through networks by recognizing the environment. This kind of method is such a technology to make our life better by being adapted to environment surveillance system, economic activities like physical distribution and production, welfare services, medical services.

Moreover, there is a limit such as electric power, computing capability and memory because each sensor of USN is small and it is very difficult to forecast the topology among sensors because there are lots of sensors scattered randomly in field and the topology of sensor network is easily changed by either adding or deleting sensor nodes. The characteristics of the USN provide convenience and automation for our lives by performing the process of information which is sensed through sensors. However, the more the dependence on system is higher, the more the danger due to this is higher [6, 7].

In this paper, home infrastructure has been constructed to realize the USN by developing the USN module based on Tiny OS [8] developed at UC Berkeley in NGHE.

MPEG-21 IPMP (Moving Picture Experts Group-21 Intellectual Property Management and Protection) is a standard about representation system of the IPMP to manage Digital Item (DI) safely when the DI is on the process of production, alteration, delivery, consumption on network which is defined in the MPEG-21.

The schema structure of the MPEG-21 IPMP Components is designed to represent IPMP_Component, IPMP_Resource, IPMP_Item associated with the protected DIDL (Component, Resource, Item) according to the Digital Item Declaration (DID) structure and the existing DID standard has been described as a concept including the IPMP [9, 10]. The DIA indicates fundamental notions of the DI based technology considering user environment. The DIA uses the DI and descriptor created from the environmental descriptor of user who consumes digital item and the general tools related to the environment as inputs. The DIA is a technology that guarantees satisfactory consumption and integrity use no matter what user's preference is or the environmental changes of contents consumer are. The DIA is used for the main request framework by being connected with a real time video transcoding technology and plays an important role in providing customized contents to the 2nd device in home.

3 HIMSF

3.1 HIMSF Design

In this paper, the design of the proposed HIMSF is divided into 3 parts roughly as following.

Total Architecture Design of the HIMSF: The overall HIMSF consists of home external elements including packager, CA server, LMS, CMS and home internal elements including HIMSF-STB (Set-Top Box), IPMP client module, content play device (Refer to Figure 1).

- *Packager*: It performs packaging function of the original contents with metadata, right. The packager module encodes the original contents mainly and it also transmits the information of right to LMS (License Management Server) and the information of contents to CMS (Contents Management Server) when the package process is working.
- *CMS*: The CMS manages metadata about contents and contents. The related DB stores contents which are packaging and provides multimedia service through the interface of VOD (Video On Demand) or streaming server.
- *LMS*: The LMS manages right information and plays a role in issuing a license of the related contents after users purchase contents through billing system. If the consumption about contents is processed on/offline, the related details will be managed or sent by callback channel.

- *CA Server*: The CA receives certificate issuing request from either users or devices and provides authentication function about the certificate issuing request. In addition, when devices request to verify the certificate for device authentication, it reports whether the certificate is right or wrong to devices.
- *HIMSF-STB*: It helps MPEG-2 TS contents which have been received from external work flexibly in device platform by transcoding of the contents. Furthermore, it manages the devices and users which belongs to target domain, and provides home environment with multimedia service by gathering and processing information.
- *IPMP Client Module*: It decodes encoded contents and plays a role in processing both the validate duration of license and spending lists through parser.
- *Content Play Device (Portable Device)*: It is a portable device that receives various multimedia services from HIMSF-STB. It means USN module embedded devices in Notebook or PDA.

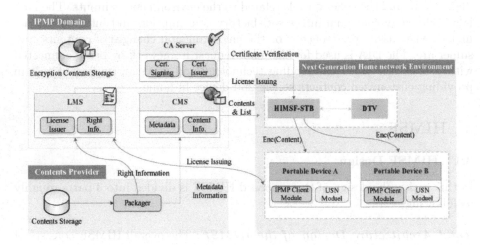

Fig. 1. Total Architecture of the HIMSF

Transcoder Design: In the proposed HIMSF, transcoder design implements the improved video trans cod-ing to play the customized contents by considering the capacity of the secondary devices in NGHE.

The proposed transcoder transforms MPEG-2 bit stream into MPEG-4 bit stream and it is designed to work well in space(pixel) area like Figure 2. Therefore, it includes both MPEG-2 decoder and MPEG-4 encoder and it also includes DS (down sampling) filter to perform resolution conversion.

In addition, the proposed transcoder includes the following blocks to maintain picture quality or decrease the calculating works. Because the supported modes for each macro block of MPEG-2 and MPEG-4 and the uses of each macro block are different, the mode de-cosion block, which is mapping the existing MPEG-2 macro block mode into MPEG-4 macro block mode efficiently, is needed.

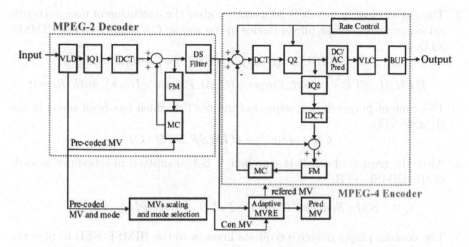

Fig. 2. Block Diagram of the transcoder on the HIMSF

Furthermore, because specification of the MPEG-2 and MPEG-4 can not use the existing moving vector due to difference of the supported frame formats and structures, the moving vector scaling block which performs the correction is needed. In that case the corrected moving vector is used, the picture quality is getting lower according to images. Therefore, block of the re-estimation find to optimize moving vector to avoid these picture quality deterioration is required. In the transcoder, it is optimized by using SIMD (Single Instruction Multiple Data) instruction because it is needed to calculate blocks like DCT, IDCT, MVRE (Motion Vector Re Estimation), MC (Motion Compensation) repeatedly through loop.

Architecture Design of the HIMSF Client: After client is authenticated through CA server, it communicates with agent server. Finally, the information of contents lists and the licenses are sent to play contents.

3.2   Protocol of the HIMSF

In this Subsection, we discuss protocol of the multimedia contents delivery and right delivery between the HIMSF-STB and the content play devices.

The device information has been sent through the USN module which is suitable for each content play device and it assumes that real-time multimedia transcoded contents are prepared in the HIMSF-STB.

1. The content player device requests both users and device authentication to the HIMSF-STB.

$$Cont\_Player \rightarrow HIMSF\_STB : Cont\_Player.Auth\_Req$$

2. The authentication result is transmitted after the execution of user authentication and the content player device in the simple CA module of the HIMSF-STB.

$$SCAM : (Cont_{player} \vee User).Auth,$$

$$HIMSF\_STB \rightarrow Cont\_Player : (Cont\_Player \vee User).Auth\_Result$$

3. The content player device requests contents list which has been saved in the HIMSF-STB.

$$Cont_{player} \rightarrow HIMSF\_STB : CL$$

4. After the requested result is searched, it is transmitted in scheduler module of the HIMSF-STB.

$$SM : SearchCL, HIMSF_STB \rightarrow Con\_Player : CL$$

5. The content player device e requests licenses to the HIMSF-STB to play the contents.

$$Cont\_Player \rightarrow HIMSF\_STB : Cont\_Player.License$$

6. The LMS module of the HIMS STB issues the license to the content player device.

$$HIMSF\_STB \rightarrow Cont\_Player : Cont_{player}.License$$

7. The IPMP client module of the content player device plays contents after de-packaging through the contents licenses.

$$Dec_{Cont\_Player.License}(Cont), Play(Cont)$$

3.3   Prototype Implementation of the HIMSF

HIMSF STB: The HIMSF-STB uses B/C Company [12]'s BCM7038 chip as a HD decoder and BCM3125 is used as a MAC chip of DOCSIS.

The BCM3125 consists of module and de-module tranciver having Physical (PHY) Layer transmission function that is MCNS Compliant and 1-Chip device that MAC(Media Access Controller) has been added based on DOCSIS [13] standard. Those devices provide the functions which are needed to transmit HD video in home using network and data transmission and receiving of the high speed.

Furthermore, the single chip device has those functions as following; analog-to-digital conversion, downstream QAM demodulation, adaptive equalization, MPEG termination, Forward Error Correction (FEC), upstream FEC encoding, upstream QPSK/16-QAM modulation, digital to analog conversion for RF output, upstream and downstream MAC control. The downstream physical layer supports 64/256 QAM transfer rate and channel decoding is executed through ITU-T J.83 ANNEX A and B. Figure 3 shows H/W function block picture of the HIMSF-STB.

Table 1. Notations

Notation	Meanings
Cont_Player	The content play device embedding USN module in NGHE
IPMP_CM	The IPMP client module in the content play device
HIMSF_STB	Set-To-Box loading HIMSF
SCAM	The simple certificate authority module in the HIMSF-STB
Sch_M	Scheduler module in the HIMSF-STB
LMSM	License management service module
VOD_SM	VOD service module
A.B_Req	Action 'B' is requested by device/User 'A'
A.Auth	Device/User A's Authentication
$A \rightarrow B$	Request/Transmission from A to B
$A : B$	Device/Action 'B' execution in module 'A'
$A \vee B$ /Play (Cont)	A or B/Play the contents
A_Result	The result of the action 'A'
CL/Usr	Contents list/User in NGHE
A.License	Device/the license of user 'A'
Cont	Multimedia content
Dec_A (Cont)	De-packaging of the contents with key 'A'

License Management Module: It reduces the number of counter after decides whether it is possible to use contents, and then it uses the contents through the license parsing. Finally, it performs the re-request function when the validate duration of the licenses expires. The following code shows the part of license control modules which is used for the proposed framework.

The part of the license control module

```
/* This part decides whether it is possible to
   be used in the license files */
...
    <validityInterval> Limit the term of validity.
        <notBefore>2006-05-26T16:34:20</notBefore>
        <notAfter>2006-05-30T16:34:20</notAfter>
    </validityInterval>

    <sx:exerciseLimit> Limit the possible number of uses
        <sx:count>4</sx:count>
    </sx:exerciseLimit>
...
```

Content List Control Module: It receives possible contents lists from server which can be played and shows them on screen. It also performs contents list and parsing internally. The contents list processes the search according to user's preference.

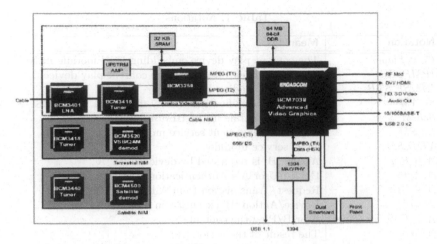

Fig. 3. H/W function block picture of the HIMSF-STB

Content control / Meta-data Control Module: After de-packaging the contents, it executes parsing works the contents detail information files and then play multimedia.

4   Conclusion

In this paper, we have proposed the hybrid intelligent multimedia service framework for secure and intelligent multimedia services suitable for NGHE. It is mixed with application technologies like home infrastructure and multimedia protection / management in NGHE. In addition, it provides multimedia interoperability among different devices for multimedia services. Furthermore, The aim of the multimedia service in ubiquitous environment, "One source, Multi-users / devices", has been prepared through the design and implementation of the transcoder in the HIMSF. Finally, super distribution has been implemented in NGHE.

In the near future, multimedia devices would play an important role in extending to ubiquitous environment. The related management / protection, and inter-operability among different devices would be a prerequisite issue. Therefore, this paper has an enough value to study as an intermediation, and additional studies would be needed for users' privacy issues.

References

1. Weiser, M.: The Computer for the 21st Century, pp. 94–104. Scientific American (1991)
2. Weiser, M.: Hot topic: Ubiquitous Computing, pp. 71–72. IEEE Computer, Los Alamitos (1993)

3. Stajano, F.: Security for Ubiquitous Computing. Wiley, Chichester (2002)
4. Park, J.H., Lee, S., Hong, S.H.: C-iUMS: Context based Smart and Secure Multimedia Service in Intelligent Ubiquitous Home. In: Zhou, X., Sokolsky, O., Yan, L., Jung, E.-S., Shao, Z., Mu, Y., Lee, D.C., Kim, D., Jeong, Y.-S., Xu, C.-Z. (eds.) EUC 2006. LNCS, vol. 4097, pp. 660–670. Springer, Heidelberg (2006)
5. Park, J.H., Song, J., Lee, S., Koh, B.-S., Hong, I.-H.: User centric intelligent IPMPS in Ubi-Home. In: Gavrilova, M., Gervasi, O., Kumar, V., Tan, C.J.K., Taniar, D., Laganà, A., Mun, Y., Choo, H. (eds.) ICCSA 2006. LNCS, vol. 3983, pp. 245–254. Springer, Heidelberg (2006)
6. Lee, D.H.: Technology Trends of the Information Security in USN, ITFIND Mailzine, vol. 1212 (2005)
7. Akyildiz, I.F., Su, W., Sankarasubramaniam, Y., Cayirci, E.: A survey on sensor networks. IEEE Communication Magazine 40(8), 102–114 (2002)
8. Tiny OS Community Forum, http://www.tinyos.net
9. MPEG-21 Overview v.5, ISO/IEC JTC1/SC29/WG11 N5231 (2002)
10. MPEG-21 IPMP Extensions Overview, ISO/IEC W6338, MPEG Munchen Meeting (2004)
11. MPEG-21 Digital Item Adaptation Draft v0.6, ISO/IEC JTC1/SC29/WG11/N5353 (2002)
12. BRCM Inc. Website, http://mwww.broadcom.com
13. DOCICS standard, Cable Modem Lab Website, http://www.cablemodem.com

Integration of Artificial Market Simulation and Text Mining for Market Analysis

Kiyoshi Izumi, Hiroki Matsui, and Yutaka Matsuo

AIST, 2-41-6 Aomi, Koto-ku, Tokyo 135-0064, Japan
{kiyoshi.izumi,hiroki.matsui,y.mastuo}@aist.go.jp

Abstract. It is important to understand and to provide a rationale for the actions of users in financial markets. Simulations of artificial financial markets are one means by which to address these needs, and this paper describes how to integrate the technique of text mining with a simulation of an artificial financial market to enhance the value and usefulness of the simulation. The procedure that is proposed consists of extracting the economic trends from the text data that is circulating in the real world and to input these economic trends into the market simulation. We show how this was experimentally tested as a decision support system for an exchange rate policy and how it suggested that the applied combination of interest rate and intervention operations was effective for stabilization of the yen-dollar rate in 1994 and 1995.

Keywords: Artificial market, multi-agent simulation, text mining, decision support systems, complex systems.

1 Introduction

In recent years, artificial market studies obtained some success in market analysis. They revealed new mechanisms that are allegedly responsible for market phenomena as financial bubbles that previous models could not explain as well. Artificial market research drew attention as a useful tool with which to test existing economic theories, e.g., the efficient market hypothesis. Many models, however, were too simple or too abstract to be useful representations of actual markets. This shortcoming motivated multi-agent simulation of social phenomena to build models of actual social phenomena on computers, and to use these for supporting actual actions since a model needs to reflect the real world so that its simulation results can be considered to be reliable.

This paper proposes an approach that we have called: SEMAS (Socially Embedded Multi Agent Simulation). This integrates an artificial market simulation with text mining by taking real world information into the simulation. By this means, it can support its user's decision in an actual market. As a consequence of this interaction, the user's action can also affect the real world and the model can reflect the changes of economic trends in the real world. In the SEMAS approach it is submitted, therefore, that an artificial market simulation is embedded into a feedback loop.

M.S. Szczuka et al. (Eds.): ICHIT 2006, LNAI 4413, pp. 404–413, 2007.

2   Framework of SEMAS

The SEMAS approach consists of the two stages as depicted in Fig. 1. First, economic trends are extracted from various text data such as news articles and webs sites using the text mining technique. Second, and without loss of generality, the extracted trend is put into a multi-agent model of a foreign exchange market, and its computer simulation is carried out. Based on the simulation results, the system can support a user in determining his/her behavior in the market.

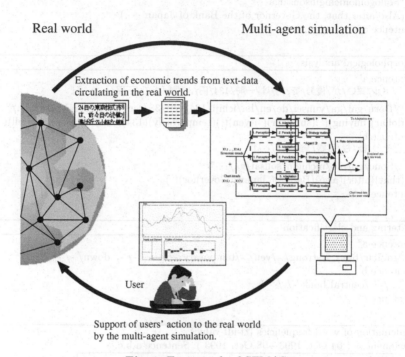

Fig. 1. Framework of SEMAS

3   Extraction of Information from the Real World

The procedure of the economic trend analysis at the first stage of SEMAS is shown in Fig. 2. A feature vector is first calculated from text data related to the given week's economic conditions (Fig. 2a-d). Economic trends are estimated by the decision tree classification method C4.5 using the feature vector (Fig. 2e).

The original data are market reports distributed by JCIF (Japan Center for International Finance, http://www.jcif.or.jp/) every week to its members such as professional dealers. Each report contained economic news relevant to trends of financial markets. It was written in Japanese using about 1,000 characters. We used the JCIF documents from 1992 to 1995 (207 weeks).

Weekly feature extraction: As preprocessing, morphological analysis, filtering, and classification are carried out. Each sentence is first divided into morphemes,

(a) Original text data

Document A (04 Oct. 1993 - 08 Oct. 1993)

Sentence a
「ドル買いが優勢で円は一時 134 円 38 銭まで下落した。」

"Dorugaigayuuseideenhaichiji134en38senmadegerakushita."

(The dollar buying was superior and yen fell to 134 yen 38 sen temporarily.)

Sentence b
「その後も日銀総裁が …」

"Sonogomonichiginsousaiga…"

(Also after that, the Governor of the Bank of Japan ….)

Sentence c
　　　　　　　　⋮

(b) Morphological analysis

Sentence a'
「/ドル/買い/が/優勢/で/円/は/一時/134/円/38/銭/まで/下落/した/./」

"/Doru/gai/ga/yuusei/de/en/ha/ichiji/134/yen/38/sen/made/geraku/shita/./"
(dollar)(buying)( )(superior)( )(yen)( )(temporarily)(134)(yen)(38)(sen)(to)(fell)( )

Sentence b'
「/その/後/も/日銀/総裁/が/…」

"/Sono/go/mo/nichigin/sousai/ga/…"
(that)(after)(also)(Bank of Japan)(Governor)( )

Sentence c'
　　　　　　　　⋮

(c) Filtering and classification

Sentence a"
"/dollar/buy/ -/strong/ -/yen/ -/temporary/-/ -/ -/ -/ -/ down/ -/ -/ "
Sentence b
"/-/-/-/central_bank/-/-/-/…"
Sentence c"
　　　　　　　　⋮

(d) Calculation of word frequencies (tf-idf value)

Document A (04 Oct. 1993 - 08 Oct. 1993): Sentence a,b,c,⋯

{dollar 0.041978}, {strong 0.018933}, {yen 0.081418}, ⋯

(e) Estimation of economic trends

Determination of decision trees about 14 economic factors by C4.5(J4.8).

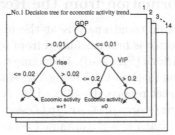

Fig. 2. Procedure of text mining

Table 1. Example of a keyword list

Keywords	Morpheme 1	(Part-of-speech),	Morpheme 2 (Part-of-speech), $\cdots$
yen	en	!noun-suffix-numerative	
	(yen)		
rise	joushou,		kyuushin, $\cdots$
	(go up)		(jump)
large	ookii,		tairyou, $\cdots$
	(large)		(massive)

The part-of-speech condition "!noun-suffix-numerative!" means that a morpheme is not a numerative noun suffix like "134-*yen*".

the smallest meaningful units, using the Japanese language morphological analysis system ChaSen [1] (Fig. 2b). Each morpheme has information about its basic form and the part-of-speech. Next, each morpheme is checked against a list of keywords as shown in Table 1. The list consists of pairs of keywords and morphemes, which relate to economic trends. There are 183 keywords containing nouns like "yen", verbs like "rise", and adjectives or adverbs like "large". Morphemes which do not appear in the list are deleted. Morphemes with the same meaning are classified into the same keyword according to the list[1] (Fig. 2c).

After preprocessing, a feature vector of each week is calculated using tf-idf (term frequency inverse document frequency) values based on frequencies of appearance of 143 keywords[2]. A tf-idf value of keyword k in week t is as follows.

$$\text{tfidf}_{k,t} = \text{tf}_{k,t} \times \log\left(N/\text{df}_k\right), \tag{1}$$

where $\text{tf}_{k,t}$ is (a frequency of keyword k in week t) / (a total of frequencies of the 143 keywords in week t). $N(= 207)$ is a number of all weeks, and df_k is a number of the weeks when the keyword k appeared at least once. About a keyword that appears many times only in this document, its tf-idf value becomes large. A feature vector of week t is as follows.

$$\mathbf{D}_t = (\text{tfidf}_{1,t}, \text{tfidf}_{2,t}, \cdots, \text{tfidf}_{183,t}). \tag{2}$$

Estimation of economic trends: As shown in Fig. 2e, the feature vector $\mathbf{D}_t$ is put into decision trees, and economic trends of week t are estimated with respect to 14 kinds of categories in Table 2. The categories follow classification in books on financial markets. The trees were determined by J4.8[3] using the training data. The training data were coding values with seven levels $\{0, \pm1, \pm2 \pm 3\}$[4] that we classified each document into with respect to the 14 categories.

We tested the decision trees acquired by J4.8 using percentages of correctly estimated documents. As shown in Table 2, the decision trees correctly estimated

---

[1] Our filtering method is based on [2,3].

[2] See [4] as another example of economy news analysis with tf-idf values.

[3] J4.8 is implementation of C4.5 in data mining software WEKA [5,6].

[4] A positive (negative) value means a trend for stronger (weaker) yen.

Table 2. Evaluation of decision trees (Number of documents N= 207)

	Categories	% of correct classification Training data	% of correct classification 10-fold cross-validation
1	Economic activity	89.4	64.2
2	Price	92.8	81.2
3	Interest rate	83.6	54.1
4	Money supply	97.6	97.6
5	Trade balance	96.1	78.3
6	Employment	87.4	63.3
7	Personal consumption	90.8	87.9
8	Intervention	88.4	65.7
9	VIP announcement	80.2	44.4
10	European trend	87.9	49.3
11	Goods market	99.0	99.0
12	Political condition	86.5	67.1
13	Stock market	91.3	74.9
14	Bond market	93.7	72.9
	Average	90.3	71.4

economic trends of 90.3% of the 207 documents in estimation of the training data, and 71.4 % in 10-fold cross-validation test[5]. The percentages reached high levels enough for practical use, although there were differences among the categories. The multi-agent model in the SEMAS approach corresponds to the real world by using the estimated economic trends.

4   Multi-agent Model of Financial Market

As the second stage of SEMAS, a multi-agent simulation is performed using the economic trends extracted from the real world in the first stage. This section describes a multi-agent model of a foreign exchange-rate market, AGEDASI TOF[6]. AGEDASI TOF is an artificial market with 100 agents, as illustrated in Fig. 3. Each agent is a virtual dealer which has dollar and yen assets and changes positions in the currencies for the purpose of making profits. Each week of the model consists of the following five steps.

4.1   Step 1: Perception

Each agent first receives coding data x_i of the 14 economic trends and 3 chart trends. The data of the 14 economics trends are produced from the feature vector of text data $\mathbf{D}_t$ and the decision tree functions $F_i(\cdot)$.

$$x_{i,t} = F_i(\mathbf{D}_t), \quad i = 1, \cdots, 14. \tag{3}$$

---

[5] In this test, 9/10 of data is used for learning and 1/10 is for the test.
[6] GEnetic-algorithmic Double Auction Simulation in TOkyo Foreign exchange market.

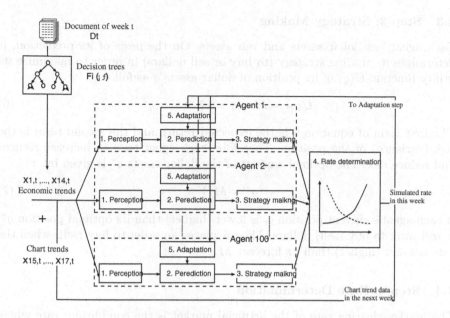

Fig. 3. Framework of multi-agent model

The 3 chart trend data are a short-term trend (a change of the exchange rate in the last week), a change of short-term trend, and a long-term trend (change through five weeks). These data are coded into 7 levels from -3 to $+3$ like the 14 economic trend data, by normalization with their standard deviations.

4.2 Step 2: Prediction

Each agent assigns its own weights, $w_i = \{0, \pm0.1, \pm0.5, \pm1.0, \pm3.0\}$ to the 17 trend data and uses the weights to predict the rate fluctuation for the next week. The mean of its forecast, M, is the weighted average of x_i as follows[7].

$$M = trunc\left(\sum_i w_i x_i\right) \times 0.02. \tag{4}$$

The variance of its forecasts, V, is calculated from the difference between the stronger-yen factors and the weaker-yen factors, as follows

$$V = \left(\sqrt{\left(\sum_{w_i x_i > 0} w_i x_i\right)^2 - \left(\sum_{w_j x_j < 0} w_j x_j\right)^2}\right)^{-1} \tag{5}$$

The variance is inversely proportional to the coherence of forecast factors. Hence, the larger the variance, the lower the confidence of the forecast.

---

[7] $trunc(\cdot)$ is a truncation function. 0.02 is a scaling coefficient. It was calculated from a ratio of the average of $\Sigma w_i x_i$ and rate changes.

4.3   Step 3: Strategy Making

Each agent has dollar assets and yen assets. On the basis of its prediction, it determines its trading strategy (to buy or sell dollars) in order to maximize its utility function $U(q)$ of its position of dollar assets q as follows.

$$U(q) = M \times q - 0.5 \times V \times q^2 \tag{6}$$

The first term of equation 6 is the expected return and the second term is the risk (variance) of the position. Therefore, each agent tries to increase returns and reduce risk. The optimal amount of the dollar assets q^* is given by.

$$q^* = M/V \tag{7}$$

If each agent's current position q_t is lower (higher) than its optimal position q^*, it will want to buy (sell) dollars. Then it places an order to buy (sell) when the rate is lower (higher) than its forecast M.

4.4   Step 4: Rate Determination

The market-clearing rate of the artificial market is the equilibrium rate where the quantities of demand and supply are equal. Agents who submit orders to buy at rates above the market-clearing rate and agents who submit orders to sell at rates below the market-clearing rate can exchange their assets for the optimizing purposes. Assets of other agents remain the same as in the previous week.

4.5   Step 5: Adaptation

Each agent improves its prediction method by referring to the prediction methods of other agents. The change of the prediction methods in the market is described by the following three operations of simple GA [7]. An individual in GA is a string of all 17 weights of one agent $\{w_i\}$ and a fitness value of each individual is calculated using a forecast error, a difference between the mean of its forecast M and the market-clearing rate. (1) *Selection*: A certain ratio G of agents replace their prediction methods by others' prediction methods with higher fitness on the probability proportional to the fitness values. The selection operator is interpreted economically as the propagation of successful opinions about forecast factors. (2) *Crossover*: A pair of agents sometimes exchanges parts of their weights. It is submitted that the crossover operator has analogy to a communication between the agents. (3) *Mutation*: Each agent has a small probability of changing each weight. This mutation operator might resemble an independent change to the prediction method of each agent. The crossover and mutation operators produce new combinations of weights[8].

After the *Adaptation* Step, we proceed to the next week's *Perception* Step.

---

[8] The crossover (mutation) occurred at a certain probability *pcross* (*pmut*). We used the following parameter sets: $pcross = 0.3, pmut = 0.003, G = 0.8$. These values were determined from forecast tests [8].

5   Support of Users' Action to the Real World

Our artificial market model can support its users such as central bank staff, policy makers, and authorities in deciding exchange rate policies. In this section our model is used as a decision support system to stabilize the yen-dollar rate in 1994-1995. The users try to decide how to operate three *control factors* (interest rates, intervention, and announcement) during the two years. First, based on simulation results using data until 1993, the users select several *condition factors* from the 17 factors. The condition factors are the factors to which many agents attached large weights. The users will operate the control factors according to the trends of the condition factors. Second, seven candidates of the users' action were prepared according to the kind of the control factors to operate. These candidates were estimated by standard deviations of rate changes in the simulation, and the best action option was suggested.

5.1   Selection of Condition Factors

First, we trained the 100 agents in the artificial market using sample data until 1993 by the following procedure. (1) *Initialization*: The initial population included a hundred agents whose weights were generated randomly and the position of each agent is square. (2) *Training*: We trained our model using data of the 17 factors and the actual rates from 1992 to 1993. During the training we skipped the rate determination step, and in the adaptation step, the fitness in GA was the cumulative value of differences between the forecast means of each agent and the actual rates.

As a result, bond market, money supply, and European trend factors got the biggest weights from the agents on the average of the 100 simulation runs. That is, the artificial market was sensitive to these three condition factors in this period. Therefore, if there is a large change of the three factors' trends, the artificial market tends to become unstable.

5.2   Decision of the Best Action Option

Second, seven candidates of the users' action (a)-(g) were prepared according to the kind of control factor to operate: (a) When there is a change of the condition factors' trends ($|x_{4,t} + x_{10,t} + x_{14,t}| > 0$), a value of the *interest rate* factor $x_{3,t}$ is set to $+3$ (if $x_{4,t} + x_{10,t} + x_{14,t} < 0$) or -3 (if $x_{4,t} + x_{10,t} + x_{14,t} > 0$), (b) the *intervention* factor is operated similarly; (c) the *announcement* factor; (d) the *interest rate* and *intervention* factor; (e) the *interest rate* and *announcement* factor; (f) the *intervention* and *announcement* factor; (g) the all three factors.

Each exchange policy options was input into the artificial market simulation and results were compared. We generated 100 simulation paths for each option by repeating the following procedure a hundred times; (1) *Initialization*, (2) *Training*, and (3) *Test*: We conducted the extrapolation simulations for the period from 1994 to 1995. In the test period, our model forecasted the rates in the rate determination step using only the 14 economic trend data. Each option

was used as input data. The 3 chart trend data and the fitness in GA were
calculated from the simulated rates in the rate determination step. Each option
is evaluated using a standard deviation of rate change in 1994 and 1995 on the
average of 100 simulation runs.

As a result, the action based on the option d (combination of the *interest rate*
and *intervention* operation) could reduce the rate fluctuation the most (Fig. 4).
The standard deviation of the simulated rate without operation was 0.8528 on
the average of 100 runs, and it was close to the value 0.7453 of the standard
deviation of the actual rate from 1994 to 1995. Comparing with this case, combination of the *interest rate* and *intervention* operation was able to reduce 47%
of the rate change on the average of 100 runs. In the option b, 34% of rate
change could be reduced only by the interest rates operation, and the reduction
effect increased in combination with the intervention operation. However, the
operation by all the three factors (the option g) made the rate unstable rather,
because its effect was too large.

The above results showed that the following exchange policy option was effective for the stabilization of the yen-dollar rate from 1994 to 1995.

When news about bond market, money supply, or European trend comes
to the market, a user should simultaneously operate intervention and
interest rate in the direction opposite to the news.

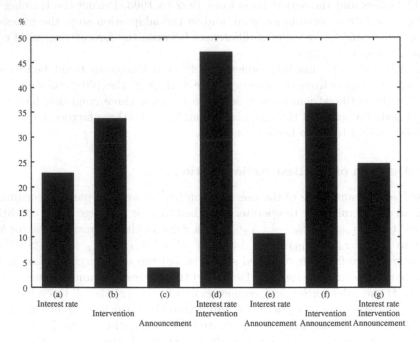

Fig. 4. Reduction rates of rate changes in comparison with the case without operation

6   Conclusion

Our approach is a method of coupling analysis and action very closely. SEMAS may offer a useful social simulation. If users act upon the real world according to information that is provided by the simulation, their action affects the dynamics of the actual market. The consequent change in the economic trend that might ensue would then become extracted by the text mining component and in this way find its way into SEMAS. Such a feedback loop is rather novel and potentially useful to track the financial market. The described technique requires, surely, many further enhancements. The news items and other text to be mined can suffer from media sensationalism. Also it is necessary to understand the true nature of the feedback loop and its convergence or divergence characteristics.

Acknowledgments. This research was partially supported by the Ministry of Education, Science, Sports and Culture, Grant-in-Aid for Scientific Research (B), no. 16300047.

References

1. Chasen (2006), http://chasen.naist.jp/
2. Carley, K., Diesner, J., Tsvetovat, M.: An integrated approach to the collection and analysis of network data. In: Proceedings of NAACSOS 2004 (2004)
3. Ahmad, K., Gillam, L., Cheng, D.: Textual and quantitative analysis: Towards a new, e-mediated social science. In: Proc. of the 1st International Conference on e-Social Science (2005)
4. Seo, Y.W., Giampapa, J.A., Sycara, K.: Financial news analysis for intelligent portfolio management. Technical Report CMU-RI-TR-04-04, Carnegie Mellon University (2004)
5. Weka (2006), http://www.cs.waikato.ac.nz/~ml/weka/
6. Witten, I.H., Frank, E.: Data Mining: Practical Machine Learning Tools and Techniques, 2nd edn. Morgan Kaufmann, San Francisco (2005)
7. Goldberg, D.: Genetic algorithms in search, optimization, and machine learning. Addison-Wesley Publishing Company, Reading (1989)
8. Izumi, K., Ueda, K.: Phase transition in a foreign exchange market: Analysis based on an artificial market approach. IEEE Transactions on Evolutionary Computation 5(5), 456–470 (2001)

Agent-Based Intelligent Decision Support for the Home Healthcare Environment*

Louie Cervantes, Yong-Seok Lee, Hyunho Yang,
Sung-hyun Ko, and Jaewan Lee

School of Electronic and Information Engineering, Kunsan National University,
68 Miryong-dong, Kunsan, Chonbuk, 573-701, South Korea
{lfcervantes,leeys,hhyang,blackpc,jwlee}@kunsan.ac.kr

Abstract. This paper brings together the multi-agent platform and artificial neural network to create an intelligent decision support system for a group of medical specialists collaborating in the pervasive management of healthcare for chronic patients. Artificial intelligence is employed to support the management of chronic illness through the early identification of adverse trends in the patient's physiological data. A framework based on software agents that proxy for participants in a home healthcare environment is presented. The proposed approach enables the agent-based home healthcare system to identify the emergent chronic conditions from the patterns of symptoms and allows the appropriate remediation to be initiated and managed transparently.

1 Introduction

Medical care problems are quite complex, and the solution of a problem involves the coordination of the efforts of different individuals with different skills and functions. This study focuses on the coordination of a group of specialists collaborating as a medical team for a patient with multiple chronic conditions. In critical situation such as healthcare management, the risk of making incorrect decision based on incomplete or outdated information is very high. Providing decision support to a medical group having potentially conflicting views in a time-critical situation represents a very important and innovative area for research. The use of artificial intelligence to support decision making have been demonstrated successfully in previous studies [1].

This paper brings together mobile software agents and artificial neural network into a real-time decision support system for pervasive healthcare management. Specifically, we propose the use of mobile agents that automate the process of obtaining data from physiological sensors and use the neural network in a decision support system to coordinate a group of medical specialists. We choose the multi-agent platform as the key enabling technology because it offer a

---

* This research was supported by the Program for the Training of Graduate Students in Regional Innovation which was conducted by the Ministry of Commerce Industry and Energy of the Korean Government.

M.S. Szczuka et al. (Eds.): ICHIT 2006, LNAI 4413, pp. 414–424, 2007.

general framework in which distributed, real-time decision support applications can be implemented more efficiently [2], [3].

The literature indicates that agent-based applications in the medical domain have recently become popular [3]. In other studies intelligent home healthcare systems showed promising results in improving the delivery of care to chronic patients particularly those receiving care in their homes. A recent medical study reported that the average age in most societies is growing higher with more people belonging to higher age groups nowadays [4]. These findings highlight the urgent need to develop better ways to deliver home health care particularly to the elderly [5].

Remediation is time-critical due to the fragile health of chronic patients. The emergent adverse condition needs to be identified as quickly and accurately as possible. In this paper we use the artificial neural network (ANN) to recognize classes of medical conditions from the symptom patterns. Our goal is to enable the system to notify the appropriate medical specialist and provide the relevant patient data when the adverse condition arise. Intelligent software agents are used in a decision support system to coordinate the group of medical specialists. By coordinating the emergency medical care, the proposed system can potentially improve the patient's safety and avoid additional costs that arise from unnecessary trips to the emergency room that result from false alarms. The use of a home healthcare system can increase the number of patients that can be safely assigned under the care of a specialist [6]. This approach could save medical funds as well as alleviate the shortage of medical personnel in some hospitals.

2   Related Works

2.1   Intelligent Decision Support System

Intelligent decision support system enables medical practitioners to obtain information more quickly and make more accurate diagnosis and treatment decisions. Foster [7] presented an overview into the current research on agent-based decision support system (DSS) in the medical domain. The advantage of intelligent DSS over plain DSS is the use of artificial intelligence that gives it a wider range of decisions including those that involve uncertainty. IDSS in medicine includes diagnosis assistants, treatment recommendation systems and patient history examination systems[7]. Frize [8] identified some important applications of ANN to various medical problems with a particular focus on the Intensive Care Unit (ICU). The study reported significant improvements in the management of resources and the quality of the medical services in a hospital. The current work extends the use of IDSS as diagnosis assistant into a coordination assistant for a group of medical specialists. The proposed healthcare system is intended for patients that require continuous monitoring - whose health conditions are too grave that rapid recognition of the medical condition from the symptom patterns obtained from medical sensors is vital to the survival of the patient.

2.2   Multi-agent Systems

Multi-agent system (MAS) is based on autonomous agents that interact with each other and their environments. A MAS is a loosely coupled network of problem solvers that interact to solve problems beyond the capabilities or knowledge of the individual problem solver. The software agents are autonomous and usually heterogeneous. In the multi-agent platform, the individual agent has a limited view point with incomplete information or capabilities for solving the problem, there is no global control, data are decentralized, and computation is asynchronous.

Moreno [6] listed the critical areas in the health care domain where the capabilities of multi-agent systems can be particularly useful: distribution of knowledge; coordination of medical personnel; and the immediate delivery of diagnosis and treatment.

2.3   FIPA and JADE

The Foundation for Intelligent Physical Agents (FIPA) promotes the establishment of standard specifications in agent technology [9]. The Agent Communication Language (ACL) is the main output of FIPA. Common patterns of agent conversations have been formalized into interaction protocols that provide agents with a library of patterns to achieve common tasks [10]. Java Agent Developer Framework (JADE) is a middleware for the development of distributed multi-agent applications based on the peer-to-peer communication architecture that complies with FIPA standards. The intelligence, initiative, information, resources and control of agents can be fully distributed on mobile terminals as well as on computers in the fixed network. We refer the reader to [10], [11] and [12] for the details of the platform services in JADE.

2.4   Homecare Monitoring Systems

There are many commercial solutions related to homecare monitoring systems mentioned in [13]. These solutions usually rely on medical call centers, transmit their data through telephones and are based on proprietary technology. Several examples of these systems are discussed in [7]. The physiological data that can be monitored include blood pressure, pulse, oxygen saturation, temperature, ECG, respiration among others.

In [14] a home health care system consists of several diagnostic peripherals such as a blood pressure monitor, glucose meter, spirometer and pulse oxymeter. Clinical data from the unit is streamed to a centralized workstation. The coordination is manually performed by a nurse who reviews the data, performs assessment of the patient's health and coordinates the necessary intervention including sending alerts and relevant medical data to the appropriate specialist involved in the collaborative care of the patient.

According to [4], the top five chronic conditions are diabetes, coronary heart disease, congestive heart failure and chronic obstructive pulmonary disease. Patients of various ages particularly the elderly tend to have multiple conditions cutting across many individual diagnostic categories. Up to 45% of the highest risk segments have five or more distinct diagnoses, each of which can be the focus of condition-specific disease management. For the current work, we limit our definition of medical sensors to include only those that can detect the abnormal medical condition or symptom from the physiological signals.

2.5 Neural Networks in Healthcare

Artificial neural network (ANN) is a computational system consisting of a set of highly interconnected processing elements, called neurons that process information in response to external stimuli. A neuron contains a threshold value that regulates its action potential. Neural networks have been applied in the medical domain for clinical diagnosis [15], image analysis and interpretation [16], and drug development [17]. In [18] neural networks were used for automatic detection of acoustic neuromas in MR images of the head. Neural networks were utilized and supported by more conventional image processing operations. The prototype system developed as a result of the study achieved 100% sensitivity and 99.0% selectivity on a dataset of 50 patient cases. A comprehensive summary of various research works on the uses of neural networks in the medical domain is presented in [14].

3   Agent-Based Intelligent Home Health Care System

The current work focused on the coordination and decision support features of the healthcare system. We use a software agent as proxy for each participant in the system: medical sensors, the patient and the medical specialists. The proxy agents are of three types: sensor agent, patient agent and doctor agent respectively. The sensor agents handle retrieval of information from health sensing devices. These agents perform the required analysis on the sequential data from the sensors. The framework supports the dynamic configuration of sensor agents according to the symptoms that need to be monitored. The doctor agent is a mobile agent that migrates from the native environment in the doctor's clinic computer to the local environment of the multi-agent healthcare system. The patient agent coordinates the interaction of the other agents. It invokes the sensor agent to check the physiological data for the presence of abnormal condition or the symptom that the sensor agent reports back as a binary value. The patient agent contains the neural network code that enables it to recognize the medical condition and send notifications and relevant medical information to the appropriate medical specialist.

3.1 Software Agent as Proxy

Figure 1 presents the framework of the proposed multi-agent home healthcare system. The patient and sensor agents are local agents in the home health care system while the doctor agents are mobile that migrate from the clinic computer of the respective specialists. Our goal is to make the network more efficient by enabling the software agents to process the data locally and send only the results to their host systems thus avoiding unnecessary transmission of raw sensor data. Our proposed healthcare system aims to support a collaborating medical team that consists of individual specialists assigned to one or more medical conditions that requires continuous monitoring.

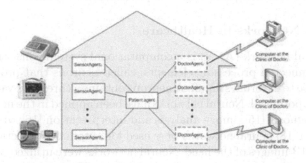

Fig. 1. Framework of the Multi-Agent Home Health Care System

We used the JADE platform for the implementation of the agent-based home healthcare system. The sensor agent registers a *medical-sensing* service that enables the patient agent to locate and subscribe to the service. Figure 2 shows the architecture of the proposed healthcare system based on the JADE platform. Each agent has a communication module which handles the exchange of data with other agents. The sensor agent uses the symptom detection module to determine the state of physiological data. The patient agent uses the neural network to classify the symptoms into the specific medical condition. The doctor agent uses the remediation management module to perform tasks such as sending the medical data to the approriate specialist.

Figure 3 shows the details of the data flow and individual functions of the component agents. Each sensor agent sends the medical data in the form of an ACL message to the patient agent. The sensor agent performs the analysis to detect the symptom based on the data from the medical sensor. *Symptom* refers to the state of the physiological data, such as heart rate which is not within the threshold considered to be normal. Each sensor agent need to be initialized with the threshold value for the particular physiological data to be able to detect the symptom. Various approaches to process the sensor data to detect the symptom are proposed in other studies [19], [20], [21]. To implement the proposed healthcare system, this study assumes that a suitable method for detecting the symptom is selected from among those presented in other studies. The patient agent provides the intelligent decision support for the medical specialists by performing the following tasks:

1. Trigger sensor agents to test for the presence of symptoms;
2. Process the inputs from sensor agents using the neural networks to recognize the medical condition from the symptom patterns; and
3. Send alerts and relevant data to the appropriate medical specialist when a medical condition is found.

The doctor agent is a mobile agent that transports the medical data from the local node of home health care environment to its host computer in the doctor's clinic. It performs other tasks such as handle the communication between the doctor and the patient and manages other remediation tasks such as the adjustments in the dosage of medication.

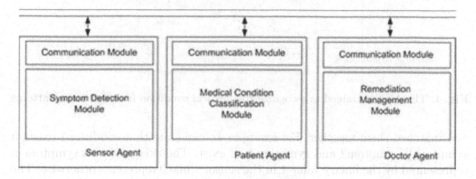

Fig. 2. Architecture of the Home Healthcare System

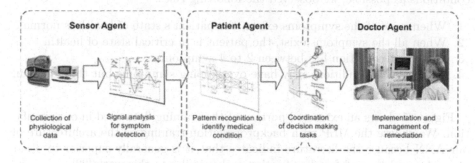

Fig. 3. Data Flow and Functions of the Component Software Agents

4   Recognizing Chronic Conditions from Symptom Patterns

The simulation environment of the proposed system consists of multi-agents created in the JADE agent platform. A single PC was used as environment for the agents to approximate the expected actual deployment configuration of the health care system. We programmed the ANN in Java and put the code in a JADE agent. Fig. 4 is a simplified model of the multi-layer neural network showing the neurons in the input, hidden and output layers. The input signal from the individual sensor agent is a binary value where the value 1 indicates that the symptom is detected by the sensor.

Synthetic classes of medical conditions were created and paired with symptom patterns to form the input-output sets. We labeled conditions and symptoms

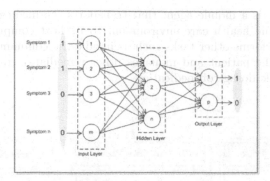

Fig. 4. The ANN is trained to recognize the medical condition from symptom patterns

and matched them together. For example we can state that condition1 exists if symptom1, symptom2 and symptom3 all exist. The existence of a symptom is represented by the binary value 1 in the sensor while 0 represents otherwise. The goal is to demonstrate that ANN can correctly recognize the chronic condition from the symptom patterns. To keep the models as close to the real medical conditions as possible, we observed the following rules:

1. When none of the symptoms exists, the patient's state of health is normal.
2. When all the symptoms exist, the patient is in critical state of health.
3. A medical condition has between 2 to 4 symptoms.
4. Two medical conditions may have overlapping symptoms but they differ in at least one.

Figure 5 shows an extracted portion of the training data used in our simulation. We choose the MLP with backpropagation training as the architecture of our ANN because the literature indicates that it is generally superior to other algorithms when used for classification tasks similar to this work [22].

	Input Signals											Target				
	V(1)	V(2)	V(3)	V(4)	V(5)	V(6)	V(7)	V(8)	V(9)	V(10)		T(1)	T(2)	T(3)	T(4)	T(5)
S(1)	1	1	1	0	0	0	0	0	0	0	C(1)	1	0	0	0	0
s(2)	1	1	1	1	0	0	0	0	0	0	C(1)	1	0	0	0	0
s(3)	1	1	1	0	1	0	0	0	0	0	C(1)	1	0	0	0	0
s(4)	0	0	1	1	1	0	0	0	0	0	C(2)	0	1	0	0	0
s(5)	0	1	1	1	1	0	0	0	0	0	C(2)	0	1	0	0	0
s(6)	0	1	0	1	1	0	0	0	0	0	C(2)	0	1	0	0	0
s(16)	0	0	0	0	0	0	0	0	0	0	C(6)	0	0	0	0	0
s(17)	1	1	1	1	1	1	1	1	1	1	C(7)	1	1	1	1	1

Fig. 5. Extracted portion of the training data

The action potential of a neuron is determined by the weight associated with the neuron's inputs (Equation 1), a threshold modulates the response of s neuron to a particular stimulus confining such response to a pre-determined range of values. Equation 2 defines the output y of a neuron as an activation function f of the weighted sum of $n+1$ inputs. The threshold is incorporated into the equation as the extra input. The output produced by a neuron is determined by the activation function. This function should ideally be continuous, monotonic and differentiable. The input data consists of the presence (or absence) of a symptom in a physiological data encoded as a binary value thus the applicable function is the step (Equation 3) or the binary sigmoid (Equation 4). If the desired output is different from the input, it is said that the network is hetero-associative, because it establishes a link or mapping between different signals; in autoassociative network the desired output is equal to the input.

In supervised learning, a feedforward ANN is trained with pairs of input-output examples. The accuracy of the response is measured in terms of an error E defined as the difference between the current and desired output (Equation 5) Weights are adjusted to minimize the overall output error.

$$z = \sum_{i=1}^{n} x_i w_i \tag{1}$$

$$y = f\left(\sum_{i=1}^{n} x_i w_i\right) \tag{2}$$

$$f(x) = \begin{cases} 1 \text{ if } \sum_{i=1}^{n} x_i w_i > 0 \\ 0 \text{ if } \sum_{i=1}^{n} x_i w_i \leq 0 \end{cases} \tag{3}$$

$$f(x) = \frac{1}{1 + e^{-x}} \tag{4}$$

$$f(x) = \frac{1}{2} \sum_{j} (t_{pj} - o_{pj})^2 \tag{5}$$

The error E is propagated backwards from the output to the input layer. The appropriate adjustments are made by slightly changing the weights in the network by a proportion d of the overall error E. After weights are adjusted, the inputs are presented again and the error is calculated, weights are adjusted and this procedure is repeated until the current output is satisfactory or the network performance cannot improve further. We used the backpropagation learning algorithm using the procedure in [14]:

1. Present the input-output pair p and produce the current output O_p.
2. Calculate the output of the network.
3. Calculate the error for each output unit for a particular pair using Equation 6.

$$\delta_{pj} = (t_{pj} - o_{pj}) f'(net_{pj}) \tag{6}$$

4. Calculate the error by the recursive computation of δ for each of the hidden units j in the current layer (Equation 7). Where w_{kj} are the weights in the k output connections of the hidden unit j, δ_{pk} are the error signals from the k units in the next layer and $f'j(net_{pj})$ is the derivative of the activation function. Propagate backwards the error signal through all the hidden layers until the input layer is reached.

$$\delta_{pj} = \sum_k \delta_{pk} w_{kj} f' (net_{pj}) \tag{7}$$

5. Repeat step 1 through 4 until error is acceptably low.

5   Experimental Evaluation

We trained the ANN using empirical models of chronic conditions and the matching sets of symptom patterns. We created 5 chronic conditions and each one is paired with a set of target outputs. We defined 2 additional conditions: the first, a normal healthy status when the outputs were all zero and the other, when all the outputs were 1, indicates the patient is in a critical state because all the symptoms were triggered. The training data consisted of 17 pairs of input and desired output patterns. Each chronic condition is matched with a set of 3 symptom patterns which were similar because they have similar symptoms but unique because they differ by at least one. This matching approximates real medical conditions that share some common symptoms. We created the test data consisting of 25 pairs of patterns that included the original 17 patterns used during training and new patterns that are different from the patterns already presented to the network.

We performed three runs of the simulation using higher epochs in successive runs to determine the performance of our ANN. An epoch is a single pass of the network through the full set of training data. Table 1 shows the performance of the network.

The ANN recognized all 17 patterns presented during training with 100% accuracy. At 5000 epochs, 22 patterns out of 25 were correctly identified. When we used higher epochs, the ANN recognized more test patterns correctly. The peak performance of the network was found at 96% generalization accuracy.

Table 1. Performance of the ANN across training epochs

Training Epochs	Training Accuracy	Generalization Accuracy
5000	100%	88%
25000	100%	92%
50000	100%	96%

6   Conclusions

This work contributes improvements to home healthcare systems particularly for critical-care patients that require continuous medical monitoring by a collaborating group of medical specialists. A framework based on software agents that proxy for participants in a healthcare environment was proposed. Neural networks were used to make the patient agent capable of intelligent decision support by being able to recognize symptom patterns that characterize certain chronic conditions. In this manner, we made the healthcare system capable of sending alerts and providing relevant medical data to the appropriate specialist responsible for the medical care of a specific chronic condition. Our simulation result indicates that ANN can recognize chronic conditions with very high accuracy. The implication is that future home healthcare systems can be enhanced with intelligent decision support in situations that require constant evaluation of the health status of a patient in critical care and the task of identifying the chronic condition from the symptoms of pattern can be safely delegated to a software agent. Using such systems, medical specialists can service more patients over a greater geographical area and be able to collaborate with colleagues from a broader spectrum of medical expertise.

Our future works will focus on improving the pattern recognition using other neural network algorithms or other adaptive systems. We will use domain-specific knowledge to perform the experimental evaluation using real-world data. Our platform that combines the multi-agent systems with the pattern recognition capabilities of neural networks will be used in other applications in other distributed environments.

References

1. Augusto, J., et al.: Pervasive Health Management. New Challenges for Health Informatics. Journal of Universal Computer Science 12(1), 1–5 (2006)
2. Heiki, J., Laukkanen, M.: Efficient Agent Communication in Wireless Environments. Software Agent-based Applications, Platforms, and Development Kits. Whitestein Series on Software Technologies, pp. 307–330 (2005)
3. Della Mea, V.: Agents Acting and Moving in Healthcare Scenario: A Paradigm of Telemedical Collaboration. IEEE Transaction on Information Technology in Biomedicine, 10–15 (2001)
4. Coughlin, J., et al.: Old Age, New Technology and Future Innovations in Disease Management and Home Health Care. Home Health Care Management and Practice 18(5), 196–207 (2006)
5. Celler, B., Lovell, N., Basilakis, J.: Using Information Technology to Improve the Management of Chronic Disease. Medical Journal of Australia 179(5), 242–246 (2003)
6. Moreno, A.: Agent-based Applications in Health Care. Medical Applications of Multi-Agent Systems. In: 1st International Joint Conference on Autonomous Agents and Multi-agent Systems, Bologna, Spain (2002)

7. Foster, D., et al.: A Survey of Agent-Based Intelligent Decision Support Systems to Support Clinical Management and Research. In: 1st Intl. Workshop on Multi-Agent Systems for Medicine, Computational Biology, and Bioinformatics (MAS*BIOMED 2005), Utrecht, Netherlands (2005)
8. Frize, M., et al.: Clinical Decision Support System for Intensive Care Units using Artificial Neural Networks. Medical Engineering and Physics 23, 217–225 (2001)
9. Bellifemine, F.: Jade - A White Paper (2003), http://jade.cselt.it/papers/2003/WhitePaperJADEEXP.pdf
10. Calisti, M.: Abstracting Communication in Distributed Agent-Based Systems. Concrete Communication Abstractions of the next Distributed Object Systems. In: Magnusson, B. (ed.) ECOOP 2002. LNCS, vol. 2374, Springer, Heidelberg (2002)
11. Bellifemine, F., Caire, G., Trucco, T.: Jade Administrator Guide (2006), http://jade.tilab.com/doc/administratorsguide.pdf
12. Bellifemine, F., Caire, G., Trucco, T.: Jade Programmer's Guide (2005), http://jade.tilab.com/doc/programmersguide.pdf
13. Hein, A., et al.: SAPHIRE Intelligent Healthcare Monitoring based on Semantic Operability Platform - The Homecare Scenario. In: 1st European Conference on eHealth (ECEH 2006), Fribourg, Switzerland (2006)
14. Sordo, M.: Introduction to Neural Networks in Healthcare. OpenClinical (2002), http://www.openclinical.org/docs/int/neuralnetworks011.pdf
15. Brause, R.: Medical Analysis and Diagnosis by Neural Networks. In: Crespo, J.L., Maojo, V., Martin, F. (eds.) ISMDA 2001. LNCS, vol. 2199, pp. 1–13. Springer, Heidelberg (2001)
16. Antkowiak, M.: Artificial Neural Networks vs. Support Vector Machines for Skin Diseases Recognition. Master's thesis, Department of Computing Science, Umea University, Sweden (2006)
17. Weinstein, J., Kohn, K., Grever, M.: Neural Computing in Cancer Drug Development: Predicting Mechanism of Action. Science 258, 447–451 (1992)
18. Dickson, S.: Investigation of the use of Neural Networks for Computerized Medical Image Analysis. PhD Thesis, Department of Computer Science, University of Bristol (1998)
19. Zamarron, C., et al.: Utility of Oxygen Saturation and Heart Rate Spectral Analysis Obtained From Pulse Oximetric Recordings in the Diagnosis of Sleep Apnea Syndrome. American College of Chest Physicians 123, 1567–1576 (2003)
20. Cotofrei, P., Stoffel, K.: Rule Extraction from Time Series Databases using Classification Trees. In: Proceedings of Applied Informatics, Innsbruck, Austria (2002)
21. Zeger, S., et al.: On Time Series Analysis of Public Health and Biomedical Data. Johns Hopkins University, Dept. of Biostatistics Working Papers. Working Paper 54 (2004)
22. Pandya, A., Macy, R.: Pattern Recognition with Neural Networks in C++, pp. 73–146. CRC Press, Florida (1996)

An Aware-Environment Enhanced Group Home: AwareRium

Hideaki Kanai[1], Toyohisa Nakada[2], Goushi Tsuruma[2], and Susumu Kunifuji[2]

[1] Center of Knowledge Science, Japan Advanced Institute of Science and Technology,
Ishikawa 923-1292 Japan
hideaki@acm.org
[2] School of Knowledge Science, Japan Advanced Institute of Science and Technology,
Ishikawa 923-1292 Japan
{t-nakada,g-tsuru,kuni}@jaist.ac.jp

Abstract. We have constructed "AwareRium", an enhanced group home with ubiquitous technology to establish person-centered care. The group home is a facility that provides care for Alzheimer's and dementia patients. The group home is effective in curbing the progress of Alzheimer's dementia. A small number of elderly persons with no relations in need of care, support, or supervision, such as Alzheimer's and dementia patients, can live together in the group home. In this paper, we describe the design concept of AwareRium and the sensing and projection devices installed therein. We present two support systems for position awareness using these devices in AwareRium: a system for finding lost objects and a system for identifying and noticing dangers in order to prevent dangers in the group home.

1 Introduction

Currently, Japan is rapidly aging. The Ministry of Health, Labour and Welfare of Japan reports that the percentage of elderly people whose ages are at least 65 years is about 20 percent in 2006, and this percentage will increase up to about 30 percent in 2020 [1]. The population of the elderly will be about 34 million in 2020. The elderly in need of care will be about 3 million in 2020.

As one of the measures toward this situation, many group homes for people with dementia are in use in Japan (Figure 1). A group home is a type of home-based care service for the elderly with dementia. It is a facility where the elderly with dementia live together and receive nursing care services such as meals and baths on a 24-hour basis. A group home provides more user friendly housing for the elderly and attempts to curb the progress of Alzheimer's dementia.

There are some problems on caring by group homes, particularly, the burdens on the caregivers. One caregiver cares for six or seven elderly people with dementia, and a few caregivers share the care duties of a house round the clock. They must provide support for any activity in an inhabitant's daily life. We consider that a reduction in these burdens is one of the most significant issues regarding group homes. To address this problem, we are pursuing the development of a group home enhanced by computing technology.

M.S. Szczuka et al. (Eds.): ICHIT 2006, LNAI 4413, pp. 425–436, 2007.

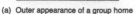

(a) Outer appearance of a group home   (b) Corridor in a group home   (c) Inhabitant's room

Fig. 1. Group Home in Japan

With the progress in ubiquitous sensing and recognition technology, several smart and ubiquitous environments including a house[2], living room[3], and kitchen[4] are being developed. These environments are installed with ubiquitous sensing and recognition devices for the daily activities performed in space. Using the sensed information, the environments will provide users with various support applications for their daily activities. We consider that group homes will be enhanced with these technologies. We are developing an aware-environment-enhanced group home "AwareRium". AwareRium can be aware of the context in which its inhabitants and caregivers act and the state of the objects therein. AwareRium offers assistance to the inhabitants and caregivers depending on the awareness.

In this paper, we outline our efforts that have gone into building AwareRium and present two applications related to the sensed position data of its inhabitants and objects. This paper is organized as follows: Section 2 describes the design concept of AwareRium that simulates a group home in the real world with the aspects of structure and the sensing and awareness functions based on the features of the elderly with dementia. Section 3 explains a position-awareness support platform in AwareRium, the position sensing devices, and the awareness-illumination devices. Section 4 presents two applications implemented on this platform.

2   Aware-Environment Enhanced Group Home: AwareRium

AwareRium[1] is an experimental environment for investigating various support systems in group homes. AwareRium is shown in Figure 2. Its size is approximately 50 square meters (4.5 m × 11 m). The installed sensors can detect and monitor the location information of the inhabitants and objects in AwareRium.

2.1   Design Concept

The design concept of AwareRium is to easily simulate a group home environment and mount various sensors in an unostentatious manner.

---

[1] The AwareRium is a combination of the English words "Aware", and the suffix "-Rium" meaning "space" and "building". It implies that space promote awareness.

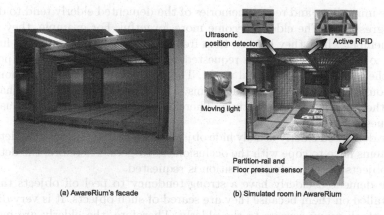

(a) AwareRium's facade   (b) Simulated room in AwareRium

Fig. 2. AwareRium

The structural features of AwareRium are the Old Japanese building style and a flexible floor plan. The former is known as the Sukiya style in Japanese. The design and shape of the group home are not uniform such as those of a hospital or a nursing home. The group home was made similar to the environment where the inhabitants had lived, namely the Old Japanese house style, in order to make the inhabitants feel at home easily. Therefore, we adopted one of the old Japanese house styles, the Sukiya style, for making the residents feel at home (Figure 2(a)). The latter is used to simulate various room arrangements in the group home. The group home has various room arrangements depending on the constitutions of the inhabitants. AwareRium simulates a layout of various room arrangements of the group home with partition rails placed on the floor in a grid layout (Figure 2(b)).

2.2 Sensing and Information Presentation Function

Similar to the current ubiquitous environments, AwareRium also has a sensing function, which is to acquire information of the actual world, and an information presentation function, which is to show information in the actual world. These functions in AwareRium are designed according to the symptoms of senior citizens.

Sensing function: The requirements of the sensing function in AwareRium are examined from a viewpoint of the characteristics of the demented elderly. The specifications are enumerated as follows:

– The features of the demented elderly differ largely across individuals depending on the progression of dementia. Dementia progresses at different rates in different elderly people. The care of demented elderly people in the group home should be on an individual basis. The items necessary for daily life also depend on the patients. Therefore, the sensors in AwareRium have the function of identifying and tracing individuals and objects.

- The immediate and recent memories of the demented elderly tend to decline progressively. The elderly become more forgetful. For example, they fail to remember where they keep an item frequently in the short term. To find the lost object, the sensors are requested to detect a highly accurate position of the lost object in the real world. The inhabitants in the group home are encouraged to walk around the rooms. The sensors can detect what happens to the inhabitants in real time. Therefore, a high tracking performance is requested.
- The demented elderly usually hide objects in cabinets. The position detection systems have to cope with the occlusions of targets. Therefore, the detection of objects in the occlusion position is requested.
- The demented elderly have a strong tendency to peel off objects that are installed on them because they are scared of such objects. It is very difficult to attach many sensors to the elderly. Therefore, the elderly are made to wear as few sensing devices as possible.

We have introduced sensing devices that fulfill the abovementioned requirements as follows:

- Ultrasonic position detection system (Furukawa Industrial Machinery System Co., LTD)
- Active RFID system (RF Code Co., LTD)
- Floor pressure sensor system (Furukawa Industrial Machinery System Co., LTD)

The ultrasonic position detection system can locate any object attached with ultrasonic tags in the 3D position with an error of approximately 5 cm. The system detects the position of an object at intervals of about 20 milliseconds. Each tag is assigned an ID. Therefore, the system can detect a position with high accuracy in real time and identify objects. 150 ultrasonic receivers are installed in the ceiling of AwareRium to maintain the location accuracy of approximately 5 cm because the number of ultrasonic receivers depends on the ceiling height. Further, the ceiling height of AwareRium is only 2 m, which is lower than that of standard Japanese houses. If the height is about 2.5 m (standard Japanese house), about 50 receivers are sufficient.

We use the active RFID system in order to detect the positions of objects in the occlusion status. In the case that an active RFID receiver can receive a signal emitted by an active RFID tag, we assume that there is a tag around the receiver and the position is detected. In other words, the position accuracy depends on the reception range. 8 active RFID receivers are installed in the ceiling of AwareRium. The system detects the position with an error of approximately 2.5 m. It should be noted that the error depends on the number of receivers installed in AwareRium.

In order to realize a position detection system that can locate hidden objects with high accuracy and in real time, we utilize a particle filter[5]. That is, the measured position data by these devices is used as inputs to the filter, and the filter estimates new position data with a maximum error of approximately 2.5 m. The details of this system are described in Section 3.2.

The floor pressure sensor system consists of multiple sensor blocks. Each block has 4 units. A unit is square-shaped and 25 cm in height and width. The unit detects functions as an ON/OFF switch by a certain pressure loading, and the sensitivity is about 10 kg/unit. The position of a block when activated as ON corresponds to that of a measurement object. The blocks are installed in the entire floor of AwareRium.

The tag-based position detection systems such as the ultrasonic position detection system treat a measurement object as a point. The floor pressure sensor system detects an object as a plane. For example, in the case that an elderly person falls on the floor, the tag-based position detection systems do not recognize whether the elderly person has fallen or is sitting on the floor. Alternatively, the floor sensor system can recognize when an elderly person has fallen on the floor because of the sensing of the body of the elderly person with multiple floor sensor blocks.

Information presentation function: The requirements of the information presentation function in AwareRium are examined from the viewpoint of the characteristics of the demented elderly as follows:

- The cognitive-functional performance of the elderly depends on the severity of dementia. The methods to show information are requested according to the severity. Particularly, it is important to determine who should be provided the awareness to and which sense organs are used in order to promote awareness of information. In the case of the elderly person with severe dementia, the awareness must be offered to the caregivers as well as the elderly. In the elderly, auditory perception declines earlier than visual perception. Therefore, in order to realize the information presentation function for the elderly, the use of sight is more effective than that of hearing.
- The demented elderly have a strong tendency to peel off objects that are installed on them because they are scared of such objects. It is very difficult for the elderly to keep any information presentation devices such as PDAs and cellular phonesD

We present some information to people using light. We utilize two types of lighting systems as the information presentation devices fulfilled the abovementioned requirements. These systems are computer-controllable lighting systems: a moving light system (MARTIN Co., Ltd) and a moving projector (Active Vision Co., Ltd). They can illuminate the light and visual sources in any position by controlling the pan and tilt of these bodies. They are typically used as stage lighting instruments. In the moving light system, various types of light (colors and shapes of the spot, light effects) can be controlled according to the information presented. Through the moving projector, visual sources from a PC or videos can be projected in any position. The projector can present richer visual content than the moving light system. In AwareRium, the abovementioned devices are used to implement the sensing function and the information presentation function.

We have developed some systems to support the daily life of the demented elderly and caregivers in the group home. We describe "finding lost objects" and "identifying and noticing dangers" with position awareness in Section 4.

3   Position Awareness Support Platform in AwareRium

3.1   Position Awareness

Position-awareness applications help the elderly and caregivers to get an insight on personalized information and services related to improving care supports depending on any position of AwareRium. The position awareness support platform of AwareRium is shown in Figure 3. The platform has the sensing function and information presentation function as described in Section 2.2. The platform consists of a position sensing unit, an awareness illumination unit, and application units. The position-sensing unit traces a person and objects in the AwareRium in real time and stores the histories in a database. One of the application units, for example, a support system for finding lost objects, searches the position information of lost objects from the database and decides the types of awareness illumination, timing, and so on. The awareness illumination unit shines light on the lost objects. Both the position detection unit and the awareness illumination unit are described as follows.

3.2   Position Sensing Unit

In AwareRium, the ultrasonic position detection system and the active RFID system are used as position detection systems. By integrated processing the position information both these systems detect, even measured objects in the occlusion status can be located with high accuracy. The estimated positions of

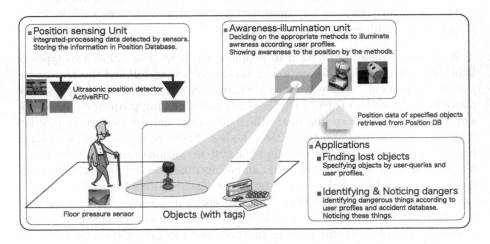

Fig. 3. Position awareness support platform

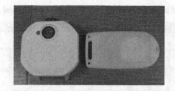

Fig. 4. Ultrasonic Tag (left) and ActiveRFID Tag (right)

the objects are stored in the position database after each updating of the position data.

Each system locates a specific tag for each system. Two types of tags must be fixed on each measured object. The tags[2] are shown in Figure 4. The left one is a tag for the ultrasonic position detection system, and it has a width of 4.2 cm and a height of 5 cm. The right one is a tag for the active RFID system, and it has a width of 6 cm and a height of 3 cm. The main objective of using the particle filter in this study is to integrate the position data measured by different position detection systems into estimated data with a confidence coefficient. The use of the particle filter is a method for representing continuous probabilistic distributions by a number of discrete particles. Since a continuous probabilistic distribution is translated to a discrete one by this method, the computation complexity can be reduced. The sensed value by a device is represented as a probabilistic distribution in the particle filter. The probabilistic distribution depends on a feature of the device. In the case of the ultrasonic position detection system that senses the 3D position (x, y, z) with an error, an input of the particle filter is a normal distribution $(mean = x, y, z \ and \ sd = error)$. In the case of the other device, the active RFID system that outputs whether an active RFID tag is in the neighborhood of an active RFID reader or not, an input of the particle filter is a uniform distribution $(P(x) = 1/(b-a) : a < x < b, P(x) = 0 : others)$. The particle filter computes the positions of all particles by locating particles during random walks and resampling the particles by the probability distribution. That is, all the particles are located in a target space at random. Every time a position detection system senses a target object, all the particles are resampled depending on the probability distribution of the system. The positions are fixed gradually. As the results, the average of the positions is the estimated position, and the standard deviation is the confidence coefficient of the estimated position. Figure 5 shows the estimated locations of three tags sensed by two types of position detection systems. Each circle in Figure 5 represents the confidence coefficient of each position.

3.3   Awareness Illumination Unit

The appropriate methods to illuminate awareness should be on an individual basis. The profile of the elderly with dementia is used to decide the appropriate

---

[2] We are developing both the tags such that they are integrated into one tag.

Estimated position by Ultrasonic position detector and Active RFID

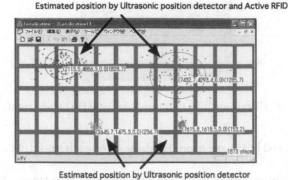

Estimated position by Ultrasonic position detector

Fig. 5. Estimated position with the confidence coefficient by the particle filter

methods. As one of elements in the profile, the severity of dementia rated by Clinical Dementia Rating Scale (CDR)[6] is used. The CDR characterizes six domains of cognitive and functional performance, and the overall score is derived by 5-point scale for each domain.

The awareness illumination unit operates the awareness illuminators – the moving light system and the moving projector system – and shows awareness embodied by light and visual content on the positions specified in the position sensing unit. The types of awareness depend on the applications.

To calculate the controlled variables of the pan and tilt of an awareness illuminator, the calculation can be treated as a robot inverse kinematics problem[7]. That is, given the desired position of the illuminated surface and the distance from the position to the illuminator's position, what must be the pan and tilt of the position of an awareness illuminator ? Here, the calculation is equal to the control computation of a 5-axis robot arm taking into consideration three calibration factors for the control elements (two rotations and one parallel translation). The calculation method is shown as follows:

T_l and T_m represent the axis of the position detection system and that of the moving light or moving projector system. Equation (1) shows the relationship between these two axes.

$$T_l = A_{ry} \cdot A_{rz} \cdot A_t \cdot A_{rp}^{-1} \cdot A_{rt}^{-1} \cdot T_m \tag{1}$$

Here, A_{ry}, A_{rz}, and A_t are the translations detected with the calibration task. A_{rp}^{-1} and A_{rt}^{-1} are the inverse translations of the pan and tilt. From Equation (1), the relationship of the position (P_l) in the T_l axis and the same position (P_m) in the T_m axis is shown in Equation (2).

$$P_m = A_{rt}^{-1} \cdot A_{rp}^{-1} \cdot A_t \cdot A_{rz} \cdot A_{ry} \cdot P_l \tag{2}$$

Here, A_{ry}, A_{rz}, and A_t are detected with the calibration task. P_m and P_l are obtained from the position detection system; therefore, the pan and tilt can be calculated from Equation (2).

4   Applications

We present two applications of the position awareness support platform in AwareRium. One is a support system for finding lost objects, and the other is a support system for identifying and noticing dangers. These supports are requested based on interviews with caregivers about the burdens of caring for patients with dementia in a group home that we have a collaborative relationship with. We consider that these applications are needed substantially for providing assistance in care practices.

4.1   Support for Finding Lost Objects

Inhabitants usually lose their personal belongings such as their wallet, glasses, etc., because of the decline of their immediate and recent memories. Caregivers have to search for the lost objects with the inhabitants on each occasion in order to avoid problems among the inhabitants. The caregivers have to devote a significant amount of time and care to find these objects. Thus, we are developing a support system for finding lost objects to reduce the burdens of caregivers[8].

Figure 6 shows the usage of the support system. The user specifies an object to be found by an input device "paper". On the front side of the paper, there are some pictures of the objects that an inhabitant needs and has lost many times on previous occasions. There is a passive RFID tag for each picture on the backside. The system is requested to find a lost object by putting a PDA embedded in a passive RFID reader on the picture. According to the specified object data, the system retrieves the position data of the object from the position database. The system can identify the position in real time because the position database automatically synchronizes all newly entered position data. The system illuminates the position by the awareness illuminator. The system makes the user aware that the missing object exists. The size of the illuminated area depends on the accuracy of the detected position.

4.2   Support System for Identifying and Noticing Dangers

The aim of a support system for identifying and noticing dangers is to make the inhabitants and caregivers aware of the dangerous objects and situations in the group home. The assistance provided to avoid the dangers that could occur in the group home leads to ensuring the safety of the inhabitants and a reduction in the caregiver's burden. The inhabitants tend to develop a fracture during a fall due to the diminishing of their spatial awareness. For example, the inhabitants often fall over slight steps on the floor and obstacles under their foot without noticing them. It is difficult to assess all these risks and eliminate them in advance due to the diversity of inhabitants and living environments in the group home. Namely, the risks differ among inhabitants, and each group home does not have a uniform environment as a hospital room. The caregivers are not able to observe the inhabitants all the time because they have many tasks to perform in the group home.

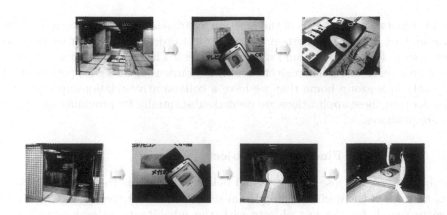

Fig. 6. Usage of a support system for finding lost objects

The system determines whether an inhabitant would be at risk (in danger) from the sensed information from the sensing units and a case database on the past accidents at the group home. In the case of a dangerous situation, the system provides the inhabitants and caregivers with an awareness of the possible danger to the inhabitant using awareness illuminators.

The case database stores cases with past accidents as if-then rules. Consider the following example: an accident happened with an elderly person tripping over a newspaper on the floor. As a rule, this accident represents "if an elderly person comes close to a newspaper on the floor, then there is a risk of the elderly person falling" in the database. To expand the possible application of rules, the semantics of "the elderly" and "a newspaper" are described in the database. Further, in the case that an elderly person closes a magazine on the floor, the above rule could be applied.

The semantic descriptions are implemented by the ontologies of the semantic web technology, OWL[9]. An ontology consists of a finite list of terms and the relationships between these terms. The terms denote the important concepts (classes of objects) of the domain. In our system, an ontology consists of objects, the victims of an accident, and the causes of an accident. For example, an ontology of "a newspaper" is represented as the "paper" class and the "tripping over" class of the "accidents" domain.

The current prototype system[10] deals with the accidents and dangerous conditions caused by the spatial relationships between inhabitants and objects and focuses on the risk of falling in the group home. Figure 7 shows an example of the system execution. The system illuminates the region around the dangerous factor (a power cable) to prevent a subject from getting entangled with the cable.

When a person with dementia approaches a power cable, the moving projector illuminates it.

Fig. 7. Usage of a support system for identifying and noticing dangers

5 Related Work

There are some researches on assisted living facilities for the elderly. The Gator Tech Smart House[11] is designed to assist older persons and enhance the quality of life and develops various applications. The Smart Medical Home[12] focuses on "health in the home" and develops personal healthcare systems for self-care, mobile care, and home care. On the other hand, AwareRium is designed to assist the elderly with dementia and caregivers in the group home. The support systems focus on a reduction in the burdens of care practices on the caregivers of the group home.

6 Conclusion

We are building an aware-environment-enhanced group home, AwareRium, to reduce the burdens of inhabitants and caregivers in the group home. This paper outlines our efforts to build AwareRium, presents a platform of support for position awareness, and introduces two applications on the platform: a support system for finding lost objects and another for identifying and noticing dangers.

We show the requirements for the sensing and information presentation functions depending on the characteristics of the elderly with dementia in Section 2.2. The requirements for the sensing function are individual identification, high tracking performance, robust measurement against occlusion, and fewer attached sensors. The requirements for the information presentation function are methods to show awareness based on the severity of dementia, the use of sight rather than hearing and the use of no PDAs. According to these requirements, we install some position detection devices and awareness illuminators in AwareRium. We use a particle filtering method for the integrated processing of the information input by the various sensors. As a result, we implement a robust high accuracy position detection system as described in Section 3.2. We introduce two applications that are needed for providing assistance in performing the care practices of the group home, as described in Section 4.

As a future work, we will conduct a field test on the above applications and improve them for practical use in the group home. We are pursuing the development of various applications for the position awareness support platform.

Acknowledgement. Our research is partly supported by a fund from the Ministry of Education, Culture, Sports, Science and Technology of Japan under the name of Cluster for Promotion of Science and Technology in Regional Areas. We would like to thank Architect Shohei Matsukawa, who is the presidency of an architectural design office, 000studio (*http* : //*www.000studio.com*), and the staff for design and construct of AwareRium.

References

1. The Japanese Ministry of Health, Labour and Welfare. White paper on aging society 2005 (2005)
2. Kidd, C., Orr, R., Abowd, G., et al.: The Aware Home: A Living Laboratory for Ubiquitous Computing Research. In: Streitz, N.A., Hartkopf, V. (eds.) CoBuild 1999. LNCS, vol. 1670, pp. 190–197. Springer, Heidelberg (1999)
3. Brumitt, B., Meyers, B., Krumm, J., Kern, A., Shafer, S.: EasyLiving: Technologies for Intelligent Environments. In: Hand held and Ubiquitous Computing 2000. LNCS, Springer, Heidelberg (2000)
4. Siio, I., Mima, N., Frank, I., Ono, T., Weintraub, H.: Making Recipes in the Kitchen of the Future. In: Extended Abstracts, Conference on Human Factors in Computing Systems (ACM CHI 2004), p. 1554 (2004)
5. Hightower, J., Brumitt, B., Borriello, G.: The Location Stack: A Layered Model for Location in Ubiquitous Computing. In: Proceedings of the 4th IEEE Workshop on Mobile Computing Systems & Applications (WMCSA 2002), pp. 22–28 (2002)
6. Hughes, C.P., Berg, L., Danziger, W.L., Coben, L.A., Martin, R.L.: A new clinical scale for the staging of dementia. Br. J. Psychiatry 140, 566–572 (1982)
7. Paul, R.: Robot Manipulators: Mathematics, Programming, and Control (Artificial Intelligence). MIT Press, Cambridge (1981)
8. Nakda, T., Kanai, H., Kunifuji, S., Support, A.: System for Finding Lost Objects using Spotlight. In: Proceedings of 7th Human-Computer Interaction with Mobile Devices and Services (MobileHCI 2005), pp. 19–22. ACM Press, New York (2005)
9. Antoniou, G., Harmelenm, F.: A Semantic Web Primer. MIT Press, Cambridge (2004)
10. Tsuruma, G., Kanai, H., Nakda, T., Kunifuji, S.: Dangerous Situation Awareness Support System for Elderly People with Dementia. In: The Second IASTED International Conference on Human-Computer Interaction (IASTED-HCI 2007) (to appear)
11. Helal, S., Mann, W., El-Zabadani, H., King, J., Kaddoura, Y., Jansen, E.: The Gator Tech Smart House: A Programmable Pervasive Space. IEEE Computer 38(3), 50–60 (2005)
12. http://www.futurehealth.rochester.edu/smart_home/

The Situation Dependent Application Areas of EPC Sensor Network in u-Healthcare

Yoonmin Hwang[1], Garam Park[1], Eunji Ahn[1], Jaejeung Rho[1],
Jonwoo Sung[2], and Daeyoung Kim[2]

[1] Auto-ID Lab Korea, School of IT business, Information & Communications
University, 119, Munjiro, Yuseong-gu, Daejeon, Republic of Korea
{ymhwang,jogme,ejahn,jjrho}@icu.ac.kr
[2] Auto-ID Lab Korea, School of Engineering, Information & Communications
University, 119, Munjiro, Yuseong-gu, Daejeon, Republic of Korea
{jwsung,kimd}@icu.ac.kr

Abstract. Electronic product code (EPC) sensor network is a collection of objects for sensing data. It is crucial to ubiquitous society. It can provide an application service based on situation dependency with its properties. The situation dependency is an emerging concept which can collect location-based and personalized information. With the situation dependency, many industries can serve ubiquitous service for independent users. u-Healthcare is one of ubiquitous service to provide seamless medical treatment. The concept of situation dependency is applied in u-Healthcare with EPC sensor network technology. Due to specialized four properties of EPC sensor network which are driven from this paper, the situation dependency is well-established in u-Healthcare service. In this paper, we defined value and architecture of u-Healthcare service and we analyzed application areas of u-Healthcare.

Keywords: u-Healthcare, EPC Sensor Network, Situation dependency.

1 Introduction

Healthcare service requires to have very detailed medical treatment for patients. The patients are very sensitive to body temperature, humidity, and other environmental condition. For instance, a diabetic has to consider on their blood-sugar level by period. A few foods may cause a serious condition. However, it is difficult to check condition of all patients regularly. Thus, the patients fear about situation that they cannot receipt sufficient medical treatment at the right time which causes a serious condition for them. In other words, the patients are not convenient and comfortable when they suffer from significant disease. The healthcare service, however, is licensed and registered labor intensive which means it strongly needs the qualified people to serve [1]. To build convenient environment for patients, the healthcare service offers a ubiquitous healthcare service also know as u-Healthcare service. u-Healthcare service is combination of ubiquitous and healthcare which indicates a seamless medical treatment and management

M.S. Szczuka et al. (Eds.): ICHIT 2006, LNAI 4413, pp. 437–446, 2007.

of healthcare service. Due to its importance, many of researches are continuing on u-Healthcare.

Sensor network is a new class of distributed systems which communicate among sensors [2]. It is dynamic to the data availability, sources and application quality of service requirements. EPC Sensor Network is a new concept which solves limitations of Sensor Network with properties of EPC Network. It is an integrated model which contains both auto identification technique from EPC Network and communication sensing information from Sensor Network. In the EPC sensor network, sensor nodes which are attached to an object, sense condition of the object and at the same time sense object information with tag device, such as RFID (radio frequency identification) tag. This property is fit into the u-Healthcare service. Thus, this paper generally finds the impact of EPC sensor network on u-Healthcare that can solve present problems in u-Healthcare.

2   u-Healthcare Service

2.1   Introduction to u-Healthcare Service

In the past, healthcare service was a simply medical treatment for a disease after patients got the disease. It further expands to medical treatment and prevention before patients got the disease. The paradigm is shifting to a more positive attitude for better medical service to patients. To handle u-Healthcare efficiently, the hybrid technology and knowledge are required. u-Healthcare is not limited to bio technology area only. It covers other advanced high-technology including IT (Information technology) and NT (Nano technology). u-Healthcare has sensor network to collect various patient dynamic bio-information.

Research trend of u-healthcare has several properties including automated capture, context awareness, central processing, and actual interface. Automated capture is continuously monitoring for patient body condition and context. It can monitor biological signal, room temperature, light exposure, noise level, and person visited automatically and invisibly with a mobile device, such as a wrist phone. With this information, it can detect patients emergency status. The context awareness understands enough of a patients current health conditions and supports suitable services, resources, or information relevant to the situation. The central processing anticipates future healthcare status with collected data from patient to early identification of deviation from normal. In addition, it can order actual interface with decision on medical treatment or operation. Actual interface is treatment and control of interfacing device which performs the fastest actions form the emergency connection to emergency team or the notification to the family members. Additionally, a continuous and steady feedback control of attached therapeutic devices, warning sign or information feedback is including in actual interface.

The major research trend is concentrated on context awareness. It requires sensor technology and other fusion technologies, such as Nano and Bio technology.

2.2   Ubiquity and Situation Dependency

A concept of ubiquity comes from ubiquitous. Ubiquity allows users to access networks from anywhere at any time, and in turn, to be reachable at any place and any time [3]. In the ubiquity environment, users can receive situation dependent service.

Situation dependency is a concept related to the spatial, personal, and temporal context in which the user accesses a service. That is, the situation dependency determines the whole context in which a user accesses a service [4]. For example, it can be applied in mobile commerce. A cellular phone which is used only for communication between caller and receiver can be expanded to a ubiquitous tool which provides different application services depend on situation information of user. This evolution enables for the technique which can identify individual with unique phone number and wireless telecommunication network with mobility. It can be applied in the u-Healthcare service with EPC Sensor Network.

The situation dependency is quiet different than context-awareness. Context-awareness is strongly related to users preference and information; whereas, situation dependency is based on situation context which refer to location. Then, use of this information, situation dependency suggests appropriately service [4]. The location-based service is also included in situation dependency [6] [7] [8]. The situation dependency has ubiquity for handling information. For instance, if a patient faints at a moment, then this situation information is sent to the server to provide sustainable service.

3   EPC Sensor Network

3.1   Introduction to EPC Sensor Network

To enable the Internet of Things, Auto-ID center and EPCglobal designed EPC-global Architecture Framework which utilizes RFID tags inside the Internet. The framework is a collection of standardized efforts for web-based interface among logical Internet components: it includes various tags, readers, ALE (Application Level Event) middleware, EPC-IS (information sharing) and accessing applications [9]. However, EPC architecture framework can not support more functions of RFID tags because it is initially designed to save and read information for utilizing of object information only. To maximize the benefits from using various and smart RFID tags network architecture should take into account advanced functions which future RFID tags may have.

Sensor Network senses object and the sensing information communicate to other sensor node with ad hoc technology. Nevertheless, it has a limitation which is difficulty on sensing for enormous volume of objects and shares with users without incompatibility; because there is no de facto standard system of sensing mass objects with identification.

EPC Sensor Network is the new architecture framework which can support both EPC Network and Sensor Network properties. It solves problems and limitations of both. This architecture is designed on the top of current EPC architecture framework so that it can support most of current usages without breaking any interoperability of diversified RFID tags.

The most novelty of EPC Sensor Network is that ultimately it will support sensor network as well as other pure legacy RFID tags together. Sensor Networks are mainly used to monitor physical phenomena from environments. The typical objectives of the Sensor Network are effectiveness, relative to design goals, reliability, accuracy, flexibility, cost effectiveness and ease of deployment [10].

EPC sensor network is the first real infrastructure to support RFID and Sensor Network for global scale integration. The sensing data networks through the EPC network that has high interpretability. It described in figure 1 below.

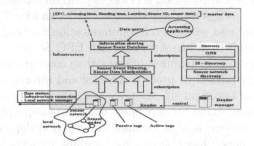

Fig. 1. EPC Sensor Network Architecture

By this model is introduced in figure 1, EPC Sensor Network has four properties: identification, networking, sensing, and communication. The four properties become value for EPC Sensor Network and make situation dependent application services in various areas.

3.2   Situation Dependency in EPC Sensor Network

The concept of situation dependency can be applied features specific to application areas of EPC Sensor Network; because, it is possible to identify object with EPC sensor tag through EPC Sensor Network. In addition, the status change gets sensed by the sensor tag. Communication among the EPC sensor tag is also possible for sensing variation of object context due to the characteristic of EPC Sensor Network. With advanced technology, mobile RFID which is mobile device containing RFID reader, can collect situation dependent information from the object. Like this, the situation dependency goes further by determining the whole context of EPC Sensor Tagging object [4].

Situation dependency may be conceived as a three dimensional space in case of EPC Sensor Network with object identity, sensing position, and sensing time as its axes. To realize situation-dependent services, four process steps are running sequentially as shown in figure 2. The service request comes from the user. The situation determination is obtained the three situation components such object identity, sensing position, and sensing time. The context computation generates further information. It means that identity, position, and time are known; and in the following computation, more semantic information is derived on the basis of additional information pools. These information pools provide data that can be accessed using the situation components as keys for data queries. Finally, the service is presented at the EPC Sensor Network [4]. With this sequence, EPC Sensor Network can collect and apply object status and context information at any time and any where to provide various services. At this moment, the object status information has dynamic information with sensing time and place.

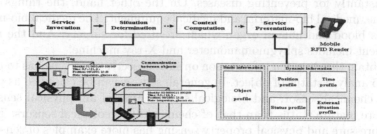

Fig. 2. The situation process of EPC Sensor Network

4 The Situation Dependent Application Areas of EPC Sensor Network in u-Healthcare

In u-Healthcare, there are objects which acquire situation information through EPC Sensor Network; and subjects which make use of the information. This paper reviews the objects and subjects in order to analyze application areas of EPC Sensor Network according to the situation.

In this regard, this paper considers what is situation information of the objects, which is provided by EPC Sensor Network and in some situations whether the information satisfies needs of users or not? Above all, this paper discusses the object in u-Healthcare service areas and situation information of object acquired by EPC Sensor Network. Then, the types of subjects and situations are derived from u-Healthcare service areas. Finally, this paper proposes the application areas based on situation information of the objects which could satisfy situation needs of these subjects. The figure 3 shows analysis framework of this section.

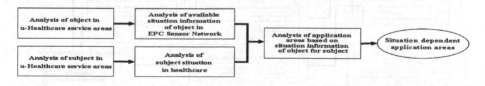

Fig. 3. The Analysis Framework of Situation Dependent Application Areas

4.1   Analysis of Situation Information of Objects in u-Healthcare

In order to analyze object situation information, deduction of objects from u-Healthcare service areas by EPC Sensor Network is required. The objects are mainly divided into human and thing. Firstly, human that includes healthcare service providers including medical doctor, nurse, and pharmacist. In addition, there are healthcare service customers who are patients receiving medical treatment according to a disease and healthy persons needed to check safeness in life constantly for preventing diseases. On the other hand, the things include medicine, medical bio-material and medical equipment. The medical bio-material includes blood and human organ which are used for surgery. And the medical equipment includes sphygmomanometer and X-ray machine.

To obtain the situation information on such object, sensor type has to be identified to analyze status of object. In general, sensors can be classified as physical sensor, chemical sensor, and biological sensor. First of all, physical sensors are in a more mature status than that of chemical and geological sensors. Temperature, pressure and physical property sensing has more examples of sensors and has a large industrial base. Secondly, chemical sensors detect and quantify chemicals that may present in gas, liquid, and solid phases. Finally, biological sensors are analytical devices that use biological molecules to detect other biological molecules or chemical substances. Typically, the detector molecule must be connected to a sensor that can be monitored by a computer which converts the biological response into an electrical and optical signal. The types of biological sensors include glucose sensor, urea sensor, and cholesterol sensor.

Among these sensors, biological sensor is the most essentially used in sensing changes of patient, medical, and bio-material status. As well as, the chemical sensor such as oxygen sensor is used to administer the patients environments. The physical sensor is used to recognize the positions of medicine and medical equipment.

EPC Sensor Network obtains both static and dynamic information. The static information is object identity data whereas the dynamic information is sensing place, sensing time, status of object, and external object situation information.

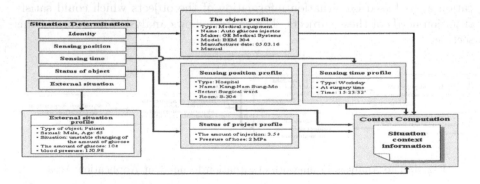

Fig. 4. Context Computation and Profiles with EPC Sensor Network in u-Healthcare

In the table 1, this paper arranges situation information of the objects according to u-Healthcare service areas. Here, u-service area is classified into inside of healthcare institution and outside of healthcare. The inside of healthcare includes hospital, healthcare center, and pharmacist. The outside of healthcare institution comprises home and office. As mentioned in table 1, the analysis of object situation information is possible through obtaining object identity, position, and time data.

Table 1. Available object situation information by EPC Sensor Network in u-Healthcare

Object / Service areas	Human		Thing		
	Healthcare service provider (Medical doctor/Nurse/ Pharmacist etc.)	Healthcare service customer (Patient/Healthy person)	Medicine	Medical bio-material	Medical equipment
Inside of healthcare institution (Hospital/ Healthcare center/ Pharmacy etc.)	*①Medical doctor/nurse situation information -Identity, Position, Time	*②Patient situation information -Identity, Position, Time -Status: Patient body (glucose, blood pressure, body heat etc.). Environment of patient (temperature, humidity, etc.)	*⑤Medicine situation information -Identity, Position, Time -Status: Environment of medicine (temperature, humidity, etc.)	*⑦Medical bio-material situation information -Identity, Position, Time -Status: Environment of bio-material (temperature, humidity, etc.)	*⑧Medical equipment situation information -Identity, Position, Time -Status: Level of X-rays radiation, pressure of ringer, height & slope of patient bed, etc.
Outside of healthcare institution (Home/ Office etc.)	-	*③Patient/ healthy person situation information -Identity, Position, Time -Status: Patient/healthy person body (glucose, blood pressure, body heat, etc.), Environment of patient (temperature, humidity, etc.) *④Emergency patient situation information -Identity, Position, Time -Status: Patient (glucose, blood pressure, body heat, etc.)	*⑥Medicine situation information -Identity, Position, Time -Status: Environment of medicine (temperature, humidity, etc.)	-	*⑨Medical equipment situation information -Identity, Position, Time -Status: Pressure of ringer, height & slope of patient bed etc.

4.2   Analysis of Subject Situation in u-Healthcare

In healthcare subjects which need situation information of objects of EPC Sensor Network comprise healthcare service provider and healthcare service. Additionally, as a communication of between objects through ad hoc technology, medical equipment can also be included in subject utilizing situation information of other objects. For analyzing application areas that are dependent on situation, the following situation is considered in which these subjects require situation information objects.

In healthcare, situations in which subjects are placed are classified in diagnosis and treatment. Healthy person receives a specialized diagnosis from healthcare service provider when developing symptoms of disease. At this time, the provider gets more information of status of patients health by using medical equipments and then infers accurate disease and the progress of the situation in comparison with symptoms of existing disease based on the information.

By referring patients characteristics such as cases of established treatment and status and history of disease, the healthcare service provider seeks treatment methods against the patients disease. Then the provider gives medical treatment including surgery and medication. These treatments are not performed just once; but undertaken constantly by making a continuous diagnosis of patients status and controlling medication type or quantity. Its sequence is shown in figure 5.

4.3   The Situation Dependent Application Areas of EPC Sensor Network in u-Healthcare

In this section, beyond the analyzed themes of the previous section, we offer application areas of situation information of objects. Firstly, in case of diagnosis situation of healthy person outside healthcare service institution, where hold medical equipment for health persons diagnosis; it has been difficult to grasp more clearly and rapidly symptom of disease. Thus, there are many cases where patients go to the healthcare service institution in spite of the irredeemable status of diagnosis. However, by attaching EPC Sensor tag to a healthy person, in a home or office, acquisition of information about status of health allows healthcare service provider to detect diagnosis in early stage.

A particular time of existing initial diagnosis are done in healthcare service institution could be advanced through EPC Sensor Network. Especially, taking accurate and fast treatment is very important for emergency patient. By attaching EPC Sensor tag for the where of the emergency patients accident to ambulance, sensed information of patient status is able to reduce the time required for taking initial diagnosis and enhance the accuracy.

Treatment tools such as medicine, medical biomaterial, and medical equipment can be managed more efficiently. Deterioration of treatment tools is prevented by sensing temperature and humidity; also the identification and tracking of treatment tools offer efficient management of medical equipments life cycle and anti-counterfeiting medicine. As a result, these applications increase safety of patients.

Information of medicine situation helps medication treatment of patients, automatically checking and alarming whether patient takes medicine. Moreover, an automated medical service, such as controlling medicine dosage according to

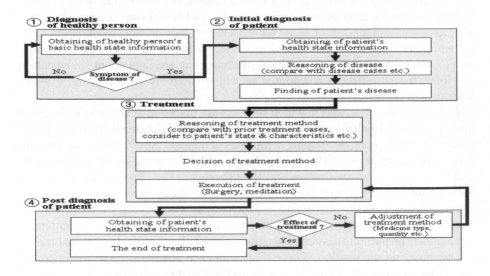

Fig. 5. Basic Subject Situation in Healthcare

the change of patient status could be served by communicating with each EPC Sensor tag attached to the medical equipment and patient.

In post diagnosis, ability to get information through EPC Sensor Network in home or office can reduce unnecessary hospital charges without hospitalization of patient for monitoring continuously. Also, in case of medicine dosage sensitive diagnosis such as melancholia and hyperpiesia, this ability provides efficient treatment adjustment without visiting a hospital.

Finally, based on the initial and post diagnosis of an infectious case, situation information enable medical doctors to track routes of an infectious disease. Besides these applications, according to the various situations, more applications are put in practice.

Table 2. Application areas of EPC Sensor Network in u-Healthcare

Subject / Service areas	Human		Thing
	Healthcare service provider (Medical doctor/Nurse/Pharmacist etc.)	Healthcare service customer (Patient/Healthy)	Medical equipment
Inside of healthcare institution (Hospital/ Healthcare center/ Pharmacy etc.)	* Tracking infected patient trace (2)-(4):①,② * Medicine, medical bio-material deterioration management based medicine, medical bio-material situation information (3):⑤,⑥,⑦ * Medical equipment life cycle management based medical equipment situation information (1)-(4):⑧,⑨ * Anti-counterfeiting of the purchasing medicine based medicine situation information (3):⑤	* Automatic check for medicine dosage (3):⑤,⑥	* Automatic fitness with dynamic situation information of patient (ex. intelligent radiator) (3):②,⑧
Outside of healthcare institution (Home/ Office etc.)	* Remote diagnosis based the patient/healthy person situation information in a home/office (1):③ᵃ * Preliminary diagnosis of the emergency patient situation information before arrival in hospital/healthcare center (2):④	* Patient/healthy person own basic diagnosis with own situation information (1),(4):③	* Automatic fitness with dynamic situation information of patient (ex. intelligent blood sugar injector) (3):③,⑨

ᵃ (1):③: In the diagnosis situation of healthy person at Fig. 5, the healthy person situation information at Table 1 is applied to remote diagnosis

5 Conclusion

This paper has considered application areas of EPC Sensor Network in u-Healthcare. EPC Sensor Network which has properties of both EPC Network and Sensor Network appropriately including identifying, networking, sensing, and communication is well-applied in u-Healthcare service which requires situation dependent information to provide a seamless medical treatment service.

Above of all, as information dependent on situation is acquired and provided application service relevant to more efficient and safety diagnosis or treatment will be served. This research suggests suitable application areas in needs of u-Healthcare to leverage the benefits of traditional healthcare service. Hereafter, application and effectiveness of cases of diseases put into practice together with technical research of EPC Sensor Network will be progressed for future study.

Acknowledgments. This work was supported in part by "Development of Sensor tag and Sensor node technology for RFID/USN" project of ETRI through IT Leading R&D Support Programs of MIC, Korea.

References

1. Fried, B.J., Johnson, J.A.: Human Resources in Healthcare: Managing for Success. AUPHA, pp. 19–40. Health Administration Press (2002)
2. Raghavendra, C.S., Sivalingam, K.M., Znati, T.F.: Wireless Sensor Networks, pp. 3–20. Kluwer Academic Publishers, Dordrecht (2004)
3. Junglas, I.A.: U-Commerce: An Experimental Investigation of Ubiquity and Uniqueness. Doctoral Dissertation, University of Georgia (2003)
4. Figge, S.: Situation-dependent services? a challenge for mobile network operators. Journal of Business Research 57, 1416–1422 (2004)
5. Hofer, T., Schwinger, W., Pichler, M., Leonhartsberger, G., Altmann, J.: Context-Awareness on Mobile Devices-the Hydrogen Approach. In: Proceedings of the 36th Hawaii International Conference on System Science (2002)
6. Kirlik, A., Strauss, R.: Situation awareness as judgement I: Statistical modeling and quantitative measurement. International Journal of Industrial Ergonomics 36, 463–474 (2006)
7. Yau, S.S., Wang, Y., Karim, F.: Development of Situation-Aware Application Software for Ubiquitous Computing Environment. In: Proceeding of the 26th Annual International Computer Software and Applications Conference (2002)
8. Figge, S.: Stefan Figge: Situation-dependent services-a challenge for mobile network operators. Journal of Business Research 57, 1416–1422 (2004)
9. Leong, K.S., Ng, M.L., Engels, D.W.: EPC Network Architecture. Auto-ID Labs Research Workshop (2004)
10. Akbar, A.A.: Pay-per-use Concept in Healthcare: A Grounded Theory Perspective. In: Proceedings of the 36th Hawaii International Conference on System Sciences (HICSS 2003) (2002)

Ubiquitous Healthcare System Using Context Information Based on the DOGF

Chang-Sun Shin[1], Dong-In Ahn[2], Dong-Seok Kim[2], and Su-Chong Joo[2]

[1] School of Information and Communication Engineering,
Sunchon National University, Korea
`csshin@sunchon.ac.kr`
[2] School of Electrical, Electronic and Information Engineering,
Wonkwang University, Korea
{`ahndong,loveacs,scjoo`}`@wonkwang.ac.kr`

Abstract. This paper proposed the Ubiquitous Healthcare System using context information for the personalized healthcare services in a home network environment. The context information is generated by location, health, and home environment information collected from sensors/devices equipped in home. This system is designed on the Distributed Object Group Framework (DOGF) supporting the functions which manage context information, applications and devices as one or more logical units in healthcare home environment. Especially, the system provides the continuous healthcare multimedia service by generating the context information based on resident's location through the Mobile Proxy and the Context Provider as components of the DOGF. For verifying the execution of our system, we implemented the seamless multimedia service, the prescription/advice service and the schedule notification/alarm service according to a healthcare scenario in home. Finally, we showed the execution results of healthcare home services by using service devices existed in the residential space.

1 Introduction

The healthcare services for human's well-being life have been provided by medical institutions like hospital. But, with the advance of medical technology as healthcare device and health diagnosis, and the appearance of ubiquitous computing, the ubiquitous healthcare service providing anytime and anywhere emerges as a new paradigm [1]. The sensors and the smart devices in ubiquitous computing environment recognize the context information from user's surrounding, and the applications provide the suitable context information service for user. Especially, it is possible that the healthcare home service in ubiquitous environment can lead well-being life by collecting and processing the information needed for healthcare via home network.

The personalized healthcare services in home must be provided without influencing the resident's location. The mobile proxy technology has been studied for supporting above mobility. The RAPP [2] and the MobiWare [3] based on

M.S. Szczuka et al. (Eds.): ICHIT 2006, LNAI 4413, pp. 447–457, 2007.

the OMG's CORBA are the representative researches for the mobile proxy. But, these researches are insufficient condition to provide the integrating management of sensors, devices and distributed resources and the context information service using collected information. Hence this paper proposes the Ubiquitous Healthcare System supporting personalized healthcare by using context information generated by location, health and home environment information collected from sensors/devices equipped in home. For verifying the system's executability, we simulate the healthcare home environment and show the execution results of healthcare context information services through the output devices as monitor or speaker installed in home. Our system targets a patient or an aged person and maintains a good health by checking the health condition, healthcare schedule, and prescription for oneself.

This paper is organized as follows; Section 2 describes the architecture and the functions of the mobile proxy and the DOGF as related works. Section 3 explains the components and the supporting services of the Ubiquitous Healthcare System. And Section 4 shows the execution results of healthcare home service using context information via healthcare application in our system. We conclude in Section 5 with summaries and future works.

2   Related Works

2.1   Mobile Proxy Technology

The resident in home moves the residential spaces as bedroom or living room frequently. The mobile proxy technology supporting the mobility service is necessary for anytime and anywhere healthcare service.

The RAPP observes the standard protocol of the OMG's CORBA and uses the CORBA adapter and the implemented object. The developing goal of the RAPP is to guarantee the QoS using the requested bandwidth by applying a proxy object in stream data and by implementing the specific stream filter. This research only uses the simple location information of device executing tasks, and not uses the user's location information. And it is insufficient to process task real-timely by adapting to user's environment. The MobiWare is based on the open programming methodology supporting the adaptive mobile service, and its control and management. But, due to focusing on the mobility support of host, this research has a restriction to manage the QoS of service devices without considering the context-awareness.

For solving the restrictions of the above researches, this paper adopts the location based mobile proxy technology supporting user's mobility.

2.2   Distributed Object Group Framework (DOGF)

The DOGF is a software architecture technology which can provide the distributed transparency among physical resources through the group management of distributed resources such as objects and sensors/devices and the construction of the logical distributed environment. And this framework consists of the group management and real-time supporting components. For group management, the DOGF consists of the Group Manager, the Security module, the Information Repository, and the Dynamic Binder. Also, the framework includes the Mobile Proxy and the Context provider supporting the mobility and the context information. For supporting real-time service, this DOGF consists of the Real-Time Manager and the Scheduler. The Group Manager executes the management functions like the group registration/withdrawal for distributed resources. The Security module is responsible for authentification and authorization for accessing the distributed application. The Dynamic Binder improving availability and reliability of distributed application including replication resources supports the binding to resource with the optimal executability when requesting service. The Real-Time Manager and the Scheduler are used to guarantee real-time according to the property of distributed application. The Mobile Proxy supports the continuity of service based on user's location. The Context Provider provides the context-awareness and reasoning technology from hardware devices/sensors. Detailed architecture and functions of the DOGF are referred to [4].

This paper proposes the healthcare context information system supporting the location based mobility by grouping sensors and devices that provide the context information in home through the interaction of the DOGF's components.

3   Ubiquitous Healthcare System Using Context Information

3.1   System Architecture

The Ubiquitous Healthcare System supports intelligent healthcare by collecting the various context information referring to resident's location in home. Figure 1 shows the architecture of the ubiquitous healthcare system based on the DOGF.

Our system is divided into 3 layers. The physical layer includes groups of healthcare sensors/devices. And the framework layer is responsible for communication between physical devices and application and provides resident with the optimal healthcare service by the group managing service. The DOGF is located on this layer. The DOGF supports the healthcare context information service by managing the healthcare resources as a logical group. The application layer includes distributed applications supporting healthcare home service. This paper studied the functionality and the supporting services of the DOGF connecting application with physical layer.

3.2   DOGF: A Ubiquitous Healthcare Framework

The Group Manager is responsible for group management of the distributed healthcare applications and the sensors/devices. The Security module executes authentification and authorization for accessing the healthcare resources. For improving availability and reliability of healthcare application, the Dynamic Binder supports the binding to resource with the optimal executability when requesting service. The Mobile Proxy guarantees the location based service execution by supporting the resident's mobility in home and provides continuous healthcare stream service. The Context Provider executes the context-awareness service using context information. The detailed properties and functions of the DOGF are shown in Figure 2.

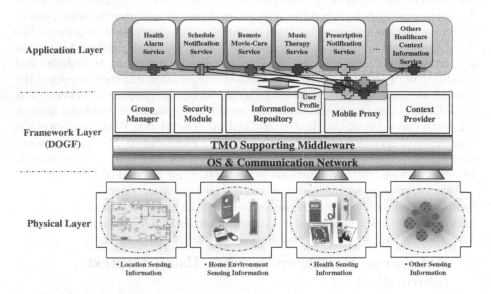

Fig. 1. Architecture of ubiquitous healthcare system based on the DOGF

3.3   Healthcare Supporting Service

The Ubiquitous Healthcare System supports not only the group and security service providing the personalized management of healthcare distributed resources, but also the healthcare context information service using the Mobile Proxy for the continuous stream service based on resident's location.

Group and Security Service of Healthcare Application.   The group and security service of healthcare application manages the healthcare application and the context information providing arbitrary residential space in home as a group. The procedures of this service are as follows. The client, healthcare interface in home, requests healthcare service to the DOGF. And the DOGF checks access right for the requested healthcare service group through the Group Manager, the Security module, and the Information Repository. Then, the framework returns a reference of healthcare service resource permitted to use by the client. After this, the client takes the appropriate service by connecting to the healthcare service group. Figure 3 shows the interaction of the framework's components for the group and security service.

Healthcare Context Information Service.   The healthcare context information service provides the moved space with the appropriate healthcare service

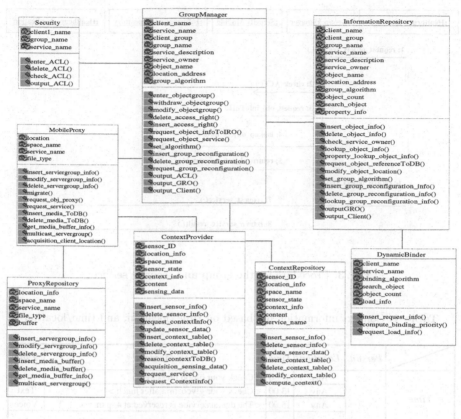

Fig. 2. Class diagram of the DOGF supporting healthcare

according to the change of resident's location in home. We show the W4H context information model which is the generating method of context information in our system and the event rule based on the model in Equations 1 and 2 [5].

$$Context(r_1\langle who\rangle, r_2\langle where\rangle, r_3\langle when\rangle, r_4\langle what\rangle, r_5\langle how\rangle) \qquad (1)$$

$$EventRule(r_1 \wedge r_2 \wedge r_3) \qquad (2)$$

This model generates the context information based on time and location using input parameters. The $\langle who\rangle$ means the resident's name. The $\langle where\rangle$ is the location information occurring the event, and the $\langle when\rangle$ is the event occurring time. The $\langle what\rangle$ is the provided context information. Finally, the $\langle how\rangle$ is the output type of service. That is, the context information is activated by the $EventRule(r_1 \wedge r_2 \wedge r_3)$ which a resident defines. Table 1 defined the context information list for executing the healthcare context information service referred to Equation 1 and Equation 2.

The healthcare notification/alarm service using text or voice is executed by applying context information based on the above time, location, and time/location

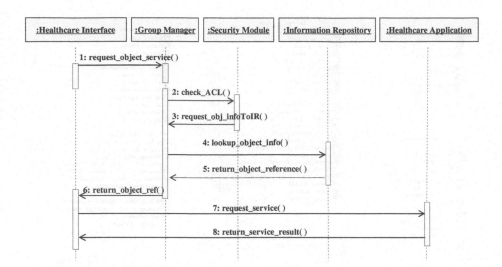

Fig. 3. Procedure for the group and security service

Table 1. Context information list based on time, location, and time/location

	Person	Location	Time	Message	Type
Time		Any	07:30	< Check your blood pressure >	Voice
		Any	13:00	< Check your glycosuria after lunch >	Text
		Any	15:00	< The dental service is reserved at 4 p.m. >	Voice
		...	...	...	...
Location	Any	Heater	Any	< Temperature of living room is 18 ℃ and humidity is 37% >	Text
		Bathroom	Any	< Water temperature 20 ℃, Appropriate temperature is 30 ℃ >	Text
		Refrigerator	Any	< Today is a fast day. You will examine an endoscope tomorrow >	Voice
		...	...	...	...
Time + Location		Kitchen	08:00	< Recommended menu is beanpaste soup and boiled rice >	Voice
		Kitchen	21:00	< Check gas valve. >	Voice
		Porch	22:00	< Temperature is very low. Stay-at-home >	Voice

to the event rule. For example, this service provides a resident with the health-care notification information with voice at the fixed time according to the event rule "EventRule(Any, Any, 07:30)" from the context information "Context(Any, Any, 07:30, ⟨Check your blood pressure⟩, Voice)".

Let us explain procedures of the healthcare context information service. First, when resident's location is changed, the healthcare interface existing in each space notifies resident's movement to the Group Manager. The Group Manager requests the Mobile Proxy to provide the continuous healthcare service to the resident by moving into the healthcare interface. After this, the Mobile Proxy requests the context information to the Context Provider. Then, the Context Provider transfers the context information created by the Equation 1 to the Mobile Proxy. According to the context information, the Mobile Proxy provides the appropriate healthcare service to the resident. Figure 4 is the executing procedures of the context information service.

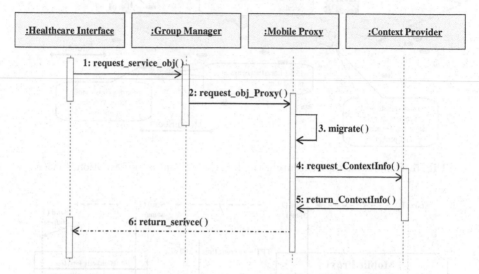

Fig. 4. Procedure for the healthcare context information service

4   Implementation of the Healthcare Context Information Service

4.1   Scenario for the Healthcare Context Information Service

For verifying the executability of the healthcare context information service, we show the healthcare services in simulation model according to the scenario like Figure 5. The healthcare context information service in this scenario provides with the prescription/advice and the schedule notification/alarm service, and the music therapy service. A resident moves among rooms. Monitors and speakers, healthcare interfaces, exist in each space.

4.2   Implementing Environment of Healthcare Service

For the healthcare context information service, we implement healthcare application on the DOGF. Figure 6 describes the interaction among the healthcare application components using the Mobile Proxy. With the implementing technology of application for processing multimedia stream, we use the Time-triggered Message-triggered Object (TMO), which is a real-time object scheme [6].

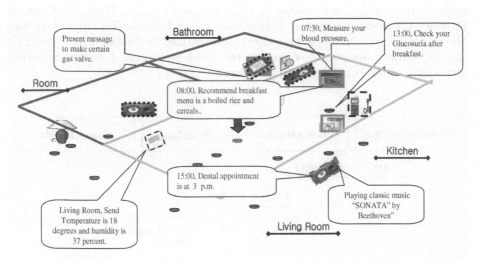

Fig. 5. Home model for executing the healthcare context information service

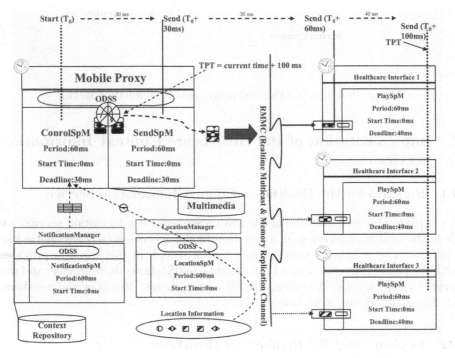

Fig. 6. Executing environment of healthcare application using the Mobile Proxy

The Mobile Proxy includes two Spontaneous Methods (SpMs) with periodic event characteristic. The Control_SpM received the context and the location information triggered by 60ms from the Notification Manager and the Location

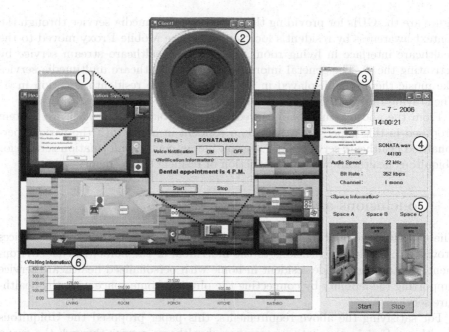

Fig. 7. GUI for the healthcare context information service

Manager. The Mobile Proxy moves to the healthcare interface equipped in the space where the resident is located. The Control_SpM transfers the context and the stream information by storing the queue buffer to the Send_SpM. The Send_SpM sends multimedia stream stored in the queue buffer to the healthcare interface. And the healthcare interface executes the schedule notification/alarm service and the prescription/advice service.

With the physical environment for the Ubiquitous Healthcare System proposed, we use 6 distributed systems. First, the healthcare context information system connected with the On/Off sensor and switch board collecting the location information of a resident is executed on one system. And we deploy the context information monitoring systems for the location tracking and the healthcare monitoring of a resident on two systems. Also, for executing the healthcare context information service, the healthcare interfaces to move the Mobile Proxy present on three systems.

4.3   Execution Results of the Healthcare Context Information Service

The moving location of a resident in this system is recognized by the received signal from the On/Off sensor and switch board equipped in home, and its result is expressed on the GUI. Figure 7 shows the monitoring GUI of the healthcare context information service provided to the resident. And, we can confirm that our system provides the resident with the defined healthcare service according to resident's location and time information. The ①, the ②, and the ③ in the figure are the GUIs for providing the healthcare multimedia service through the context-awareness by resident's location. Now, the Mobile Proxy moved to the healthcare interface in living room provides the healthcare stream service by activating the ②. The general information of the healthcare multimedia service like music therapy is displayed in ④. We can see the resident's location in ⑤. The staying time on each room is showed in ⑥.

From the above results, we showed that the Ubiquitous Healthcare System proposed in this paper could provide a resident in home with the various personalized healthcare services, and the continuous context information service for ubiquitous healthcare without restriction of resident's movement.

5 Conclusions and Future Works

The ubiquitous computing environment extends the existing healthcare services provided in medical institutions to the individual or the home. In this environment, we have to provide resident in home with personalized healthcare service supporting the mobility by connecting the physical home space with the healthcare service.

For satisfying the above requirements, this paper proposed the Ubiquitous Healthcare System supporting continuous healthcare service through the group management and the context-awareness of the various context based information by using the Mobile Proxy which is a component of the DOGF. Also, for verifying the executability of our system, we implemented the seamless multimedia service based on resident's location and the prescription/advice and the schedule notification/alarm service as healthcare applications in home. And we showed the executing results of healthcare home service by using service interfaces existed in the residential space according to the healthcare scenario. Hence, we confirmed that our proposed system could support the Ubiquitous healthcare by considering location, health, and home environment without consciousness of the resident.

In the future, we will study the automatic and adaptive healthcare service technology by adopting the intelligent context-awareness, and develop the healthcare integrating service system which provides users with anytime and anywhere healthcare in mobile environment.

Acknowledgements. This work was supported by the Korea Research Foundation Grant funded by the Korean Government (MOEHRD) (the Center for Healthcare Technology Development, Chonbuk National University, Jeonju 561-756, Republic of Korea).

References

1. Berler, A., Pavlopoulos, S., Koutsouris, D.: Design of an interoperability framework in an regional healthcare system. Proceedings of Engineering in Medicine and Biology Society 2, 3093–3096 (2004)

2. Seitz, J., Davies, N., Ebner, M., Friday, A.: A CORBA-based Proxy Architecture for Mobile Multimedia Applications. In: Proceedings of the 2nd IFIP/IEEE International Conference on Management of Multimedia Networks and Service(MMNS 1998), Versailles, France (1998)

3. Campbell, A.T.: QoS-aware Middleware for Mobile Multimedia Communications. Multimedia Tools and Applications 7(1/2), 67–82 (1998)

4. Shin, C.S, Kang, M.S., Jeong, C.W., Joo, S.C.: TMO-based Object Group Framework for Supporting Distributed Object Management and Real-Time Services. In: Zhou, X., Xu, M., Jähnichen, S., Cao, J. (eds.) APPT 2003. LNCS, vol. 2834, pp. 525–535. Springer, Heidelberg (2003)

5. Truong, K.N., Abowd, G.D., Brotherton, J.A.: Who, what, when, where, how: Design issues of capture & access applications. In: Proceedings of the International Conference on Ubiquitous Computing, pp. 209–224 (2001)

6. Kim, K.H., Liu, J.: Distributed Object-Oriented Real-Time Simulation of Ground Transportation Networks with the TMO Structuring Scheme. In: Proceedings of the IEEE CS 23nd International Computer Software & Applications Conferences, pp. 130–138 (1999)

7. Kopetz, H.: Real-Time Systems: Design Principles for Distributed Embedded Applications. Kluwer Academic Publishers, Boston (1997)

Load Balancing Using Dynamic Replication Scheme for the Distributed Object Group[*]

Romeo Mark A. Mateo[1], Marley Lee[2], and Jaewan Lee[1]

[1] School of Electronic and Information Engineering, Kunsan National University, 68 Miryong-dong, Kunsan, Chonbuk, 573-701, South Korea
{rmmateo,jwlee}@kunsan.ac.kr
[2] School of Electronic and Information Engineering, Chonbuk National University 68 664-44, DeokJin-dong, Jeonju, Chonbuk 561-756, South Korea
mrlee@chonbuk.ac.kr

Abstract. CORBA is the most widely used middleware for implementing distributed application. Currently, object implementations facilitate object group model to organize the object services. In addition to the complexity of designing the object groups, researchers also seek to improve the quality of service (QoS) by various means such as implementing load balancing to the system. This paper deals with the proposed load balancing service of distributed object groups. The proposed load balancing service uses the dynamic replication scheme which is mechanized by flow balance assumption (FBA) that derives from the arrival and service rate to execute request forwarding to new objects. The proposed on-demand replication scheme adjusts the number of replicated objects based on the arrival rate to minimize the waiting time of clients. It consists of procedures such as intercepting the request and executing on-demand activation of objects. The result of the simulation shows the improvement of the total mean client request completion time of the system as compared to other load balancing schemes.

1 Introduction

Replication is a key to achieve high availability and fault tolerance in distributed systems [1]. It is often achieved in CORBA-based distributed systems by using replication of services [2]. Replication is a technique for enhancing services and it is done by assigning the workloads to each replicated objects which makes the services more available. Fault tolerance is also achieved in replication by making a copy of the object so the data can be accessed at an alternative server if the default server fails or become unreachable. According to these necessities, we need to manage the replicated data by using object group models. Object group modeling is one concentration of research in recent studies on object services and various approaches are used to implement effective object groups. Improving the

---

[*] This research was supported by grant R01-2006-000-10147-0 from the Basic Research Program of the Korea Science and Engineering Foundation.

M.S. Szczuka et al. (Eds.): ICHIT 2006, LNAI 4413, pp. 458–468, 2007.
© Springer-Verlag Berlin Heidelberg 2007

membership schemes by binding other objects to a group is defined through IDL
for accessibility of the object services [3]. The management of object group is
critical for coordination of objects.

Load balancing is important to promote QoS in object group models. Recent
research studies are focusing on adaptive [4] and evolutionary [5] algorithms
but these are more complex to implement. Round robin and minimal dispersion
methods are commonly used for load balancing. However, the schemes are only
efficient on handling the queues for the requests but have no dynamic replication
of objects to support the large number of requests efficiently.

This paper proposes a load balancing service using dynamic replication scheme
for the distributed object groups. Dynamic replication scheme for load balancing
is based on the flow balance assumption that derives from the arrival and service
rate. The number of replicated objects is adaptive based on the arrival rate
of clients. These client's inter-arrivals is predicted based on the patterns from
the history to calculate the average of the next arrival periods and compare
to the threshold. The proposed load balancing service consists of intercepting
the request, on-demand activation of objects to replicate objects and executes
forwarding of request. The result of the simulation shows an improvement of the
total mean client request completion time of the system by implementing our
scheme.

2   Related Works

Implementing load balancing and fault tolerance based on replication schemes
to the distributed objects promotes scalability and dependability of the system. Othman introduced the adaptive on-demand load balancing [6] and tested
the performance of round robin and dispersion load balancing schemes. He also
carefully discussed the challenges of implementing the load balancing service and
illustrated the components which are necessary. However, the scheme is only efficient in handling queues for requests. There are also schemes that use threshold
value for load balancing policy [7]. A threshold-based load balancing policy is
based on a real-time data exchange mechanism between the home agents. The
load balancing is implemented to other available home agent. To implement an
efficient dynamic replication, the number of replicas must be adaptive to the
number of clients requesting for the objects in the system. Arrival rate scheme
can be used to calculate the number of replicas needed in a period of time by
using the flow balance assumption [11]. The main issue from the scheme is to
predict the next arrival period and calculate the average of the arrival periods
then compares it to the mean service time of the system. A neural network [8] is
used to predict the packet arrival rate. However, this approach of data training
and predicting consumes more time and it is more complicated to implement.
Our proposal for dynamic replication scheme is adaptive to the number of clients
requesting for the objects. The idea of threshold value to adjust the number of
replica and predicting the patterns from inter-arrival periods is presented more
details in Chapter 4.

3   Architecture of Load Balancing the Distributed Object Groups

The architecture of the load balancing the distributed object groups is shown in Figure 1. Heterogeneous hardware like desktop PC, notebooks and PDA are interconnected by a network and software components communicate through Object Group Services (OGS) by the Object Request Broker (ORB). The portable object adapter and portable interceptor are CORBA specifications defined in OMG's standards which is used as tools to implement our proposed load balancing service. Each OGS manages by group managers (GM) and consists of a load balancing service.

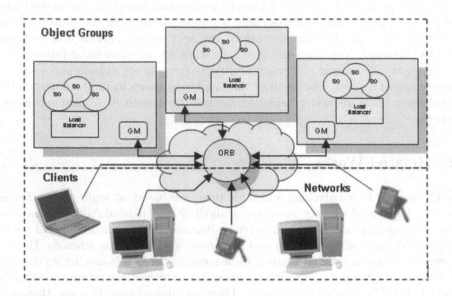

Fig. 1. Architecture of load balancing for distributed group objects

Figure 2 presents the interaction of the objects inside the group. This communication is managed by the group managers or GM. The information of objects is stored in the object information service. GM processes the request to the security service for authorization of clients to access the service objects. After the process, it directs the request to the appropriate service. GM has a function of mapping each group manager from the other groups by adding the collection of object services information to its database. GM also initializes the grouping of objects for efficiency of service. Replica manager (RM) manages the replication of object services to support load balancing and fault tolerance. RM communicates to the object factory to create additional objects. Specifically, RM replicates the healthcare expert service (HES) within the system in our proposal. The healthcare manager (HM) manages the invocation of clients to process the data and

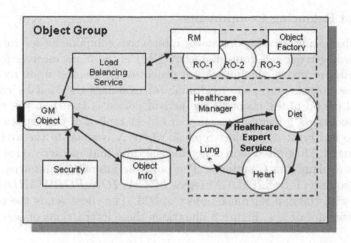

Fig. 2. Object groups for home healthcare service

returns a consultation result. The HES consists of collaborating object services to produce reliable information to customers or patients. In Figure 2 the services of lungs, heart and diet collaborates to process the data of clients. After processing, information is collected by the HM and return the result to the client. Our work also considers the QoS function from this service by providing a dynamic replication and load balancing in case of large number of request is processed in the system.

3.1 CORBA Specification Tools

Portable Object Adapters. Load balancing service features uses object replication to enhance the service on multiple requests of clients. The proposed architecture used the portable object adapter (POA) as basis of object replication from the object group. On-demand activation also uses POA. In case of replication to a different group or server, object agreements are required. The replication of objects from the original server considers the concurrence transparency. If the service object needed to change value after it has been requested by the client then the other replications of the object changes transparently. To be able to do this, the object must agree that it is a replica of the original object.

Portable Interceptor. Portable interceptors are one of the useful components from the CORBA specifications. A request interceptor is used to intercept the flow of a request or reply at specific interception points so that services can transfer to clients and servers. These can either reside on the client request or server invocation to monitor the request or provide an exception in case of failure of request. In this research, this is used to intercept the request and forward the request to the appropriate object replica and thus provide load balancing.

3.2   Load Balancing Components

The load balancing service consists of interactive components for intercepting and forwarding request. Group manager (GM) is the main receiver for client's request. After receiving, the interceptor processes the request upon receiving its arrival time. In this stage, the load analyzer examines the load by calculating the arrival period of the client. If the arrival period is less than the threshold, then the request is forwarded to another object replica of the replica manager (RM). The RM creates N object replicas where N is equal to the arrival rate's object demand. More discussion of getting the threshold and arrival rate's object demand is explained in Chapter 4. The portable interceptor informs GM that ForwardRequest() is occurred. GM issues $LOCATION\_FORWARD()$ reply to the client with the new reference object of RM. The client sends the request to the new reference object. Figure 3 illustrates these interactions of services.

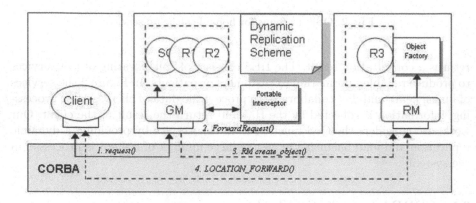

Fig. 3. Load balancing components interaction

Figure 3 visualizes the interaction of each component of object group to support the load balancing service. The communication of the object groups use the ORB core. RM is only informed for creating new replicas if the average arrival period is less than the threshold. This is done to minimize queues of request to the system and implement an efficient service. On the other hand, RM will terminate the created replica if the average arrival period is greater than the threshold and we consider it to be on normal state. The interaction of the components is also presented in Event Trace Diagram (ETD) for a clearer view of the interaction that is in Figure 4.

Figure 4 presents the ETD of the load balancing service. The procedures are grouped into two phases. The first phase consists of procedures from 1 to 8 which analyzes the current arrival time. After determining that the $t > Ap$ is true then the second phase is operational. This phase is consisted of procedures from 9 to 11 which forward the request. The deactivation of the object replica from the replica manager is done if the flow of the arrival rate is normalized for a defined period of time.

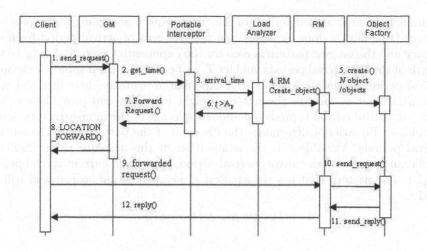

Fig. 4. Load balancing components interaction

4 Dynamic Replication Scheme

Our research implements the dynamic replication scheme using the round robin algorithm for assigning the request to the replicated objects. The number of the replicated objects to predict adds latency and it is so more suitable to use the round robin because it has lesser time to assign the request compared to dispersion scheme [6]. The scheme is based on the flow balance assumption which states that if the period is sufficiently long, the arrival rate is equal to the service rate [10], [11]. This is determine by calculating the number of arrivals observed as well as the number of completion ($\lambda = X$), where arrival rate is equal to arrival of clients over period of time ($\lambda = A/T$) and throughput rate is equal to total completion of request over period of time ($X = C/T$). Dynamic replication of objects and forwarding of request occurs if the threshold is greater than the average arrival period. Equation 1 illustrates the threshold which is determined by calculating the mean service time (μs) of the object replicas divided by the number of replica (r) in the system. After calculating the threshold, the average arrival period is calculated. A_p is the average value of all arrival period ($A_n - A_n - 1$) from 1 to n where n indicates the number of arrival period. A single arrival period is calculated by the difference of the current request's arrival time and the previous request's arrival time based on the n count. Equation 2 presents the average arrival periods.

$$t(s) = \frac{\mu s}{r} \tag{1}$$

$$A_p = \frac{\sum_1^n A_n - A_{n-1}}{n} \tag{2}$$

The arrival periods in Equation 2 is predicted to calculate its average for determining the number of object replica. The exponential average is calculated to predict the next arrival periods of the client request according to the previous history [12]. This is done by comparing the patterns of arrival period from the history and the nearest pattern is used for the exponential average. Let t_n be the length of the nth arrival periods, and let T_{n+1} be the predicted value for the next arrival period. Equation 3 defines the exponential average where $\alpha, 0 \le \alpha \le 1$. The parameter α controls the relative weight of recent and past history. The next nth arrival period is predicted and the average of the inter-arrival period is calculated. Equation 4 determines the distance of the pattern from the current arrival periods. Variable c is the summation of the absolute value from the Euclidian distance of the history arrival period (H_{AP}) and current arrival period (C_{AP}). The pattern that has the smallest c means it is the nearest and will be used.

$$T_{n+1} = \alpha t_n + (1 - \alpha)T_n \tag{3}$$

$$c = \sum_{1}^{m} \|H_{AP} - C_{AP}\| \tag{4}$$

The $t(s)$ and A_p is compared in Equation 5. If $t(s) > A_p$ is true then RM assigns the request to the new replica object. $LOCATION\_FORWARD()$ is sent back to client and informs to forward the request. This procedure assumes more available replica objects to serve the clients and removes the request from the queue. On the next arrival, $t(s)$ decreases the value because of the number of replica serves the system. Equation 5 presents the procedure of comparing the threshold and forwarding the request.

$$\text{if } t(s) > A_p \text{ then } R_m createobject() \text{ and } LocationForward() \tag{5}$$

The number of the new object replica is determined by dividing the threshold by the arrival period and get its approximate value. Equation 5 occurs in a certain period of time equal to the number of arrival period used in prediction. The number of created replicas is called the arrival rate's object demand. Equation 6 shows the calculation of the arrival rate's object demand. The maximum number of replicated objects that can be created is set by the max value. When the maximum number of requests is exceeded, any additional request are kept in the system queued.

$$N = t(s)/A_p \tag{6}$$

The system also assumes that in a certain period of time when Equation 5 is not true then the N replica objects that were created by the RM will be terminated. A default value of $t(s)$ is compared in Equation 5 to determine that the system has a normal rate of arrivals of request. Figure 5 shows the codes for portable interceptor. It receives and processes the arrival time. The analyzer method gets the previous value of the object service to compute the arrival period and threshold. Values are compared and it returns a forward if threshold is greater than the arrival period.

```
public void receive_request(ServerRequestInfo ri)
    throws ForwardRequest {
    analyzer(arrival_time(), assigned_object());
    if (Location_Forward == true) {
        throw new ForwardRequest(RM); //forward to RM object
    }
}
private void analyzer(int r, double ta, objectreplica[] obj) {
    double tp;
    while (n != 10) { //10 is a number of the history for prediction
        temptp = obj[r].arrivals(n) - obj[r].arrivals(n-1)
        tp = tp + temptp;
        n++;
    }
    tp = tp / n;
    if (threshold()>tp) {Location_Forward = true;}
}
private double threshold (objectreplica[] obj) {
    double ts = obj[r].meanservice()/r; //ts =  s/r
    return ts;
}
```

Fig. 5. Portable interceptor codes for dynamic replication scheme

5   Experimental Evaluation

The object group and the load balancing service were developed in Java Visibroker. The proposed load balancing service used the portable object adapter and portable interceptor. The simulation consisted of 18 clients where each client produces 10 loops of requests and 3 object replica to serve the request. The mean service time of each service was 30 ms. Each client had an arrival mean of 10 ms. In every arrival, the clients overlapped their arrival time and on this stage the arrival and service rate scheme was operational. We obtained the estimated simulation time by multiplying the number of clients and the mean service then the result was divided by the number of object replica. The expected simulation time will be approximately 18 second to process all requests ((180 * 0.30) / 3). The system also supported queuing of multiple requests. It produced maximum of three additional object replicas from the replica manager.

5.1   Performance Analysis

In measuring the performance, the loads of each replica in a given period of time are monitored. Othman used the comparison of different schemes in load balancing and implemented the load balancer service for TAO CORBA [6]. Different benefits of load distribution are observed by two algorithms. The round robin algorithm minimizes the latency of sending message but does not provide equal distribution of loads. The dispersion algorithm provides load normalization by transferring demand to the least loaded replica but the scheme has a high over-head latency. In our research, the round robin algorithm is implemented with dynamic replication scheme. The schemes are evaluated by adding all client request completion $(C(s))$ and divided by the number of clients (C).

$$C(s) = ST + TT(T_{server} + T_{client}) \tag{7}$$

Equation 7 shows the calculation of $C(s)$ where the transfer time of message request and service time are added. The transfer time is consisted of the time of transferring the request message from the client to the server and time of transferring the result message from the server to client. The mean client request completion $(\mu C(s))$ is given in Equation 8. The capacity of the system to process the request is measured by Equation 8. Processing all clients request in a lesser time means the scheme is more efficient. The efficiency of the scheme is determined by increasing the number of clients and calculates the $\mu C(s)$ on each additional load.

$$\mu C(s) = \frac{1}{C} \sum_{i=1}^{n} C(s) \tag{8}$$

5.2   Result

Figure 6 is the graphical result of the total free time and waiting time of on-demand replication scheme (ORS) and round-robin (RR) in discrete simulation. The free time refers to the time of service in idle or not serving clients and waiting time refers to the time where the request of client is in the queue. We used the same parameter values discussed in the later sections. In the simulation, ORS has 4 replica objects while RR has 3 replica objects. An inverse effect of the two variables that is observed in both algorithm. The ORS has approximately averaging of 9 seconds of waiting time which minimizes the waiting time of clients compared to the RR has an average of 20.4 seconds. Also ORS was more consistent on the value of waiting time compared to the RR where the highest peak of waiting time could reach until 30 seconds. The total mean client request completion (CRC) time of ORS is approximately 2.7 seconds only while the RR is 3.7 seconds in which lots of requests are queued on the system. The waiting time of each request is minimized on using the proposed algorithm. It is also assumed that the initialization of the replicated objects can have an overhead but because of the distribution of loads in the additional objects, it can serve a large number of clients.

The result of the graph in Figure 7 shows both algorithms with the same CRC for lower client loads of 3 or less. The switching time of each scheme have latency that is added to the actual service time of the request and is added at the CRC. At higher client loads, ORS outperforms the RR as indicated by lower CRC time of the system. The same latency is also caused by initialization of the replica manager but we can consider that it has less significant effect on the overall performance of the system.

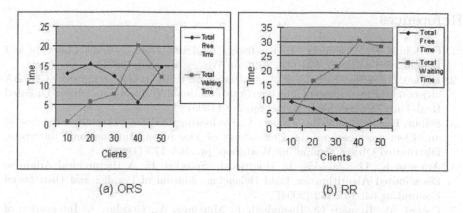

(a) ORS   (b) RR

Fig. 6. Free time vs. waiting time of clients over the number of clients

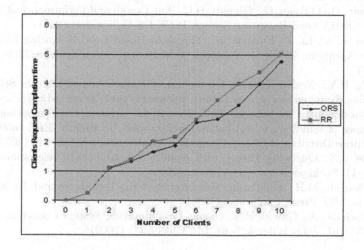

Fig. 7. Waiting time of clients over the simulation time

6   Conclusion

In this paper, a dynamic replication scheme for load balancing the distributed object groups is proposed. CORBA portable object adapter and portable interceptor specifications are used to design and implement the load balancing service component. The dynamic replication scheme based on flow balance assumption (FBA) is used. The number of replicated objects is adaptive based on the arrival rate of clients. These client's inter-arrivals is predicted based on the patterns from the history to calculate the average of the next arrival periods and compare to the threshold. The result of the simulation shows that ORS minimizes the total waiting time of clients and improves the completion time of clients' request by implementing our proposed scheme.

References

1. Coulouris, G., Dollimore, J., Kindberg, T.: Distributed Systems Concepts and Dessign, 3rd edn., pp. 553–606. Addison Wesley, Reading (2001)
2. Defago, D., Felber, P., Schiper, A.: Optimization techniques for replicating CORBA objects. In: Proceedings of The Fourth International Workshop on Object-Oriented Real-Time Dependable Systems, pp. 2–8 (1999)
3. Felber, P., Guerraoui, R., Schiper, A.: Evaluating CORBA portability: the case of an object group service. In: Proceedings of The Second International Enterprise Distributed Object Computing Workshop, pp. 164–173 (1998)
4. Antonis, K., Garofalaskis, J., Mourtos, I., Spirakis, P.: A Hierarchical Adaptive Dis-tributed Algorithm for Load Balancing. Journal of Parallel and Distributed Computing 64, 151–162 (2004)
5. Castro, M., Román, G., Buenabad, J., Martínez, A., Goddar, J.: Integration of Load Balancing into a Parallel Evolutionary Algorithm, International School and Symposium on Advanced Distributed Systems, pp. 219–230 (2004)
6. Othman, O., O'Ryan, C., Schmidt, D.C.: The Design and Performance of an Adaptive CORBA Load Balancing Service. IEEE DS Online 2(4) (2001)
7. Vasilache, A., Li, J., Kameda, H.: Threshold-Based Load Balancing for Multiple Home Agents in Mobile IP Networks. Telecommunication Systems 22(1-4), 11–31 (2003)
8. Zadeh, H.Y.: Neural Network Estimation of Packet Arrival Rate in Self-Similar Queuing Systems, www.ece.uci.edu/~hyousefi/publ/sssdm.pdf
9. Balasubramanian, J., Schmidt, D.C., Dowdy, L., Othman, O.: Evaluating the performance of middleware load balancing strategies. In: Eighth IEEE International Enterprise Distributed Object Computing Conference, pp. 135–146 (2004)
10. Daigle, J.N.: Queueing Theory with Applications to Packet Telecommunication, pp. 1–11. Springer Science, Heidelberg (2005)
11. MacDougall, M.H.: Simulating Computer Systems Techniques and Tools, pp. 16–17. The MIT Press, Cambridge (1983)
12. Silberschatz, A., Galvin, P.B., Gagne, G.: Operating System Concepts, 6th edn., pp. 158–161. John Wiley & Sons, Inc, Chichester (2002)

Design and Implementation of a Performance Analysis and Visualization Toolkit for Cluster Environments

Tien-Hsiung Weng[1], Hsiao-Hsi Wang[2], Tsung-Ying Wu[3], Ching-Hsien Hsu[4], and Kuan-Ching Li[1]

[1] Parallel and Distributed Processing Center
Dept. of Computer Science and Information Engineering
Providence University, Shalu, Taichung 43301, Taiwan
{thweng,kuancli}@pu.edu.tw

[2] Parallel and Distributed Processing Center
Dept. of Computer Science and Information Management
Providence University, Shalu, Taichung 43301, Taiwan
hhwang@pu.edu.tw

[3] Grid Operation Center
National Center for High Performance Computing
Taichung City, Taichung 40767, Taiwan
alex@nchc.org.tw

[4] Dept. of Computer Science and Information Engineering
Chung Hua University, Hsinchu 300, Taiwan
chh@chu.edu.tw

Abstract. The low cost and wide availability of PC-based clusters have made them excellent alternatives to supercomputing. However, while Network of Workstations are readily available, there is an increasing need for performance tools that support these computing platforms in order to achieve even higher performance. Strategies that may be considered toward such performance achievement we may list are: performance data analysis, algorithm design, parallel program restructuring, among others. Introduced in this paper is a toolkit that generates performance data and graphical charts of pure MPI, pure OpenMP, as well as hybrid MPI/OpenMP parallel applications, reflecting to its sequence of execution over time and cache behavior, with the use of DP*Graph representation, a parallel version of timing graph. That is, parallel applications have their execution sequence in a cluster system platform shown through graphical charts composed by sequential codes, parallel threads, dependencies and communication structures, symbols defined in DP*Graph. It is discussed the implementation of this toolkit, as also some of its features, together with experimental use of the toolkit on parallel applications such as matrix multiplication (parallel implementation using MPI) and SPICE3 (parallel implementation using OpenMP).

Keywords: OpenMP, MPI, Cluster Systems, Program Representation, SPICE3.

M.S. Szczuka et al. (Eds.): ICHIT 2006, LNAI 4413, pp. 469–479, 2007.
© Springer-Verlag Berlin Heidelberg 2007

1 Introduction

Advances in networking, high-end computers and middleware capabilities in recent years have resulted in a new computing paradigm: Networks of Workstations (NOW), or PC-based clusters. The potential of this computing infrastructure has attracted attention from computing industry, since this technology depends solely on commodity components. Furthermore, such platforms have been widely used to improve the performance of applications with intensive demands for computational power.

In merely a few years, computer clusters have become one of the most convenient and cost-effective tools for solving many complex computational problems, such as THE GRAND CHALLENGES. These problems are fundamental in science and engineering with broad scientific and economic impact, which solution can definitely be advanced with high-performance computing.

Its popularity is mainly due to scalability, ability to provide significant cost effective computing relying on commodity technology, and efficiently support both single processor interactive processing and large batch parallel processing. The PCs or workstations are typically interconnected through a high-speed network, such as Gigabit Ethernet, SCI, Myrinet, or Infiniband, and they run commodity and open source distribution operating systems, such as Linux.

In any performance methodology or software tool, high-level abstraction of an application plays an important role. Based on the distributed and shared memory programming paradigm, a class of timing graphs named DP*Graph (Distributed Processing Graph) [4,5,6] has been designed. The objective of this class of timing graphs is to describe the sequential and parallel computations and threads, as well as the communication and synchronization relationships of parallel applications.

In this research paper, we present the architecture design of a toolkit featured to provide application developers parallel timing graph representation of his parallel application, as also technical data such as cache miss. The former representation shows to application developer the sequence of execution of his parallel application, while latter information is provided solely when the computing system is a cluster of SMPs or cluster of multicore CPUs inside computing nodes.

The remainder of this paper is structured as follows. In section 2, motivation for the design of this toolkit, as also some backgrounds for this research. The timing graph DP*Graph is introduced in section 3, while in section 4 the implementation of this toolkit is briefly discussed. In section 5, experimental results using this toolkit are presented, and finally, some conclusions and future activities of this research in section 6.

2 Background and Motivation

Writing a parallel program is quite a difficult task, particularly if the programmer plans to efficiently describe a parallel algorithm. Additionally, the execution of a parallel program depends on several factors which interact in complex ways. Basically, a parallel algorithm can be defined as a sequence of local computations, interleaved with stores/loads, device I/O operations, and network communication steps, that is, all types of operations are allowed to be executed concurrently among themselves.

To identify and understand which would be the most effective structure to be chosen when writing a parallel program, it is important also to understand what factors affect the performance of this parallel program. That is, it is needed a representation model, where computation, network communication, and local I/O times are evaluated in orthogonal way.

Sequential graph-based representation and computation of maximum task execution time have widely been discussed, e.g., multi-level graph partitioning [2] and task graph scheduling [3]. Though, there is a need for parallel version of such timing graph representation. New representation terminologies for a parallel version of timing graph have been introduced in [4,5,6]. Basically, it is a class of timing graph symbols that can represent not only serial programs, but also parallel threads interleaved with communication and synchronization operations.

3   DP*Graph Representation

In distributed processing, synchronization and communication operations among computing nodes are fundamental operations, while in concurrent processing, the representation of parallel threads where parallelism occurs.

Figure 1 shows DP*Graph timing graph elements used to represent parallel programs [4,5]. On the left, figure 1-(A), it is shown components to represent distributed processing, as a segment of sequential code (1), the stream of execution, represented by the edge (4), node (5), and one-to-one communication primitives send (2) and receive (3), and all-to-all communication primitive, represented in (6). On the right side of this same figure 1, figure 1-(B), a representation given to the execution of multiple threads in a computing node, if this node contains multiple CPUs or multi-core CPUs capable of real execution of multiple threads in that computing node.

An instance of how parallel applications are represented is presented as follows. In Figure 2, it is shown a sample parallel code and a visualization representation of this parallel code using the DP*Graph. Note that we discuss the visualization for parallel applications instrumented with MPI only, since the visualization for parallel applications with OpenMP is quite simple, and similar to Fig. 1(B).

In figure 2, ▼n means that the computing node is receiving a message sent by computing node n, while ▲n means that computing node sends a message to processing node n.

The sample parallel code in figure 2 performs the following: all nodes calculates a simple vector sum in a loop of size 1000, and then, computing node 0 sends the data to computing node 1, while computing node 3 send data to computing node 2.

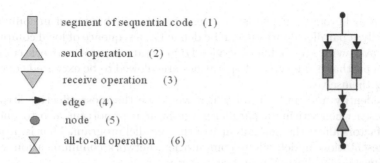

segment of sequential code   (1)

send operation   (2)

receive operation   (3)

⟶   edge   (4)

●   node   (5)

all-to-all operation   (6)

Fig. 1. DP*Graph elements

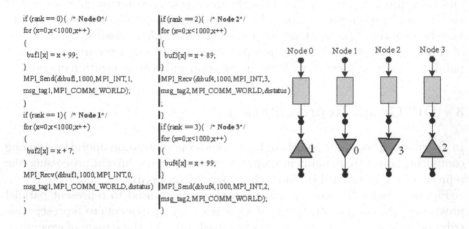

```
if (rank == 0){ /* Node 0 */
for (x=0;x<1000;x++)
{
  buf1[x] = x + 99;
}
MPI_Send(&buf1,1000,MPI_INT,1,
msg_tag1,MPI_COMM_WORLD);
}
if (rank == 1){ /* Node 1 */
for (x=0;x<1000;x++)
{
  buf2[x] = x + 7;
}
MPI_Recv(&buf1,1000,MPI_INT,0,
msg_tag1,MPI_COMM_WORLD,&status)
;
}
```

```
if (rank == 2){ /* Node 2 */
for (x=0;x<1000;x++)
{
  buf3[x] = x + 89;
}
MPI_Recv(&buf4,1000,MPI_INT,3,
msg_tag2,MPI_COMM_WORLD,&status)
;
}
if (rank == 3){ /* Node 3 */
for (x=0;x<1000;x++)
{
  buf4[x] = x + 99;
}
MPI_Send(&buf4,1000,MPI_INT,2,
msg_tag2,MPI_COMM_WORLD);
}
```

Fig. 2. Parallel code listing and DP*Graph representation

4   Proposed Toolkit Architecture

The input to the proposed toolkit is a parallel application that belongs to one of the following classes: sequential, pure MPI, pure OpenMP, hybrid (mixed with MPI and OpenMP constructions), while the output generated by this toolkit is DP*Graph parallel timing graph, in addition to performance data such as cache miss, in case the computing node involved in the computation either is a SMP or contains multi-core CPU. Figure 3 shows briefly our proposed toolkit.

We have started with the implementation of this toolkit using SUN Java J2SE (version 5.0) [10]. Figure 4 presents the execution steps including input and output to the toolkit.

As first step, the toolkit reads as input a sequential / parallel application into it. This application is then parsed, as also to capture the number of computing nodes involved in the computation. Later, not only is detected if the application is sequential or parallel, parallel and message passing constructions involved in the computations are identified and then saved into data file Rank-Status.dat.

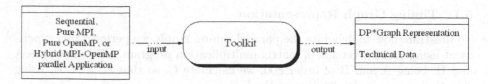

Fig. 3. Input and output of the proposed toolkit

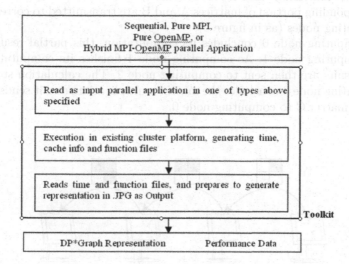

Fig. 4. Proposed Architecture of Performance Analysis and Visualization Toolkit

As next step, the application is compiled and executed in the specified computing nodes of a cluster computing platform, and all types of execution times are registered in data file CPU-time.dat. The times we keep in this data file are: execution of sequential code, data exchange and communication, and maximum time among concurrent execution of parallel threads (if occurs). As output, DP*Graph parallel timing graph is generated, based on performance data previously saved in files, in addition to a performance data table presenting cache miss results, if any of computing nodes involved in the computation is either a SMP or that particular computing contains a multicore CPU. Figure 4 provides visualization on the sequence of execution inside the toolkit.

5 Experiments and Preliminary Results

In this section, we present some experiments performed using our proposed research toolkit. In subsection 5.1, we show the execution process to generate a post-execution DP*Graph timing graph, while in subsection 5.2 we show performance data that are captured and their interpretation.

5.1   Timing Graph Representation

To illustrate how to generate the parallel timing graph, we perform this experiment using parallel version of matrix multiplication program. Given matrices A and B, being A and B of order 800, we calculate C, so that $C = AxB$, using standard matrix multiplication method.

The calculation using 4 computing nodes was performed in the following way:

- values of matrices A and B are initialized,
- corresponding portion of matrices A and B are transmitted to corresponding computing nodes (as in figure 5),
- as computing node 0 concludes its computation, this partial result is sent to computing node 1. As computing node 1 finishes its computation, partial results are then sent to computing node 2. The calculation stops when computing node 3 concludes the matrix multiplication, and it sends the final result matrix C to computing node 0.

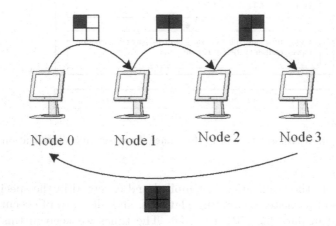

Fig. 5. Proposed Architecture of Performance Analysis and Visualization Toolkit

Since 4 computing nodes have been involved in the computation, there are 4 rows in our DP*Graph, ranked from bottom to top, starting from 0. This is easily observed in figure 6. Therefore, computing node 0's sequence of execution is presented in first row, next to X-axis. Sequential code executions are represented by a number of rectangles, while one-to-one communications are represented in a number of triangles. In fact, the parallel version o matrix multiplication program has a number of message passing send and receive constructions. In addition, the total execution time of the application is presented in X-axis, while Y-axis represent activities involved by each computing node.

The matrix multiplication program was written with the intention to leave other computing nodes idle, to make easier visualization of what in fact each computing node was doing during the global execution of a parallel application.

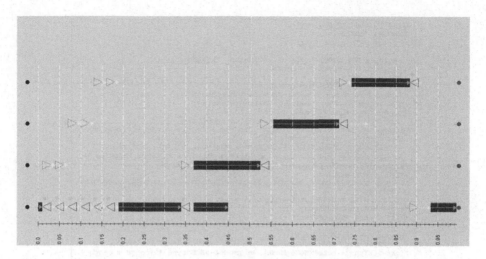

Fig. 6. DP*Graph representation for execution of Matrix Multiplication using 4 computing nodes

5.2   SPICE3 (Parallel Version of SPICE Instrumented with OpenMP)

SPICE3 is a general purpose circuit simulation program for DC, transient, linear AC, pole-zero, sensitivity, and noise analysis developed by UC Berkeley [9,10,11,12] and written in C. Several commercial codes are based on SPICE. It is used to simulate circuits for various applications from switching power supplies to SRAM cells and sense amplifiers. By doing so, it is required the simultaneous solution of a number of equations that capture the behavior of electrical/electronic circuits. The number of equations can be quite large for a modern electronic circuit with transistor counts from several hundred thousands to millions, and thus the simulation of circuits has become complex and quite time-consuming. Thus, a shared memory parallel program version is needed to achieve cost-effective performance.

Figure 7 reproduces a compact example code to demonstrate the SPICE3 simulation in its OpenMP implementation [13]. The SRAM circuit consists of many instances of the MOSFET devices with MOS (Meta Oxide Semiconductor) level 3 model. Therefore the time-consuming part of the original sequential routine was the MOS3load function, which is the device-loading routine in SPICE3. The actual size of the source code of the loop is approximately 1.3K LOC (Line of Code). The size of iteration is depending on the size of the circuit, the number of devices such as transistor, capacitor, etc. simulated may vary widely.

We run experiment based on different versions of SPICE3 circuit simulator programs simultaneously instead of running sequentially one by one on cluster of 2 CPUs inside computing nodes. The results of execution from all nodes are collected, computed, and plotted. Each version of the programs labels "Source", "Capload", "MOS3load", "Capload+MOS3load", and 3 versions

```
int MOS3load(inModel,ckt)
GENmodel *inModel;
register CKTcircuit *ckt;
{  ......
   register MOS3model *model = (MOS3model *) inModel;
   register MOS3instance *here;
   ......
   MOS3instance **MOS3instanceArray;
   MOS3instanceCount = model->MOS3instanceCount;
   MOS3instanceArray = model->MOS3instanceArray;
#pragma omp parallel default(none) shared(ckt,
   CONSTKoverQ, MOS3instanceCount, MOS3instanceArray)
#pragma omp for private(vt,Check, SenCond,EffectiveLength,DrainSatCur,
   SourceSatCur,GateSourceOverlapCap,GateDrainOverlapCap, \
   GateBulkOverlapCap,Beta,OxideCap,vgs,vds,vbs, vbd,vgb, vgd,xfact, \
   vgdo,delvbs,delvbd,delvgs,delvds,delvgd,cbhat,cdhat,tempv, cdrain, \
   capgs,capgd,capgb,von,evbs,evbd,vdsat,cdreq,xrev,xnrm,
   ceqbd,ceqbs,ceqgb, \
   ceq,geq,vgs1,vgd1,vgb1,arg,sarg, sargsw,error,gcgs,ceqgs,gcgd,ceqgd,gcgb
   ,model,here)
for( i = 0; i < MOS3instanceCount; i++) {
   here = MOS3instanceArray[i];
   model = here->MOS3modPtr;
   ......

#pragma omp critical(lockA)
{ // Right hand side of Ax = b
*(ckt->CKTrhs + here->MOS3gNode) -= (model->MOS3type * (ceqgs + ceqgb +
ceqgd));
   ......
// Sum of contributions for the element of matrix A
*(here->MOS3DdPtr) += (here->MOS3drainConductance);
*(here->MOS3GgPtr) += ((gcgd+gcgs+gcgb));
   ......
} /* end critical */
} /* end of for loop */
return(OK);
} /* end of MOS3load() */
```

Fig. 7. MOS3load function which is Part of SPICE3 OpenMP source code

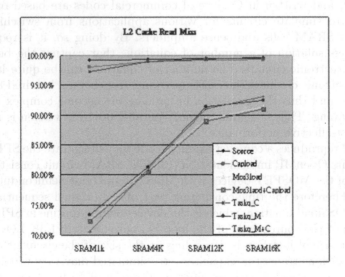

Fig. 8. Cache read miss information

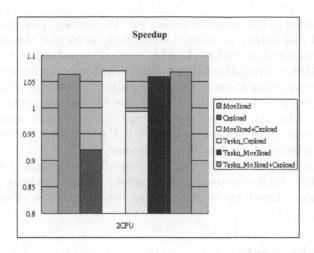

Fig. 9. Performance of SPICE3 application

label with beginning word *"Taskq"*. They correspond to original sequential program of SPICE3, parallelized *Capload()* function alone, parallelized *MOS3load()* function alone, parallelized both *Capload()* and *MOS3load()* functions using `omp parallel for` and synchronization directives, 3 versions each parallelized using `omp task queue` respectively. Each version takes input of simulation data of 1, 4, 8, and 16K SRAM based on MOS level 3 model simulation running on (each node) a DELL PowerEdge SC1420 Intel Xeon Processor 3.0GHz/2M, Em64T 800Mhz FSB *2 and 1GB DDR400 memory. Figure 8 and 9 show the L2 cache read miss and the performance speedup respectively.

6 Conclusions and Future Work

We showed in this paper the viability of implementing a toolkit that brings to developers parallel representation of his sequential or parallel applications under development, easing the hand-made load balancing process. Based on the representation and performance data the developer may obtain from this toolkit, he can perform "what-if" analysis on his application, as also performance tuning, with the goal to achieve higher performance.

Preliminary analysis and evaluation of performance results are obtained from the performance data of shared-memory parallel programs. It is crucial not only because to its cost-effectiveness in developing parallel application using OpenMP, but also to benefit the application developers for performance tuning of their applications. Again, this discussion makes sense only when the cluster platform contains SMP or multi-core CPU as computing node.

As future work, several directions of this research are ongoing. The first activity is to extend the implementation of DP*Graph to other message passing constructions, such as one-to-all and all-to-all constructions, since they have not yet been included in the proposed toolkit yet. In addition, race conditions, task splitting and dispatching in selected computing nodes will also be considered for implementation into toolkit. Another topic of investigation is to couple in this toolkit efficient scheduling policy toolkit, in order to distribute computations fairly among computing nodes in a cluster computing platform with efficacy, i.e., to detect if there presence of SMPs or multi-core CPU as computing node in cluster platform, and assign heavier threads computations to those computing nodes. One efficient scheduling method to be used is Parallel Loop Self-Scheduling, or general task scheduling for distributed systems.

Semi-automatic performance tool that captures performance data such as functions within their execution time, synchronization overheads, load imbalance, inefficient thread execution are also listed in future direction of this research.

Acknowledgements

This paper is based upon work supported by National Science Council (NSC), Taiwan, under grants no. NSC95-2745-E-126-002-URD, NSC95-2221-E-126-006-MY3, and NSC95-2218-E-007-025, and National Center for High-Performance Computing (NCHC), Taiwan. Any opinions, findings, and conclusions or recommendations expressed in this material are those of the authors and do not necessarily reflect the views of the NSC or NCHC.

References

1. El-Rewini, H., Lewis, T.G., Ali, H.H.: Task scheduling in parallel and distributed systems. Prentice-Hall, NJ, USA (1994)
2. Karypis, G., Kumar, V.: Analysis of multilevel graphs partitioning, Technical Report 98-037, University of Minnesota, USA (1998)
3. Kwok, Y.K., Ahmad, I.: Benchmarking and comparison of the task graph scheduling algorithms. Journal of Parallel and Distributed Computing 59, 381–422 (1999)
4. Li, K.-C., Gaudiot, J-L., Sato, L.M.: Performance measurement and prediction of parallel programs for NOW environments using P3MP. In: NPDPA 2002 IASTED International Conference on Networks, Parallel and Distributed Processing, and Applications, Tsukuba, Japan (2002)
5. Li, K.-C., Gaudiot, J-L., Sato, L.M.: Performance prediction methodology for parallel programs with MPI in NOW environments. In: Das, S.K., Bhattacharya, S. (eds.) IWDC 2002. LNCS, vol. 2571, Springer, Heidelberg (2002)
6. Li, K.-C., Chang, H.-C., Yang, C.-T., Sato, L.-M., Yang, C.-Y., Wu, Y.-Y., Liao, H.-K., Hsieh, M.-C., Tsai, C.-W., Pel, M.-Y.: On construction of a visualization toolkit for MPI parallel programs in cluster environments. In: AINA 2005 The 19th IEEE International Conference on Advanced Information Networking and Applications, Taipei, Taiwan, vol. II (2005)
7. Lumetta, S., Mainwaring, A.M., Culler, D.E.: Multi-Protocol active messages on a cluster of SMP's. In: Proceedings of Supercomputing 1997 The International Conference for High Performance Computing, Networking, Storage, and Analysis, San Jose, USA (1997)

8. Nagel, L.W.: SPICE2 - A Computer program to simulate semiconductor circuits, University of California at Berkeley, ERL. Memo ERL-M520 (May 1975)
9. Quarles, T.L.: Analysis of performance and convergence issues for circuit simulation, University of California at Berkeley, ERL. Memo ERL-M89 (April 1989)
10. Quarles, T.L.: Adding devices to SPICE3, University of California: Berkeley, UCB/ERL M89/45 (1989)
11. Vladimirescu, A., Liu, S.: The simulation of MOS integrated circuits using SPICE2, University of California: Berkeley, UCB/ERL M80/7 (1980)
12. Vladimirescu, A.: The Spice Book. John Wiley & Sons Inc, New York, NY (1994)
13. OpenMP Architecture Review Board, Fortran 2.0 and C/C++ 1.0 Specifications, http://www.openmp.org
14. SUN Java SE overview, https://java.sun.com/javase/index.jsp

Enterprise Application Framework for Constructing Secure RFID Application*

Hyundong Lee, Kiyeal Lee, and Mokdong Chung

Dept. of Computer Engineering, Pukyong National University,
599-1 Daeyeon-3Dong, Nam-Gu, Busan, 608-737, Korea
{win4class,zestgame}@hanmail.net, mdchung@pknu.ac.kr

Abstract. In the ubiquitous environment, anyone could easilyaccess all shared informations which means it also has many serious drawbacks, such as security problems. Therefore the ubiquitous environment should provide a security service. This paper suggests an Enterprise Application Framework(EAF) which includes a security module as well as a business process module for constructing secure RFID application. The security module includes user authentication mechanism, key sharing mechanism, and authorization mechanism. Thus, this framework is expected to provide more secure management in the ubiquitous environments such as RFID applications.

Keywords: RFID, Application Framework, RBAC, Security.

1 Introduction

Researches and developments for ubiquitous computing environment which is human centered computing paradigm are widely spread. One of them is the research and development of RFID(Radio Frequency Identification) technology.

In the ubiquitous environment, anyone could access easily to all shared information. It also, however, has many serious side effects, such as unauthorized information usage due to cracking, virus circulation, computer crime, privacy infringement, and copyright infringement[8]. Therefore, it should provide security service such as authentication, data protection and authorization to prevent these side effects [4,5].

The EPCglobal Network is a secure means to connect servers containing information related to items identified by EPC numbers. The servers, called EPC Information Services of EPCIS, are linked via a set of network services. Each participant in the EPCglobal Network will store relevant information related to specific EPC numbers in their own EPCIS servers. When user submits a query

* This work was supported by the Regional Research Centers Program (Research Center for Logistics Information Technology), granted by the Korean Ministry of Education & Human Resources Development.

M.S. Szczuka et al. (Eds.): ICHIT 2006, LNAI 4413, pp. 480–489, 2007.

to the EPCglobal Network, it will send the query to the registries, which would return the address of the various EPCIS containing the requested information [1].

This paper suggests an Enterprise Application Framework (EAF) which includes a security module as well as a business process module for constructing secure RFID application. The security module includes user authentication mechanism, key sharing mechanism, and authorization mechanism.

The structure of the paper is as follows. Section 2 describes related work, section 3 deals with Enterprise Application Framework (EAF) architecture, session 4 shows EAF security model for RFID application, session 5 suggests an application of Enterprise application scenario, and section 6 discusses the conclusion and the further work.

2   Related Work

2.1   EPCglobal ALE (Application Level Events)

EPCglobal network shows a new standard, called Application Level Events (ALE), which is developed from the concept of a middleware, called Savant. The role of the ALE is to provide means to process the event data which have been collected by the RFID reader and to deliver them to the higher-level applications [1,7].

On looking into the structure and components of the EPCglobal, RFID reader delivers identified tag data to the middleware, ALE Engine. Middleware is trying to filter out various overlapped tag data, and transmit accumulated/filtered tag data index to EPCIS or applications. The EPC capturing application stores received tag data and relating business information in the EPCIS and is ready to response to the hereafter query from the application.

ALE, a kind of an interface, defines API (Application Programming Interface) regarding on accumulation, filtering, counting and logging the transferred tag data from the RFID readers. Event cycle - the smallest unit of interaction between the ALE interface and application - delivers the EPC list, which has been collected during every event cycle to the application.

ALE interface provides API for defining Event cycle, which sets up the beginning and the end of Event cycle and determines the conditions of filtering and grouping, and reporting format. ALE application defines its own Event cycle, registers it through API, and receives the EPC index which might be used in an application through subscription/publication API.

2.2   CBD (Component-Based Development)

Software components are binary units of independent production, acquisition, and deployment that interact to form a functioning system. Composite systems composed of software components are called component software [6]. Abstractions, such as procedures, classes, modules, or even entire applications could form components, as long as they are in a 'binary' form that remains composable. The benefit of component software is as follows: components are the way to go because all other engineering disciplines introduced components as they became mature [6].

2.3 RBAC (Role-Based Access Control)

Access control is one of the most important aspects of information security. It has a great impact on integrity and confidentiality and also a considerable one on availability. A process is needed to support subject-based security policies, such as access based on competency, conflict-of-interest rules, or access based on a strict concept of least privilege. Supporting such policies without disregarding the organizational structure requires the ability to restrict access based on a user function or role within the enterprise. A solution to meet these needs was proposed in 1992 by Ferraiolo and Kuhn, integrating features of existing application-specific approaches in a generalized role-based access control (RBAC) model [2,3].

3 Enterprise Application Framework Architecture

3.1 Enterprise Application Framework

EAF is a framework for RFID application development consists of Data Manager for the communication with other systems and EPC event transmission, Security Manager for security issues, Business Event Manager for management EPC event and for management business process, and GUI for client.

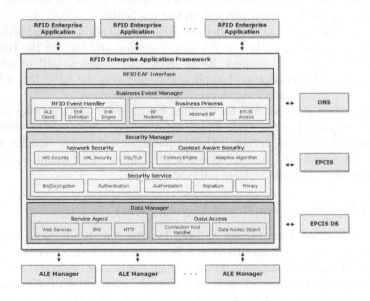

Fig. 1. Enterprise Application Framework

This allows clients to develop and use the RFID application based on ALE efficiently and easily. Also, it offers diverse communicating environments and various platforms to the standard, and can get the contextual information on EPC data which is related to EPCIS, ONS, and EPCIS DS. Figure 1 shows EAF overall structure.

3.2 Architecture of EAF

3.2.1 Data Manager
Data Manager offers function to access outer systems through database and web services. It includes Data Access module, which is in charge of input and output processing of data by using databases, and flexibility to the change of outer system by Web Service module's delegating the interaction between the outer systems. It also supports various networking protocols such as JMS and Socket. It delivers defined EPC event(ALE: ECSpec) to middleware through Even Definition of Event Manager, transmits it, in turn, to Event Handler of Event Manager in order to receive and process the result(ALE: ECReport), and has charge of communication with EPCIS or other legacy systems.

3.2.2 Security Manager
Security Manager provides appropriate authentication model to the distributed service environment connected to the network. In RFID platform, authentication mechanism to the entire platform, not the authentication to each system, is required; moreover, every service in the platform should be available by one authentication. Since the device computing capacity of data and distributing environment can be deficient due to the properties of the RFID platform environment, it is difficult to apply a complex computing algorithm, such as public-key encryption algorithm. Consequently, we utilize ID/PW-based security model currently, and later we will extend it to a lightweight PKI. In addition to this, flexibility and effective authentication, and authorization granting environment are provided by applying RBAC for the authorization and authentication.

3.2.3 Business Event Manager
Business Event Manager provides two modules, called RFID Event Handler and Business Process, which define the EPC event sent by ALE Middleware, and transform it into higher level business process which will be processed in the application.

Business Process module implements the businesses of application which includes business rules regarding on business logic, and processes business contexts regarding on granting appropriate contexts to the business processes. Business processes are defined by the flow of XML schema according to the requirement of user, and indicate logical events to represent meaningful expression in the applications.

3.2.4 The Characteristics of Enterprise Application Framework (EAF)
The Characteristics of EAF is that we can construct RFID application fast and efficiently with low cost through general, reusable, and extensible API. EAF utilizes Web Services, thus provides higher interoperability, and can apply to any sort of applications owing to the component based architecture.

4   EAF Security Model for RFID Application

4.1   Security Model

4.1.1   Security Manager

Security Manager is the component which is responsible for access control of external system (EPCIS) and RFID application service, user authentication, and data cryptography.

Various technologies like JCA, JCE, JSSE, JAAS, XML Security, and WS-Security are being used on this component. Security programmer, however, may have heavy burden to select appropriate techniques among all these various technologies. To solve this difficulty, EAF provides security API which has all the technologies mentioned above, and is designed to select and use them more easily. Thus, we can reduce the developing time and maintenance effort.

4.1.2   User Authentication and Authorization Mechanism

Table 1 shows the user authentication and Authorization protocol between user and AS (Authentication Server).

Table 1. User authentication and Authorization protocol

Notiations	
IDu	User Identifation
PWu	User password
Kas	Secret Key of the AS (Authentication Server)
EKas[M]	Encrypted Message M Using Kas
LKas	The span of life of the Message M
R	Authority Information to use Service
Detailed Protocol	
1: $U \rightarrow AS$: IDu || PWu	
2: $AS \rightarrow U$: IDu || PWu || EKas[M] || LKas	
3: $U \rightarrow AS$: IDu || EKas[M] || LKas	
4: $AS \rightarrow U$: IDu || R	

Authentication and Granting Authority consist of four steps. Transmission information and processing method in each step are as follows.

(1) Requesting User Authentication (U → AS): When user wants to use an external service, he or she should be authenticated by AS(Authentication Server) at first. User sends his or her ID and password to AS.

(2) Processing User Authentication (AS → U): AS sends an encrypted message and a span of life of the message to the user for granting authority after completing user's authentication. The encrypted data and the span of life of the message for granting authority is simple method to authenticate the user. Later we will replace ID/PW authentication by lightweight PKI which would be appropriate for the wireless environment.

(3) Requesting Service Authority (U → AS): When user needs to get his or her authority for service, he or she sends an encrypted message and a span of life of the message to AS.

(4) Transmitting Service Authority (AS → U): AS checks a validation of the received data through AS's secret key. If it conforms, AS sends an authorization information to user.

4.2   Design of Security Manager

4.2.1   User Authentication and Authorization

Authentication in the EAF is based on the ID/Password processing, and access control is based on RBAC. It decides AccessTarget, AccessGrant, and SecurityLevel during log-in procedure, and provides corresponding security service.

- AccessTarget: approaching external system (Manufacturer EPCIS, Distributor EPCIS, Retailer EPCIS, ONS, etc)
- AccessGrant: access authorization (Select, Insert, Update, Delete, etc)
- SecurityLevel: Security Algorithm (RSA, DSA, DES, etc)

Figure 2 shows the structure of database tables being utilized authorization and access control (RBAC).

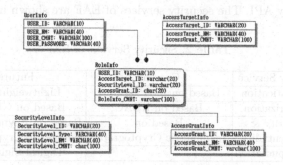

Fig. 2. The structure of database tables based on RBAC

4.2.2   Data Cryptography

To do the cryptographic process, we uses key agreement based on the Diffie-Hellman as a standard cryptography method for EAF. Also, it provides flexible design architecture for selecting other methods.

Figure 3 shows use case diagram for data encryption and decryption.

EAFKeys class is used to save a session that provides a secure communication between EAF application and the external object. EAFKey agreement is used to make a session key. Cryptography's classes using factory method pattern are used to use asymmetric key cryptography.

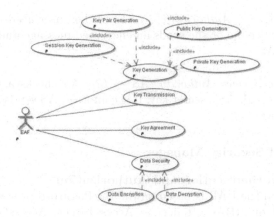

Fig. 3. Use-case diagram for data encryption and decryption

4.3   Characteristics of EAF Security Model

Security Model of EAF consists of ID/password-based authentication mechanism, key sharing mechanism and authorization mechanism, and provides authentication, granting authority (Security Model based on RBAC), integrity, and confidentiality (algorithm based on the public key).

EAF helps to reduce the developing time and the maintenance effort by suggesting security API. The security services of EAF are shown in Table 2.

Table 2. Security Services of EAF

Security Service	Current	Future
Authentication	Based on ID/Password	Lightweight PKI
Authorization	Based on RBAC	Based on GRBAC
Integrity & Confidentiality	Based on Asymmetric Cryptography	Based on Asymmetric & Symmetric Cryptography
Non-Repudiation	-	Digital Signature

Figure 4 shows EAF Security Service Sequence.

1. User create a member instance to register his or her ID/password information.
2. EAF application sends member instance to LoginAgent for login service.
3. LoginAgent sets up user's authority to member instance using security protocol
4. After login service, EAF application matches to the session key of destination that EAF application wants to communicate
5. EAF Application and the destination encrypt and decrypt data using session key that is matched in the fourth step

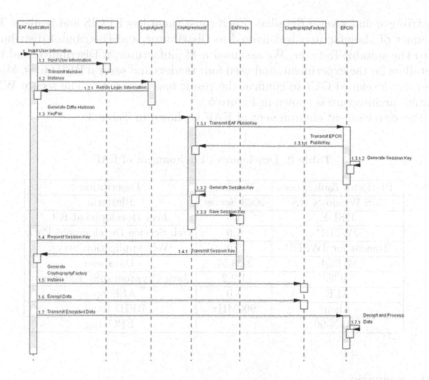

Fig. 4. EAF Security Service Sequence

5   Constructing Secure WMS System Using EAF

5.1   Implementation Environment

This section describes an RFID application based on EAF framework, secure WMS (Warehouse Management System), which consists of Manufacturer,

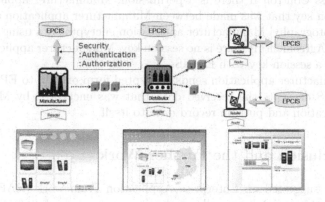

Fig. 5. Secure WMS System Architecture

Distributor and several Retailers. Each constituent has EPCIS and readers. The product of Manufacturer is delivered to Distributor, and Distributor distributes it to the suitable Retailer. We assumed a Manufacturer, a Distributor and two Retailers for the experiment, and used four readers and several RFID tags. Moreover we developed GUI to confirm the result based on EAF. The secure WMS system architecture is shown in Figure 5.

The development environment of EAF is shown in Table 3.

Table 3. Development Environment of EAF

Platform/Tools/Spec	Version	Description
MS Windows OS	2003 Server	Platform
J2SDK	1.5	Java Development Kit
JWSDP	1.6	Web Service Development Pack
Tomcat for JWSDP	5.0	Web Application Server
MySQL	3.23.57	Database
JCE	1.2.2	Java Cryptography Extension
ALE	1.0	ALE Spec
Alien	900MHz	RFID Reader
GTIN-96	-	EPC Tag

5.2   Scenario

1. [Authentication] User who wants to use Manufacturer application sends ID and password information.
2. Manufacturer application gets information on login and authority from databases.
3. User puts data of product on EPCIS.
3-1. [Access Control] If there is no permission to use EPCIS, putting data on EPCIS would be denied by Manufacturer application.
3-2. [Access Control] If there is a permission, Manufacturer application finds session key that was made between Manufacturer application and EPCIS.
4. [Cryptography] Manufacturer application encrypts data using session key.
4-1. [Key Agreement] If there is no session key, Manufacturer application would make a session key with EPCIS.
5. Manufacturer application sends encrypted form of data to EPCIS.
6. EPCIS decrypts the received data that was encrypted by Manufacturer application and puts on record data to itself.

6   Conclusion and the Further Work

This paper suggested an Enterprise Application Framework (EAF) which includes a security module as well as a business process module for constructing secure RFID application. The security module in EAF proposed suitable security requirements for RFID application environment to consider efficiency, and usefulness about the user and service provider of logistics. Also this paper suggested ID/PW and RBAC-based security model which satisfies these security requirements. We focus on user authentication mechanism, key sharing mechanism, and authorization mechanism in the suggesting model.

In the future, we will extend security model to the lightweight PKI which would be appropriate for the wireless environment. Also we need to accept GR-BAC and Context-aware technology to propose stronger security model.

References

1. EPCglobal, The Application Level Events (ALE) Specification, Version 1.0 (2005)
2. NIST, http://csrc.nist.gov/rbac/NIST
3. Sandhu, R.S., Coyne, E.J., Feinstein, H.L., Youman, C.E.: Role-based access control models. IEEE Computer 29(2), 38–47 (1996)
4. Securing RFID Data for the Supply Chain, http://www.verisign.com/epc
5. Stallings, W.: Cryptography and Network Security, Pearson Education, Inc. Prentice-Hall, Englewood Cliffs (2003)
6. Szyperski, C.: Component Software. Addison-Wesley, Reading (1998)
7. Verisign, http://www.verisign.com
8. Yoon, J., Chol, J.: Ubiquitous, 21th Books (2003)

A GDB-Based Real-Time Tracing Tool for Remote Debugging of SoC Programs*

Myeong-Chul Park[1], Young-Joo Kim[1], In-Geol Chun[2],
Seok-Wun Ha[1,**], and Yong-Kee Jun[1,**]

[1] Division of Computer Science, Gyeongsang National University
660-701 Jinju, South Korea
{africa,akates,swha,jun}@gnu.ac.kr
[2] Embedded S/W Research Division,
Electronics and Telecommunications Research Institute (ETRI)
305-700 Daejeon, South Korea
igchun@etri.re.kr

Abstract. Since embedded systems based on System-on-a-Chip(SoC) have limited resources, debugging programs in such systems requires a remote debugging system that has enough resources. However, existing JTAG based remote debugging system that uses GDB in Linux environment does not provide tracing function, so it is hard to monitor the executions of SoC program in real time. This paper adds a tracing facility to existing GDB remote debugging system to provide a real time monitoring tool. To demonstrate a real time tracing of synthetic program, Intel a Xscale PXA series processor based target system is used.

Keywords: SoC Program, GDB, Remote Debugging, JTAG, Real-time Tracing.

1 Introduction

Embedded system has been used in industrial control devices for long time and has been requiring enhanced functions due to compact and integration of microprocessor. In order to develop appropriate SoC(System-on-a-Chip) program [5,8,15,19], we need remote debugging host tool that is high in resources [3,7,9,17,18]. There are Remote type [3,11,14] and Non-remote [4,16] type in order to debug SoC program. Remote type has advantage of debugging in host system that is high in resources than non-remote type that is done in resource limited embedded system. Regular SoC program debugging uses remote type. Also supporting of multi equipment and tool compatible remote connection adapter that is essential in remote type is not sufficiently developed yet [10].

Existing GDB [13] remote debugging tool [1,2,9,12] in Linux environment configures *Breakpoint* to obtain registry or memory information in any point during

---

* This paper was supported by ETRI.

** In Gyeongsang National University, he is also involved in the Research Institute for Computer and Information Communication(RICIC).

M.S. Szczuka et al. (Eds.): ICHIT 2006, LNAI 4413, pp. 490–499, 2007.

program running and temporarily stops the program that enables interference or control of the program. But this is not appropriate in SoC program debugging that had time restrictions and also it is hard to visualize progress of the program. Also hardware emulator tools [1,2,9] are easy for debugging environment configuration but they require designated adapters resulting high cost for set up. Software emulator tools [12] are economical in set up but they have restriction of not having designated adapter.

In this study, we have added tracing function in existing GDB based tool [12] to propose real time tracing tool *EKDebugger* that enables logging of program status. For remote debugging environment low cost adapter was used [10]. For host system, *Tracepoint Server* that is tracing function added GDB based remote debugging tool *Jelie* [12]. For target system, *Debug Handler* that uses *Hot-Debug* [7] solution based in JTAG(Joint Test Action Group) [6] through mini-IC to debug Intel Xscale PXA core is used. *Tracepoint Server* has three main modules. First is *Event Parser* that interprets commands from GDB and converses into remote debugging symbols. Second is *Tracing Preprocessor* that delivers SoC address information using *Testcase* file that is delivered from GDB and Tracepoint configuration information. Lastly, *Trace Logger* that logs trace information by address information. GDB and *Tracepoint Server* in connected through RSP(Remote Serial Protocal) that is TCP/IP base and the environment is configured with USB from host system and JTAG interface from target system. Proposed tool can be adapted to remote visualization of SoC program and monitoring tool development and also effective in establishing economical debugging environment. We used PXA series [8,9] processor based target system for testing in this study to demonstrate tracing of synthetic program in real time.

In chapter 2, existing real time remote debugging tool is introduced using GDB for SoC program as a background of this study. In chapter 3, real time tracing tool that is main point in this study is introduced and then testing environment and real time remote debugging test are introduced. Lastly in chapter 4, we proposed conclusion and future assignments.

2 Background

In this chapter, SoC program debugging techniques are introduced. Also we will examine problems of existing GDB remote debugging in Linux base using JTAG interface. For Soc program, it is inefficient to use debugger in target system that is limited in resources. So there are debugging methods that are host based, On-Chip based or remote based method.

First, host based debugging method is conducted in host system using target systems command group simulator when target system that Soc program will run is not developed. It is easy to develop command group simulator but it has problem of inaccurate forecast of program run time.

Second, On-Chip based method uses debug kernel that is provided as a part of thee chip to monitor system resources by transferring delivering information to

Table 1. Comparison of remote debugging tools

Debugger	Emulator	Adapter	Tracepoint
Aiji GDB[2]	OPENice32-Axxx	O	X
dbi GDB[1]	BDI2000	O	X
winIDEA[9]	IC Series	O	X
Jelie GDB[12]	Jelie	X	X

execute control host or designated emulator when processor cannot communicate with other parts of the system. Three standard interface protocols in debug kernel are BDM(Background Debug Mode), IEEE 1140.1 JTAG, and IEEE-50001 ISTO(Nexus).

Third, remote debugging method uses debugger that is distributed to target system and host system due to limited resources of target system. The communications are done through serial or Ethernet channel. There are hardware emulator based and software emulator based for remote debugging method. [1] uses BDI series interface based in GDB using JTAG and Ethernet connection of target system. [2] that uses OPENice series interface has special feature that can communicate with USB of host system. But all these hardware emulator based methods requires high cost debugging environment. The most common tool for software emulator is [12]. It uses software emulator *Jelie* to provide environment to communicate with target system JTAG. Also internal configuration to interface host system parallel port and USB is available. It has advantage of economical debugging environment but real time SoC debugging is unavailable due to deprived tracking function. In Table 1 comparison of existing debugging tools are shown. There are two methods for existing tool that supports remote debugging using JTAG interface of target system based on GDB in Linux environment for information exchange between target system and host system. One is using designated emulator and the other is using adapter that can only provides signal conversion between target and host system. Tool to use designated emulator has disadvantage of high cost on debugging environment. And tool to use adapters like *Jelie* has advantage of software only debugging but it deprives designated adapter. Also existing GDB based remote debugging tool has only *Breakpoint* function. It is hard to monitor logical procedures of program and changed memory value of variables in *Breakpoint* function. On the other hand, *Tracepoint* provides logging function of process information in real time without controls or interferences. Existing tools do not support this function.

3   A Real-Time Tracing Tool

This chapter introduces *Tracepoint engine* design using remote debugging tool *Jelie* based in GDB of existing adapter [10]. This engine has added process routine and *Tracepoint* related commands in GDB and *Jelie*. Related commands to *Breakpoint* can use without modification in GDB. First, described about design

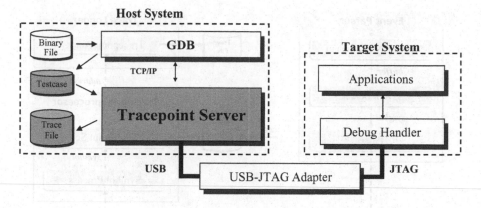

Fig. 1. A structure of EKDebugger

of tracing tool. This tool is GDB-based real-time tracing tool for remote debugging of SoC programs. Second, we also explain for implementation of tool using GDB and *Jelie*. Finally, we verify the result by experiment of remote debugging with this tool.

3.1   Tool Design

Real time tracing tool *EKDebugger* uses USB-JTAG signal conversion adapter to communicate between host system and target system in Linux environment. The structure is shown in Fig. 1.

In Fig. 1 GDB receives user specified binary program and produces in *Testcase* file type. *Testcase* file contains monitoring information of compiled binary file in debugging mode. GDB is connected to *Tracepoint Server* RSP based on TCP/IP. Server transfers *Tracepoint* related command processes and results to GDB. Also creates files that has log of trace information created by tracing engine in *Debugging Server*. Target system is composed of *Debug Handler* and *Applications*. *Debug Handler* analyzes commands from host system and controls SoC program in target system to transfer process status information to host system. *Applications* are SoC program that are transferred from host system in binary format. *Tracepoint Server* is composed of *Buffer Handler*, *Tracing Engine* and *Port Controller*. *Buffer handler* is compose of *Socket Handler* that opens TCP/IP port and stands by for input from port developers, *Character Reader* that saves received information in variables and delivers to *Input Handler*, and *Input Handler* that saves transferred information in history and transfers to *Tracing Engine*.

In this study, the most essential *Tracing Engine* module is configured in Fig. 2. First of all, translate commands from GDB and transfers in buffer(*Debug-Mode Select*). Event Select uses contents in buffer to analyze event to transfer appropriate symbol. And grant meanings on symbol to transfer to *Symbol Process*

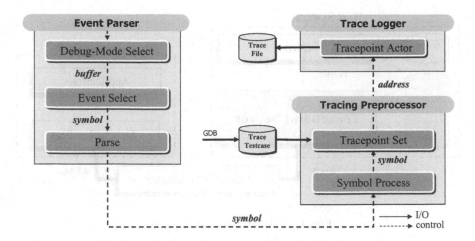

Fig. 2. Design of Tracing Engine

(*Parse*). It is a protocol between GDB and *EKDebugger* in host system in order to conduct *Tracepoint* function.

Symbol Process processes symbols and delivers *Tracepoint Set* for *Tracepoint* configuration. It uses already created *Testcase* file to transfer address within SoC to *Tracepoint Actor* to configure *Tracepoint*. It uses this address information to log corresponding trace information to trace file.

Port Control module related to adapter is composed of *Reset Routine* that initializes USB port, *JTAG Register Routine* that handles approaching to JTAG of target system, and *USB Port Routine* that conducts actual signal transmitting.

3.2 Tool Implementation

The real time tracing tool introduced in this study has added process routine and *Tracepoint* related commands for GDB and *Jelie*. Added commands are related to creation of *Testcase* in *Tracepoint* configured point, mode setting/modification to converse to *Tracepoint* function, and *Tracepoint* execution/stop. *Tracepoint* related command routine that is modified *Breakpoint* function that supports source level debugging method was added to *Jelie*.

Fig. 3 is proposed tool showing "parseValidPacket()" that is function to process symbols. These are functions to correspond symbols that are separated from *Event Parser* to remote debugging commands. Symbols starting with 'Q' calls for "Prepare_ tracepoint()" to create tracing file that is *Tracepoint* initial process and also adds newly configured *Tracepoint*. Symbols starting with 'c' calls for "cont()" to compare address of *Program counter* with program configured *Breakpoint* or *Tracepoint* address. *Breakpoint* and *Tracepoint* use same *Continue* command but executions are different. Execution of *Tracepoint* is logging (without stopping the execution) variables or memory values in real time

```
void GdbRemote::parseValidPacket(){
    int startOfPacket = ptr;
    switch(current()) {
        case 'H': setThread(); break;
        case 'd': remote_debug = !remote_debug; break;
        case 'q': query(); break;
        ...
        case 'M': writeMem(); break;
        case 'c': cont(); break;
        case 'Q': prepare_tracepoint(); break;
        case 'T': debug_tracemode(); break;
        default : printf("command %s not understand\n",
                    current());
                    message[0] = 0;
        ...
```

Fig. 3. parseValidPacket Function

```
#include <stdio.h>              @trfunc_1_pre
int temp;                       LOCAL_TRACE n  fffffff0 4
int main(void) {                LOCAL_TRACE f1 ffffffec 4
    int n=64, f1=0, f2=1, f3;   LOCAL_TRACE f2 ffffffe8 4
    int i;                      LOCAL_TRACE f3 ffffffe4 4
    for(i=0;i<n-2;i++) {        LOCAL_TRACE i  ffffffe0 4
        f3 = f1 + f2;           @end
        f1 = f2;
        f2 = f3;
    }
    return 0;
}
```

Fig. 4. Synthetic program and Testcase file

when configured memory value of *Tracepoint* is same with memory value of SoC program on same location.

Fig. 4 shows *Testcase* file that is created by user synthetic program GDB for Fibonacci sequence. *@trfunc_1_pre* indicates order of *Tracepoint*. There are four types of trace type *LOCAL_TRACE*, *GLOBAL_TRACE*, *REGISTER_TRACE* and *MSGTRACE*. This time *Testcase* file of 5 variables that are *LOCAL_TRACE* are created. Also address and size of each variable is indicated. *Debug Handler* used handler module provided by *Jelie* using *Hot-Debug* solution for Intel Xscale. Fig. 5 shows structure of *Debug Handle*. *WaitForCommand* method module analyzes commands from host system to call corresponding module. When certain module is operated, command inputs are always lined up in this module. Command methods are *SetBreakpoint* to configure *Breakpoint*, *Continue* to continue program execution, *SendToHost* to send information to JTAG to deliver to host

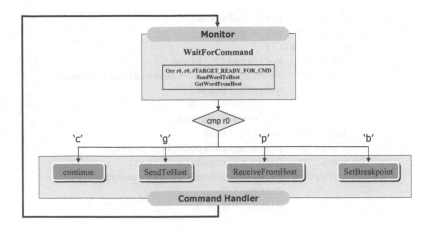

Fig. 5. Debug Handler

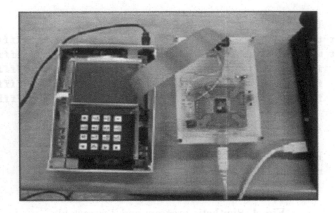

Fig. 6. Environment of Tracing Tool

system, and *ReceiveFromHost* method to download data to memory of target system from host system.

3.3 Tool Experimentation

Environment to test the tool is composed of kernel version 2.4.18 Linux Red-Hat 8.0 host system and Xscale PXA255 Palm Tynux Box for target system. For connection between host system and target system AN2131QC Chip from Cypress is used and adapter [10] to connect USB port of host system and JTAG port of target system is used. Fig. 6 shows the overall environment.

Fig. 7 is execution screen of *EKDebugger* showing RSP connection base on TCP/IP and GDB after successful connection with target system. In Fig.7, first rectangle box(**1**) identifies type of the target by recognizing *IDCODE* of target

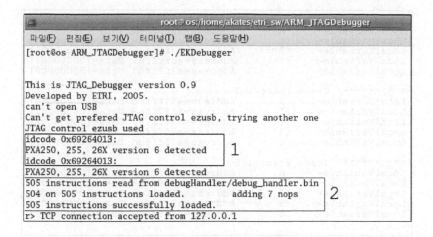

Fig. 7. The screen for a normal operation confirmation

```
0x000000f4 in ?? ()
(gdb) load fibo
Loading section .text, size 0x100 lma 0xa0000000
Start address 0xa0000000, load size 256
Transfer rate: 2048 bits/sec, 128 bytes/write.
(gdb) break main
Breakpoint 1 at 0xa0000038: file fibo.c, line 8.
(gdb) trace 33
Tracepoint 1 with trfunc_1_pre, @(trfunc_1,i) at 0xa00000d0: file fibo.c, line 33.
(gdb) action 1
Enter actions for tracepoint 1, one per line.
End with a line saying just "end".
> collect $local
> end
(gdb) trmake -c testcase
mv: cannot stat `testcase_trfunc.cpp': 그런 파일이나 디렉토리가 없음
echo "tracepoint file"> testcase_trfunc
trmake...O.K.
(gdb) trsetmode TRACE
(gdb) tstart
sendingO.K.
(gdb) break 37
Breakpoint 2 at 0xa00000f0: file fibo.c, line 37.
(gdb) continue
Continuing.

Breakpoint 1, main () at fibo.c:8
8          int f1 = 0;
(gdb) ▮
```

Fig. 8. Result of operation by specifications

system. In second rectangle box(**2**), we can confirm downloaded *Debug Handler* in target *mini-IC*. Fig. 8 shows actual testing screen.

Fig. 9 shows traced Logger File by using Testcase file from SoC program and GDB in target system. After determination of Trace File created by above process, it is confirmed that it is possible to debug SoC program running in target system in real time.

```
*********** Trace Point : 1
LocalVariable=      n        Address=a3ffffed Data=[00000064]
LocalVariable=      f1       Address=a3ffffe9 Data=[00000001]
LocalVariable=      f2       Address=a3ffffe5 Data=[00000001]
LocalVariable=      f3       Address=a3ffffe1 Data=[00000001]
LocalVariable=      i        Address=a3ffffdd Data=[00000000]

*********** Trace Point : 1
LocalVariable=      n        Address=a3ffffed Data=[00000064]
LocalVariable=      f1       Address=a3ffffe9 Data=[00000001]
LocalVariable=      f2       Address=a3ffffe5 Data=[00000002]
LocalVariable=      f3       Address=a3ffffe1 Data=[00000002]
LocalVariable=      i        Address=a3ffffdd Data=[00000001]

*********** Trace Point : 1
LocalVariable=      n        Address=a3ffffed Data=[00000064]
LocalVariable=      f1       Address=a3ffffe9 Data=[00000002]
LocalVariable=      f2       Address=a3ffffe5 Data=[00000003]
LocalVariable=      f3       Address=a3ffffe1 Data=[00000003]
LocalVariable=      i        Address=a3ffffdd Data=[00000002]
. . .
```

Fig. 9. Verification of Trace file

4 Conclusion

In this study a real time tracing tool is introduced that enables logging of SoC status on execution of specific command. This tool uses *Tracepoint* to trace execution information in real time to monitor execution progress. Also GDB is used for low cost debugging tool and used existing graphic user interface that supports GDB without modification. This tool can be used in visualization of program execution progress or development of monitoring tool by using program execution information. But this tool is limited to Intel PXA series for target system. The future study would be developing independent debugging tool and appropriate visualization tool for target system.

References

1. Abatron, A.G.: bdiGDB JTAG interface for GNU Debugger User Manual, Manual Ver 1.07 for BDI2000, ABATRON AG Switzerland (2004)
2. AIJI System: OPENize & Spider User's Manual, 2F Samho-Tower. 1122-10, Ingye-dong, Paldal-gu, Suwon-city, Gyeonggi-do, Korea (2004)
3. ARM: RealView Debugger: Extensions User Guide (2004)
4. Cantrill, B.M., Shapiro, M.W., Leventhal, A.H.: Dynamic Instrumentation of Production Systems. In: 2004 USENIX Annual Technical Conference, Boston, MA, USA (June 2004)
5. Pospiech, F., Olsen, S.: Embedded Software in the SoC World. How HdS Helps to Face the HW and SW Design Challenge. In: IEEE 2003 Custom Integrated Circuits Conference (2003)
6. IEEE: Standards Board, Standard Test Access Port and Boundary-Scan Architecture, Std 1149.1-2001 (2001)
7. Intel: Hot-Debug for Intel Xscale Core Debug: White Paper, 2200 Mission College Blvd. Santa Clara, CA 95052 USA (May 2005)

8. Intel: Intel XScale Microarchitecture for the PXA255 Processor: User's Manual, 2200 Mission College Blvd. Santa Clara, CA 95052 USA (March 2003)
9. ISYSTEM Inc.: Intel XScale Family On-Chip Emulation, Carl-Zeiss-Strasse 1. Schwabhausen, Germany (May 2001)
10. Koo, G., Park, M., Ha, S., Jun, Y., Lim, C.: A USB-JTAG Adapter for Remote Debugging Tool of SoC Programs. In: Proc. of the 24th KIPS Fall Conference, KIPS vol. 12(2), pp. 1449–1452 (2005)
11. Lee, K., Kim, J., Lim, C., Kim, H.: A Development of Remote Tracepoint Debugger for Run-time Monitoring and Debugging of Timing Constraints on Qplus-P RTOS. In: IEEE Workshop on Software Technologies for Future Embedded Systems, Hakodate, Hokkaido, Japan, pp. 93–96 (May 2003)
12. Pilet, J., Magnenat, S.: Jelie: manuel de l'utilisateur, EPFLI&CISIM LAP. INF 136 Station 14 CH-1015 Lausanne Switzerland (2003)
13. Stallman, R., Pesch, R., Shebs, S., et al.: Debugging with gdb. Free Software Foundation (2003)
14. Akgul, T., Kuacharoen, P., Mooney, V.J., Madisetti, V.K.: A Debugger RTOS for Embedded Systems. In: 27th EUROMICRO Conference, Warsaw, Poland, pp. 264–269 (September 2001)
15. Whitney, T., Neville, G.: SoC Software Hardware NIGHTMARE or Bliss. ACM QUEUE (April 2003)
16. Prasad, V., Cohen, W., Eigler, F.Ch., Hunt, M., Keniston, J., Chen, B.: Locating System Problems Using Dynamic Instrumentation. In: 2005 Linux Symposium, Ottawa, Canada (2005)
17. Vink, G.: Trends in Debugging Technology, Embedded System Conference, Chicago (March 1998)
18. Gatliff, W.: Implementing a Remote Debugging Agent Using the GNU Debugger (2001)
19. Yaghmour, K.: Building Embedded Linux Systems. O'Reilly & Associate (2003)

A Novel Buffer Cache Scheme Using Java Card Object with High Locality for Efficient Java Card Applications*

Won-Ho Choi[1,**], Ha-Yong Jeon[2,***], Rhys Rosholt[3,†], Gwang Jung[3], and Min-Soo Jung[4,‡]

[1,2,4] Department of Computer Engineering, Kyungnam University, Masan, Korea
{hoya9499,hayongj,msjung}@kyungnam.ac.kr
[3] Department of Mathematics and Computer Science, Lehman College,
The City University of New York, Bronx, NY
{rhys.rosholt,gwang.jung}@lehman.cuny.edu

Abstract. Java Card technology enables smart cards and other devices with very limited memory to run small applications. It provides users with a secure and interoperable execution platform that can store and update multiple applications on a single device. However, a major difficulty with Java Card is its low execution speed caused by hardware limitations. In this paper, we propose a novel scheme about how to improve the execution speed of Java Card. The key idea of our approach is a buffer cache scheme that uses RAM instead of EEPROM to improve the execution speed of Java Card. The proposed scheme reduces I/O count, especially EEPROM writing. Our scheme is based on the high locality of Java Card objects and the use of RAM that is several magnitude faster than EEPROM.

Keywords: Java, Java Card, Java Card objects, Buffer Cache.

1 Introduction

Java Card technology enables smart cards, SIM, USIM and other devices with very limited memory to run small applications and has advantages such as a platform independence and dynamic downloading [1, 2, 3, 15]. For these reasons, Java Card technology is an accepted standard for smart card and SIM devices [15].

Java Card is essentially an Integrated Circuit Card with an embedded Java Card Virtual Machine. Its CPU can access three different types of memory. These are ROM which usually contains a basic card operating system (COS) and the greatest part of the Java Card runtime environment, EEPROM which can be used to store code and data, and RAM in which applications are executed [4, 7].

Most Java Card applications generally use RAM and EEPROM. The difference between these memories is that writing operations to EEPROM are typically more

---

* This work is supported by Kyungnam University Research Fund, 2006.
** Post Doc. of Kyungnam University.
*** Ph. D. student of Kyungnam University.
† Assistant Professor of Lehman College, the City University of New York.
‡ Professor of Kyungnam University.

M.S. Szczuka et al. (Eds.): ICHIT 2006, LNAI 4413, pp. 500–510, 2007.
© Springer-Verlag Berlin Heidelberg 2007

than 10,000 times slower than to RAM. In a traditional Java Card application, the EEPROM is used to support Java Card objects, atomic data and transactions [1, 3]. This makes the Java Card operate more slowly. Too many EEPROM write operations cause a drop in the execution speed of Java Card [4].

The primary cause of many of the EEPROM writes is that the traditional Java Card uses a low-level EEPROM write with a page-buffer. The size of the page-buffer depends on the platform, such as ARM, Philips and so on. This page-buffer is designed to write blocks as small as one byte or any number of consecutive bytes up to the size of the page-buffer into EEPROM. However, this page-buffer generally is designed without regard to the locality of Java Card objects [5, 7]. The second cause of many of the EEPROM write operations is the resolution of indirect references during the download of a new application. The Java Card Installer performs these resolutions, changing indirect references to real physical addresses in the Java Card API [2].

In this paper, we suggest two methods to improve the speed of Java Card. One of methods is a new buffer cache scheme based on the high locality of those Java Card objects that are stored in heap area. The other is a new Java Card Installer that employs resolution using RAM technology to resolve indirect references into direct references in RAM during the download of new applications.

This paper is organized as follows. Section 2 describes the Java Card memory system and the Java Card Installer as related works in detail. Section 3 gives a design of a new buffer cache scheme and Java Card Installer with resolution in RAM to improve the execution speed of Java Card. Section 4 explains algorithms for implementation of these methods. Section 5 gives the performance results of these methods. Finally, the conclusion and possible future work is presented in section 6.

2 Java Card Environment

2.1 Java Card Memory System

Smart card hardware provides very limited storage capabilities. The memory resources typically consist of ROM, RAM and EEPROM. EEPROM is used to store long-lived data, Java Card objects, and atomic data. In contrast, RAM loses its contents after a power loss and is only available for temporary storage.

In Figure 1, a typical Java Card system places the Java Card virtual machine, System API classes, and other software in ROM. Applet code can also be stored in ROM. RAM is used for temporary storage. The Java Card runtime stack is allocated in RAM. Method parameters, native methods and local variables are put on the stack. Persistent data such as Java Card objects and downloaded applet classes are stored in EEPROM [1, 3, 4, 7].

The applet instance and associated persistent objects of an application must survive during a session. Therefore they are placed in the non-volatile EEPROM storage on the card. EEPROM provides read and write access similar to that of RAM. However, the difference between the two forms of memory is that write operations to EEPROM are typically more than 10,000 times slower than to RAM [4].

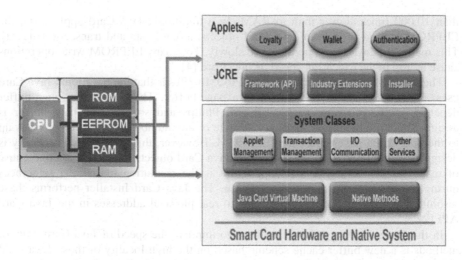

Fig. 1. Java Card Memory System that consists of ROM, RAM, EEPROM and its contents

2.2 Java Card Installer

Java Card supports dynamic downloading of an applet, called post-issuance. Using one of the ports of Java Card, the installer loads a CAP file sent from the CAD (Card Acceptance Device) and creates applets using the loaded CAP file. The installer enables secure downloading of software and applets onto the card after the card is made and issued to the card holder. The installer cooperates with off-card installation

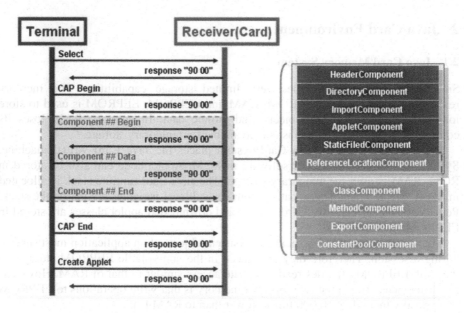

Fig. 2. Java Card Installer and off-card installation program

programs. Together they accomplish the task of loading the binary contents of CAP files. The installer is an optional JCRE (Java Card Runtime Environment) component. Without the installer, all card software including applets, must be written into card's memory during the card manufacturing process [1, 2, 3].

The installer APDU protocol follows a specific sequence of events in the transmission of Applet Protocol Data Units as shown in the following figure.

3 Java Card Using RAM Instead of EEPROM

3.1 EEPROM Writing Based on New Buffer Cache Scheme

3.1.1 EEPROM Writing Mechanism in Traditional Java Card
In the Java Card environment, a one-page buffer in RAM is used to write data into the EEPROM. The size of a page-buffer depends mainly on the platform. Generally, it is between 128 and 256 bytes. The chip with CalmCore16 that we used to test our technologies had a 128 byte page-buffer size. Using it, the Java Card can write any number of consecutive bytes from 1 byte up to 128 bytes to EEPROM at one time. For example, suppose there are two objects generated sequentially by Java Card. Each object has 5 bytes. EEPROM addresses of these objects are sequentially 0x87175 and 0x87170, although both addresses are within 128 bytes included the size of one page buffer in RAM. Java Card will first write one object data in 0x86005 through the page-buffer, and then, after the page-buffer is clear, the other object data will be written in 0x86000.

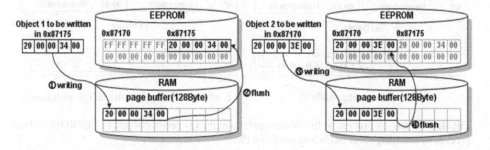

Fig. 3. EEPROM writing mechanism in traditional Java Card using an inefficient one-page buffer when objects are written into EEPROM by Java Card

Figure 3 describes the one-page buffer algorithm of a traditional Java Card. This page buffer is used to write continuous data to EEPROM. However, it erases the contents of the page buffer if a different address of an object is generated, because it does not have a function for caching.

3.1.2 Java Card Objects with High Locality
When an applet is executed on Java Card, if information such as objects and class data that the applet writes are close to each other, the total number of EEPROM writes would be reduced by adding a caching function to the page buffer. To do this,

the write address of objects and data created by Java Card must have high locality. That allows the number of EEPROM writes to decrease and the hit ratio of the caching function to increase.

We investigated the general tendencies of write operations as compared with the EEPROM addresses. We discovered that Java Card has an internal rule about the locality of EEPROM write addresses. Consequently, the locality of Java Card objects and data is considerably high.

```
public class Wallet extends Applet{
  int balance;
  int withdraw;
  OwnerPIN pin;
  wallet () {
    pin = new OwnerPIN(3, 8); // OwnerPIN(trylimit, Pinsize)
  }
  initialize() {
    balance = 90;
  }
  withdraw() {withdraw = 50;
    balance = balance - withdraw;
  }
}
```

a. A sample code about the locality of EEPROM writing address

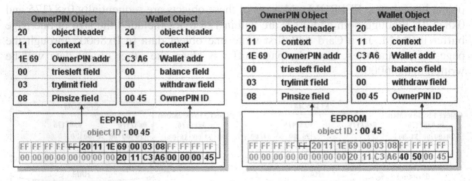

b. after executing wallet()   c. after executing initialized() & withdraw()

Fig. 4. The creation process of the Wallet applet and the OwnerPIN object in EEPROM and the process of changing local fields and rewriting them

As illustrated in Figure 4, when the Wallet class is created by the install() method, the Wallet object that has 3 fields is first written in EEPROM. The value of the wallet object is 2011C3A600000000. And the OwnerPIN object that assigned 0045 as an objectID is created and written in EEPROM. The value of the OwnerPIN object is 20111E69000308. After the OwnerPIN object is created, Java Card writes the objectID as a pin reference field of the Wallet object. After the Wallet applet is created, a method such as initialize() and withdraw() generally would be invoked. Initialize() method is to change the value of balance field into 100. After this operation, the content of the wallet object is 2011C3A690000045. The withdraw() method also changes the field value of withdraw and balance into 50 and 40 separately. At this time, the content of the wallet object is 2011C3A640500045.

If Java Card just performs these processes by using the one-page buffer mentioned above, it might spend much time writing and changing localized data like the above example.

3.1.3 The Proposed Buffer Cache Scheme Using High Locality of Java Card Objects

As mentioned earlier, a traditional Java Card has only a one-page buffer in RAM to write data into EEPROM. The page buffer has a function for buffering of continuous bytes. In this paper, we suggest an object buffer that performs buffering and caching to improve the execution speed of Java Card. The most important requirement for adding a caching function to Java Card is a high hit ratio for the caching buffer.

3.2 Java Card Installer with Resolution in RAM

A CAP file has a compact and optimized format, so that a Java package can be efficiently stored and executed on a Java Card. Among several components in a CAP file, the ConstantPool component and the Method component employ various types of constants, including method and field references which are resolved when the program is linked or downloaded to a Java Card.

The ConstantPool component has lots of constants that must be resolved during the downloading of a CAP file. Before constants are resolved to real addresses, they are parsed into tokens. The token of each constant is resolved to a real address in the Java Card API. After the resolution of constants, Java Card performs the resolution of indirect references into constants that are already resolved if bytecodes in the Method component have indirect references as an operand. This linking operation of the Method component is executed when a ReferenceLocation component is finally sent to the Java Card.

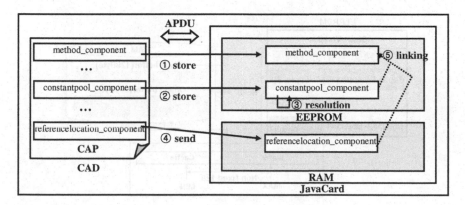

Fig. 5. The procedure for downloading a CAP file with resolution in EEPROM in a traditional Java Card

Figure 5 shows the procedure for downloading a CAP file. First, both the Method component and the ConstantPool component are saved in the heap area in EEPROM. Second, the installer performs the resolution of constants for the ConstantPool component that is already downloaded. Next, the ReferenceLocation component is sent to RAM. This component has lists of offsets that must be replaced into constants in the ConstantPool component among bytecodes in the Method component. Finally, the installer replaces bytecodes in the Method component into resolved constants by using the offset data of the ReferenceLocation component.

Therefore the size of ReferenceLocation represents the number of bytecodes that must be replaced in the Method component. While the ReferenceLocation component is downloading, Java Card continually changes an operand of bytecodes in the Method component tracking the changes of size of ReferenceLocation. This makes the Java Card Installer operate more slowly.

In this paper, we suggest a new installer that is very simple. The key idea of our Installer is to use RAM as much as possible by resolving indirect references in RAM, not EEPROM. The important difference between these memory types is that writing to RAM is typically more than 10,000 times faster than writing to EEPROM.

4 The Proposed Enhanced Java Card Using RAM

4.1 The Proposed Buffer Cache Scheme with a High Locality

In Section 3, we explained how to write data in EEPROM by using the one-page buffer in a traditional Java Card. We discovered that all objects and data that the Java Card creates during execution have high locality. This means that an additional caching function would make the number of EEPROM writes decrease. For these reasons, we developed a new Java Card with a two-page buffer in RAM. One, called the page buffer, is the existing page buffer for non-heap area static variables. The other, called object buffer, is for the heap area in EEPROM. The heap area is where objects created by the Java Card are allocated.

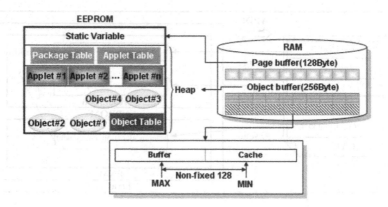

Fig. 6. The proposed buffer cache scheme consists of buffer and cache. (The data between Min and Max can be written to EEPROM all at once.).

Figure 7 below shows the main algorithm using the object buffer and the page buffer. Writing of the non-heap area is performed with the existing page buffer. Writing of the heap area is executed with the object buffer. When the Java Card

writes data related to Java Card objects into the heap area of the EEPROM, the first
operation is to get 128 bytes lower than the address that will be written and to copy
them to the cache area of the object-buffer. Next, the buffer area (128 bytes) of the
object buffer is cleared. Two points, Max and Min, have the highest and lowest points
that are written after the Java Card gets the new 256 bytes to the object buffer. The
gap between Max and Min is continually checked in order to write the object buffer to
EEPROM. Max and Min are non-fixed points, allowing an increase in the efficiency
of the object buffer. The reason why the gap between Max and Min is 128 bytes is
that our target chip, CalmCore16, supports the EEPROM writing of 128 bytes all at
once.

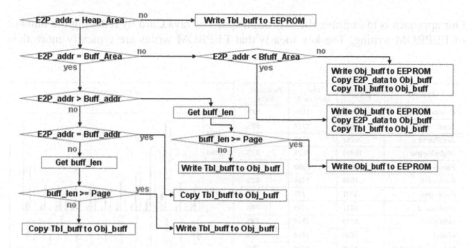

Fig. 7. The Object-buffer algorithm that checks continually the Min and Max points to write the
object-buffer to EEPROM when Java Card writes data to heap area

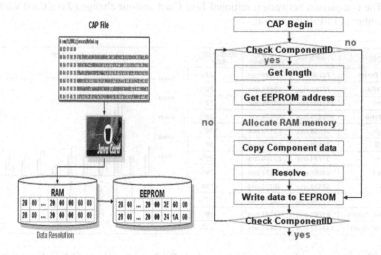

Fig. 8. The resolution and linking procedure of our changed installer using a RAM Area

4.2   The Proposed Java Card Installer with Resolution Using RAM

Our changed installer is more flexible than a traditional one. Especially, the size of Method component is not always fixed. This means that the size of the component can exceed the remaining size of RAM. For this reason, when downloading a ReferenceLocation component, the installer must first check the total size of the Method component to calculate the proper block-size for resolution. Figure 8 shows the resolution and linking procedure of our changed installer using a RAM Area.

5   Evaluation of Our Approach

Our approach is to improve the execution speed of Java Card by reducing the number of EEPROM writing. The key idea is that EEPROM writes are typically more than

Applets	Traditional	Our approach	Reduced Rate
Channel Demo	7552	1596	79%
JavaLoyalty	7291	1322	82%
JavaPurse	22712	4537	80%
ObjDelDemo	16416	3025	82%
PackageA	9685	2000	79%
PackageB	7698	1406	82%
PackageC	3497	745	79%
Photocard	6737	1409	79%
RMIDemo	6119	1261	79%
Wallet	5641	1190	79%
EMV small Applet	6721	1419	79%
EMV Large Applet	11461	2433	79%

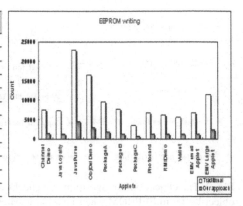

Fig. 9. The comparison between traditional Java Card and our changed Java Card with respect to the number of EEPROM writings

Applets	Traditional	Our approach	Reduced Rate
Channel Demo	76140 ms	55406 ms	27%
JavaLoyalty	72703 ms	52079 ms	28%
JavaPurse	232109 ms	169250 ms	27%
ObjDelDemo	159420 ms	112016 ms	30%
PackageA	90530 ms	63172 ms	30%
PackageB	74859 ms	52703 ms	30%
PackageC	32734 ms	23188 ms	29%
Photocard	64608 ms	46172 ms	29%
RMIDemo	57328 ms	40437 ms	29%
Wallet	57140 ms	41844 ms	27%
EMV small Applet	61766 ms	43187 ms	30%
EMV Large Applet	119812 ms	88671 ms	26%

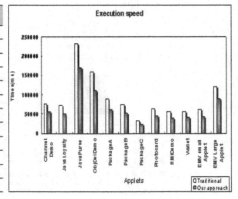

Fig. 10. The comparison between traditional Java Card and our changed Java Card with respect to the execution speed

10,000 times slower than writes to RAM. Results of analyzing the traditional Java Card are that the one-page buffer writes data to EEPROM without regard to the high locality of Java Card objects and that the installer uses a resolution in EEPROM. For this reason, we developed a new buffer cache scheme and new Java Card Installer.

We experimented with the CalmRISC16 MCU and the SAMSUNG microcontroller for smart card. Figure 9 and Figure 10 show the comparison between a traditional Java Card and our enhanced Java Card in regard to the number of EEPROM writes and the execution speed. The number of EEPROM writes is reduced by about 80%. The execution speed of Java Card is improved by 30%.

6 Conclusion and Future Work

Java Card technology is already a standard for smart cards and SIM cards [11, 15]. In this paper, we have proposed a method to reduce the number of EEPROM writes with a new buffer cache scheme using both the high locality of Java Card and also the new installer. It makes both the downloading time and also the execution time of Java Card quicker. With our approach, the number of EEPROM writes decreased by 80% and the downloading speed increased by 30%. Our approach also enables an application to be downloaded more quickly in the case of an application sent to a mobile phone via the GSM network.

References

1. Sun Microsystems, Inc. JavaCard 2.2.1 Virtual Machine Specification. Sun Microsystems, Inc. (2003), http://java.sun.com/products/javacard
2. Sun Microsystems, Inc. JavaCard 2.2.1 Runtime Environment Specification. Sun Microsystems, Inc. (2003), http://java.sun.com/products/javacard
3. Chen, Z.: Java Card Technology for Smart Cards: Architecture and programmer's guide. Addison Wesley, Reading, Massachusetts (2001)
4. Rankl, W., Effing, W.: Smart Card Handbook, 3rd edn. John Wiley & Sons, Chichester (2001)
5. Gosling, J., Joy, B., Steele, G., Bracha, G.: The Java Language Specification, 2nd edn. Addison-Wesley, Reading (2001), http://java.sun.com/docs/books/jls/index.html
6. Oestreicher, M., Krishna, K.: USENIX Workshop on Smartcard Technology, Chicago, Illinois, USA (May 10–11, 1999)
7. Oestreicher, M., Ksheeradbhi, K.: Object Lifetimes in JavaCard. In: Proc. Usenix Workshop Smart Card Technology, Usenix Assoc., Berkeley, Calif, pp. 129–137 (1999)
8. Baentsch, M., Buhler, P., Eirich, T., Höring, F., Oestreicher, M.: IBM Zurich Research Laboratory. Java Card From Hype to Reality (1999)
9. Hartel, P.H., Moreau, L.: Formalizing the safety of Java, the Java virtual machine, and Java card. ACM Computing Surveys (CSUR) 33(4), 517–558 (2001)
10. Oestreicher, M.: Transactions in JavaCard. In: Proc. Annual Computer Security Applications Conf., Los Alamitos, Los Alamitos (to appear)
11. Kim, J.S., Hsu, Y.: Memory system behavior of Java programs: methodlogy and analysis. In: Proceedings of the ACM Java Grande 2000 Conference (June 2000)
12. OTA White Paper. Gemplus (2002), http://www.gemplus.com
13. The 3rd Generation Partnership Project.: Technical Specification Group Terminals Security Mechanisms for the (U)SIM application toolkit. 3GPP (2002)

14. MCULAND, http://mculand.com/e/sub1/s1main.htm
15. Leroy, X.: Bytecode verification for Java smart card. Software Practice & Experience, 319–340 (2002)
16. SAMSUNG, http://www.samsung.com/Products/Semiconductor
17. Beckert, B., Mostowski, W.: A program logic for handling Java Card's transaction mechanism. In: Pezze, M. (ed.) ETAPS 2003 and FASE 2003. LNCS, vol. 2621, pp. 246–260. Springer, Heidelberg (2003)
18. Jeon, H.S., Noh, S.H.: An Efficient Buffer Cache Management Algorithm based on Prefetching. Journal of KISS 27(5), 529–539 (2000)

Design and Implementation of the Decompiler for Virtual Machine Code of the C++ Compiler in the Ubiquitous Game Platform*

YangSun Lee, YoungKeun Kim, and HyeokJu Kwon

Dept. of Computer Engineering, SeoKyeong University
16-1 Jungneung-Dong, Sungbuk-Ku, Seoul 136-704, Korea
{yslee,ykkim,hjkwon}@skuniv.ac.kr

Abstract. The ubiquitous game platform implemented by our team is composed of a C++ compiler, a java translator, and a virtual machine. The EVM (Embedded Virtual Machine) is a stack-based solution that supports object-oriented languages such as C++ and java. It uses the SIL (Standard Intermediate Language) as an intermediate language, which consists of an operation code set for procedural and object-oriented languages. The existing C++ compilers are used to execute programs after translating them into a target machine code. The downside of this method is its low practicality, along with its platform-dependency. To resolve this matter, we developed a C++ compiler that generates virtual machine codes based on platform-independent stacks that are not target machine codes. This paper presents a decompiler system that converts a C++ compiler generated intermediate language, namely SIL, to a representation of a C++ program. This method optimizes the simulation needed for the generation of exacted SIL code, and a solution that can verify the SIL code generation through a C++ program represented in the decompiler. Furthermore, the ease of extracting the meaning of a program, as opposed to assembly-structured SIL codes, allows much more convenience in changing the software structure and correcting it to improve performance.

Keywords: Decompiler, Virtual Machine Code, C++ Compiler.

1   Introduction

Mobile devices, set-top boxes, and digital TVs all contain EVMs (Embedded Virtual Machine) that downloads and executes applications independently of their hardware platform. The SIL (Standard Intermediate Language), an intermediate language of the EVM, is designed to make content developments much easier, since it can support both C, a procedural language, and C++ or java, object oriented languages.

The existing C++ programs need to be translated into target machine codes through a platform-dependent compiler. A major defeat of this method is the fact

---

* This research is accomplished as the result of the research project for culture contents technology development supported by KOCCA.

M.S. Szczuka et al. (Eds.): ICHIT 2006, LNAI 4413, pp. 511–521, 2007.

that every platform requires a compiler to translate the target code according to its own platform. Hence, reusability and portability become problems because every platform translates different codes. In order to resolve such difficulty, we developed a stack-based virtual machine, the EVM, and a C++ compiler that generates virtual machine codes, the SIL, for the EVM. Our C++ compiler translates a C++ source program to virtual machine codes, the SIL, and the translated SIL codes are then executed in an EVM-installed system. This method increases the portability and the reusability of codes because they do not need to be translated when a program is executed in different platform environments [2],[6],[7],[8],[19],[20].

In this paper, we designed and implemented a decompiler that translates an intermediate SIL code into a represented C++ program, a quadruple format. Likewise, the execution of a decompiler-generated C++ program can be examined for whether or not the SIL code has been correctly generated and the meaning can be extracted from a program at the source level for easier analysis. By executing a decompiler-represented C++ program, this system supports the verification of a correctly generated SIL code and provides a simulation needed for optimal SIL code generation. Moreover, the meaning of a program can be extracted by this system with less effort than by assembly-type SIL codes, which, consequently, permits easier structural transformation and correction that will enhance performance.

2   Related Studies

2.1   Decompiler

One of studies on translators is about an intermediate language translator that generates a native code of the target machine from bytecode for improving the execution speed of Java applications. It, however, can't be used in other computer environments because of its dependency on the target machine [9],[10],[11].

There are various types java decompilers ranging from the publicly used ones to the professional ones used commercially. Among the publicly used decompilers, Mocha[12] was the very first decompiler to be introduced and Jad[14] is the most widely used one at present. Commercial decompilers include DejaVu[13], which is provided as a part of 'OEW for Java', a Java development environment of Innovative S/W, and WingDis[15] made by Wing Soft. In terms of the .NET decompilers, some examples are Salamander[16] made by Remote Soft and Dis#[17] made by .NET Decompiler.

Although translations are performed correctly, errors are likely to be detected in most cases: Mocha translates boolean data into integers and the initial statement of an array is sometimes translated to an invalid code. Meanwhile, Jad has a defeat in that it cannot decompile when attempting to use internal and external break/continue statements in an internal loop statement that is overlapped with a break/continue of a loop statement. Furthermore, some exceptions cannot be handled and they will only be notified by messages. In the case of DejaVu, it is impossible to analyze the flow while WingDis often mistranslates pre-increment operators into post-increment operators. Dis# also has a disadvantage of not utilizing the meaning of the source code due to the auto-generated local variable names. Except for Jad, the decompilers mentioned above are designed to run under Java runtime environments. Therefore, they are said to be dependent on the runtime environment.

The SIL-to-C++ decompiler implemented in this study, similar to the Jad and .NET decompiler, can be run regardless of runtime environments. Moreover, it supports a variety of languages without being limited to a certain language unlike how the Java decompiler is limited to Java. This is because the SIL code is designed to accept C++ and Java as an intermediate language for the embedded system thereby translating them into diverse source code types. Therefore, although the SIL-to-C++ decompiler translates codes into a C++ program, they can always be translated into other languages.

2.2 EVM (Embedded Virtual Machine)

An EVM, commonly installed in mobile devices, set-top boxes, and digital TVs, is an embedded virtual machine that downloads and executes software programs without relying on the hardware. Figure 1 illustrates the system structure for the ubiquitous game platform [18],[20],[21].

The ubiquitous game platform is composed of three key parts. The first part consists of a compiler that converts a high-level language program, such as C/C++ or Java programs, into a stack-based virtual machine code. The second part is an assembler that translates virtual machine codes (SIL) into an EVM file format that can be executed in virtual machines. The final part, namely the virtual machine, is directly installed in the actual embedded hardware and it executes EVM files. The ubiquitous game platform has a pyramid structure, which minimizes the burden on the targeting process [19],[20].

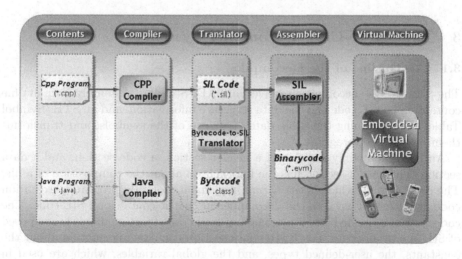

Fig. 1. System Configuration for the Ubiquitous Game Platform

2.3   SIL (Standard Intermediate Language)

The SIL, the virtual machine code of the EVM, is designed as a standardized model of the virtual machine code for common embedded systems. Consisting of a union of stack-based commands, the SIL is hardware and platform-independent and is based on the existing virtual machine codes such as java bytecode, .NET MSIL, and other intermediate code so that it is compatible with a range programming languages. Moreover, it consists of operation codes that support both procedural and object-oriented languages.

The SIL is composed of operation codes that correspond to actual instructions and pseudocodes that indicate the process of a specific task, such as a class declaration. The operation codes are written in an abstract form that does not necessarily restrict itself to any specific hardware of source language and is codified in easy-to-decipher mnemonics by applying a consistent name rule to make the debugging of assembly language code smoother. It also has short form operation codes for optimum environment. The 418 instructions of the SIL are classified into 7 categories, which are themselves divided into sub-categories. Figure 2 shows the categories of the SIL operation code [19],[20],[21].

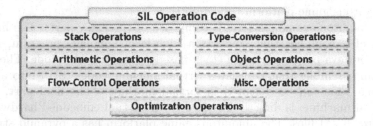

Fig. 2. SIL Operation Category

3   The SIL-to-C++ Decompiler System

3.1   An Outline of the Decompiler

The SIL-to-C++ decompiler receives an SAF (Standard Assembly Format) file containing an SIL code and the data structure information, and an STB (Symbol Table) file containing the table data structure of the symbols, and transforms them into a C++ program.

An SAF file mainly consists of a header section, a code section, and a data section. The header section has the information about the composition of a file. The code section, which is the actual implementation of a program's execution code, is extremely vital to the decompiler in that it contains an SIL code. The code section is divided into functional units and each functional unit is composed of SIL operation codes. Lastly, the data section contains information about the constants, the user-defined types, and the global variables, which are used in the code section. The decompiler lexically and syntactically analyzes these three sections to generate an AST (Abstract Syntax Tree) and decompiles it after analyzing the tree in the code translation phase. Figure 3 shows the structural schematic of the decompiler's execution process.

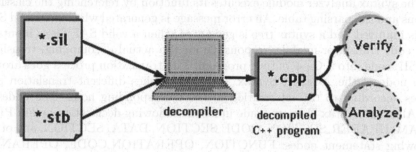

Fig. 3. Outline of Decompiler System

3.2   System Configuration

The decompiler is largely broken down into a front-end and a back-end. The front-end, consisting of a lexical analyzer module and a syntax analyzer module, is responsible for generating an Abstract Syntax Tree(a type of an intermediate tree) by receiving an SAF file as input and then partitioning it into a minimal unit of tokens. Moreover, the front-end receives an STB file as input and reconstructs the symbol table data structure that is later used in the back-end. In the back-end, a whole code translator module performs the actual decompiling, accomplishes the actual searching and translating of an abstract syntax tree into the corresponding code.

The data structure that the decompiler often references contains the API translation table and the runtime stack, which approaches the statement part and the symbol table that are required for managing the declaration part during the decompiling process. The symbol table saves information such as user-defined types and variables declared in a C++ program, which are designed to be referenced in the statement part. The runtime stack is intended to be a type of stack that is used for saving the intermediate result value of handling the right-hand side of the process, which translates the SIL code into a 3-address code format. The API translation table is a data structure used for handling program statements for library functions defined by the standard C++ regulation.

The SIL-to-C++ decompiler system described in this paper is implemented in Visual C++ 6.0 in Windows XP. Figure 4 illustrates the architecture and the development environment of the decompiler system.

3.3   SIL-to-C++ Decompiler

The decompiler is implemented in standardized module units. The lexical analyzer module obtains the lexical information and the parsing table by analyzing SIL instructions and designing LALR(1) type of syntax, which is then inputted into the Parser Generating System. The SIL code of the inputted SAF file is determined according to the type of token shown in the generated lexical information.

The syntax analyzer module executes its function by referencing the classified tokens and the parsing table. An error message is generated when an invalid SAF file is inputted and a syntax tree is generated when a valid SAF file is inputted. The code translator module, responsible for the actual decompiling, translates the SIL code into a C++ output program. The translation process goes around each node of the Abstract Syntax Tree and applies different translation processes according to the information of the corresponding node. The nodes of the Abstract Syntax Tree are made up of the following declaration nodes: PRO-GRAM, HEADER_SECTION, CODE_SECTION, DATA_ SECTION, and of the following statement nodes: FUNCTION, OPERATION_CODE, OPERANDS. Figure 5 illustrates the translation algorithm of the decompiler.

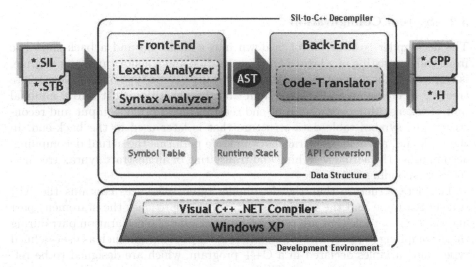

Fig. 4. Architecture of the Decompiler System

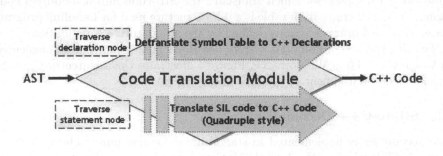

Fig. 5. Translation Algorithm of the Decompiler

The PROGRAM node of the declaration part outputs the variable declared in the global scope, the user-defined type, and the 3-address code as a quadruple format while also processing the output of the temporary variables. The CODE_SECTION node handles the output of the used standard library and the function prototype.

The FUNCTION node of the statement part stores the definitions of the functions. It outputs the name, the return type, and the parameters of the corresponding function after translating them to C++ statement formats. The OPERATION_CODE, a node that stores the SIL code instructions and the instruction parameters, has OPCODE and OPERANDS nodes beneath. The OPCODE node contains one of the SIL's 418 instructions and it performs different translation processes according to this instruction. The push and the pop of the runtime stack decide whether the Assemble module, the Execute module, or both are called in translation. The instruction parameter of the SIL code is referenced by the OPERANDS node during the translation process. The OPERANDS node contains information such as the variable's base level, the offset value, and the label. The scope of the variable is determined by the base and the offset information. After this, the information of the corresponding variable is obtained from the symbol table, translated to the name of the variable, and then represented. The label information indicates the location of the flow control and the label of the flow control statement is translated to a code that moves into the location of the flow control.

4   Experimental Results and Analysis

The following example describes how to generate an SAF (SIL Assembly Format) file of a C++ program by using our C++ compiler, and to translate the SAF file back to a represented C++ source program using the decompiler described in

```
#define _WIN32
#define _WCHAR_T_DEFINED 0
#include "C:\Program Files\Microsoft Visual
Studio\VC98\Include\stdio.h"
#define MAXDISC 64
char pinname[ ] = {'a', 'b', 'c'};
enum pin {a, b, c};
class Hanoi {
private:
  int stack[3][MAXDISC] ;
  int top[3] ; /* stack pointer */
public:
  void move(int, int);
  void transfer(int, int, int, int);
  void hanoi();
};
```

```
...
void Hanoi::hanoi() {
  int i, n;
  printf("*** Enter the number of discs : ");
  scanf("%d",&n);
  for (i = 1; i <= n ; i++)
    stack[a][i] = n + 1 - i;
  top[a] = n; top[b] = top[c] = 0;
  transfer(a, c, b, n);
}
void main() {
  Hanoi h;
  h.hanoi();
}
```

Fig. 6. Original C++ Program

%%HeaderSectionStart	ldc.i 1
...	lod.i 1 8
%%HeaderSectionEnd	call &Hanoi::transfer$2
%%CodeSectionStart	ret
%FunctionStart	.opcode_end
.func_name &Hanoi::hanoi$3	%FunctionEnd
.func_type 2	...
.param_count 0	%%CodeSectionEnd
.opcode_start	%%DataSectionStart
proc 12 1 1	%LiteralTableStart
ldp	.literal_start @1 0 3
ldc.p @0x0000	0x25,0x64,0x00
ldc.i 4	.literal_end
call &printf	...
ldp	%LiteralTableEnd
ldc.p @0x0023	%InternalSymbolTableStart
lda 1 8	.var_start $pinname 0 3 3 1 (1)
ldc.i 8	0x61,0x62,0x63
call &scanf	.var_end
...	%InternalSymbolTableEnd
ldp	%ExternalSymbolTableStart
lod.p 1 0	.evar_decl $_iob
ldc.i 0	%ExternalSymbolTableEnd
ldc.i 2	%%DataSectionEnd

Fig. 7. Compiled SAF File

#include "h.cpp.saf.h"	tmp_int21 = (int)n;
void Hanoi::hanoi()	tmp_ptr11->transfer(0,2,1,tmp_int21);
{	}
// [LOCAL VARIABLE DECLARATION]	...
int i;	void main()
int n;	{
// [FUNCTION STATEMENT]	// [LOCAL VARIABLE DECLARATION]
tmp_int0 = printf(" *** Enter the number	Hanoi h;
of discs : ");	// [FUNCTION STATEMENT]
tmp_int1 = scanf("%d", &n);	h.hanoi();
...	}
tmp_ptr11 = this;	

Fig. 8. Decompiled C++ Program

this paper, and finally to execute it. Figure 6 is the original code of an example C++ program, Figure 7 is the compiled SAF file, and Figure 8 is the final C++ source program regenerated by the decompiler.

Table 1 shows the result of performance evaluation between C++ source programs and decompiled C++ source programs. As it is evident in the table, the performance time is about the same for a source program and a decompiled program.

Fig. 9. Execution Result(Left : Original C++ Program, Right-down : Decompiled C++ Program)

Table 1. The Result of Performance Evaluation

Test Program	Original C++ Program	Decompiled C++ Program	SIL Code
Palindrome Number	101 ms	98 ms	114 ms
Perfect Number	119 ms	111 ms	125 ms
Prime Number	101 ms	100 ms	123 ms
Heap Sort	100 ms	100 ms	158 ms
Marge Sort	100 ms	97 ms	103 ms
Quick Sort	98 ms	98 ms	131 ms
Magic Square	105 ms	108 ms	121 ms

5 Conclusion

In this paper, we present the SIL-to-C++ decompiler that translates an SIL code, which is an intermediate language used in stack based virtual machines, into a C++ source program. The strength of this decompiler, which consists of the front-end of a syntax analyzer, the back-end of a code translator, and a lexical analyzer that uses the standardizing method of a compiler, is in its effective reusability of module units. The C++ program represented by the decompiler is executed after being translated into a target code by the Visual C++ compiler. By running and analyzing C++ programs generated by such decompiler, the correct generation of an intermediate code, namely SIL, can be verified and it can be used as a simulation tool for generating optimized intermediate codes. Moreover, losses that occur due to the nature of the embedded virtual machine in the form of content download during program transfer or software malfunctioning can be easily corrected and analyzed at the source level as opposed to the assembly format code.

References

1. Stroustrup, B.: The C++ Programming Language. Addison-Wesley, Reading (2000)
2. Galles, D.: Modern Compiler Design. Addison-Wesley, Reading (2004)
3. International Standard, Programming Language-C++ 14882:1998(E), ISO/IEC (1998)
4. Microsoft, MSIL Instruction Set Specification (November 2000)
5. Microsoft, The IL Assembly Language Programmer's Reference (October 2000)
6. Lindin, S.: Inside Microsoft.NET IL Assembler. Microsoft Press, Redmond (2002)
7. Venners, B.: Inside the JAVA Virtual Machine, 2nd edn. McGraw-Hill, New York (2000)
8. Lindholm, T., Yellin, F.: The Java^{tm} Virtual Machine Specification, 2nd edn. Addison-Wesley, Reading (1999)
9. Hsieh, C.A., Conte, M.T., Johnson, T.L.: Java Bytecode to Native Code Translation: the Caffeine Prototype and Preliminary Results. In: Proceedings of the IEEE 29th Annual International Symposium on Microarchitecture (December 1996)
10. McGhan, H., O'Conner, M.: PicoJava: A Direct Execution Engine for Java Bytecode. IEEE Computer, 22–30 (1998)
11. Veldema, R.: Jcc, A Native Java Compiler, Vrije Universiteit Amsterdam (July 1998)
12. Van Vliet, H., Mocha, http://www.brouhaha.com/~eric/computers/mocha.html
13. Innovative Software, Dejavu, http://www.isg.de/OEW/Java
14. Kouznetsov, P., Jad, http://www.kpdus.com/jad.html
15. Wingsoft, Wingdis, http://www.wingsoft.com
16. Remotsoft, Salamender, http://www.remotesoft.com/salamander/index.html
17. Netdecompiler, Dis#, http://netdecompiler.com/index.html
18. Son, M.-S., Bae, S.-K., Lee, Y.-S.: Design and Implementation of a Symbol Table Detranslator for C++ Compiler. In: Korea MultiMedia Society Conference, vol. 8(2), pp. 181–184 (2005)
19. Oh, S.-M., Lee, Y.-S., Ko, K.-M.: Design and Implementation of the Virtual Machine for Embedded System. Journal of Korea MultiMedia Society 9(8), 1282–1291 (2005)
20. Kim, Y.-K., Kwon, H.-J., Lee, Y.-S.: Design and Implementation of Decompiler for Generating C Program from EVM SIL. In: Korea Information Processing Society Conference, vol. 12(1), pp. 549–552 (2005)

21. Kim, Y.-K., Kwon, H.-J., Lee, Y.-S.: Java Bytecode-to-SIL Translator using an Abstract Syntax Tree. In: Korea Information Processing Society Conference, vol. 1(11), pp. 519–522 (2004)
22. Park, J.-K., Kim, Y.-K., Kwon, H.-J., Lee, Y.-S.: Translator System for Embedded Virtual Machine. In: Korea MultiMedia Society Conference, vol. 8(2), pp. 180–183 (2005)
23. Tonella, P.: Alessandra Potrich: Reverse Engineering of Object-Oriented Code. Springer, Heidelberg (2005)
24. Eilam, E.: Reversing: Secrets of Reverse Engineering. Wiley Publishing, Chichester (2005)

Mobile Pharmacology

Patrik Eklund, Johan Karlsson, and Annica Näslund

Umeå University, Department of Computing Science, SE-901 87 Umeå, Sweden
{peklund,johank,dva98and}@cs.umu.se

Abstract. In mobile usage scenarios, patient care involving point of care considerations build upon increasingly complex information and information structures. Furthermore, communication of information must be enabled regardless of space and time. Decision support and guidelines are expected to communicate with other subsystems, such as those provided by information sources and hardware devices. As also these subsystems may build upon intelligence and involve their own usage scenarios, this implies further complications also the knowledge representation and software engineering tasks. In these developments, public health problems provide case studies with potentially rather huge impacts. Examples are provided e.g. by guidelines involving pharmacological treatment. Knowledge and reasoning need to interact with information management, and often involves utility of various devices. Well organized databases for pharmacological information are necessary for successful engineering of mobile extensions in these case studies. The public is now also one of the driving forces in these developments. As electronic prescriptions and generic substitutes are appreciated by the public, we experience how knowledge in its various forms provide success stories in this field.

1 Introduction

It is important not to focus on devices. A device is, from a programmer's view, basically a processor and an (open) operating system that helps you execute software. We require suitable software development kits together with emulators for given devices. Having said that, we should note how the overall software development process, as identified over the last decade, in fact remains largely the same. Since the introduction of user interface oriented operating systems, new types of devices enter the market and thereby corresponding operating systems are updated including their respective software development kits. Once in a while major changes and developments in devices require major software and operating system amendments. Software development in these situations should pay attention not to bind software structure altogether to state-of-the-art of certain present time development kits. Software development for mobile devices is always to be seen as a living process where maintenance needs to survive device changes and upgrades. However, once hardware and software environments mature, as they eventually do and stay that way for a significant amount of years, software development is as comfortable as it ever was.

M.S. Szczuka et al. (Eds.): ICHIT 2006, LNAI 4413, pp. 522–533, 2007.

In this paper we will treat communication physiology as a secondary issue. Indeed, communication protocols and media do not, and should not, interfere with implementations and utility of knowledge structures such as those involving pharmacologic information. In any case, communication is technically implemented using various well specified communication software components. In the end, the communication media, be it telephony or internet, or some dual mode combination of the two, must be independent from the application and its usage.

2   Related Work

Health care professionals have realized the benefits of using mobile devices and handheld computers. Various surveys show a usage in diverse areas. The usage of handheld computers in U.S. family practice recidency programs was shown in [9] to be widespread. Mobile clinical applications in the programs included medical reference applications, such as drug information, and, in a few cases, integration with the electronic medical record (EMR). One important factor that limited the scope of handheld usage was the fact that, at that time, only 22% of the programs had an EMR.

An earlier study [2] presents a survey among physicians and nurses concerning suitability of mobile access to clinical information. The results reveal an acceptance of mobile communication, even though technological issues such as network problems and the relatively large size of the mobile prototype caused some resistance among physicians. Physicians found access to drug information especially useful in the mobile prototype.

In [7], a rule-based electronic drug prescribing system is presented that uses information concerning drug interactions. This system and the rules within it are however more specialized for drugs associated with renal diseases. In their evaluation, professional users indicated that the system provided useful information such as warning messages for possible drug interactions. The system highlights advantages of the mobility and direct access to information through mobile devices. It also demonstrates the utility of a direct connection to the patient record information about previously prescribed drugs.

Several drug reference databases exists for handheld computers, see [8]. One of the more spread is the ePocrates Rx drug information guide. A survey of users of Rx was done in [19], where it was found that such a tool saved time and contributed to decisions on drug prescriptions. However, Rx does not have a publicly available software development kit that allows it to be integrated with other applications.

Apart from pharmacological examples, an application for wireless transfer and image processing on a mobile device of medical images from a Picture Archiving and Communication System (PACS) is presented in [3]. In an application intended for consultation purposes rather than image processing, authors show that even by reducing the image color depth, images are still of sufficient diagnostic quality.

However, mobile devices do suffer from inherent hardware and software problems. Studies [5] and [15] highlight some of these problems such as cumbersome text entry and bad connectivity. In both studies, users used docking of the devices to update data. This process requires the device to be physically connected to a host computer on a regular basis in order to update and synchronize data on the mobile device. A preferable solution is to connect through a local wireless network [11], such as done in [3,12,6,17], or to use the mobile phone system [22].

Additionally, screens are generally speaking a limitation on mobile devices [5]. Surprisingly, users participating in [15] did not consider the limited screen size of the devices as a major problem. Instead, they focused more on positive aspects such as the portability of the devices used. Differences in results can partly be explained by diversity in devices and application areas. In [15,19] users had positive attitudes to mobile devices, mainly due to earlier experience with mobile technology. A possible solution to the limited screen problem of user interfaces can be found in [12]. Haptic interfaces make use of human sense of touch to add levels of interaction.

Complexity of drug prescribing naturally grows as the number of available drug options (articles on the market) increase. Also, while the prescribed drug may be the proper one, the dosage can either be too low (not resulting in the desired effect) or be too high (toxicity). Potential interactions with other, previously prescribed, drugs increase the complexity of prescribing drugs. Interactions between two drugs cannot only change the effect of one or both drugs but also cause undesired combined effects [1].

Authors in [4] surveyed drug treatments of elderly patients in six European countries in order to study drug-drug interactions. Patients selected to be in the study were older than 65 years, used at least four prescribed drugs and were living at home, still able to take care of themselves. The average number of drugs per patient were 7.0. Potential drug-drug interactions (DDI) were discovered in 46% of this population. Almost 10% of the potential DDIs found were classified as combinations to be avoided.

While [4] only looked at medication problems for elderly, it is nevertheless clear that carefully maintained databases with information on drug-drug interactions, drug side-effects etc. can play an important part in the drug prescribing task [21] also for other patient groups. However the frequency of drug-drug interactions could be expected to be lower than the one in [4] in most cases simply because of fewer simultaneously used drugs.

3   Mobility of Pharmaceutical Information

Pharmaceutical information, like warnings and interactions, or information on generic substitutes, should be available instantly regardless of location or time. In a home care scenario, we might need to collect information about drugs prescribed to the patient or customer visited. A mobile device including data collection facilities is then suitable, in particular if the device additionally comes with various questionnaires or even screening and decision support tools. It is possible to change a prescription of treatment for hypertension, identify generic substitutes, and even contribute to screening of cognitive disorders, all using one and the same mobile device.

The mobile device can in this way function as an important decision support tool. When we speak of the mobility of pharmaceutical information we do not only mean the mobility of devices but the integration/mobility of pharmacological information with other information systems. Health care professionals are obligated to motivate their treatment decisions. Therefore the device should be integrated in the clinical process and interact with other systems such as the electronic patient record (EPR).

Because of the abundance of EPRs used in health care today, a decision support system should not be built only for a specific system. In this way, successful application of decision support systems depend on the use of terminology systems to enable data exchange with other clinical systems. A terminology system assigns terms to concepts or objects based on the specification of these concepts or objects, thereby allowing systems to understand these concepts with considerably less ambiguity.

The need for terminology systems are no less important in pharmaceutical care. Complexity of drug prescribing naturally grows as the number of available drug options (articles on the market) increase. Also, while the prescribed drug may be the proper one, the dosage can either be too low (not resulting in the desired effect) or be too high (toxicity). Potential interactions with other, previously prescribed, drugs increase the complexity of prescribing drugs. Interactions between two drugs cannot only change the effect of one or both drugs but also cause undesired combined effects.

4 ATC

Originating from the Norwegian Medicinal Depot, the Anatomic Therapeutic Chemical (ATC) classification system [23] is now an accepted WHO standard. The main purpose of ATC is to facilitate national and international comparisons of drug utilization data. The acronym ATC tells in itself something about how the codes are constructed. The anatomical (A) part denotes the organ system for which the drug is intended, therapeutic (T) tells about for what therapy the drug is used (main indication of use), and finally chemical (C) stands for the chemical class of the drug.

To further drug utilization statistics, a unit of measurement called defined daily dose (DDD) was also developed to complement ATC. A DDD is the average dose per day for a drug that is used for its main indication when treating adults. It is however not a recommendation of dosage. Indeed, the DDD could be in the middle between two commonly prescribed dosages and as such never be an actual prescribed dosage.

To outline the ideas in ATC, we look in Table 1 at the classification of paracetamol (code N02BE01), an analgesic drug (pain reliever that does not affect consciousness) that is commonly used to relieve pain and fever.

Table 1. Classification of "paracetamol"

N	Nervous system	1st level, main anatomical group
N02	Analgesics	2nd level, therapeutic subgroup
N02B	Other analgesics and antipyretics	3rd level, pharmacological subgroup
N02BE	Analides	4th level, chemical subgroup
N02BE01	Paracetamol	5th level

The DDD for paracetamol is 3 g/day and there are three administration routes registered (oral, parenteral and rectal).

There are relations in ATC that can be said to be in the form "used to treat". As an example, the drug amoxicillin (J01CA04) can be of use against certain infections caused by bacteria, for example pneumonia. The ATC group J01 is reserved for drugs whose main indication is "antibacterials for systemic use". Since amoxicillin belongs to that group, we get an indication of what it is mainly used for.

It is important to note that the drugs in ATC are, with a very few exceptions, classified according to their main indication of use. As such, the ATC terminology system should not be used as a recommendation of therapeutic usage. Drugs in the same ATC group are not necessarily therapeutically equivalent, even if they are classified under the same chemical subgroup (4th level ATC). Drugs can and do have different dosages, drug-drug interactions or administration routes.

As a complement to ATC encoded drugs, article codes are often used. They allow each article to be uniquely identified in patient journal notes, inventory systems and so on.

5   Information Structures

In this section, we will discuss underlying data structures of pharmacologic information. Data structures and information types are described as a basis for encoding information management of various kind.

In a rather straightforward usage, e.g. with asthma or pregnancy placing restrictions on usage of pain killers, the article number, as available in the package barcode, leads immediately to required warning information about that particular drug. Devices with barcode scanners and wireless internet connections are then useful tools providing complete and immediate access to required information.

In this paper, pharmacologic information is organized using an open source database [20] (see Figure 1). Information management is supported by a application programming interface (API) supporting information structures and data types as described and used above.

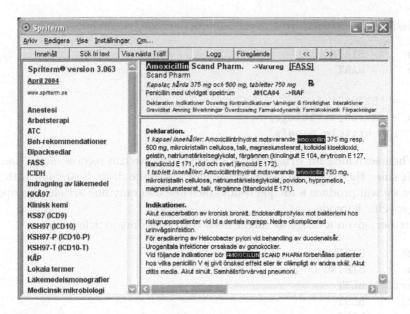

Fig. 1. Browsing for pharmacologic information, a solution by Spriterm, used by several County Councils in Sweden

We illuminate usage of these information structures by encoding search of drugs and articles using search strings as well as barcodes. Such a search leads typically to a drug name. Given a drug name it might be interesting to continue towards finding article prices or interactions.

In this context and for these examples, various handles to data or lists are useful. We need handles e.g. to articles and drugs, respectively denoted hART and hDRG, as well as to ATC codes and interactions, respectively denoted hATC and hICTS. Further, we use hLST generally to denote a handle to a list.

Listing drugs for the specific search string *amoxi* would be achieved by the following execution.

```
s       STRING
dl      hLST
d       hDRG

s = "amoxi"
dl = search_drugs(s)
loop
  d = remove_first_element(dl)
  print drug_name(d)
until empty(dl)
```

Finding a particular drug given only its barcode, is similar.

```
s       STRING
a       hART

s = "522243"
a = search_article(s)
print article_name(a)
```

When searching for drugs with a search string we can receive more than one result since there may be many articles for a specific drug. Searching with barcodes we will produce a unique result since one and only one article corresponds to a specific barcode.

Further, given a drug, we might need corresponding article prices.

```
s       STRING
a       STRING
artl    hLST
dl      hLST
d_1     hDRG
d_2     hDRG
art_1   hART
art_2   hART
atc     hATC

s = "522243"
art_1 = search_article(s)
print article_price(art_1) article_name(art_1)

d_1 = drug_family(art_1)
atc = drug_atc(d)
a = atc_code(atc)
dl = drugs_with_atc(a)
loop
  d_2 = remove_first_element(dl)
  artl = drug_articles(d_2)
  loop
    art_2 = remove_first_element(artl)
    print article_price(art_2) article_name(art_2)
  until empty(artl)
until empty(dl)
```

A more elaborate example is given for drug interactions. Note for the following two implementations

```
s       STRING
dl      hLST
```

```
d_1     hDRG
d_2     hDRG
art     hART

s = "522243"
art = search_article(s)
d_1 = drug_family(art)
dl = drug_interacting_drugs(d_1)
loop
   d_2 = remove_first(dl)
   print drug_name(d_2)
until empty(dl)
```

and

```
s       STRING
il      hLST
d       hDRG
i       hICST
art     hART

s = "522243"
art = search_article(s)
d = drug_family(art)
il = drug_interactions(d)
loop
  i = remove_first(il)
  print interaction_atc_code(i)
until empty(dl)
```

Fig. 2. A mobile device (screensize 320 × 240 pixels) with a barcode scanner

how the side effects are different. In the latter implementations ATC codes appear explicitely.

In our example, the search string *amoxi* leads to drug name *amoxicillin*. Screenshots are drawn from a mobile device having a screen size of 320 × 240 pixels (see Figure 2).

6   Client/Server Architectures

In Section 5, the device acts as a client, and the server provides the information using the programming interface to the drug information database. In this client/server architecture, a suitable communication protocol must be designed and utilized so that clients on various operating system platforms can request information from the server.

From the server point of view, various information provision services are implemented to meet client demands. In addition to managing pharmacologic information, the server can provide guideline based decision support. Decision

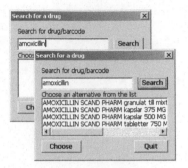

Fig. 3. Search with drug name

Fig. 4. Search with barcode

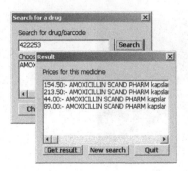

Fig. 5. Search for price using the barcode

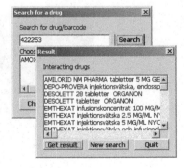

Fig. 6. Search for interactions

related to drug treatment are particularly suitable in this context, as decisions frequently are combined with requirements related to drug information. In [18] we implemented a computerized guideline for hypertension treatment. This desktop solution was later adapted for use over the internet [16,10].

Communication between clients and the server is typically as follows.

```
Client                           Server

send(open)                       receive(open)
receive(open)                    send(open)
send (search\_name)              receive(search\_name)
send (amoxi)                     receive(amoxi)
receive(nr\_hits)                send(nr\_hits)
receive(amoxicillin 375 mg)      send(amoxicillin 375 mg)
...                              ...
receive(amoxicillin 750 mg)      send(amoxicillin 750 mg)
```

Communication is tcp/ip based and as implemented within software development kits. This is usually event based, i.e. communication is observed and messages are managed in an event handling manner. The protocol thus provides synchrony like communication whereas the application developed may depart from synchrony requirements and handle messages as required. Our approach adopts the latter view. It is important also to consider that protocol levels are handled somewhat differently by software development kit vendors. This obviously calls for identification of communication patterns between clients and the server, and that implementation of these patterns are supported for respectively client and server side software.

7   Towards Platform and Device Independency

All devices do not have the necessary operating system or memory capacity to contain and run client side tools that interact with server side pharmacological databases. If end-users only require general information on a specific drug, tag-based formats involving various markup languages can be utilized.

There are several advantages to communicating the information in this way. In information presentation scenarios, no installation is needed in order to access information since an internet connection and a standard browser will suffice. Browsers exist for most platforms and devices, from desktops to handheld devices and mobile phones. Use of markup languages with mobile phones to access medical information is presented in [14] and [22].

We have implemented a server application (PharmaJ) [13] that formats information from the pharmacological database as documents encoded by a chosen markup language. A browser on the client side has no knowledge *per se* on how to display tagged information. Therefore stylesheet languages are used in order to transform markup language data to be displayed by the client browser.

Also, a more complex client application could directly parse the XML data (ready-made parsers exist for most platforms) and thereby not only display the results but also use the information for making decisions. PharmaJ really provides a more structured and interchangeable format, removing the need for a specific operating system or installation of the complete pharmacological database locally for the end-user.

8 Conclusions

The starting point of this paper was the existence and maintenance of a public application programming interface to pharmacologic information. Various software solutions have been developed to enable a broad range of applications e.g. in clinical practise and home care.

Our software developments have been focused on providing accessible protocols and APIs for the drug database used. This approach is contrary to the approach taken with Rx where the application is intended to be used as is. Our approach allows information from the database to be integrated with external patient record systems and decision support systems.

These developments clearly shows how recommendations and national initiatives have promoted usage and utility of pharmacology databases. Important additional features are possibilities to integrate with decision support and information systems related to pharmacologic treatment. The final goal is to ensure availability of drug information at the time and place of the medical decision.

References

1. American Medical Directors Association: 'White Paper on Quality Pharmaceutical Care in Long Term Care'. Technical report, American Medical Directors Association, Columbia, MD (1999)
2. Ammenwerth, E., Buchauer, A., Blaudau, B., Haux, R.: Mobile information and communication tools in the hospital. International Journal of Medical Informatics 57, 21–40 (2000)
3. Andrade, R., Wangenheim, A., Bortoluzzi, M.: Wireless and PDA: a novel strategy to access DICOM-compliant medical data on mobile devices. International Journal of Medical Informatics 71, 157–163 (2003)
4. Björkman, I., Fastbom, J., Schmidt, I., Bernsten, C.: Drug-Drug Interactions in the Elderly and the Pharmaceutical Care of the Elderly in Europe Research (PEER) Group. The annals of Pharmacotherapy 36, 1675–1681 (2002)
5. Carroll, A., Saluja, S., Tarczy-Hornoch, P.: The implementation of a Personal Digital Assistant (PDA) based patient record and charting system: lessons learned. In: Proceedings of American Medical Informatics Association Symposium, pp. 111–115 (2002)
6. Chen, E., Mendonça, E., McKnight, L., Stetson, P., Lei, J., Cimino, J.: PalmCIS: A Wireless Handheld Application for Satisfying Clinician Information Needs. Journal of the Americal Medical Informatics Association 11(1), 19–28 (2004)

7. Clark, I.R., McCauley, B.A., Young, I.M., Nightingale, P.G., Peters, M., Richards, N.T., Abu, D.: Electronic Drug Prescribing and Administration - Bedside Medical Decision Making. In: Horn, W., Shahar, Y., Lindberg, G., Andreassen, S., Wyatt, J.C. (eds.) AIMDM 1999. LNCS (LNAI), vol. 1620, pp. 143–147. Springer, Heidelberg (1999)
8. Clauson, K., Seamon, M., Clauson, A.: Evaluation of drug information databases for personal digital assistants. American Journal of Health-System Pharmacy 61, 1015–1024 (2004)
9. Criswell, D., Parchman, M.: Handheld computer use in U.S. family practice residency programs. Journal of the American Medical Informatics Association 9, 80–86 (2002)
10. Eklund, P., Eriksson, S., Karlsson, J., Lindgren, H., Näslund, A.: Software development and maintenance strategies for guideline implementation. In: Adlassnig, K.-P. (ed.) Proc. EUNITE-Workshop Intelligent Systems in Patient Care, pp. 26–34. Austrian Computer Society (2001)
11. Frenzel, J.: Data Security Issues Arising From Integration of Wireless Access Into Healthcare Networks. Journal of Medical Systems 27(2), 163–175 (2003)
12. Grasso, M.: Clinical Applications of Handheld Computers. In: CBMS 2004. Proceedings of the 17th IEEE Symposium on Computer Based Medical Systems, pp. 141–146 (2004)
13. Karlsson, J.: Interface for Accessing Pharmacological Information. UMINF-05.06, Dept. of Computing Science, Umeå University (2005)
14. Maglaveras, N., Koutkias, V., Meletiadis, S., Chouvarda, I., Balas, E.A.: The Role of Wireless Technology in Home Care Delivery. In: Patel, V.L., Rogers, R., Haux, R. (eds.) Proceedings of Medinfo 2001, pp. 835–839. IOS Press, Amsterdam (2001)
15. McAlearney, A., Schweikhart, S., Medow, M.: Doctors' experience with handheld computers in clinical practice: qualitative study. British Medical Journal 328, 1162–1166 (2004)
16. Näslund, A.: EviBase Ht Web - implementation of decision support for hypertension treatment (in Swedish), UMNAD 393/02, Dept. of Computing Science, Umeå University (2002)
17. Oyama, L., Tannas, H., Moulton, S.: Desktop and Mobile Software Development for Surgical Practice. Journal of Pediatric Surgery 37(3), 477–481 (2002)
18. Persson, M., Bohlin, J., Eklund, P.: Development and maintenance of guideline-based decision support for pharmacological treatment of hypertension. Computer Methods and Programs in Biomedicine 61, 209–219 (2000)
19. Rothschild, J., Lee, T., Bae, T., Bates, D.: Clinician Use of a Palmtop Drug Reference Guide. Journal of the American Medical Informatics Association 9, 223–229 (2002)
20. Spriterm, version 3.069 (October 2004)
21. Schiff, G., Rucker, T.: Computerized Prescribing - Building the Electronic Infrastructure for Better Medication Usage. Journal of the American Medical Informatics Association 279(13), 1024–1029 (1998)
22. Vassányi, I., Mógor, E., Szakolczai, K., Tarnay, K.: Mobile Access to a Cardiology Database. In: Patel, V.L., Rogers, R., Haux, R. (eds.) Proceedings of Medinfo 2001, p. 881. IOS Press, Amsterdam (2001)
23. WHO Collaborating Centre for Drug Statistics Methodology: ATC classification index with DDDs. Technical report, Norwegian Institute of Public Health (2005)

Wireless Control System for Pet Dogs in a Residential Environment

Ji-Won Jung and Dong-Sung Kim

Networked System Lab., School of Electronic Eng.
Kumoh National Institute of Technology, Korea
{kisstice,dskim}@kumoh.ac.kr
http://knut.kumoh.ac.kr/~nsl

Abstract. This paper concerns a wireless control system (WCS) for pet dogs using wireless sensor networks (WSNs) in a residential environment. The developed WCS is composed of a central control system, a wireless auto-feeder, a small-sized guidance robot, and wireless sensing devices. The developed system uses luminance, temperature, and sound data from a pet dog and the surrounding environment. The presented design method provides an efficient way to control and monitor the pet dog using WSNs. The implemented system can be used as a design framework of portable devices for pet dog control within a residential network.

Keywords: Wireless Control System (WCS), Wireless Sensor Network (WSN), Pet dog control system, Sensor network platform, Frequency analysis, Small-sized guidance robot.

1 Introduction

A WSN can support an intelligent operation on smart environments using multiple WSN nodes. A WSN node consists of a sensor and short range radio communication module. Each node can detect and deliver local conditions such as temperature, sound, light, or the movements of objects. Therefore, there were wide ranges of applications envisioned for WSNs.

Military operations such as enemy permeation detection and attack perception of biological or chemical weapons have been studied [1]. As a control of natural resources, WSNs have been applied to watching endangered animals [2][3][4], water containment monitoring, and forest fire and flood watch [5]. For the field of health care, WSNs have been applied to remote medical treatment systems [6]. In addition, WSNs have been applied to the building automation and industrial control fields such as factory automation and nuclear plants [7][8].

The growth of wireless networks and usage led to a potential breakthrough in ubiquitous computing possibilities for pet control. To the best knowledge of the authors, a WCS for pet dogs using WSNs has not been reported as yet in technical literature. Therefore, this research can be a useful reference for designing a pet animal control system using WSNs in a residential environment.

M.S. Szczuka et al. (Eds.): ICHIT 2006, LNAI 4413, pp. 534–545, 2007.

In this paper, WSNs are applied to a WCS for pet dogs. The implemented WSN system is composed of wireless sensing modules, a central control system and wireless devices such as a small-sized guidance robot and automatic feeder. The wireless sensor module can be attached to the pet dog. It measures the luminance, temperature and sound data and sends it to the WCS.

This paper is organized as follows. The problem analyses of wireless pet dog control are described in Section 2. Section 3 presents the structure of the WCS for a pet dog. The developed pet dog WCS and experimental results are presented in Sections 4. Finally, conclusions are drawn in Section 5.

2 Problem Analysis

Because of the demands for efficient control of pet dogs, interest in a pet dog control system has increased. However, the systems for managing pets have been used in restricted fields such as an automatic temperature controller of a tropical fish tank or food feeders.

Recently, the studies on behavior analysis between owners and pet dogs have been studied [9][10]. These studies have been focused on the analysis of the behavior of the pet dog based on utterance, and physiological signals such as the heart beat count. These results can be applied to the pet dog control system.

The major problem of a pet dog control system lies on a sound occurrence, feeding, or disposal of urine/feces. The requirements of a pet dog control system using WSNs are affected by many environmental factors such as temperature, residential types, and so on.

The first problem lies in noise generated by a pet dog. Pet dog utterances can be caused from the instinctual guarding of territory unfamiliar objects. However, it causes inconveniences between apartment residents and dog owners. Table 1 shows the audible range of a person and a dog, respectively. It shows that the dog's hearing range is four times larger than a human's. The frequency analysis of pet dog sounds can be used in the analysis of demands and health condition. Measurement of frequency distribution is shown in Fig. 1. Utterances have different frequency distributions according to the emotion of the dog.

Measured wave forms have a constant pattern according to the breed and emotional situation. This can be defined as a voiceprint. This is one of the tools for managing several dogs together. Fig. 2 shows that wave forms have constant patterns and a sequential procedure for analyzing them respectively. The second problem is that an automatic food feeder based on a timer can perform an unnecessary operation when the pet dog leaves the residence. Whenever a person

Table 1. Comparison of hearing range

Type	Utterance range [Hz]	Audible range [Hz]
Person	85~1100	20~16000
Dog	452~1080	15~60000

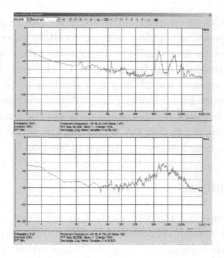

Fig. 1. Frequency graph of dog's utterance

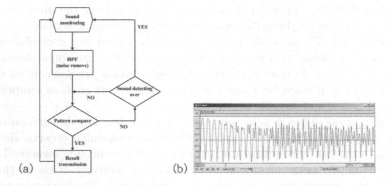

Fig. 2. Analysis of waveform: (a) Flowchart of analysis (b) Tested waveform

and dog goes outside, the feeder needs an additional manual setting for the operation. If the food feeder can detect the pet dog automatically, it can provide a more efficient way of feeding.

The third problem lies in an environment change by urine and feces. This is caused by a climate change. This can be monitored by a temperature sensor attached to the pet dog's collar. The temperature data can be used to control a fan and heater. It can be applied to health condition monitoring using body temperature checks.

The operation of a pet dog small-sized guidance robot for attracting the interest of pet dogs can be used for decreasing unnecessary noise at night. Based on these problems, the whole structure of a pet dog control system is presented in Section 3.

Table 2. Control object using sensing data

Measurement items	Control Object
Luminance	Food feeder
Temperature	Fan or heater
Sound	Guidance robot
Mixed Sensing	Health check

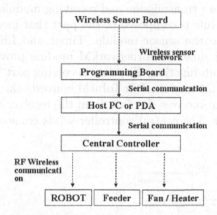

Fig. 3. Overall structure of pet control system

3   Overall Structure of the Pet Dog Control System

The developed system uses the basic module for wireless computing environments. These wireless devices include light, temperature, magnetic and acceleration sensors. These units include a microphone-based sounder system which can be used for utterance range measurements of pet dogs. It adapts the IEEE 802.15.4 communication protocol, which is standard for low-speed short-range personal area networks [11][12].

The implemented system can gather sensor information from each wireless sensor node and transmit collected data to a central control system. A control command from the central control system is transmitted to each device. Each controller's operation is designed to attract the interest of the pet dog using mixed sensing information. Table 2 shows a control object and user environment.

In operation, sensing information from the pet dog and environment invokes the small-sized pet dog guidance robot and other devices. When the pet dog approaches the wireless auto feeder, a sensor node checks the luminous intensity. Then, it activates the automatic food feeder. When a variation in room temperature is detected, a heater or electric fan can be activated.

The overall structure of implemented system is shown in Fig. 3. The host PC or PDA receives the measured data from each sensor node and transmits through the wireless links. All components are controlled via wireless/wired links.

During operation, the sensed data such as luminance, temperature, and sound are transferred to the central controller. Communications between the control system and small-sized guidance robot or automatic foot feeder are performed via RF (433Hz) radio communication.

3.1   Technical Details of Wireless Module

Wireless Sensing Module
A block diagram of data transmission and receiving module are shown in Fig. 4. The transmission module is composed of the part that measures light, temperature and sound from the sensor module. Timer and LED operation can be supported by each module. SensornetworkM module provides the function to control and manage sub-function. The data receiving part is composed of I/Os, Library and RfmToIntM module. RfmToIntM converts the received I/O data to integer type. RfmToInt receives the data from the receiver module and transfers to a central controller. The central controller sends commands to each wireless device.

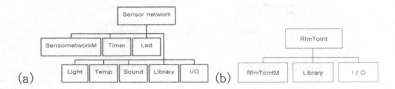

Fig. 4. Structure of communication module: (a) Transmission module (b) Receiving module

Packet Structure of Wireless Module
The transmission protocol of the WSN system is composed of 36 bytes. The frame structure and the details of data field are presented in Fig. 5. The Add and Type field includes a destination address and data type, respectively.

The Group field includes the group ID of channel. The length field includes the size of received and transmitted packet length. The Data field includes information of sensed data within 29 bytes. It allocates information of light, sound and temperature data. For periodic data transmission, data packets are transmitted periodically even if there is no sensed information.

dest addr	handlerID	groupID	msg len	source addr	counter	channel
7e 00	0a	7d	1a	01 00	14 00	01 00

readings
96 03 97 03 97 03 98 03 97 03 96 03 97 03 96 03 96 03 96 03

Fig. 5. Frame structure of WCS

Graphical User Interface (GUI)

Fig. 6 shows the implemented GUIs. It shows the screenshot of the program that is the main control part and device control module respectively. It can be preset to an automatic or manual mode by a user.

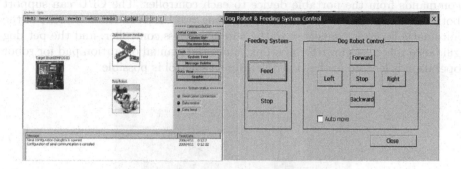

Fig. 6. System monitoring program and control box

3.2 Design and Implementation of Control Devices

Fig. 7 shows the overall system structure and its operation. The overall system is composed of a wireless sensor board, controller, robot and automatic feeder. Data communication is performed at RF communication of wireless platforms and serial communication is connected between receiver and controller. A portable device is connected to the controller via serial link. Controller, automatic feeder, electric fan or heater, and robot are connected via RF wireless links. The wireless controller creates control data according to the measured data and delivers the data to each controller. Based on the measured data, guidance robot, automatic feeder and electric fan/heater can be operated.

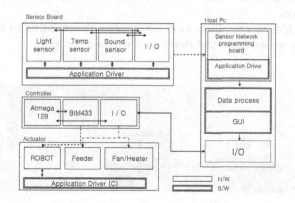

Fig. 7. Overall system structure and operation flow

Wired / Wireless Controller

Fig. 8 shows the wireless sensor node and controller. The wireless transceiver on the pet dog is combined with a sensor board. The wireless transceiver system transmits measured data using wireless communication. The wireless sensor node is attached to the collar of the pet dog. The wireless controller delivers received commands from the portable device to each controller. The CPU can support both serial and wireless communication using two UARTs. The controller of the automatic feeder is connected to the wired/wireless controller, and the pet dog guidance robot uses an RF radio link. Since the manual operation pad for robot operation is located in the controller, direct control is possible.

Fig. 8. Wireless sensor node and controller

Wireless Pet Guidance System

Fig. 9 shows the wireless pet guidance system. If the sensor board detects a pet dog's sound, the central controller transmits an activating command to the pet dog guidance robot. At that time, the pet dog guidance robot is operating according to the predefined scenarios. It can move in any direction or be controlled using manual mode by the user.

The operation of the pet dog guidance robot can be tested in various cases based on the given conditions. We tested various situations using the robot's movements (circular, forward, backward and complex) and recorded the user's voice commands.

An automatic feeder supplies food automatically according to luminance condition which is sensed from the developed wireless device.

Fig. 9. Wireless pet guidance system

Connection with Residential Network

If a user wants to connect a control system from the outside the house, it must connect with a residential network server. The pet control system could be used via a cellular phone or PDA from outside.

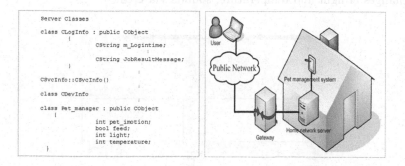

Fig. 10. Pet control system in a residential network

A residential network server operates data processing of an oriented object, and pet dog control system may have transmit data of class form to the residential network server [13]. Fig. 10 shows the overall structure of the residential network system.

4   Experimental Test

4.1   Test Environment

The test environment set for the developed system is as follows. We assume that the environment is an indoor and apartment residential type in the daytime.

(1) Light above 90 % is intercepted in the place where the pet dog eats food.
(2) When temperature changes ±2 ° C, the temperature maintenance system is set to operate.
(3) If sound measurement comes up to 75% of maximum value, a central controller activates.

In an experimental test, we tested the operation of the automatic feeder by dog approach or specific behaviors. The pet dog guidance robot starts to operate when sound measurement comes up 75% of maximum value. The robot's various movements (circular, forward, backward and complex) and the recorded voice of a person are tested for attracting the pet dog's interest. In the case of temperature changes over ±2 ° C, the electric fan or heater is automatically activated to maintain indoor temperature.

4.2   Experimental Test Using Pet Dog

The user monitoring function is implemented for measuring data from the sensor board. The measurement window of sensor data is shown in Fig. 11. Each graph displays the received light, temperature and sound data. The user can monitor the changes of light and temperature, sounds via GUI.

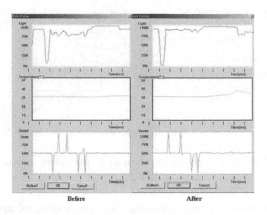

Fig. 11. Experimental measured sensor data

Light Processing
When the developed system detects light changes by pet dog movements, the wireless sensor board transmits sensing status. Received data is displayed through GUI at the central control system. The system controller sends the activation command to the automatic feeder. The automatic feeder supplies food to the pet dog automatically. The test is conducted by using a small test pet dog. According to the test results, even if there are some differences in size, the automatic feeder worked without error.

Temperature Processing
In the developed system, the wireless automatic control module can control the fan and heater according to the measured temperature.

Indoor temperature was set as 30 ° C in the actual test. We changed the surrounding temperature by using the heater and fan. If temperature changed, the command to start or stop the fan and heater was given by the wired/wireless system controller.

If the desirable temperature is set, the commands (Fig. 12) are given to the fan and heater. As shown in Fig. 12, when current temperature is 30 ° C, and the user wants to adjust the desirable temperature up to 34 ° C, the fan is stopped and heater is activated.

If the desirable temperature is changed to 26 ° C again, the fan starts to operate. When there is difference of 2 ° C between the indoor temperature and desired temperature in the controlled item, heater and fan are activated. A further part

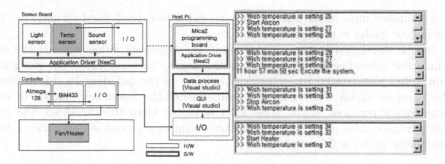

Fig. 12. Temperature maintenance module

can be developed to expand the system that can control and alarm and clean the air automatically when the pet dog spoils the atmosphere.

Operation Procedure

The operation procedure of the implemented WCS is shown in Fig. 13. When the dog's sound and temperature are sensed in the control module, it transmits data to the programming interface. The central controller has the role of transmitting command data into the wireless system controller. The wired/wireless system controller uses RF communication to operate the pet dog guidance mini-robot and fan/heater.

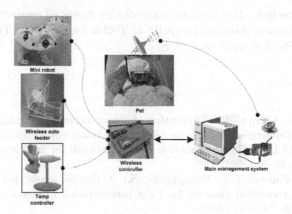

Fig. 13. Operation scenario of WCS

In the developed system, if the sound measurement value exceeds the set value, the central controller recognizes the dog as barking. Then, the guidance mini-robot is activated to attract the interest of the pet dog. In this system, the reaction signal is set with the value within the frequency domain of the pet dog's utterances. If the noise level is higher than the threshold, the pet dog guidance mini-robot will operate according to a predefined scenario or the automatic feeder will be activated.

In the developed system, if the sound measurement value exceeds the set value, the central controller recognizes the dog as barking. Then, the guidance mini-robot is activated to attract the interest of the pet dog. In this system, the reaction signal is set with the value within the frequency domain of the pet dog's utterances. If the noise level is higher than the threshold, the pet dog guidance mini-robot will operate according to a predefined scenario or the automatic feeder will be activated.

5   Conclusion

In this paper, we describe a WCS for pet dogs in a residential environment using WSNs. The developed WCS is composed of a wireless control system, a wireless auto-feeder, a small-sized guidance robot, and wireless sensing devices. Three sensed data types from a pet dog and the surrounding environment are used for the WCS.

The presented design method using sensed data provides an efficient way to control and monitor a pet dog using wireless links. The implemented system can be used as a design framework for portable devices for pet dog control.

As a future direction, one may consider an wireless operation modules using differences in utterance frequency and physiological signals such as the heart beat of a pet dog.

Acknowledgments. The authors acknowledge financial support from the Research Fund, Kumoh National Institute of Technology, Korea. They also thank the reviewers for their time and effort.

References

1. Nemeroff, G., Hampel, D.: Application of sensor network communications. In: International Conference on Military Communications, vol. 1, pp. 336–341 (2001)
2. Kumagai, J.: Life of birds - wireless sensor network for bird study. IEEE Spectrum 41(4), 42–49 (2004)
3. Yihan, L., Panwar, S.S., Burugupalli, S.: A mobile sensor network using autonomously controlled animals. In: First International Conference on Broadband Networks, pp. 742–744 (2004)
4. Sikka, P., Corke, P., Overs, L.: Wireless sensor devices for animal tracking and control. In: IEEE International Conference on Local Computer Networks, pp. 446–454 (2004)
5. Lorincz, K., Malan, D.J., Nawoj, A., Clavel, A., Shnayder, V., Mainland, G., Welsh, M., Moulton, S.: Sensor networks for emergency response: challenges and opportunities. IEEE Pervasive Computing 3(4), 16–23 (2004)
6. Timmons, N., Scanlon, W.G.: Analysis of the performance of IEEE 802.15.4 for medical sensor body area networking. In: IEEE conference on Sensor, Ad Hoc Communications and Networks, pp. 16–24 (2004)

7. Hsieh, T.T.: Using sensor networks for highway and traffic applications. IEEE Potential 23(2), 13–16 (2004)
8. Aakvaag, N., Mathiesen, M., Thonet, G.: Timing and power issues in wireless sensor networks - an industrial test case. In: International conference on Parallel Processing, pp. 419–426 (2005)
9. Vas, J., Topál, J., Gácsi, M., Csányi, V.: A friend or an enemy? Dogs' reaction to an unfamiliar person showing behavioral cues of threat and friendliness at different times. Applied Animal Behavior Science 94(1-2), 99–115 (2005)
10. Clara, P., Emanuela, P.P., Marina, V.: Heart rate and behavioral responses of dogs in the Ainsworth's Strange Situation: A pilot study. Applied Animal Behavior Science 94(1-2), 75–88 (2005)
11. Akingbehin, K., Akingbehin, A.: Alternatives for short range low power wireless communications. In: First International Workshop on Self-Assembling Wireless Networks, pp. 320–321 (2005)
12. Lynch, C., O'Reilly, F.: PIC-based TinyOS implementation. In: Second European Workshop on Wireless Sensor Networks, pp. 378–385 (2005)
13. Kim, D.S., Lee, J.M., Kwon, W.H.: Design and Implementation of Residential Network Systems using UPnPMiddleware for Networked Appliances. IEEE Transactions on Consumer Electronics 48(4), 963–972 (2002)

Intelligent Embedded Real-Time Software Architecture for Dynamic Skill Selection and Identification in Multi-shaped Robots

Laxmisha Rai and Soon Ju Kang

School of Electrical Engineering and Computer Science,
Kyungpook National University, Korea
laxmisha@ieee.org, sjkang@ee.knu.ac.kr

Abstract. This paper presents an intelligent embedded, modular software and hardware architecture for multi-shaped robots using real-time dynamic skill identification and selection. It is a layered architecture with reusable and reconfigurable modules, which can embed in an expert system as both hardware and software modules and demonstrated with snake robot and physically reconfigured four-legged robot as examples. The intelligent dynamic selection and synchronization of selected behaviors enable the mobile robot to perform many tasks in complex situations. The architecture proposed is applicable to multi-shaped robots, for dynamic selection of behaviors during reconfiguration, where the hardware and software modules can be reused during reconfiguration. Related videos of these robots can be viewed at: http://rtlab.knu.ac.kr/robots.htm

1 Introduction

Mobile robots such as snake robots and legged robots have the ability to locomote over variety of surfaces, narrow gaps and in unexpected physical conditions. Such robots are playing a critical role in many applications. In such applications, multi-shaped robots are highly desirable to perform multiple tasks by changing its shape and behavior dynamically rather than using different robots to perform different tasks. Design of such robots demands reusable software and hardware architecture. In this paper we propose layered software architecture, for dynamic skill selection and identification in multi-shaped robots. Proposed architecture has 5 layers: reasoning, behavior, skill, skill primitive and sensor/actuator driver layers. In addition to these layers, the architecture also provides flexibility to map it to software and hardware modules directly.

The paper is organized as follows. The section 2 describes the related works and section 3 describes the requirement analysis. The section 4 describes conceptual modeling; section 5 explains the design architecture supporting dynamic behavior identification and selection. The experimental evaluation is described in the section 6 and finally the conclusion in section 7.

M.S. Szczuka et al. (Eds.): ICHIT 2006, LNAI 4413, pp. 546–556, 2007.

2 Related Works

Over the years, there are many software frameworks for robots were proposed. Some of these are MIRO [1], ActivMedia's ARIA [2], Player/Stage/Gazebo (PSG) [3], ORCOS [4] and CARMEN [5]. MIRO reduces software development times and costs by providing data structures, functions, communications protocols and synchronization mechanisms specific to robots. ActivMedia robots are managed by a micro-controller that implements the server. PSG is based on client-server design and to offer an abstract interface of the hardware of the robots, not to offer common blocks. CARMEN is an open source collection of robot control software which provides basic set of primitives for robotics research. OROCOS is a software framework for robotic control software. It aims to integrate low-level real-time services provided by operating systems (OSs) and provides a high-level software framework for robotic control software development. Also, researchers have studied and proposed various approaches related to robot behaviors [6,7]. The paper [8] presents an approach to implementing these AI-level concepts into Behavior-Based Systems (BBS), without compromising BBS's key properties. In [9], researchers attempt to construct a high-level action decision process for a cognitive robot, and have developed a modular control system that uses knowledge of its individual behaviors to predict the outcome under a task context.

However, many of these architectures focus on reusable software blocks rather than both software and hardware blocks and fail to realize 1-1 mapping between hardware and software modules exactly. And one of the weak points in earlier architectures is: they lacks the ability of integrated high-level reasoning modules for intelligent behavior generation. Our major contribution in this paper is to develop embedded real-time software and hardware architecture for dynamic skill-selection and identification in multi-shaped mobile robots integrated with rule-based expert system. The architecture can also be extended to support real-time communication protocol, CAN. We believe that the proposed framework is extremely useful not only for students studying embedded real-time software but also in AI applications.

3 Requirements Analysis

The requirements include the following basic objectives:

- To develop an intelligent embedded real-time application, where the real-time expert system modules can be used in the higher layer to decide the combination of behaviors or skills during reconfiguration. The reasoning layer must be capable of managing, allocating and adjusting different behavioral tasks dynamically.
- The architecture must support to generate complex behaviors from existing behaviors, and high-level behaviors must be formed dynamically with the support of intelligent reasoning layer. The hardware and software modules

must be reusable to develop complex applications and adapt in dynamic environments.

- Third requirement is to develop reusable hardware blocks using the existing infrastructure (such as LEGO blocks) and integrate with real-time embedded software; real-time communication protocol (CAN) and rule based expert system.
- We also aim to develop virtual prototyping tools to reduce the time and complexity of the development task in complex robots such as snake or legged with reconfigurable abilities with minor modifications. As many research groups not affordable to buy expensive robot tools and simulators, the toolkit developed in our lab intended to support robotic developers by allowing real-time performance.

4   Conceptual Modeling

4.1   Modular and Reusable Behavior Modeling

To realize our proposed approach, we took case studies of two complex robots: snake and legged robot. We have selected these because of their multi-shaped and shape-reconfigurable behavior.By understanding the locomotion of snake and legged robots we can come to many conclusions. First, snakes move by pushing their body against environment, to achieve this, the different sections (Fig. 1) of snake's body must generate serpentine locomotion or lateral undulation. But, in legged robots there is a need to synchronize the movements of legs to generate purposeful locomotion. In both cases, the generating purposeful movement is a challenging task. To generate serpentine locomotion, the robot must generate left and right movements in robot units. As shown in Figs. 1 and 2, the snake robot and four legged robots are divided into smaller modules called units. Many units put together form a section. A unit is a primitive individual element of a robot. In general a snake may have N sections ($1 \leq k \leq N$) and each section may have M units ($1 \leq p \leq M$). The total number of units $= M \times N$. Similarly, a N-Legged Robot may have maximum of N sections (($1 \leq k \leq N$) and each section may have M units ($1 \leq p \leq M$). So, the total number of units required for the

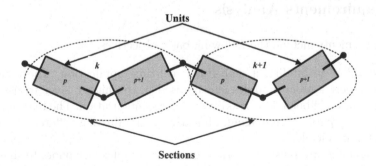

Fig. 1. Schematic of snake robot sections and units

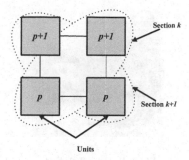

Fig. 2. Schematic which explain 4-legged robot's sections and units

movement is $M \times N$, which is similar as in the snake robot. The "KNU-Snake Robot" is the name of the snake robot developed in our lab. It has 4 modules and we reconfigured the snake robot into a four legged robot without the software change. (The changing shape of a robot dynamically is out of scope of this paper and we only have a plan as next research topic). This implies that, it is easy to reconfigure 6 module snake robot into a six legged robot rather than 4 or 8.

4.2 Reusable Module Based Hardware Reconfiguration

The physical prototyping of snake robot is developed using LEGO based embedded real-time toolkit [10]. We represent each unit of the robot as hardware module with a motor and an angle sensor embedded in it. The angle sensors are used to co-ordinate the sequence of sine wave movements in snake robot and legged-gait in four-legged robot. Fig. 3 shows the role of reusable hardware module in snake robot and in four-legged robot with minor modifications. We also designed reusable hardware module with CAN protocol [11] to support real-time communication performance to connect actuators and sensors as shown in Fig. 4. Fig. 5 shows the actual physical prototype of snake robot and four-legged robot, which is physically reconfigured from snake robot.

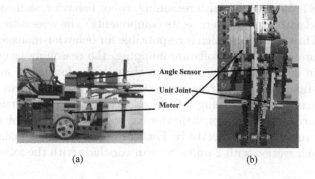

Fig. 3. View of a reusable robot module in (a) snake Robot, which is reconfigured in (b) Four-legged Robot

Fig. 4. The reusable robot module integrated with CAN

(a)   (b)

Fig. 5. (a) KNU-LEGO snake robot and (b) Four-Legged Robot

5   Design Architecture Supporting Dynamic Behavior Identification and Selection

5.1   Layered Architecture for Behavior-Based Reasoning

In order to perform high-level goals the robot must be capable of autonomous reasoning about the information it has about the outside world. In such cases a flexible reasoning mechanism that is dedicated to operate in practical problem domain is essential for proper functioning of the mobile robot. Fig. 6(a) describes the high-level layered model, with reasoning, robot behavior, skill, skill-primitive and sensor-actuator driver layers as its components. The reasoning layer details are described in section 5.2, which is responsible for behavior-management during robot locomotion. In case of software mapping, the reasoning layer mapped to expert system modules, behavior layer to threads (or functions) and skill-layer to thread or functional primitives respectively. The expert system modules are designed and implemented using CLIPS expert system tool, version 6.23 [12]. In case of hardware mappings, behavior layer mapped to robot sections and skill layer to robot units respectively. For example, in mobile snake robot with 3 sections, each section with 2 units, we can conclude with the following related parameters.

No. of behaviors generated =3, No. of sections=3, No. of Skills=2, No. of robot units=2.

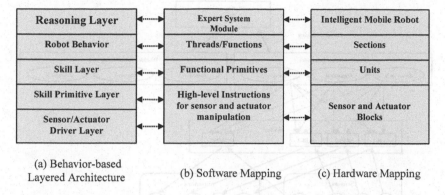

(a) Behavior-based
Layered Architecture   (b) Software Mapping   (c) Hardware Mapping

Fig. 6. The layered behavior based software architecture and 1-1 software and hardware mappings

These parameters may change depending on expected behavior of the robot behavior. In the above example, the number of behaviors generated is equal to 3 indicates that, 3 behaviors can be generated concurrently. The basic behavior of robot is modeled as a thread or function, and these robot behaviors are synchronized to generate purposeful locomotion. In our earlier work, we proposed multi-thread based synchronization of locomotion control in snake robots [13], where sinusoidal locomotion behavior of each robot section is mapped to thread directly.

5.2 Behavior Reasoning Manager

As shown in Fig. 7, the Skill reasoning manager is a decision engine that selects what types of locomotion is needed (i.e sinusoidal or legged). Depending on the decision results, the behavior is enabled (in software terms, various thread or program modules are executed) or disabled. The Disabled Robot Behaviors block indicates, behaviors which are not currently active and Enabled Robot Behaviors block indicates active behaviors. These modules are responsible for generating purposeful locomotion. Each behavior block lists all the required skills needed by the mobile robot. The basic building block of behavior-based control is a Skill. Combinations of skills generate required behavior. The new set of behaviors are stored in the knowledge base and called by the reasoning manager. In the snake robot, these skills are move-forward, move-backward, move-left, move-right, move-head and turn-tail-left, etc. The selection of these skills is based on the robot context and required behavior. This architecture adds the flexibility of choosing different skills and behaviors dynamically as the robot operates. This architecture has two main advantages. First, there is flexibility in robot configuration (from snake robot to four-legged robot) and different types of behaviors can be chosen dynamically (i.e., backward to forward).

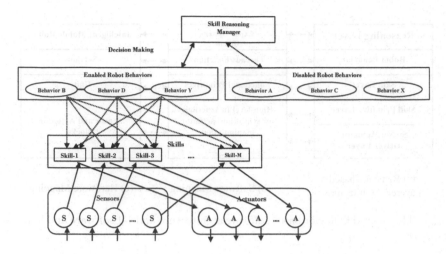

Fig. 7. Intelligent software architecture for dynamic skill selection and identification

6   Experimental Evaluation

6.1   Implementation

A hardware interface is developed in our lab to support many sensors and actuators of LEGO kit [10]. An embedded computer system with LEGO sensors and actuator interfacing can be shown as shown in Fig. 8. The KNU embedded real-time toolkit(Fig. 9) supports LEGO touch sensors, light sensors, angle sensors, motors etc. The PC104 is an embedded computer standard and commonly used with LEGO systems. Also, the PCI 104 is suitable for development of autonomous embedded robot control applications. The proposed architecture is implemented in RT-Linux based dual kernel environment. This architecture support both real time and non-real time tasks execution.

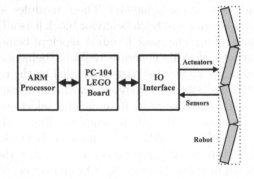

Fig. 8. An embedded computer system with LEGO sensors and actuators

Fig. 9. KNU-Embedded real-time system toolkit

6.2 Integrated CLIPS Environment

As a rule-based shell, CLIPS stores the knowledge in rules, which are logic-based structures. The rules are defined by using defrule constructs. These rules control the high-level decisions and can call on C-compiled procedures. The apply_skill() is the external module in our implementation and it is used to generate various behavior dynamically using the rules. The apply-skill() is a generic skill module which can be applied with different parameters. It has 5 parameters as shown in Table 1. The apply_skill() module can be inserted as a fact and can be applied to robot as shown below:

(apply_skill 1 2 10 1 11).

The above fact turns the motor in the unit 2 of section 1 with 45 degrees in clockwise direction with motor speed value of 11. This fact can be applied with different parameters dynamically to generate different behavior for the robot with different parameters. In practice, skills are executed based on various conditions. For example, if the robot is facing towards an obstacle, then it has to change the direction of its movements. In such cases, the reasoning layer decides

Table 1. apply_skill() module parameters

apply_skill() parameters	Description
Section	Robot Section Number, For example 1, 2, 3... etc
Unit	Robot Unit Number, For example 1, 2, 3 etc
Angle	Angle Sensor Value,to turn the particular unit of robot left-right in case of snake robot and backward-forward in case of four-legged robot,by an angle value of Φ. For every angle sensor value (AS Value)of 5,the robot turns with 22.5 degrees
Dir	Direction,to run the motor embedded in a particular unit of a particular section with clockwise (forward) or anticlockwise (backward)directions.If the value is 1, then the motor turns in clockwise and if it is 0,then the motor turns in anti-clockwise direction
Speed	Speed, the speed of the motors in each unit of robot

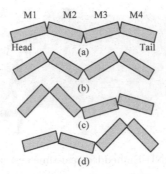

Fig. 10. Different types of snake locomotion generated with different angle sensor values (a) Rectilinear (b) Serpentine (c) and (d) Concertina

to move backward, by firing appropriate rules. The following rule makes the robot to move single step, when the fact (robot_step single) is asserted.

This rule is to test, how the robot generates the purposeful locomotion in case of snake robot or four-legged robot.

(defrule move_onetime (robot_step single)=>
(apply_skill 1 1 10 1 10) (apply_skill 1 2 10 0 10) (apply_skill 2 1 10 1 10)
(apply_skill 2 2 10 0 10) (apply_skill 1 12 0 0 10) (apply_skill 1 2 20 1 10)
(apply_skill 2 1 20 0 10) (apply_skill 2 2 20 1 10) (apply_skill 1 1 20 1 10)
(apply_skill 1 2 20 0 10) (apply_skill 2 1 20 1 10) (apply_skill 2 2 20 0 10)
(apply_skill 1 1 10 0 10) (apply_skill 1 2 10 1 10) (apply_skill 2 1 10 0 10)
(apply_skill 2 2 10 1 10))

6.3   Testing and Results

We have tested the movement of snake robot for sine-wave motion. The robot is tested to exhibit real snake behavior.The robot generated the different kinds of snake movements as shown in Fig. 10, including rectilinear, serpentine and concertina movements. The different movements (a, b, c, d) are generated with different angle sensor values. Table 2 illustrates various robot movements in a robot with 4 units, generated by setting different angle sensor values to motors (M1 to M4). In legged robot, the movements of legs are forward-backward instead

Table 2. Angle sensor values for different snake robot movements

	M1	M2	M3	M4
AS Value (10.a)	3	3	3	3
AS Value (10.b)	9	9	9	9
AS Value (10.c)	12	9	6	3
AS Value (10.d)	3	9	9	12

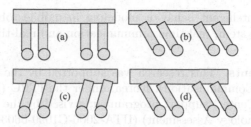

Fig. 11. Sequence of movements of four-legged robot

Table 3. Angle sensor values of robot movement in four-legged robot

	M1	M2	M3	M4	Position
AS Value (11.b)	14	14	14	14	Forward
AS Value (11.c)	26	26	26	26	Backward
AS Value (11.d)	26	26	26	26	Forward
AS Value (11.a)	14	14	14	14	Initial

of left-right as in snake robot. The Fig. 11 explains the sequence of leg movements in four-legged robot. Fig. 11(a) shows robot legs is in its initial condition, in 11(b), the robot legs moves in forward direction with greater angle sensor value (AS Value) and in 11(c) the robot legs moves backward , to balance itself against gravitational forces with lesser AS Value. Table 3 shows the various angle sensor values in robot units and resulted movements.

6.4 Advantages

The proposed architecture has several advantages including (a) Flexibility in dynamic reconfiguration and skill selection (b) Execution of different skill-sets with different priorities and (c) Combining different actuators and sensors using real-time communication protocol, CAN. Different priorities to different modules are necessary during complex situations such as obstacle detection and avoidance. For example, in a snake robot, the head-section module can be given higher priority than tail-section module to generate more realistic behavior. In case of obstacle avoidance, the tail section skill can be given more priority to remove the obstacle than the other robot sections.

7 Conclusion

This paper presents a modular software and hardware architecture for multi-shaped robot using real-time dynamic skill identification and selection. Modular and reusable software architecture combined with expert system modules provides more flexible and convenient architecture for robot control and operation. The proposed architecture is a good example for embedded real-time software architecture for intelligent behavior modeling in mobile robots by combining real-time expert system, real-time communication and real-time system.

Acknowledgements. This research was supported by the MIC (Ministry of Information and Communication), Korea, under the ITRC (Information Technology Research Center) support program supervised by the IITA (Institute of Information Technology Assessment) (IITA-2006-C1090-0603-0020).

References

1. Utz, H., Sablatnog, S., Enderle, S., Kraetzschmar, G.: MIRO - middleware for mobile robot applications. IEEE Transactions on Robotics and Automation 18(4) (August 2002)
2. ActivMedia: ARIA Reference Manual (v. 1.1.10), (2002), http://www.users.cs. umn.edu/stergios/classes/csci5551/Aria-Reference.pdf
3. Gerkey, B., Vaughan, R.T., Howard, A.: The Player/Stage Project: Tools for Multi-Robot and Distributed Sensor Systems. In: Proceedings ICAR 2003, pp. 317-323 (June 2003)
4. Orocos: Open Robot Control Software, http://www.orocos.org/
5. CARMEN: Carnegie Mellon Robot Navigation Toolkit, http://www.2.cs.cmu. edu//~carmen/
6. Abdalla, H.H.K., Karam, Z., Farsi, M.: A Knowledge Managed Multi-Controller Structure Using CLIPS. In: Proceedings of the Third IEEE Conference on Control Applications, vol. 2, pp. 1343-1344 (1994)
7. Brooks, R.A.: A robust layered control system for a mobile robot. IEEE Journal of Robotics and Automation RA-2, 14-23 (1986)
8. Nicolescu, M.N., Mataric, M.J.: A Hierarchical Architecture for Behavior Based Robots. In: Proceedings of the First International Joint Conference on Autonomous Agents and Multi-Agent Systems (2002)
9. Ratanaswasd, P., Dodd, W., Kawamura, K., Noelle, D.: Modular behavior control for a cognitive robot. In: Proc. of ICAR, Seattle WA (July 2005)
10. Jung, G.H., Kim, D.H., Park, S.H., Kim, O.G., Kang, S.J.: Experimental Software Engineering Course for Training Embedded Real-Time Systems. In: SERP 2002, pp. 410-416 (2002)
11. Etschberger, K.: Controller Area Network-Basics, Protocols, Chips and Applications (2002) ISBN 3-0000-7376-0
12. CLIPS, http://www.ghg.net/clips/CLIPS.html
13. Rai, L., Kang, S.J.: Multi-Thread based Synchronization of Locomotion Control in Snake Robots. In: Proceedings of The Eleventh IEEE RTCSA 2005, pp. 559-562 (2005)

The Accurate Performance Evaluation of Time Hopping UWB Systems with Pulse Based Polarity

Jang-Woo Park[1], Kyung-Ryoung Cho[1], Nam-Hong Jo[2], and Sung-Eon Cho[1],*

[1] Dept. of Inform. & Comm. Eng., Sunchon National University,
Sunchon-shi, Chonnam, 540-472, Korea
{jwpark,jkl,chose}@sunchon.ac.kr
[2] Zenocom. co. LPD
itnhjo@naver.com

Abstract. The bit rate performance of the time hopping impulse radio UWB system with the pulse based polarity is analyzed. It is well known that the pulse polarity helps reduce the spectral spike appearing in the conventional impulse radio system. This paper provides the method for accurately modeling the multiple access interference(MAI) in the system with the pulse polarity. The characteristic function is used to consider the MAI. We also show the MAI can be simplified as Gaussian random variable when the number of pulses representing an information symbol or the pulse rate becomes large. It is obtained directly from approximating the Characteristic Function of the MAI in case of the large number of pulses. Some results have been shown to prove validity for our method. The results also show with the total processing gain fixed, increasing the pulse rates proves the system performance. But the system without the pulse polarity does not.

Keywords: Ultra Widband, MAI.

1 Introduction

Communication systems employing UWB(ultra-wideband) technology have been strongly investigated since FCC(US Federal Communications Commission) approved the use of this technology. Its robustness to severe multipath conditions and low cost and low power implementation make the UWB technology a viable solution for high-speed indoor short range wireless communication systems. PPM(pulse position modulation) and PAM(pulse amplitude modulation) were introduced as a modulation method for UWB signals. To avoid catastrophic collisions among different users, UWB signals employed TH(time hopping) method.

These conventional impulse radios send several pulses representing an information symbol with the same polarity. Recently, pulse-based polarity randomization was proposed, where each pulse has a random polarity code [1,2]. The

---

* Corresponding author.

M.S. Szczuka et al. (Eds.): ICHIT 2006, LNAI 4413, pp. 557–565, 2007.
© Springer-Verlag Berlin Heidelberg 2007

use of polarity codes help optimize the spectral shape abiding by FCC specifications by eliminating the spectral lines that are inherent in IR systems without polarity [1] and also can provide additional robustness against MAI(multiple access interference).

The spectral shape and performance analysis for IR systems with a random polarity code were investigated by Piazzo [1]. He called these IR systems DS-IR(direct sequence impulse radio). Gezici et al. [2] have investigated performance of IR UWB systems with pulse based polarity randomization, including the effects of interframe interference and multi-access interference in the Gaussian channel and multipath environment. They proved that when the number of pulses representing the information pulses goes very large, the MAI can be approximated to the Gaussian random variable. Fishler and Poor [3] explained the tradeoff between two types of processing gain, where the one type is the pulse rate or the number of pulses for the one information symbol, denoted N_f and the other is called the pulse spreading gain, N_c which is defined as the ratio of the average total time between two consecutive transmissions and the actual transmission time. Gezici et al. also explained the same problem considering the timing jitter [4].

In this paper, we perform the BER calculation for UWB IR systems with pulse-based polarity randomization. Because it is well known that modeling MAI as a Gaussian random variable leads to much discrepancy from the exact result, we are treating MAI with characteristic function method, which has been widely used in explaining MAI effects in CDMA [5] and also in IR UWB systems [6,7]. The results from this paper have proved the modeling method proposed in this paper is very accurate. When the number of pulse repetitions becomes large, we show the fact that the MAI can be approximated as a Gaussian distribution from the results of the accurate characteristic function. This paper analyzes the system performance with the pulse polarity with the SNR(signal to noise ratio), two types of processing gain and the number of asynchronous users.

2   System Model

In this section, we consider a TH-IR(time hopping impulse radio) system of $K+1$ asynchronous users with BPSK(binary phase shift keying) in an AWGN(additive white Gaussian noise) channel. The received signal can be described as

$$r(t) = \sum_{k=0}^{K} S_k(t) + z(t) \tag{1}$$

for $-\infty < t < \infty$. The second term is a zero mean AWGN process with a two-sided noise power spectral density of r N_0. The first term represents the sum of the received signals from each user. The signal of k-th user can be expressed as

$$S_k(t) = \sqrt{\frac{E_k}{N_f}} \sum_{j=-\infty}^{\infty} d_j^{(k)} b_{\lfloor j/N_f \rfloor}^{(k)} p\left(t - jT_f - c_j^{(k)} T_c - \tau_k\right) \tag{2}$$

where $p(t)$ is the transmitted unit-energy UWB pulse with duration $T_p \leq T_c$, E_k is the bit energy of user k, N_f is the number of pulses for one information symbol and $b_{\lfloor j/N_f \rfloor}^{(k)} \in \{-1, 1\}$ is the information bit transmitted by user k. $d_j^{(k)}$'s, the random polarity codes, are binary random variables taking ± 1 with equal probability, satisfying that $d_i^{(l)}$ and $d_j^{(k)}$ are independent for $(i, l) \neq (j, k)$. In order to allow the channel to be exploited by many users and to avoid catastrophic collisions, a long pseudorandom sequence $\{c_j^{(k)}\}$, such that $c_j^{(k)}$ is an integer taking one of the values in $[0, 1, \ldots, N_c - 1]$ is assigned to each user. Usually referred to as TH sequence, each sequence provides an additional time shift $c_j^{(k)} T_c$ seconds to the jth pulse of the kth user. Tc is the chip interval which is chosen to satisfy $T_c \leq T_f/N_c$ for preventing the pulses from overlapping.

Defining the new sequence as follows [3]

$$a_j^{(k)} = \begin{cases} d_{\lfloor j/N_c \rfloor}^k, & j - N_f \lfloor \frac{j}{N_c} \rfloor = c_{\lfloor j/N_c \rfloor}^k \\ 0, & \text{otherwise} \end{cases} \tag{3}$$

the received signal can be described by the following model

$$S_k(t) = \sqrt{\frac{E_k}{N_f}} \sum_{j=-\infty}^{\infty} a_j^{(k)} b_{\lfloor j/N_f \rfloor} p(t - jT_c - \tau_k) \tag{4}$$

where $N = N_f N_c$ represents the total processing gain of the system. The sequence $\{a_j^k\}$ can be regarded as a pulse rate/chip rate spreading sequence. Throughout the paper, we denote $k = 0$ to be the user of interest and $1 \leq k \leq K$ for the interfering users. It means the receiver is perfectly synchronized with user 0. To facilitate the analysis in subsequent sections, it is helpful to express the interfering signals in (4) as

$$S_k(t) = \sqrt{\frac{E_k}{N_f}} \sum_{j=-\infty}^{\infty} \hat{a}_n^{(k)} p(t - nT_c - T_k) \tag{5}$$

for $k \in [1, K]$, where $\hat{a}_n^{(k)} = a_{\lfloor n - l_k \rfloor} b_{\lfloor (n - l_k)/N \rfloor}$, $l_k = \lfloor \tau_k/T_c \rfloor$ and $T_k = mod(\tau_k, T_c)$ respectively and $\tau_k = l_k T_c + T_k$. The term $T_k \in [0, T_c)$ is referred to as the chip delay and T_k's are i.i.d. uniform RVs. Gezici et al. [2] have shown the MAI from user k given T_k has asymptotic Gaussian distribution with zero mean and variance expressed as only a function of T_k regardless of l_k. $\hat{a}_n^{(k)}$ can be effectively referred to as an i.i.d. random sequence for user k having the same pmf as $a_n^{(k)}$.

In eq. (4), the chip sequence $a_j^{(k)}$ can be modeled as random ternary with probability mass function(pmf) [8]

$$p_a\left(a_j^{(k)}\right) = (1 - 2\alpha)\delta\left(a_j^{(k)}\right) + \alpha\delta\left(a_j^{(k)} - 1\right) + \alpha\delta\left(a_j^{(k)} + 1\right) \tag{6}$$

where $0 < \alpha \leq 1/2$. In any given chip interval, there is probability $1 - 2\alpha$ that no pulse exists and probability 2α that ± 1 occurs. Especially, if $N_c = T_f/T_c$ is assumed, $\alpha = 1/2N_c$.

In the receiver having the matched filter(MF), matched to $p(t)$, the MF output is sampled at every chip and is then multiplied by the chip sequence $a_i^{(0)}$, carrying out the integration

$$\left(y_i|a_i^{(0)} \neq 0\right) = \sqrt{E_0}b_0^{(0)} + \sum_{k=1}^{K}\sqrt{\frac{E_k}{N_f}}\sum_{l=-\infty}^{\infty}a_i^{(0)}\hat{a}_l^{(k)}\rho\left((l-i)T_f + T_k\right) + \eta_i \quad (7)$$

where $\rho(\tau) = \int_{-\infty}^{\infty}p(t)p(t-\tau)dt$ is the continuous-time auto-correlation function of the pulse. Note that in (7) the subsequence $\left(y_i|a_i^{(0)} = 0\right) \equiv 0$ and we now assume y_i, $i = c(0), c(1), \ldots, c(N_f - 1)$ is subsequence conditioned on $a_i^{(0)} \neq 0$, where $c(i) = N_c i + c_i^k$. The noise term $\eta_i = a_i^{(0)}w_i$ is conditionally Gaussian when $a_i^{(0)}$, where $w_i = \int_{-\infty}^{\infty}z(t)p(t - iT_c)dt$. Also $x = a_i^{(0)}\hat{a}_l^{(k)}$, conditioned on $a_i^{(0)} \neq 0$, has the same pmf as $a_j^{(k)}$, given by (6).

The decision variable for one information bit is obtained by summing this N_f chip sequence

$$y = \sum_{i=0}^{N_f-1} y_{c(i)} = \sqrt{E_0 N_f}b_0^{(0)} + M + n \quad (8)$$

In (8), M is the contribution of the multiple access interference as

$$M = \sum_{k=1}^{K}M^{(k)} \quad (9)$$

$$M^{(k)} = \sqrt{\frac{E_k}{N_f}}\sum_{i=0}^{N_f-1}\sum_{l=-\infty}^{\infty}a_{c(i)}^{(0)}\hat{a}_l^{(k)}\rho((l-c(i))T_c + T_k) \quad (10)$$

As the function $\rho(\tau)$ has time support of $\tau \in [-T_c, T_c)$, (10) can be rewritten as

$$M^{(k)} = \sqrt{\frac{E_k}{N_f}}\sum_{i=0}^{N_f-1}\sum_{l=-1}^{\infty}a_{c(i)}^{(0)}\hat{a}_{c(i)+l}^{(k)}\rho(lT_c + T_k). \quad (11)$$

3   Analysis of the MAI

This section drives the characteristic function(CF) of MAI component. The CF of the MAI is derived as $\phi_M(\omega|T) = E[\exp(j\omega M)]$. Conditioned on $T = \{T_1, T_2, \ldots, T_k\}$, the CF of MAI is expressed as

$$\phi_M(\omega|T) = \prod_{k=1}^{K}[\phi_{M_k}(\omega|T_k)] \quad (12)$$

$$\phi_{M_k}(\omega|T_k) = E\left[\exp\left(-j\omega M^{(k)}\right)|T_k\right] \quad (13)$$

$$= \prod_{i=0}^{N_f-1}\prod_{l=-1}^{0}E\left[\exp\left(j\omega\sqrt{\frac{E_k}{N_f}}a_{c(i)}^{(0)}a_{c(i)+l}^{(k)}\rho(lT_c + T_k)\right)\right]$$

$\phi_{M_k}(\omega|T_k)$ is the MAI contributed from user k and can be easily obtained with the pmf of $a_{c(i)}^{(0)} a_{c(i)+l}^{(k)}$ in case of $a_{c(i)}^{(0)} = 1$.

$$\phi_{M_k}(\omega|T_k) = \left[\prod_{l=-1}^{0} \left((1 - 2\alpha) + 2\alpha \cos\omega\sqrt{\frac{E_k}{N_f}} \rho(lT_c + T_k) \right) \right]^{N_f} \tag{14}$$

To simply the CF of (14), the terms in parenthesis will be approximated as

$$1 - 2\alpha + 2\alpha \cos\omega\sqrt{\frac{E_k}{N_f}} \rho(iT_c + T_k) \simeq \exp\left(-\omega^2 \alpha \frac{E_k}{N_f} \rho^2(iT_c + T_k) \right) \tag{15}$$

where we use the Taylor series expansion of the cosine function. Applying (15) to (13) gives the simple CF,

$$\phi_{M_k}(\omega|T_k) = \exp\left(-\omega^2 \alpha E_k \sum_{l=-1}^{0} \rho^2(lT_c + T_k) \right) \tag{16}$$

With $\alpha = 1/2N_c$, (16) becomes

$$\phi_{M_k}(\omega|T_k) = \exp\left(-\omega^2 \frac{E_k}{2N_c} \sum_{l=-1}^{0} \rho^2(lT_c + T_k) \right) \tag{17}$$

(17) means $M^{(k)}$ is Gaussian random variable having zero mean and variance of

$$\sigma_{m,k}^2 = \frac{E_k}{N_c} \sum_{l=-1}^{0} \rho^2(lT_c + T_k) \tag{18}$$

As a result, we can express the CF of the MAI as

$$\phi_M = \prod_{k=1}^{K} \phi_{M_k} \tag{19}$$

$$\phi_{M_k} = \frac{1}{T_c} \int_0^{T_c} \phi_{M_k}(\omega|T_k)dt. \tag{20}$$

4 Probability of Bit Error Calculation

This section shows a BER, P_e expression based on the CF method. The BER for the system is given by

$$P_e = \Pr(y \leq 0 | b_0^{(0)} = 1) \tag{21}$$
$$\text{or} \quad P_e = \Pr(\sqrt{E_0 N_f} + M + n \leq 0)$$

With the expressions for the CF given in previous section, the BER expression can be obtained from the CF method [5,6,7]

$$P_e = \frac{1}{2} - \frac{1}{\pi} \int_0^\infty \frac{\sin\left(\omega\sqrt{E_0 N_f}\right)}{\omega} \phi(\omega) e^{-\omega^2 N_0 N/4} d\omega \qquad (22)$$

From (22), it is known that the calculation of the BER for the UWB TH systems needs a double integration, which is performed numerically but is not so difficult. We use the composite Simpson's rule for computing a double integration [6].

Fig. 1 is the results of the bit error rate with the SNR(signal to noise ratio) of the IR UWB systems with the pulse based polarity. In this calculation, total processing gain is 32 and two different number of pulses representing an information symbol, N_f are considered, which means two situations have the same bit transmission rate. The results from Gaussian approximation and simplified modeling (eq.(16)) show the same BER regardless of N_f when there is the same total processing gain. But simulation and accurate results have shown that even if the system has the same total processing gain, the increase in N_f results in the reduction in the BER, which explains more pulses with smaller power results in better performance than the system using fewer pulses with the big power. Monte-Carlo simulation results are very close to the results from the accurate modeling. And it also can be known that with N_f increased, the simplified results become the good approximation.

Fig. 2 shows the BER with the number of users. In this case, we also consider the fixed total processing gain, 60. This figure shows the modeling method results explain the reduced pulses representing one information bit makes the system

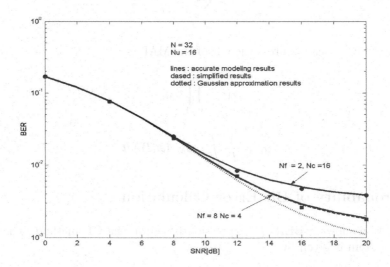

Fig. 1. Bit error rate with SNR. In this figure, the total processing gain of 32 is used and two frame numbers of 2 and 4 are used.

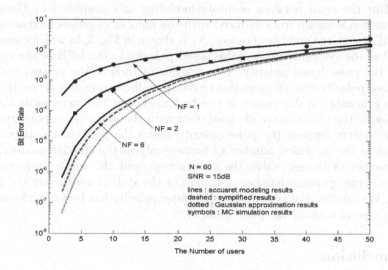

Fig. 2. This figure shows the BER with the number of users. Our modeling results is very close to the simulation results and proves the fact the different number of pulses leads to the different BER of the system with the same total processing gain.

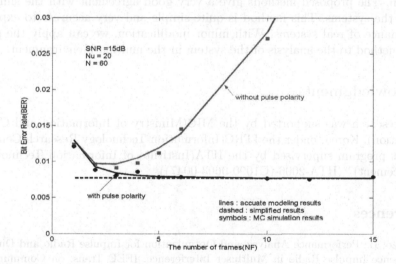

Fig. 3. BER vs. the number of frames. We compare the system performance with the pulse polarity with that without the pulse polarity. The results from the proposed method make good agreement with the simulation results.

worse so that the many pulses for one bit are in favor for the system, which cannot be proved with the simplified method or Gaussian approximation. The simplified modeling and Gaussian approximation give always the same BER regardless of N_f. But the error between accurate modeling and simplified or Gaussian approximation is shown to be reduced with the number of pulses increasing.

The BER with the number of pulses, N_f is shown in Fig. 3, in which compares the BER of the system with the pulse based polarity to the BER of the system without the pulse based polarity. The analysis method of the system without pulse based polarity was not given due to the space limitation, but is nearly same with the procedure in this paper. It can be known that the increase in N_f can be improving the performance of the system with the pulse based polarity. But while the system without the pulse polarity shows the performance improving according to the increased number of frames only in the smaller number, the larger number of frames makes the system represent the worse performance. So, there is the optimum number of frames in the system not having the pulse polarity. We can find the system with the pulse polarity has better performance than the system without the pulse polarity.

5   Conclusion

We have proposed the performance evaluation methods for the UWB systems with pulse-based polarity randomization. Under the flat channel environments, the multiple access interference has been characterized via the characteristic function. The proposed methods give a very good agreement with the simulation of the systems. This method is quite simple and very accurate to explain performance of real systems. With minor modification, we can apply the proposed method to the analysis of the system in the multipath environment.

Acknowledgment

"This research was supported by the MIC(Ministry of Information and Communication), Korea, under the ITRC(Information Technology Research Center) support program supervised by the IITA(Institute of Information Technology Advancement)" (IITA-2006-(C1090-0603-0047)).

References

1. Piazzo, L.: Performance Analysis and Optimization for Impulse Radio and Direct-Sequence Impulse Radio in Multiuser Interference. IEEE Trans. on Communications 52(5), 801–810 (2004)
2. Gezici, S., Kobayashi, H., Poor, H.V., Molisch, A.F.: Performance Evaluation of Impulse Radio UWB systems with Pulse-Based Polarity Randomiztion. IEEE Trans. on Signal Processing 53(7), 2537–2549 (2005)
3. Fishler, E., Poor, H.V.: On the Tradeoff between Two Types of Processing Gains. IEEE Trans. on Communications 53(9), 1744–1753 (2005)
4. Gezici, S., Molisch, A.F., Poor, H.V., Kobayashi, H.: The Trad-off Between Processing Gains of Impulse Radio Systems in the Presence of Timing Jitter. In: Proc. IEEE Int. Conf. Communications, vol. 6, pp. 3596–3600 (2004)

5. Yoon, Y.C.: An Improved Gaussian Approximation for Probability of Bit-Error Analysis of Asynchronous Bandlimited DS-CDMA Systems With BPSK Spreading. IEEE Trans. on Wireless Communications 1(3), 373–382 (2002)
6. Hu, B., Beaulieu, N.C., Cho, Y.S.: Accurate Evaluation of Multiple-Access Performance in TH-PPM and TH-BPSK UWB Systems. IEEE Trans. on Communications 52(10), 1758–1766 (2004)
7. Hu, B., Beaulieu, N.C.: Accurate Performance Evaluation of Time-Hopping and Direct-Sequence UWB Systems in Multi-User Interference. IEEE Trans. on Communications 56(6), 1053–1062 (2005)
8. Sadler, B.M., Swami, A.: On the Performance of Episodic UWB and Direct-Sequence Communication Systems. IEEE Trans. on Wireless Communications 3(6), 2246–2255 (2004)

Improvement of Adaptive Modulation System with Optimal Turbo Coded V-BLAST Technique

Kyunghwan Lee[1], Kwangwook Choi[1], Sangjin Ryoo[2], Kyoungwon Lee[2],
Mingoo Kang[3], Intae Hwang[4], Taejin Jung[4],
Daejin Kim[4], and Cheolsung Kim[4]

[1] Dept. of Computer Engineering, Chonnam National University 300 Yongbong-dong,
Buk-gu, Gwangju 500-757, Korea
signal-ds@nate.com, kwangwooke@hanmail.net
[2] Dept. of Electronics Engineering, Chonnam National University 300
Yongbong-dong, Buk-gu, Gwangju 500-757, Korea
sjryoo@hanyeong.ac.kr, caesar@teramail.com
[3] Dept. of Information Science & Telecommunication, Hanshin University 411
Yangsan-dong, Osan, Kyongggi-do, 447-791, Korea
kangmg@hanshin.ac.kr
[4] Dept. of Electronics & Computer Engineering, Chonnam National University 300
Yongbong-dong, Buk-gu, Gwangju 500-757, Korea
{hit,tjjung,djinkim,chskim}@chonnam.ac.kr

Abstract. In this paper, we propose and analyze the Adaptive Modulation System with optimal Turbo Coded V-BLAST (Vertical-Bell-lab Layered Space-Time) technique that adopts the extrinsic information from MAP (Maximum A Posteriori) Decoder with Iterative Decoding as a priori probability in two decoding procedures of a V-BLAST scheme; the ordering and the slicing. Also, comparing with the Adaptive Modulation System using conventional Turbo Coded V-BLAST technique that is simply combined a V-BLAST scheme with a Turbo Coding scheme, we observe how much throughput performance can be improved. As a result of a simulation, it has been proved that the proposed system achieves a higher throughput performance than the conventional system in the whole SNR (Signal to Noise Ratio) range. Specifically, the result shows that the maximum throughput improvement is about 350 kbps.

1 Introduction

In the next generation mobile communication systems, the data throughput performance improvement will be among the hot issues. In order to fulfill the need for ultra-high speed service, active researches on multiple-input-multiple-output (MIMO) systems that uses multiple transmit antennas and receive antennas have been in progress. Generally in MIMO systems, the main schemes considered are the MIMO diversity scheme and the MIMO multiplexing scheme [1][2][3].

Also, in order to improve the throughput performance, together with a MIMO system, an Adaptive Modulation and Coding (AMC) has drawn much attention to the pioneer of the next generation mobile communication systems [4][5].

M.S. Szczuka et al. (Eds.): ICHIT 2006, LNAI 4413, pp. 566–575, 2007.

The AMC scheme adapts a coding rate and a modulation scheme to the channel condition, resulting in an improved throughput performance. And it guarantees a transmission quality. Consequently, the combination of a MIMO system and an AMC scheme could be the solution for improving the throughput performance.

Considering the complexity, in this paper, as the scheme of a MIMO system combined with an AMC scheme, we will select a V-BLAST scheme [6][7] and a Turbo Coding scheme [8][9]. The Turbo Coding scheme with Iterative Decoding is parallel concatenated recursive systematic convolutional codes, and is iteratively decoded using a posteriori probabilities (APP) algorithms for the constituent codes [10][11]. We will present the performance analysis of the Adaptive Modulation Systems with several Turbo Coded V-BLAST techniques.

The remainder of this paper is organized as follows. The structure and the characteristics of AMC, V-BLAST, and Turbo Coded V-BLAST with Iterative Decoding are described in Section 2. In Section 3, we will show the structure of a transmitter and a receiver of the Adaptive Modulation Systems with several Turbo Coded V-BLAST techniques. In Section 4, the performance of each system is verified by a computer simulation, analyzed, and compared. Finally, conclusions will be drawn in Section 5.

2 The Structure and the Characteristics of AMC, V-BLAST and Turbo Coded V-BLAST System with Iterative Decoding

2.1 The Structure and the Characteristics of an AMC Scheme

Fig. 1 shows the transmitter-receiver structure of an AMC system. The data from a BS (Base Station) is coded, interleaved, modulated and then transmitted

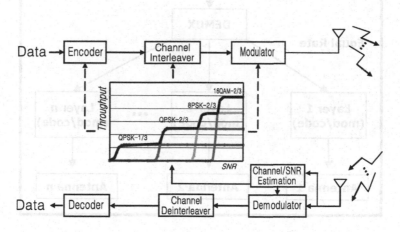

Fig. 1. Transmitter-receiver structure of an AMC System

through the channel. At the MS (Mobile Station), the estimation of the channel condition is performed using the received signal through the channel. The information about the channel condition is returned to the transmitter. The channel estimation is based on the SNR. The transmitter chooses the Modulation and Coding Scheme (MCS) level, and adapts the channel coding, interleaving, and modulation structure based on the channel information.

If the channel condition is favorable, high order modulation and coding rate (e.g. 16QAM and turbo code with coding rate 2/3) will be used. Otherwise, low order modulation and coding rate (e.g. QPSK and turbo code with coding ratio 1/3) will be used.

An AMC scheme promotes throughput performance and quality of system with appropriate MCS level considering trade-off between throughput performance and BER performance according to channel condition.

2.2   The Structure and the Characteristics of a V-BLAST Scheme

The BLAST scheme in the MIMO Multiplexing technique is classified as a D-BLAST (Diagonal-VLAST) scheme and a V-BLAST (Vertical-BLAST) scheme [12][13][14]. Although a D-BLAST scheme was proposed before a V-BLAST scheme, a D-BLAST scheme is hard to be implemented because of high complexity. In order to simplify the complexity, a V-BLAST scheme was proposed as an alternative. Fig. 2 shows the transmitter structure of a V-BLAST scheme.

The sequential input data stream, parallelized as many as the number of the transmit antenna, is modulated and coded. The layer (bit stream) is formed

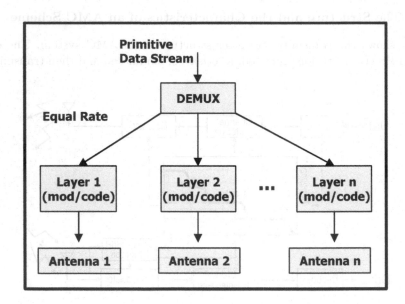

Fig. 2. Transmitter structure of a V-BLAST scheme

with the same number of the transmit antenna according to the data input. The layer signals received from the multiple antennas interfere with each other. As a result, the system performance is degraded. Consequently, the V-BLAST receiver removes the interference caused by the symbols of other layers using the nulling and canceling process.

In the nulling process, two schemes; MMSE (Minimum Mean-Squared Error) and ZF (Zero Forcing) are generally used. In the detection process, it is important to consider how to order and detect the symbols that are transmitted from several transmit antennas. Generally, the symbol, which has the highest SNR, under best channel condition is detected and then removed by the scheme. It is because the symbol under the best condition has much possibility to be exactly detected. This is done in order to minimize the error effect to other signals.

2.3 The Structure and the Characteristics of Turbo Coded V-BLAST System with Iterative Decoding

Fig. 3 shows the decoder structure of Turbo Coded V-BLAST system. It is parallel concatenated convolutional coded (PCCC) [8] V-BLAST system. In this system, the decoded output sequence is channel-interleaved and fed into the V-BLAST decoder as a priori probability, recursively. When SISO (Soft Input Soft Output) module [10] is applied as a channel decoder, it provides both the bit probability and the codeword probability. The codeword probability is used as a priori probability on a V-BLAST decoding procedure. This operation is performed iteratively.

In this recursion scheme, the codeword probabilities from the decoder1 and the decoder2 are used at both V-BLAST decoder and the other decoder as extrinsic information.

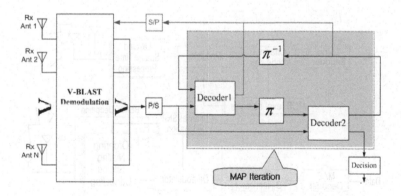

Fig. 3. Decoder structure of Turbo Coded (PCCC) V-BLAST system with Iterative Decoding

3   The Adaptive Modulation Systems with Several Turbo Coded V-BLAST Techniques

The system configurations of the Adaptive Modulation Systems with several Turbo Coded V-BLAST techniques are shown in this section. In addition, some considerable factors for the proposed system are discussed.

3.1   The Adaptive Modulation System with Conventional Turbo Coded V-BLAST Technique

Fig. 4 shows transmitter-receiver structure of the Adaptive Modulation System with conventional Turbo Coded V-BLAST technique that is simply combined a V-BLAST scheme with a Turbo Coding scheme. The information bits are transmitted using LST (Layered Space-Time) of a V-BLAST scheme after the channel encoding, interleaving, and modulation of an AMC scheme. The received signal is decoded by a V-BLAST procedure. This procedure consists of a repetition structure of ordering, nulling, slicing, and canceling.

When we apply the Adaptive Modulation System with conventional Turbo Coded V-BLAST technique, each transmit antenna has different SNR. Therefore we should consider applying to a MCS level based on which each antenna's SNR.

In this paper, a minimum SNR threshold is applied to our implemented system. In case of applying the maximum SNR or the average SNR to the threshold, even a poor channel layer selects a MCS level that has high order modulation and code rate. This leads to an increased error probability and a degraded total throughput performance.

Fig. 5 shows the operation process of the Adaptive Modulation System with conventional Turbo Coded V-BLAST technique. In this part, instead of the hard decision values, the soft decision values are inputted to a decoding stage for Turbo decoding.

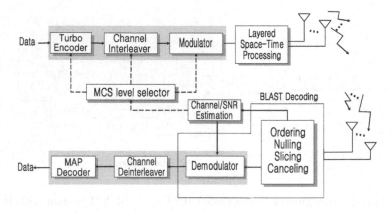

Fig. 4. Transmitter-receiver structure of the Adaptive Modulation System with conventional Turbo Coded V-BLAST technique

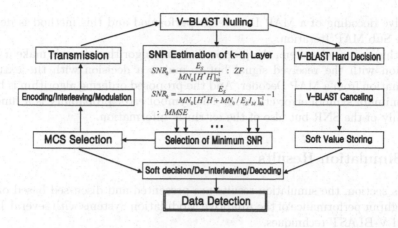

Fig. 5. Operation process of the Adaptive Modulation System with conventional Turbo Coded V-BLAST technique

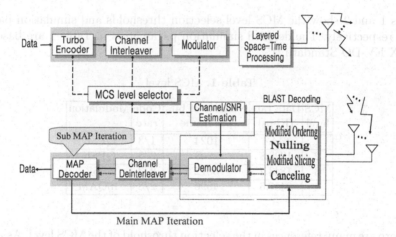

Fig. 6. Transmitter-receiver structure of the Adaptive Modulation System with optimal Turbo Coded V-BLAST technique

3.2 The Adaptive Modulation System with Optimal Turbo Coded V-BLAST Technique

In this section, the structure of Adaptive Modulation System with optimal Turbo Coded V-BLAST technique is proposed.

Fig. 6 shows the transmitter-receiver structure of the proposed system. The difference with the Adaptive Modulation System using conventional Turbo Coded V-BLAST technique in Section 3.1 is that the extrinsic information from MAP Decoder is used as a priori probability in two decoding procedures of a V-BLAST scheme; the ordering and slicing. This scheme operates iteratively and is defined as the Main MAP Iteration. Also, whenever it operates, internally an iterative decoding of a MAP Decoder is performed and this method is defined as the Sub MAP Iteration.

In this proposed system, the proposed slicing algorithm doesn't make a Hard Decision with the received signal but it makes a decision with the extrinsic information from a MAP Decoder. And the proposed ordering algorithm is based on minimum symbol error criterion. The symbol error probability is a function not only of the SNR but also of the extrinsic information.

4   Simulation Results

In this section, the simulation results are presented and discussed based on the throughput performance of the Adaptive Modulation Systems with several Turbo Coded V-BLAST techniques.

4.1   MCS Level and Simulation Parameter

Tables 1 and 2 show the MCS level selection thresholds and simulation parameters, respectively. The detailed simulation parameters in Table 1 are based on the 1X EV-DO Standard [15].

Table 1. MCS level

MCS level	Data rate (Kbps)	Number of bits per frame	Code rate	Modulation
1	614.4	1024	1/3	QPSK
2	1228.8	2048	2/3	QPSK
3	1843.2	3072	2/3	8PSK
4	2457.6	4096	2/3	16QAM

There are many references in the selection threshold of the MCS level. As an example, the threshold can be selected to satisfy the required BER (Bit Error Rate) performance and the required FER (Frame Error Rate) performance. Since we have put more emphasis on the data transmission rate, we select the threshold that maximizes the throughput performance in this paper. Accordingly, each MCS level selection threshold is based on the throughput performance cross point.

4.2   Performance of the Adaptive Modulation Systems with Several Turbo Coded V-BLAST Techniques

In Section 3.1, we have considered the Adaptive Modulation System with conventional Turbo Coded V-BLAST technique and in Section 3.2, the Adaptive Modulation System with optimal Turbo Coded V-BLAST technique has been proposed. Fig. 7 shows the throughputs of each MCS level in the Adaptive Modulation systems with several Turbo Coded V-BLAST techniques using 2 transmit and 2 receive antennas. We can see that the proposed system achieves

Table 2. Simulation parameters

Parameter	Value
Modulation	QPSK, 8PSK, 16QAM
Code rate	1/2, 2/3
Turbo Coding scheme	PCCC (Parallel Concatenated Convolutional Code)
MAP Iteration of the Adaptive Modulation system with conventional Turbo Coded V-BLAST technique	4
Main MAP Iteration of the Adaptive Modulation system with optimal Turbo Coded V-BLAST technique	4
Sub MAP Iteration of the Adaptive Modulation system with optimal Turbo Coded V-BLAST technique	2
Number of Tx. & Rx. antennas	2
Channel	Flat Rayleigh fading

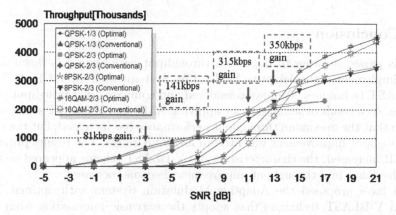

Fig. 7. Throughputs of each MCS level in Adaptive Modulation Systems with several Turbo Coded V-BLAST techniques

a better throughput performance improvement about each MCS level than the conventional system. The maximum throughput improvement of each MCS level is about 81 kbps, 141 kbps, 315 kbps, and 350 kbps, respectively.

Fig. 8 shows the throughputs of the Adaptive Modulation Systems with several Turbo Coded V-BLAST techniques using 2 transmit and 2 receive antennas. The result shows that the proposed system achieves a throughput performance improvement of more 300 kbps in the range between about 7dB~15dB

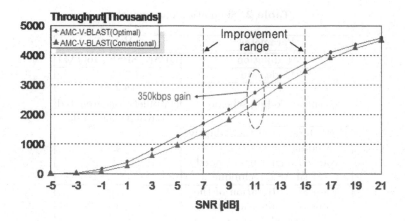

Fig. 8. Throughputs of the Adaptive Modulation System with several Turbo Coded V-BLAST techniques

SNR. Specifically, the proposed system at 11dB SNR shows that the maximum throughput improvement is about 350 kbps.

As a result of the simulation, in the Adaptive Modulation Systems with several Turbo Coded V-BLAST techniques, the proposed system has a higher through-put performance than the conventional system in the whole SNR range.

5   Conclusion

In this paper, in order to improve a throughput performance in downlink, we have implemented the Adaptive Modulation Systems with several Turbo Coded V-BLAST techniques. We have considered and compared its throughput perfor-mance. As a result of simulation, by applying AMC and V-BLAST schemes, it is shown that the maximum throughput performance has increased, but the rate of a throughput improvement was little in low SNR range. The results prove that as SNR increased, the characteristics of a V-BLAST scheme appeared regularly and the range of a throughput improvement became increased.

We have proposed the Adaptive Modulation System with optimal Turbo Coded V-BLAST technique that adopts the extrinsic information from MAP Decoder with Iterative Decoding as a priori probability in two decoding proce-dures of a V-BLAST scheme; the ordering and the slicing. We have considered the Adaptive Modulation System with conventional Turbo Coded V-BLAST technique that is simply combined a V-BLAST scheme with a Turbo Coding scheme. As a result of a performance comparison of each system, it has been proved that the proposed system achieves a superior throughput performance than the conventional system in the whole SNR range. Specifically, a simulation result shows that the maximum throughput improvement is about 350 kbps.

Acknowledgments. This study was financially supported by Chonnam Na-tional University.

References

1. Winters, J.H.: The Diversity Gain of Transmit Diversity in Wireless Systems with Rayleigh Fading. IEEE Trans. on Veh. Tech. 47(1), 119–123 (1998)
2. Alamouti, S.M.: A Simple Diversity Technique for Wireless Communications. IEEE Journal on select areas in communications 16(8), 1451–1458 (1998)
3. Foschini, G.J.: Layered Space-Time Architecture for Wireless Communication in a Fading Environment When Using Multi-Element Antennas. Bell Labs Technical Journal (Autumn 1996)
4. Bender, P., Black, P., Grob, M., Padovani, R., Sindhushayana, N., Viterbi, A.: Qualcomm, Incorporated, CDMA/HDR: A Bandwidth-Efficient High-Speed Wireless Data Service for Nomadic Users. IEEE Comm. Magazine, 70–77 (2000)
5. Goldsmith, A.J., Chua, S.G.: Adaptive Coded Modulation for Fading Channels. IEEE Trans. on Comm. 46(5), 595–602 (1998)
6. Farrokhi, F.R., Foschini, G.J., Lozano, A., Valenzuela, R.A.: Link-Optimal BLAST Processing with Multiple-Access Interference. VTC2000 Fall 1, 87–91 (2000)
7. Bhargave, A., de Figueiredo, R.J.P., Eltoft, T.: A Detection Algorithm for the V-BLAST System. In: GLOBECOM 2001. IEEE, vol. 1, pp. 494–498 (November 2001)
8. Benedetto, S., Montorsi, G.: Unveling Turbo Codes: some results on parallel concatenated coding schemes. IEEE Trans. on Inform. Theory 42, 409–429 (1996)
9. Berrou, C., Glavieux, A., Thitimajshima, P.: Near Shannon Limit Error-Correcting Coding and Decoding: Turbo Codes. In: Proc. of International Conference on Communications, vol. 3, pp. 1064–1070 (May 1993)
10. Bennedetto, S., Divsalar, D., Montorsi, G., Pollara, F.: A soft-input soft-output maximum a posteriori(MAP) module to decode parallel and serial concatenated codes, JPL, TDA Progress Rep. 42-126 (August 1996)
11. Hagenauer, J., Offer, E., Papke, L.: Iterative Decoding of Binary Block and Convolutional Codes. IEEE Trans. on Inform. Theory 42, 429–445 (1996)
12. Catherine, Z.W., Sweatman, H., Thompson, J.S., Mulgrew, B., Grant, P.M.: Comparison of Detection Algorithm including BLAST for Wireless Communication using Multiple Antennas. In: PIMRC 2000, vol. 1, pp. 698–703 (2000)
13. Beach, M.A., McNamara, D.P., Fletcher, P.N., Karisson,: MIMO-A solution for advanced wireless access? In: 11th international conference on Antennas and Propagation, vol. 480, pp. 231–235 (April 2001)
14. Wolniansky, P.W., Foschini, G.J., Golden, G.D., Valenzuela, R.A.: V-BLAST: An Architecture for Realizing Very High Data Rates Over the Rich-Scattering Wireless Channel. In: ISSSE 1998, pp. 295–300 (1998)
15. 3GPP2 C.P9010, Draft baseline text for the physical layer portion of the 1X EV specification, pp. 9–78 (August 2000)

Header Compression of RTP/UDP/IP Packets for Real Time High-Speed IP Networks

Kyung-shin Kim, Moon-sik Kang, and In-tae Ryoo

[1] Chungkang college, 37 Haewol-ri, Majang, Ichon, Kyunggi, 467-744, Korea
[2] Kangnung National Univ.,120 Daehak-ro, Kangnung, Kangwon, 210-702, Korea
[3] Kyunghee Univ., seocheon-ri, Yongin, Kyunggi, 560-759, Korea
update@ck.ac.kr, mskang@knu.ac.kr, itryoo@khu.ac.kr

Abstract. In this paper, a new header compression scheme considering BCB (Basic Compression Bits) or NCB (Negotiation Compression Bits). The header compression scheme can be used for reducing the header size by eliminated repeated fields in the packet header. Here, the efficiency of the compression of the dynamic field in RTP/UDP/IP packets is very important in real-time high-speed IP networks. Our new compression method with SN and TS fields can be applicable to IPHC (Internet Protocol Header Compression), ROHC (Robust Header Compression Protocol), and other header compression schemes. The performance of the proposed scheme is discussed via simulation results.

Keywords: Header compression, RTP, BCB, NCB, SN, TS.

1 Introduction

In this paper, we seek to present a compression scheme that will economize bandwidth by compressing the header of this RTP protocol.

The packets transferred in the RTP (Real Time protocol), when audio/video data are transferred through cable or wireless networks, are capsulated into an IP header (20 octets) or UDP (User datagram protocol) header (8 octets) or RTP (12 octets), with the total header size of 40 (60 bytes in case of IPv6) bytes[1]. RPT is the protocol for various real-time communications and multimedia services using IP Networks, such as Multi-User online games and video/audio broadcasting, etc. In addition, RTP can also be used for video/audio streaming or remote video conferencing and remote patient care. IPHC uses the packet header's characteristic of maintaining a constant size within consecutive packet streams. Header domains in the same stream (session) usually have the same value [2]. For example, the address domain and port domain in the same IP or the same UDP/TCP are the same. Also, the value of SN (Sequence Number) and TS (Time Stamp) domains of the RTP changes at a constant rate one after another, and the UDP length domain overlaps that of the link layer. Based on such pattern analysis, the method of compressing the header by eliminating the overlapping portions of the RTP/UDP/IP and transferring change values for the constantly increasing domain values is currently approved as RFC 2508 [3].

M.S. Szczuka et al. (Eds.): ICHIT 2006, LNAI 4413, pp. 576–585, 2007.

We call this process of reducing the header size by eliminating increased packet header overheads as the header compression, which includes two representative methods of IPHC and ROHC. These utilize the repetitiveness of header fields, and they enable more efficient communications. Based on this fact that the information is repetitively maintained among packets, the header compression protocol defines the Compressor and the decompressor. Such reduction of the header size to be transferred increases user throughput [4].

Consecutive packet headers of audio or video streams include static, dynamic, and random fields. Consecutive packets in the same stream in UDP will have the same source and destination addresses [5]. Furthermore, the UDP length field can be inferred from the protocol stacks of lower layers. Through such analysis, the Destination Node can reason and restore a significant part of the header field after the receipt of the stream's consecutive packets (under the assumption that the information of a previous packet is saved). Here, we describe the new compression methods of SN and TS fields with higher compression rates than that of the previous IPHC method, the representative header compression scheme. This proposed scheme reduces the dynamic fields of the RTP packet into the BCB (3-bits) or the NCB (BCB + additional bits) bits. This new compression method is applicable to IPHC, ROHC, and all other header compression schemes. In order to verify the performance of the proposed header compression scheme, we present and discuss the most appropriate compression ratio according to the video packets in IP wireless networks.

2   A New Proposed Compression Scheme

In order to study an efficient header compression method of the RTP protocol in Multi-User online game's peer-to-peer multimedia services or video streaming services using IP Wireless Networks, we have executed various experiments on the MPEG streaming data with RTP/UDP/IP packets in IP Wireless Networks.

In order to compress the header with the highest compression rate for the SN and the TS, which increases dynamically and constantly among the many fields in the RTP header, the proposed scheme has some features as follows.

First, it should be designed so that the increase in the SN and the TS among the packets sent and received between the Compressor and the Decompressor uniformly maintained in the following way; The SN increases one by one, if BCB is defined as the demanded bit value of the compression result. Also, the TS increases by the PCTSI (PiCTure clock interval) times.

Second, the uncompressed SN and TS comprise a total of 48 bits. The Compressor and the Decompressor first decide how many bits these 48 bits should be compressed.

It can be decided as the 3-bits(Basic Compression Bit). If this value is inadequate, the value of basic compression bit decides by way of adjustment between compressor and decompressor. It is the NCB(the Negotiation Compression Bit). Of course, the NCB (Negotiation Compression Bit) is equal to BCB(Basic Compression Bit) as 3-bit is adequate.

Third, the Compressor compresses the total 48-bits of the SN and the TS into n NCB bits, of which the n bits are transferred to the Decompressor. The Decompressor, calculates the SN by $SN_{n-1} + 1$ with the previous values and restores the TS as PCTSI $+ TS_{n-1}$. Finally, the 48-bits (16-bit SN and 32-bit TS) can be compressed into n bits. The detailed explanation is as follows.

2.1   BCB and NCB

Consider that the BCB of SN and the TS between the compressor and the decompressor nodes can be set as 3-bits. However, the compressor and the decompressor decides the final NCB according to the characteristics of the payload traffic transferred through the RTP, and the SN and the TS are compressed based on this result. Here, when the NCB value is determined as n, it is decided whether to compress the 48-bits of SN and the TS in 3-bits (BCB) or to compress them with an additional n bits in 3+n (NCB).

As a result, the 48-bits of SN and TS value is sent from the Compressor after being compressed in NCB $(BCB + n\ bits, 0 \leq n \leq 48)$. This means that long streamed or large video traffic can be compressed with a value larger than 3-bits, while small files or small traffic is compressed by the BCB, 3-bits. Here, the n NCB is decided upon by calculating basic and extended bits in order to not only increase the compression rate, but also to prevent cases where the decompressor itself has problems of restoring the lost consecutive packets more than $2^n - 1$.

Considering the ability to restore consecutive packet loss, that is, the Robustness of the protocol, we set the n-bit size according to depending on the negotiation results of the compressor and the decompressor. Consequently we suggested that the number of compressed and sent bits be extended to NCB(BCB+n).

In other words, when the large amount of video data are being transferred through a wireless network, which the link status is not so stable, the robustness against multiple packet loss can be provided through the diverse expansion of the compression range by compressing the SN and the TS in a large bit value (3+n) (of course, the compression rate is lowered). In other case, the compression rate can increase by configuring the bit value at the BCB. If the 48-bits of SN and TS are compressed in BCB 3-bits, the compression rate is increased, while the robustness of the protocol is lowered. However, if they are 3-bits, the compression rate is lowered, while the protocol robustness increases.

In the next section, we analyze and execute simulation experiments of these relationships and compare the relationship between the compression rate and the performance of the protocol.

2.2   SN Compression Method

An existent method for the sequence Number (SN) in the RTP is to increase the sequence numbers of the consecutive packets by one. In other words, a RTP packet with a 16-bit sequence number increases by 1 from 0 to 65535. This way, because it is easy to detect loss of packet in the part of the reception. However, in order to achieve a higher compression rate, when the nth packet's SN is 3000, the SN of the N+1th SN is $3000 + 8(2^3) = 3008$. Therefore, we can obtain the following formula for the compression and restoration of the sequence number SN.

$$SN_n = SN_{n-1} + 1(SN_n < 65535)$$
$$....$$
$$SN_1 = SN_0 + 1$$
$$SN_0 = Random\ value \tag{1}$$

Here, SN_n is the original sequence number in the RTP packet, and also the NCB is the basic 16-bit SN compression bits. Therefore, if n=3, this means that the 16-bit SN is compressed in 3-bits. When the range is expanded in the proposed method, the sequence number (SN) can be represented in equation (2), where SN_n is the RTP's nth packet's sequence and SN_{n-1} is the most recently decoded sequence number.

$$SN_n = SN_{n-1} + 1 \tag{2}$$

If the SN is compressed in such a way, the SN compression rate of the RTP header can be calculated by the following relationship.

$$C_{SN} = \frac{SN - SN_{proposed}}{SN} = 1 - \frac{SN_{proposed}}{SN} \tag{3}$$

Here, the SN is the uncompressed packet header size (byte), and the $SN_{proposed}$ is the compressed size of the SN. C_{SN} is the compression rate of the SN, namely the relative ratio of the compressed SN size. The following equation shows the calculation of the compression rate of a 16-bit Sequence Number using formula (3) (A NCB is 3 bits).

$$C_{SN} = 1 - \frac{3bits}{16bits} = 0.8125$$

2.3   TS Compression Method

As with the constitution of the SN, in order to achieve a higher compression efficiency, changes in the current RTP TS (Time Stamp) value should be set at a constant multiple value. We propose that the TS value of the RTP header increases by the PCF (Picture clock frequency) value.

In a video stream, within the same picture the difference in TS becomes 0 [7]. And the TS value for the following B or the original Intra(I) or Inter(P) pictures can be many times larger than PCF or bigger than 0. The PCF of H.261, H.263 ver 1 is 29.97Hz, and this means that the increase between the two coded pictures is 3003. The following represents why such PCTSI value was calculated. When the video is compressed using RTP, a packet may not include more than 1 picture. A single picture may be divided into two or more packets. The header compression profile regarding video information uses 90kHz as the reference clock [8].

$$PCTSI = PiCTure\ and clock\ and interval\ in\ and TS$$
$$= \frac{Video Information Reference Clock(90KHz)}{PCF(Picture Clock Frequence)}$$

$$PCTSI = \frac{90000}{PCF(29.97Hz} = 3003\ in H.261$$

$$PCTSI = \frac{90000}{25} = 3600\ in H.263$$

$$PCTSI = \frac{90000}{30} = 3000\ in MPEG-4$$

Also, the PCF of both the H.263 ver. 2 and the MPEG-4 become 25Hz and 30Hz, respectively, of which values show the increase multiple in TS of 3600 and 3000, respectively. Even if there is loss in the transfer process between the Compressor and the Decompressor, since the TS increases by the constant multiple of PCTSI, this is immediately restorable using the TS of the previous packet. Like this, if the TS value of the RTP header is configured to increase by the constant multiple of PCTSI, the TS value can be simply compressed and restored using the following formula.

$$TS_n = TS_{n-1} + PCTSI \tag{4}$$

If TS_n is the time stamp of the nth packet of the RTP, and TS_{n-1} is the most recently decoded time stamp, when the range of the time stamp is expanded by the suggested method, it can be represented by equation (4).

If the change in TS between two packets is configured to be the multiple of PCTSI, and the PCTSI value is shared between the compressor and the decompressor, the decompressor can use the previously mentioned BCB n-bit to easily restore the TS value. Using this method, the Compressor may use either 3-bits or $3 + extended\_n$ bits to compress and restore the 16-bit SN and the 32-bit TS. The following formula shows what the compression rate is in the TS field. The compression rate of the TS of the RTP header can be calculated by the following way.

$$C_{TS} = 1 - \frac{TS_{proposed}}{TS} \tag{5}$$

Here, TS is the uncompressed packet header size (byte), and the $TS_{proposed}$ is the compressed TS size. CTS is the compression rate of the TS, namely the relative ratio of the TS size. The following shows how the compression rate of a 32-bit TS is calculated using formula (5).

$$C_{TS} = 1 - \frac{0bits}{32bits} = 1.0$$

The reason why the $TS_{proposed}$ value is set at 0 is that the Decompressor can restore the TS using the NCB for the compression of the SN. Since the value of the TS field changes in multiples of PCTSI, this is restorable without sending any compression bits. In other words, the only NCB is needed to be sent when the total 48-bits of the SN and the TS are compressed.

2.4 Compressed Header Format

The existing header compression methods(VJHC, IPHC, ROHC) transmit and receive delta value of SN or TS. And then they decreased the size of them. That is to say delta value is transmitted and received by means of each SN and TS are 3-bits compressed. In this paper, the proposed method can be compressed SN and TS simultaneously. Actually, 48-bits of SN and TS can be reduced by 3-bits of them. The following shows the process of transmitted and received and type of the compressed header. First, there is the type of packet in stage of compression negotiation.

$$\boxed{\text{CID}}\boxed{\text{Seq(BCB)}}\boxed{\text{Data Type}}$$

First, determine CID, create the same context DB in compressor and decompressor. CID becomes the key of context DB. And then, Decompressor need compressed packet by means of received CID and context when compressor transmitted CID. The following shows compressor the earliest transmit full header subsequent negotiation.

$$\boxed{\text{CID}}\boxed{\text{Seq(BCB)}}\boxed{\text{Non-Compressed Total Field}}$$

The following shows the type which compressor with the best compression rate transmits and receives in normally.

$$\boxed{\text{CID}}\boxed{\text{Seq(BCB)}}\boxed{\text{Random 1}}\boxed{\text{Random 2}}....\boxed{\text{Random n}}$$

Thus, the method by means of context and NCB bits can compress header more efficiency without modification of the original upper layer protocol such as App., RTP, UDP and IP.

3   A New Proposed Compression Scheme

3.1 Simulation Model

To evaluate the performance of the proposed scheme, the data transmission is considered only between the sending node and the receiving node with the compressor and the decompressor. Both the sending node and the destination node are assumed to be connected by Cellular Link, known to have a high packet error rate and long RTT.

There are some assumptions to verify the adequacy of the suggested method. Packets transmitted in the forward data channel can be lost, and there is no special form of packet errors. The order of packets is maintained between the compressor and the decompressor. In other words, the receiver always receives

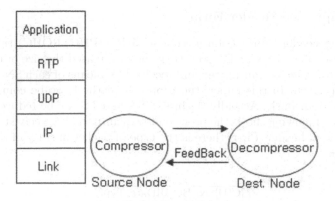

Fig. 1. Simulation Model

packets in the order sent from the transmitter. There is no strict delay or error demands in the feedback channel. The proposed algorithm does not assume any special distribution of RTT. It can dynamically adapt to any changes in RTT, and the header compression does not affect interleaving or channel coding. We executed the simulation in the following two directions in order to assess the suggested compression scheme. First, we experimented the change in abilities depending on the size of the compression bit number (BCB or NCB) regarding the consecutive packet loss in video streaming traffic, and second, we experimented on the compression bit number, error rate, and abilities analysis. For the target for experimentation, MPEG4 video traffic between the compressor and the decompressor was considered. First, as we mentioned, the SN and the TS of the RTP header both increase by a constant amount. It is assumed that the SN begins at a random value, increasing by 2^n, while the TS increases by the ratio between the reference clock and the Picture Clock frequency. In order to analyze the compression condition and the restoration of the TS and the SN, and to compare the robustness regarding compression bit number and errors, the simulation has executed using the Visual SLAM 3.0(AWESIM).

The following equation shows the compression rate when the SN and the TS are compressed in n bits.

$$C_{SNTS} = 1 - \frac{H_{compress}}{H_{SNTS}} \tag{6}$$

SN and the TS are a total of 48 bits. The following shows the compression rate when this is compressed in Basic Compression Bits (3 bits).

$$C_{SNTS} = 1 - \frac{3bits}{48bits} = 0.0625$$

We achieved the following experimental model (Table 1) by means of formula(6) in order to compare the header's compression rate using

Table 1. Comparison of Compression and Robustness

Number of NCB(Bits)	3bit	5bit	7 bit	9 bit	...	n bits
Compression Rate	0.0625	0.1042	0.1458	0.1875	...	$1 - \frac{nbits}{48bits}$
ChangeScope for Robustness	[0,7]	[0,31]	[0, 127]	[0, 511]	...	$[0, 2^n - 1]$

Table 1 shows the compression rate and the change scope of the compressed bits of the SN and the TS. Here, the factor that should be considered is the consecutive packet error bit number. The consecutive error bit number is a factor that has effect at the wrap-around point considering the compression bits and the compression step, and it has a significant effect in the decision of the compression bits. If the consecutive packet error bit number gets long, as the NCB compression bit length gets longer, error restoration rate is increased, while error restoration rate is lowered as the NCB compression bit number is lowered.

3.2   Simulation Results in Case of Video Streaming

The average data transmission speed is approximately 8Kbps, the frame rate is 10fps, and the video frame size is 100 bytes. The Maximum Transmit Unit (MTU) in the RTP layer is 4000 bits. This means that the video frames larger than 500 bytes are segmented and are transported after being divided into 2 or more RTP packets. However, in this experiment, it assumed that the RTP packets are divided in a set size, and the algorithm and the process of the MPEG4 stream being divided into RTP packets are not considered. In this experiment, only the constant size RTP packets are used as test subjects. Two channels were used between the two nodes: the data channel where data are transferred between the Compressor and the Decompressor, and the feedback channel where the ACK or the NAK signal is transmitted from the Decompressor. The results are as follows.

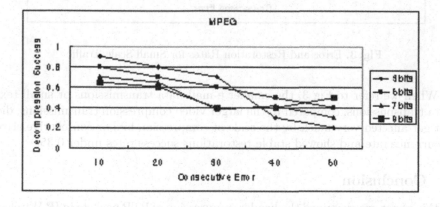

Fig. 2. Consecutive Error Rate and the Error Restoration Rate of Video Streaming

The x-axis shows the length of consecutive packet loss, and the y-axis shows the decompression success rate. The results show that, for smaller lengths of consecutive packet loss, the success rate for the 3-bit compression method is increased, while the error rate for the method of large compression bit number of the header is decreased as the length of consecutive packet loss gets longer. In addition, for 3-bit compressions, we could see that the difference between consecutive error rates of 10 and 50 was more than 8 times, while the 9-bit compression showed not much effect from the consecutive error lengths, as the error rate was maintained from 0.4 to 0.6. As previously analyzed, from this, it is inferred that it would be advantageous to do compression and restoration by increasing the compression bit numbers of the SN error rate or large consecutive packet error lengths.

3.3 Simulation Results in Case of Small Traffic

For the experiment, the average data transmission speed was assumed approximately 8Kbps, the payload of packets was assumed 30 bytes or less and the traffic generation interval was assumed at 60 seconds. Consequently, we made experimental model that has 2 communication channels(data channel and feedback channel), RTP fixed 30bytes packets only and 8Kbps bandwidth. The results are as follows.

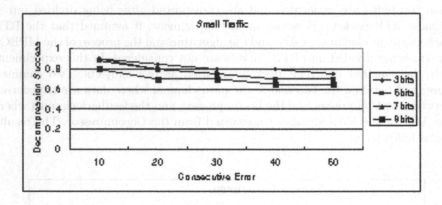

Fig. 3. Error and Restoration Rates for Small Scale Traffic

When we refer to Fig 3, the smaller e-mail data transmissions or small text data transmissions, compared to the larger video compression transmissions, did not get affected very much by the scale of compression bits or consecutive error occurrence rate and showed stable restorations success rates under 0.35.

4 Conclusion

In this paper, we examined the header compression of RTP packets in IP Wireless networks, in order to present an efficient header compression scheme of the RTP protocol for peer-to-peer multimedia services and video streaming in Multi-User online game systems using IP Wireless Networks.

The RTP header compression scheme has been used in a variety of ways, including IPHC and ROHC. Also, it is possible to execute more efficient communications by reducing the large amount of the bandwidth. We think that a header compression scheme with higher compression rate can be achieved if the proposed compression scheme is applied to a variety of other header compression applications. From the results of performance analysis with various compression methods of 48-bit SN and TS, it is extremely important to decide at how much compression bits we choose by considering the consecutive packet error rates and the restoration error rates. According to the results, in case of the 3-bit compressions, we can say that the difference between consecutive 10 error case and 50 case is more than 8 times of restoration rate of the other, while the 9-bit compression method shows not much the effect from the consecutive error lengths, as the error rate was maintained from 0.4 to 0.6. Also, it can be shown that the e-mail, with small amount of traffic, should be compressed at 3-bits, while the video traffic, if the consecutive error bit averages 30, should be compressed at an average of 5-bits where the error rate falls under 0.6. If the NCB is 5-bits, the compression step is [0, 127]. This means that the average consecutive error bit (30 bit) is restorable without problems even if it is lost due to an error. As the efficient transmission of video streams is very important in IP wireless networks with limited bandwidths, the efficient use of bandwidths and desired service quality should be provided along with an efficient header compression scheme. In particular, the header sizes for RTP/UDP/IP packets are too large, posing a significant problem for the protocol using existent headers in wireless links where bandwidth is limited. For further study, we will devote time to the research of more efficient header compression schemes to cooperate with next generation networks.

References

1. Schulzrimme, H., et al.: RTP: A Transport Protocol for Real Time Applications. IETF RFC 1889 (January 1996)
2. Degermark, M., Nordgren, B., Pink, S.: IP Header Compression. RFC 2507 (February 1999)
3. Casner, S., Jacobson, V.: Compressing RTP/UDP/IP Headers for Low-Speed Serial Links, RFC 2508 (February 1999)
4. Ertekin, E.: Internet Protocol Header Compression, Robust Header Compression and Their Applicability in the GIG. IEEE Communications Magazine (November 2004)
5. Jin, H.: Performance Comparison of Header Compression Schemes for RTP/UDP/IP Packets. IEEE Communications Society, 1691–1696 (2004)
6. Bormann, C.: Robust Header Compression(ROHC). RFC 3095 (June 2001)
7. Hoffman, F.G.: RTP Payload Format for MPEG1/MPEG2. RFC2250 (January 1998)
8. Kikuchi, V., Nomura, T.: RTP Payload Format for MPEG-4 Audio/Visual Streams. RFC3016 (November 2000)

Repetition Coding Aided Time-Domain Cancellation for Inter-Carrier Interference Reduction in OFDM Systems

Jeong-Wook Seo[1], Jae-Min Kwak[1], Won-Gi Jeon[1], Jong-Ho Paik[1],
Sung-Eon Cho[2], and Dong-Ku Kim[3]

[1] Korea Electronics Technology Institute,
68 Yatap-dong, Bundang-gu, Seongnam-si, Gyeonggi-do 463-816, Korea
{jwseo,jmkwak,jeonwg,paikjh}@keti.re.kr
[2] Sunchon National University,
315 Maegok-dong, Sunchon-si, Jeollanam-do 540-742, Korea
chose@sunchon.ac.kr
[3] Yonsei University,
134 Sinchon-dong, Seodaemun-gu, Seoul 120-749, Korea
dkkim@yonsei.ac.kr

Abstract. In this paper, an enhanced time-domain cancellation method is proposed for inter-carrier interference (ICI) reduction in OFDM systems. The conventional time-domain cancellation neglects the effect of channel variation in cyclic prefix during the time-domain cancellation and does not work well in deep fades. In order to supplement the conventional method, the simple repetition (de-)coding and the modulation order increasing are employed in the time-domain cancellation. The repetition coding provides reliable symbols in the regeneration operation, and the modulation order increasing maintains or increases the spectral efficiency. Simulation results indicate that the proposed method using 16QAM significantly improves the BER performance compared to the conventional method using QPSK, while maintains the spectral efficiency. Moreover, the proposed method using 64QAM concurrently improves both the BER performance and the spectral efficiency.

1 Introduction

Orthogonal frequency division multiplexing (OFDM) is a promising access technology for next-generation communication systems such as IEEE 802.16e and 3G long-term evolution, since it can mitigate severe effects of frequency selective fading by inserting a guard interval and using a simple one-tap equalizer [1]-[3]. However, if the time variation of a fading channel over an OFDM symbol period is occurred, a loss of subcarrier orthogonality, referred to as inter-carrier interference (ICI), results in an error floor. The performance degradation due to ICI becomes significant as the carrier frequency, the number of subcarriers, and vehicle velocity increase[4]. In order to suppress the ICI, many techniques have been proposed such as minimum mean square error (MMSE), matched

M.S. Szczuka et al. (Eds.): ICHIT 2006, LNAI 4413, pp. 586–595, 2007.

filtering, Taylor series expansion, and polynomial cancellation, etc. [4]–[6]. Recently, time-domain cancellation (TDC) was proposed in [6] to mitigate the ICI with low complexity and high bandwidth efficiency. But, it neglects the effect of channel variation in cyclic prefix during the time-domain cancellation. Moreover, it does not work efficiently, if the data symbols in deep fades are used to regenerate the signal necessary to the cancellation operation. In order to supplement the weakness of the conventional time-domain cancellation, we propose an enhanced time-domain cancellation method exploiting a simple repetition coding and derive the cancellation procedure considering the channel variation in cyclic prefix. The proposed method provides better performance compared to the conventional method, while it maintains spectral efficiency by increasing the modulation order.

This paper is organized as follows. Section 2 presents an OFDM system model in time-varying multipath channels. Section 3 describes the conventional time-domain cancellation reflecting the cyclic prefix. Section 4 derives the proposed cancellation method using the modulation order increasing and the repetition (de-)coding. Section 5 gives some simulation results to evaluate the performance of the proposed method. This paper is concluded in Section 6.

2   OFDM Systems in Time-Varying Multipath Channels

After the inverse discreter Fourier transform (IDFT) operation and the addition of cyclic prefix at the transmitter, we can obtain discrete-time baseband samples given by

$$x_n = \frac{1}{\sqrt{N}} \sum_{k=0}^{N-1} X_k e^{j2\pi nk/N}, \ -N_g \leq n \leq N-1 \tag{1}$$

where x_n is the nth time-domain sample, N is the number of subcarriers, X_k is the kth frequency-domain modulated symbol, and N_g is the number of samples in cyclic prefix. The channel impulse response (CIR) of time-varying multipath channels is represented by tapped delay line (TDL) model such as eq. (2)

$$h_n = \sum_{l=0}^{L-1} \alpha_{l,n} \delta(n - \tau_l) \tag{2}$$

where h_n is the nth CIR, L is the number of paths, $\alpha_{l,n}$ is the nth complex channel gain in the lth path, and τ_l is the lth path delay. The complex channel gains are wide-sense stationary uncorrelated scattering (WSSUS) processes with the Jakes' power spectrum and can be represented by the average term (α_l) and the variation term $(\Delta\alpha_{l,n})$ given by

$$\alpha_l = \frac{1}{N} \sum_{n=0}^{N-1} \alpha_{l,n} \tag{3}$$

$$\Delta\alpha_{l,n} = \alpha_{l,n} - \alpha_l \tag{4}$$

We assume that the length of cyclic prefix is longer than the maximum delay spread and the synchronization is perfect at the receiver. The received sample of an OFDM symbol is represented by

$$y_n = \sum_{l=0}^{L-1} \alpha_l x_{n-\tau_l} + \sum_{l=0}^{L-1} \Delta\alpha_{l,n} x_{n-\tau_l} + w_n, \quad -N_g \le n \le N-1 \quad (5)$$
$$= y_{n,1} + y_{n,2} + w_n$$

where $y_{n,1}$ is the ICI-free sample, $y_{n,2}$ is the ICI sample, and w_n is independent and identically distributed (i.i.d.) complex additive white Gaussian noise (AWGN) samples with zero mean and variance of σ_w^2. After the subtraction of cyclic prefix and the DFT operation, the mth frequency-domain symbol is obtained by

$$Y_m = \frac{1}{\sqrt{N}} \sum_{n=0}^{N-1} y_n e^{-j2\pi mn/N} = \frac{1}{\sqrt{N}} \sum_{n=0}^{N-1} (y_{n,1} + y_{n,2} + w_n) e^{-j2\pi mn/N} \quad (6)$$
$$= Y_{m,1} + Y_{m,2} + W_m$$

where $Y_{m,1}$ is the DFT of $y_{n,1}$, $Y_{m,2}$ is the DFT of $y_{n,2}$, and W_m is the DFT of w_n. $Y_{m,1}$ and $Y_{m,2}$ are represented by eq.(7) and eq.(8), respectively.

$$Y_{m,1} = \sum_{k=0}^{N-1} X_k \frac{1}{N} \sum_{l=0}^{L-1} \alpha_l e^{-j2\pi\tau_l k/N} \sum_{n=0}^{N-1} e^{j2\pi(k-m)n/N} \quad (7)$$
$$= X_m H_{m,m}$$

$$Y_{m,2} = \sum_{k=0}^{N-1} X_k \left(\frac{1}{N} \sum_{l=0}^{L-1} \sum_{n=0}^{N-1} \alpha_{l,n} e^{j2\pi(k-m)n/N} e^{-j2\pi\tau_l k/N} \right.$$
$$\left. - \frac{1}{N} \sum_{l=0}^{L-1} \alpha_l e^{-j2\pi\tau_l k/N} \sum_{n=0}^{N-1} e^{j2\pi(k-m)n/N} \right) \quad (8)$$
$$= I_m, k \ne m$$

where $Y_{m,1}$ is the ICI-free symbol, and $Y_{m,2}$ is the ICI symbol. If the multipath channel is time-invariant, the $Y_{m,2}$ would be eliminated. $H_{m,m}$, I_m, and $H_{m,k}$ are defined by eq.(9)-(11).

$$H_{m,m} = \sum_{l=0}^{L-1} \alpha_l e^{-j2\pi\tau_l m/N} \quad (9)$$

$$I_m = \sum_{k=0,k\ne m}^{N-1} X_k H_{m,k} \quad (10)$$

$$H_{m,k} = \frac{1}{N} \sum_{l=0}^{L-1} \sum_{n=0}^{N-1} \alpha_{l,n} e^{j2\pi(k-m)n/N} e^{-j2\pi\tau_l k/N} \quad (11)$$

3   Conventional Time-Domain Cancellation

In this section, we derive the procedure of the conventional time-domain cancellation, where the time-variation in cyclic prefix is considered, and address its weakness in deep-fading environment. The mth frequency-domain symbol is rewritten by

$$Y_m = X_m H_{m,m} + I_m + W_m \tag{12}$$

In order to estimate channel frequency responses (CFRs), various channel estimator can be used. For the convenience of explanation, it is assumed that the least square (LS) channel estimator is employed and provides the desired CFRs perfectly, namely, $\hat{H}_{m,m} = H_{m,m}$. Then, the output of the one-tap equalizer is represented by

$$\tilde{X}_m = \frac{H_{m,m}}{\hat{H}_{m,m}} X_m + \frac{I_m + W_m}{\hat{H}_{m,m}} = X_m + \frac{I_m + W_m}{H_{m,m}} \tag{13}$$

By the hard-decision operation, we can obtain the detected symbol given by

$$\bar{X}_m = f_{dec,HD}\left(\tilde{X}_m\right) = X_m - E_m \tag{14}$$

where $f_{dec,HD}$ is the hard-decision function related to the modulation type, and E_m represents the detection error. The detected symbol is transformed into the time-domain by the IDFT operation and the addition of cyclic prefix.

$$\bar{x}_n = \frac{1}{\sqrt{N}} \sum_{m=0}^{N-1} \bar{X}_m e^{j2\pi nm/N} = x_n - e_n, \quad -N_g \leq n \leq N-1 \tag{15}$$

If the variation term $(\Delta\alpha_{l,n})$ and the path delay (τ_l) of the CIR are estimated perfectly, the regenerated signal can be represented by

$$\bar{y}_n = \sum_{l=0}^{L-1} \Delta\alpha_{l,n} \bar{x}_{n-\tau_l} = \sum_{l=0}^{L-1} \Delta\alpha_{l,n} x_{n-\tau_l} - \sum_{l=0}^{L-1} \Delta\alpha_{l,n} e_{n-\tau_l} \tag{16}$$

The time-domain cancellation is represented by

$$y_n^d = y_n - \bar{y}_n = \sum_{l=0}^{L-1} \alpha_l x_{n-\tau_l} + \sum_{l=0}^{L-1} \Delta\alpha_{l,n} e_{n-\tau_l} + w_n \tag{17}$$

The ICI-reduced signal is transformed again into the frequency-domain by the subtraction of cyclic prefix and the DFT.

$$Y_m^d = \frac{1}{\sqrt{N}} \sum_{n=0}^{N-1} y_n^d e^{-j2\pi mn/N} = X_m H_{m,m} + \sum_{k=0,k\neq m}^{N-1} E_k H_{m,k} + W_m \tag{18}$$

If eq.(13)-(18) are performed iteratively, we can obtain the better performance since the E_m is gradually decreased.

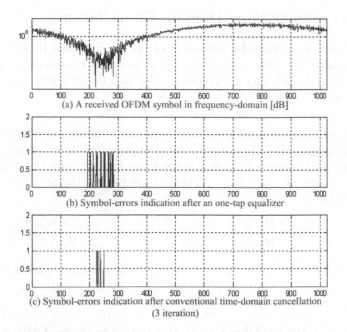

(a) A received OFDM symbol in frequency-domain [dB]

(b) Symbol-errors indication after an one-tap equalizer

(c) Symbol-errors indication after conventional time-domain cancellation
(3 iteration)

Fig. 1. Conventional time-domain cancellation under time-varying two-path channels

When the deep fades have occurred, the detected symbol in eq. (14) becomes unreliable, resulting in the bit error rate (BER) performance degradation. The weakness of the conventional time-domain cancellation under the time-varying two-path channel is shown in Fig. 1. Fig. 1.(a) represents the received signal strength through the two-path channel in [dB], and Fig. 2.(b) represents the errors after the one-tap equalizer without the time-domain cancellation. Even though the time-domain cancellation is iteratively performed three times, the errors still remain as shown in Fig. 1.(c).

4   Repetition Coding Aided Time-Domain Cancellation

In order to overcome the weakness, that the conventional method suffers from the deep-faded symbols, a repetition coding aided time-domain cancellation (RCA-TDC) is proposed. Repetition coding was adopted as a useful technique in the specification of mobile WiMAX, since it can be used to increase signal margin over the modulation and forward error correction (FEC) mechanisms [2]. In the repetition coding, the same symbol is transmitted on several different subcarriers. Even if it becomes impossible to receive the reliable symbol from the specific subcarrier, the other symbols on the different subcarriers would compensate the corrupted symbol by maximum ratio combining (MRC) [7]. While the repetition coding provides more reliable symbols for the time-domain cancellation, it decreases the spectral efficiency. For the purpose of maintaining the spectral efficiency, the modulation order is increased in our

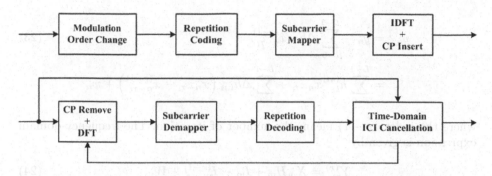

Fig. 2. Block diagram of repetition coding aided time-domain cancellation (RCA-TDC)

method. Fig. 2 shows the block diagram of our RCA-TDC method. N_s data symbols, $\{c_j | 0 \leq j \leq N_s - 1\}$ are copied to M groups, generating the group symbols $\{c_{i,j} | 0 \leq i \leq M - 1, 0 \leq j \leq N_s - 1\}$. The group symbols are allocated to the subcarriers determined by the subcarrier mapper. The allocated symbols, $\{X_k | 0 \leq k \leq N - 1\}$ are given by

$$X_k = c_{i,j}, if k = f_{map}(i, j) \tag{19}$$

where $f_{map}(i, j)$ is the mapping function from the jth symbol in the ith group to the kth subcarriers. The subcarrier demapping is represented by

$$d_{i,j} = Y_k, if (i, j) = f_{demap}(k) \tag{20}$$

where $f_{demap}(k)$ is the demapping function from the kth subcarrier to the jth symbol in the ith group. The received group symbol is represented by

$$d_{i,j} = c_{i,j} H_{i,j} + I_{i,j} + W_{i,j} \tag{21}$$

The repetition decoding or the MRC operation provides the symbol given by

$$d_j = \sum_{i=0}^{M-1} G_{i,j} d_{i,j} = c_j \sum_{i=0}^{M-1} G_{i,j} H_{i,j} + \sum_{i=0}^{M-1} G_{i,j} (I_{i,j} + W_{i,j}) \tag{22}$$

$$= c_j + \left(\frac{\sum_{i=0}^{M-1} H_{i,j}^* (I_{i,j} + W_{i,j})}{\sum_{i=0}^{M-1} |H_{i,j}|^2} \right)$$

where $G_{i,j} = H_{i,j}^* / \sum_{i=0}^{M-1} |H_{i,j}|^2$ is the weight coefficients for the repetition decoding. The same procedure of the time-domain cancellation is used for the RCA-TDC method. The time-domain cancellation of the RCA-TDC is represented by

$$\hat{y}_n^{(r)} = y_n - \sum_{l=0}^{L-1} \Delta h_{l,n} \hat{x}_{n-\tau_l}^{(r-1)} \qquad (23)$$

$$= \sum_{l=0}^{L-1} h_l^{(m)} x_{n-\tau_l} + \sum_{l=0}^{L-1} \Delta h_{l,n} \left(x_{n-\tau_l} - \hat{x}_{n-\tau_l}^{(r-1)} \right) + w_n$$

where the index $(r-1)$ means the number of iteration. The frequency-domain expression is given by

$$\hat{Y}_m^{(r)} = X_m H_m + I_m - \hat{I}_m^{(r-1)} + W_m \qquad (24)$$

where the ICI term, $\hat{I}_m^{(r-1)}$ is represented by eq. (25).

$$\hat{I}_m^{(r-1)} = \sum_{k=0, k \neq m}^{N-1} X_m^{(r-1)} \frac{1}{N} \sum_{l=0}^{L-1} \sum_{n=0}^{N-1} h_{l,n} e^{j2\pi n(k-m)/N} e^{-j2\pi k\tau_l/N} \qquad (25)$$

After the RCA-TDC operation and the repetition decoding, we can obtain the reliable symbol given by

$$\hat{d}_j^{(r)} = c_j + \left(\frac{\sum_{i=0}^{M-1} H_{i,j}^* \left(I_{i,j} - I_{i,j}^{(r-1)} + W_{i,j} \right)}{\sum_{i=0}^{M-1} |H_{i,j}|^2} \right). \qquad (26)$$

5   Simulation Results

In this section, the performance of the proposed RCA-TDC method is investigated in a time-varying multipath channel. A QPSK/16QAM/64QAM-OFDM system with N=1024 subcarriers occupies a bandwidth of 10 MHz operating in the 2.3 GHz. The sampling interval is given as T=0.1 μs. The number of samples in cyclic prefix, N_g is 128. The "Vehicular A" channel model, which is defined by ETSI for the evaluation of UMTS radio interface proposals, is considered. The velocity of the mobile terminal is 250 km/h. In order to obtain the boundary performance of our method, we assume that the average and the variation of the channel is known.

Fig. 3 shows the BER performance of the conventional time-domain cancellation. "TDC" represents the conventional time-domain cancellation, and the x in the parenthesis TDC(x) means the number of iteration. As stated above, the conventional method suffers from the deep-faded symbols which are used for the regenerated signal. Therefore, it cannot provide good performance even though the number of iteration is increased. The BER performances of the proposed RCA-TDC are shown in Fig. 4 and Fig. 5. "RCA" represents the repetition coding without the time-domain cancellation, and the x in the parenthesis RCA(x) means the number of repetition. "RCA-TDC" represents the

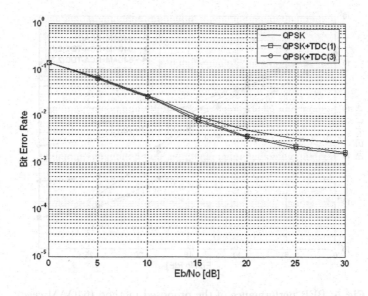

Fig. 3. BER performance of conventional time-domain cancellation

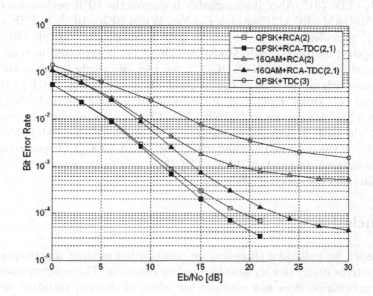

Fig. 4. BER performance of the proposed method (QPSK/16QAM case)

proposed method. The x and y in the parenthesis RCA-TDC(x, y) mean the number of repetition and iteration, respectively. In Fig. 4, QPSK/16QAM modulation and two times repetition are considered for our method. When BER=10^{-3}, "16QAM+RCA-TDC(2,1)" provides the E_b/N_0 gain more than 15 dB compared

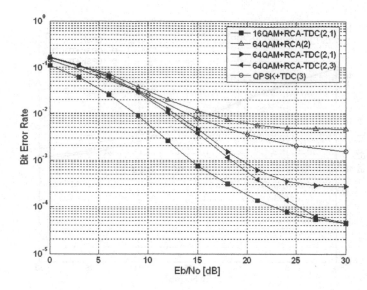

Fig. 5. BER performance of the proposed method (64QAM case)

to "QPSK+TDC(3)". Also, it remarkably improves the BER performance compared to "16QAM+RCA(2)" at high E_b/N_0s. When BER=10^{-4}, "QPSK+RCA-TDC(2,1)" provides about 2 dB E_b/N_0 gain compared to "QPSK+RCA(2)". If the proposed method, "16QAM+RCA-TDC(2,1)" is exploited instead of the conventional mehtod, "QPSK+TDC(3)", we can effectively improve the BER performance without decreasing the spectral efficiency. In Fig. 5, 64QAM modulation and two times repetition are considered for our method. "64QAM+RCA-TDC(2,1)" is superior to "QPSK+TDC(3)". As the number of iteration is increased, the BER performance is significantly improved (see "64QAM+RCA-TDC(2,3)"). If the proposed method, "64QAM+RCA-TDC(2,1)" is used instead of the conventional method, "QPSK+TDC(3)", we can improve both the BER performance and the spectral efficiency.

6   Conclusion

In this paper, an enhanced time-domain cancellation method was proposed for OFDM systems under doubly-selective fading channels. The conventional time-domain cancellation does not consider the effect of channel variation in cyclic prefix during the time-domain cancellation. It does not provide the performance improvement, when the deep fades have occurred. In order to supplement the conventional method, we employed the simple repetition (de-)coding and the modulation order increasing. The repetition coding provides reliable symbols used for the regeneration operation, and the modulation order increasing maintains or increases the spectral efficiency. When BER=10^{-3}, the proposed method using 16QAM provides the E_b/N_0 gain more than 15 dB compared to the conventional method using QPSK, without decreasing the spectral efficiency. Moreover, the proposed method using 64QAM is still superior to the conventional method. As the number of its iteration is increased, the BER performance is significantly improved. Since the proposed method can efficiently mitigate the inter-carrier interference without decreasing the spectral efficiency, compared to the conventional method, it can be applied to the wireless broadband systems under high-speed mobile environment.

References

1. van Nee, R., Prasad, R.: OFDM for Wireless Multimedia Communications. Artech House (2000)
2. IEEE Std 802.16e^{TM}-2005, IEEE standard for local and metropolitan area networks - Part 16: Air interface for fixed and mobile broadband wireless access systems (February 2006)
3. Ekström, H., Furuskär, A., Karlsson, J., Meyer, M., Parkvall, S., Torsner, J., Wahlqvist, M.: Technical solutions for the 3G long-term evolution. IEEE Commun. Mag. 44, 38–45 (2006)
4. Jeon, W.G., Chuang, K H., Cho, Y.S.: An equalization technique for orthogonal frequency division multiplexing systems in time-variant multipath channels. IEEE Tran. Commun. 49, 1185–1191 (1999)
5. Choi, Y.-S., Voltz, P.J., Cassara, F.A.: On channel estimation and detection for multicarrier signals in fast and selective Rayleigh fading channels. IEEE Tran. Commun. 49, 1375–1387 (2001)
6. Chen, S., Yao, T.: Intercarrier interference suppression and channel estimation for OFDM systems in time-varying frequency-selective fading channels. IEEE Tran. Consum. Electron. 50, 429–435 (2004)
7. Tse, D., Viswanath, P.: Fundamentals of Wireless Communication. Cambridge Univ. Press, Cambridge (2005)

On Scheduling Transmissions for Hidden Terminal Problems in Dynamic RFID Systems

Ching-Hsien Hsu[1], Jong Hyuk Park[2], and Kuan-Ching Li[3]

[1] Dept. of Computer Science and Information Engineering
Chung Hua University
Hsinchu 300, Taiwan
chh@chu.edu.tw
[2] R&D Institute
Hanwha S&C Co. Ltd.
Jangkyu-Dong, Jung-Gu, Seoul Korea
parkjonghyuk@gmail.com
[3] Parallel and Distributed Processing Center
Dept. of Computer Science and Information Engineering
Providence University
Shalu, Taichung 43301, Taiwan
kuancli@pu.edu.tw

Abstract. The problem of scheduling transmissions of dynamic Radio Frequency Identification (RFID) systems has been recently studied. One of the common problems, reader collision avoidance has instigated researchers to propose different heuristic algorithms. In this paper, we present a prime based First Come Higher Priority (FCHP) transmission scheduling method for reader collision problems that caused by hidden terminal. FCHP is a simple mechanism for coordinating simultaneous transmissions among multiple readers. A significant improvement of this approach is that FCHP prevents reader collisions by giving contention free scheduling. The second advantage of the proposed technique is that FCHP is adaptive in both static and dynamic RFID environments. The simulation results show that the proposed technique provides superior performance in both static and dynamic instances. The FCHP is shown to be effective in terms of system throughput, system efficiency and easy to implement.

1 Introduction

The Radio Frequency Identification (RFID) is an automatic technology aids machines or computers to identify, record or control individual target through radio waves. The electronics in the RFID reader use an outside power resource to generate signal to drives the reader's antenna and turn into radio wave. The radio wave will be received by RFID tag which will reflect the energy in the way of signaling its identification and other related information. In some matured system, the reader's RF can also instruct the memory to be read or write from which the tag contained.

In many applications, it will be necessary to install several RFID readers in appropriated distance to each other. Otherwise it would be interfere with each other. The interference could be caused when the frequency band is shared with other potential

M.S. Szczuka et al. (Eds.): ICHIT 2006, LNAI 4413, pp. 596–606, 2007.

users. As an RFID reader is designed to accept the tiny signal reflected from a tag. It will be particularly influenced to any relatively powerful transmissions from other readers that happen at the same time.

The distributed control mode means readers switch the communication state with each other. The centralized control mode means appropriate coordination is handled by a top reader among other readers. We combined with the centralized and distributed reactions and propose a transmission scheduling method for reader collision problems [4] that caused by hidden terminal [3,10]. By establishing communications among other readers through control channel while one is transmitting data with tag via data channel [5], a prime based First Come Higher Priority (FCHP) transmission scheduling algorithm is presented. A significant improvement of this approach is that FCHP prevents reader collisions by giving contention free scheduling. The second advantage of the present technique is that FCHP is adaptive in both static and dynamic RFID environments. The FCHP is shown to be effective in terms of throughput, system efficiency and easy to implement.

The rest of this paper is organized as follows. In Section 2, a brief survey of related work will be presented. In Section 3, we introduce the hidden terminal problem and reader collision problem. A prime based FCHP scheduling technique will be discussed in section 4. The theoretical performance analysis and simulation comparison will be given in Section 5. Section 6 briefly concludes this paper.

2   Related Work

Many research efforts for collision avoidance have been presented in the literature. Frequency Division Multiple Access (FDMA), Code Division Multiple Access (CDMA), Time Division Multiple Access (TDMA) [1] and Carrier Sense Multiple Access (CSMA) are four basic access methods to categorize MAC-layer protocols.

FDMA is functioned via frequency assignment in which the communication is applied in form of many-to-one. Since RFID tags are without a frequency tuning circuitry, the additional of such a tuning circuitry will increase the cost. CDMA uses spectrum modulation techniques that based on pseudo random codes for data transmission. The complicated technique and computation intensive bring up the cost of tags. TDMA uses time slot to avoid collision. Because there is only one code transmitted during each slot, it can successfully avoid the collision. CSMA enables individual data transmission by detecting whether the medium is busy. Collision avoidance and collision detection are two common types defined in CSMA.

Standard collision avoidance protocols like RTS-CTS [8] cannot be directly applied to RFID systems due to the collision avoidance mechanism for CTS.

Techniques for resolving RFID reader collision problems are usually proposed as reader anti-collision techniques or tag anti-collision solutions. We briefly describe the related works in these two aspects.

The Colorwave [9] is a scheduling-based approach prevents RFID readers from simultaneously transmitting signal to a RFID tag. The Colorwave is used as a distributed anti-collision system based on TDMA in RFID network. Addressing RFID network with graph coloring theory and accomplishing each reader has the smallest possible number of adjacent nodes in the same color. The protocol attempts to optimize the graph to achieve a percentage of successful transmissions. However, some limitations don't ensure the system stays in stable condition such as randomly selecting colors.

Pulse protocol [2] is referred as beacon broadcast and CSMA mechanism. A "beacon" is sent by readers periodically in separated control channels during communication with tags. The contend_back-off and the delay_before_beaconing [7] in the protocol are similar in wireless networks.

A coverage-based RFID reader anti-collision mechanism was proposed in [6]. Kim et al. presented a localized clustering coverage protocol for solving reader collision problems occurring among homogeneous RFID readers. However, this technique is not applicable in dynamic RFID environments.

From what we have learned in the literature about developed protocols, which are called medium access protocols, are defined to coordinate the use of a shared medium with certain regulation. However, the existing method under the mobility environments has several limitations.

3   Hidden Terminal Problem

Figure 1 shows the tag T1 is surrounded by two readers. Each of the readers located out of sensing range from the other one in the RFID network. Therefore, these two readers are not able to communicate each other and the reader collision might happen. The situation is mentioned as hidden terminal problem which has the following features:

- The reader that are not in each others' sensing range might interfere with tags and cause carrier sensing to work ineffective.
- When queries or transmissions from multiple readers collide on a tag, signals can be distorted and the queries might be incorrect.

Interferences in RFID system are usually classified into reader to reader frequency interference and reader-to-tag interference.

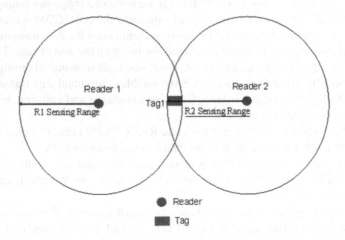

Fig. 1. Hidden terminal problem

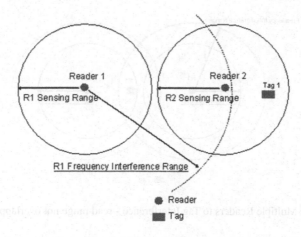

Fig. 2. Reader to Reader Frequency Interference

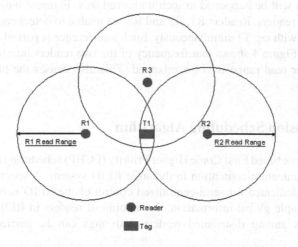

Fig. 3. Multiple Readers to Tag Interference - read range overlapped

Reader-to-reader frequency interference also calls frequency interference, occurs when readers are interfered with others from communicating with tags. Fig. 2 shows reader R2 resides in the frequency interference range of reader R1 who has wider interference signal range (the dotted line). As tag T1 response to reader R2, it might be influenced by the interference signal of reader R1. Such hidden terminal problem occurred even when the readers' range of the two readers are not overlapped.

Also defined as simply tag interference, occurs when two or more readers in the transmission zone attempts to communicate with one tag simultaneously. In this situation, each reader performs one to one communication to the tag. However, it is not

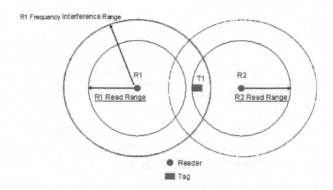

Fig. 4. Multiple Readers to Tag Interference - read range not overlapped

known by readers that the tag should responsible to multiple readers simultaneously. Reader collision will be happened in such undesired way. Figure 3 indicates the overlapping of three readers. Readers R1, R2 and R3 are unable to detect each other during communication with tag T1 simultaneously. Such interference is part of the hidden terminal problem. Figure 4 shows that frequency of the two readers interfered with each other while their read range is not overlapped. This also causes the hidden terminal problem.

4 Transmission Scheduling Algorithm

We present a prime based First Come Higher Priority (FCHP) scheduling algorithm aims to avoid communication contention in dynamic RFID system. A specific reader (say coordinator) is dedicated for semi-centralized control of the RFID network. Through broadcasting simple global information, the amount of readers in RFID network, the communications among distributed readers with tags can be carried out without contention.

Figure 5 shows scenario of four readers R1, R2, R3 and R4 reside in communication range reachable to the tag. It is expected that if the readers know existence of others, the communications with tag will not be collided. Therefore, a coordinator reader (R5) could be on demand associated with the tag which will be read by other readers. The coordinator keeps track of total amount of readers and makes the request readers be informed this information to build its own transmission schedule locally. Since the transmission schedule is distributed determined while the global information is centralized collected, the proposed technique is classified as a semi-distributed algorithm.

The FCHP has three major steps:

– The request reader sends beacon through control channel to the corresponding coordinator notifying its attempt to communicate with tag before transmitting data,

- The coordinating reader learns beacon's source address according to the received frame. Then, it informs existing readers the fact of new prime is assigned through sending beacon reply.
- In case that two or more readers transmit data with tag at the same time slot and causing communication conflict, lower prime time slot will be preempted by higher prime time slot.

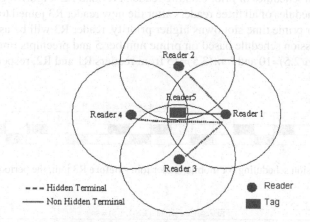

Fig. 5. Motivating scenario of the hidden terminal

Reader	Prime Number	Time Slot, the period is $lcm(2, 3) = 6$																	
		1	2	3	4	5	6	7	8	9	10	11	12	13	14	15	16	17	18
R1	2		■		■		■		■		■		■		■		■		■
R2	3			■			■			■			■			■			■

Fig. 6. Prime based scheduling mechanism, the period is $lcm(2,3)=6$

We use an example to clarify the above operation. Figure 6 shows R1 is assigned to use prime number 2 and R2 is assigned having prime number 3 as time slot. That is, R1 transmits data with tag every two time units (i.e., $t_2, t_4, t_6, t_8, \ldots$) while R2 transmits data with tag every three time units (i.e., $t_3, t_6, t_9, t_{12}, \ldots$).

4.1 Reader Join

This subsection presents the process of FCHP for handling reader join in dynamic RFID system. The FCHP has two main operations when reader joins to the network.

- Before join to an RFID network, the new reader first sends beacon message to the coordinating reader notifying its attempt to communicate with tag. The coordinating reader will inform present readers the fact of new reader join upon the received beacon message and assigns the next unused prime (p_{next}) to the new reader,

– For each prior existing reader, since it is informed that prime number p_{next} was assigned to the latest joined reader, it will adjust its communication schedule by disable time slots that is multiple of the least common multiple of p_{next} and its own prime.

Figure 7 considers again the example in previous subsection to demonstrate the situation of mobile reader join. Following the situation of Figure 6, Figure 7 shows the communication schedules of prior existing readers, R1 and R2. Figure 8 gives the communication schedules of all three readers after the new reader R3 joined to the network. Because higher prime time slot owns higher priority, reader R3 will be assigned an ordinary transmission schedule based on prime number 5 and preempts time slots that is multiple of $lcm(2,5)=10$ and $lcm(3,5)=15$ from readers R1 and R2, respectively.

Reader	Prime Number	Time Slot, the period is $lcm(2,3)=6$																	
		1	2	3	4	5	6	7	8	9	10	11	12	13	14	15	16	17	18
R1	2																		
R2	3																		

Fig. 7. Transmission scheduling for mobile reader join - **before** R3 join, the period is $lcm(2,3)=6$

Reader	Prime Number	Time Slot, the period is $lcm(2,3,5)=30$																													
		1	2	3	4	5	6	7	8	9	10	11	12	13	14	15	16	17	18	19	20	21	22	23	24	25	26	27	28	29	30
R1	2																														
R2	3																														
R3	5																														

Fig. 8. Transmission scheduling for mobile reader join - **after** R3 join, the period is $lcm(2,3,5)=30$

4.2   Reader Leave

The process of FCHP for handling reader leave in dynamic RFID system is abstracted as follows.

As one reader finished the transmission with tag, it will not further send beacon information to the coordinating reader. Before it leaves the RFID network, a "finalized" beacon message will be sent (through control channel) stating its completion of the communication. Meanwhile, the coordinating reader will inform present readers the fact of old reader (R_{leave}) leave upon the received beacon message which contains the virtual address of R_{leave} . At the same time, the virtual source address will be translated into corresponding allocated prime which will facilitate other readers to be notified and adjust its local transmission schedule.

Given a snapshot of RFID network, assume there are n readers, R1, R2, ..., Rn in the network. For these readers, the assigned prime as their transmission time slot are $k_1, k_2, ..., k_n$, respectively. Let $R_{leave} = R_i$, where $1 \leq i \leq n$. The dynamic adjustment of transmission schedule of remaining readers has three types.

For reader R_j whose index is smaller than i, i.e., $j < i$, resets all time slots that are multiples of k_n from unavailable to available. For reader R_j whose index is larger than i, i.e., j>i and j≠n, switches its time slot from k_j to k_{j-1} and disables all time slots that conflict with primes k_j, k_{j+1}, ..., k_{n-1}. For those readers R_j whose index is the largest before R_i has leaved, i.e., $j = n$, resets its time slot from k_n to k_{n-1}.

5   Performance Evaluation and Simulation Results

To evaluate performance of the proposed algorithms, we have implemented a random RFID network generator. The FCHP method was compared along with the Colorwave and the Pulse protocols. Both static and dynamic scenarios were conducted in simulations.

To simplify the presentation of the following performance analysis, some notations and terminologies are defined in Table 1. According to these definitions, two performance metrics, system throughput and system efficiency can be further defined as follows,

- System Throughput = N_{succ}
- System Efficiency (%) = N_{succ} / T_{in_use}

As defined in Table 1, there are two types of failure time slot, the collided time slots ($N_{collide}$) and the idle time slots (N_{idle}). Collided time slot represents two or more readers attempt to transmit data with tag at the same time slot. Idle time slot denotes no reader transmits data at that time slot.

Table 2 shows the parameters and simulation results of the three algorithms under static circumstance. The numbers of initial readers are set from 20 to 80. Because of the

Table 1. List of notations and terminologies

R_{init}	The number of initial readers in a given RFID network
$R^{join}_{t_m,t_n}$	The numbers of "join" readers during time period $t_m \sim t_n$
$R^{system}_{t_m,t_n}$	$R^{system}_{t_m,t_n} = R_{init} + R^{join}_{t_m,t_n}$
N_{succ}	Number of time slots upon which reader transmits data with tag successfully
N_{idle}	Number of time slots upon which no reader can transmit data through it
N_{free}	Number of time slots during which there is no reader request data from tag in the RFID network
$N_{collide}$	Number of collided time slots
N_{fail}	Number of failure time slots, $N_{fail} = N_{idle} + N_{collide}$
$T_{deadline}$	Total number of time slots set for experiment, $T_{succ} + T_{fail} = T_{deadline}$
T_{in_use}	The amount of time slots during which there exist readers request data from tag (counted by # of time slots), i.e., $T_{in\_use} = T_{deadline} - N_{free}$
$T_{complete}$	The cmpletion time for all readers finish data transmission with tag.

Table 2. Parameters and simulation results under static circumstance

Protocols	R_{init}	$R_{t_0,t_5 00}^{join}$	R^{system}	T_{in_use}	N_{free}	Throughput N_{succ}	Efficiency ((N_{succ}/T_{in_use}))
Pulse				202	298	113	55.9
Colorwave	20	0	20	350	150	112	32.0
FCHP				117	383	113	96.6
Pulse				405	95	185	45.7
Colorwave	40	0	40	460	40	124	27.0
FCHP				239	261	235	98.3
Pulse				461	39	159	34.5
Colorwave	60	0	60	479	21	83	17.3
FCHP				286	214	284	99.3
Pulse				482	18	138	28.6
Colorwave	80	0	80	500	0	59	11.8
FCHP				353	147	352	99.7

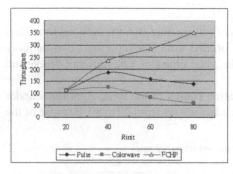 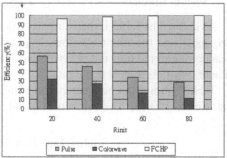

Fig. 9. System throughput and efficiency for three algorithms under static circumstance

static circumstance, we have $R^{join} = 0$ and $R^{system} = R^{init}$. The time period set for this simulation is $T_{deadline} = 500$. $T_{complete}$ is completion time for all readers finish data transmission with tag. According to definitions in Table 1, we have $N_{free} = T_{deadline}$ - $T_{complete}$. System throughput is equal to the number of time sots upon which reader transmits data with tag successfully in a given period of time; while the system efficiency is equal to the ratio of N_{succ} over T_{in_use}. From Table 2, we observe that most readers finish data transmission before $T_{deadline}$ (t_{500}) except the case of the Colorwave algorithm with 80 readers initially. Because the time period set for this simulation was 500 time slots, the completion time is denoted up to 500 in case some of the readers do not finish data transmission before deadline.

The proposed FCHP scheduling algorithm also has the highest system throughput and system efficiency; the higher reader density, the performance improvement is more significant. To facilitate the comparison, the results of system throughput and system efficiency are also depicted in Figure 9 which is identical to the last two columns listed in Table 2.

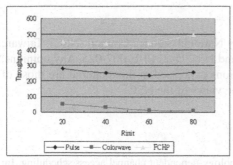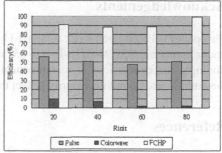

Fig. 10. System throughput and efficiency for three algorithms under dynamic circumstance

Table 3. Parameters and simulation results under dynamic circumstance

Protocols	R_{init}	$R^{join}_{t_0,t_{5}00}$	R^{system}	T_{in_use}	N_{free}	Throughput N_{succ}	Efficiency $((N_{succ}/T_{in\_use})$
Pulse				500	0	281	56.2
Colorwave	20	196	216	500	0	51	10.2
FCHP				500	0	454	90.8
Pulse				500	0	254	50.8
Colorwave	40	177	217	500	0	32	6.4
FCHP				500	0	440	88.0
Pulse				500	0	236	47.2
Colorwave	60	227	287	500	0	11	2.2
FCHP				500	0	444	88.8
Pulse				500	0	254	50.8
Colorwave	80	188	268	500	0	9	1.8
FCHP				500	0	496	99.2

The FCHP method outperforms the other two protocols in terms of system through-put and system efficiency. The results of system throughput and system efficiency are also depicted in Figure 10, which is identical to the last two columns listed in Table 3.

6 Conclusions

In this paper, we have presented a prime based FCHP transmission scheduling tech-nique for reader collision avoidance. By using the method of dual channel control, the proposed technique achieves contention free communication with tags among multiple readers. To evaluate the performance of the proposed technique, we have compared the FCHP scheduling technique along with two other algorithms. The simulation results show that the proposed technique provides superior performance in both static and dy-namic environments. The FCHP is shown to be effective in terms of system throughput and efficiency and easy to implement. In summary, the FCHP transmission scheduling is capable of scalability under high density mobile readers.

Acknowledgements

This paper is based upon work supported by National Science Council (NSC), Taiwan, under grants no. NSC94-2218-E-007-057 and NSC95-2218-E-007-025. Any opinions, findings, and conclusions or recommendations expressed in this material are those of the authors and do not necessarily reflect the views of the NSC.

References

1. Bao, L., Aceves, J.J.G.L.: Collision-free topology-dependent channel access scheduling. In: MILCOM. Proceedings of the IEEE 21st Century Military Communications Conference, pp. 507–511 (2000)
2. Birari, S.M.: Mitigating the Reader Collision Problem in RFID Networks with Mobile Readers, Master thesis, Kanwal Rekhi School of Information Technology (2005)
3. Chen, W.-T., Ho, T.-W., Chen, Y.-C.: An MAC Protocol for Wireless Ad-hoc Networks Using Smart Antennas. In: ICPADS 2005. Proceedings of the 11th IEEE International Conference on Parallel and Distributed Systems, pp. 446–452 (2005)
4. Engels, D.W., Sarma, S.E.: The reader collision problem. In: Proceedings of the 2002 IEEE International Conference on Systems, Man and Cybernetics, p. 6 (2002)
5. Jain, N., Das, S.R., Nasipuri, A.: A multichannel CSMA MAC protocol with receiver-based channel selection for multihop wireless networks. In: Proceedings of the 10th IEEE International Conference on Computer Communications and Networks, pp. 432–439 (2001)
6. Kim, J., Kim, E., Kim, S., Kim, D., Lee, W.: Low-Energy Localized Clustering: An Adaptive Cluster Radius Configuration Scheme for Topology Control in Wireless Sensor Networks. In: VTC. Proceedings of the IEEE Vehicular Technology Conference, pp. 2546–2550 (2005)
7. Konorski, J.: Solvability of a Markovian Model of an IEEE 802.11 LAN under a Backoff Attack. In: MASCOTS 2005. Proceedings of the 13th IEEE International Symposium on Modeling, Analysis, and Simulation of Computer and Telecommunication Systems, pp. 491–498 (2005)
8. Sobrinho, J.L., de Haan, R., Brazio, J.M.: Why RTS-CTS is not your ideal wireless LAN multiple access protocol. In: Proceedings of the IEEE Wireless Communications and Networking Conference, pp. 81–87 (2005)
9. Waldrop, J., Engels, D.W., Sarma, S.E.: Colorwave: an anticollision algorithm for the reader collision problem. In: Proceedings of the IEEE International Conference on Communications, pp. 1206–1210 (2003)
10. You, T., Hassanein, H., Yeh, C.-H.: PIDC - Towards an Ideal MAC Protocol for Multi-hop Wireless LANs. In: Proceedings of the IEEE International Conference on Wireless Networks, Communications and Mobile Computing, pp. 655–660 (2005)

Efficient RFID Authentication Protocol for Minimizing RFID Tag Computation*

Keunwoo Rhee[1], Jin Kwak[2], Wan S. Yi[1], Chanho Park[3],
Sangjoon Park[1], Hyungkyu Yang[4], Seungjoo Kim[1], and Dongho Won[1,**]

[1] Information Security Group, Sungkyunkwan University,
300 Cheoncheon-dong, Jangan-gu, Suwon-si, Gyeonggi-do, 440-746, Korea
{kwrhee,wsyi,sangjoon,skim,dhwon}@security.re.kr
http://www.security.re.kr
[2] Department of Information Security, Soonchunhyang University,
646 Eupnae-ri, Shinchang-myun, Asan-si, Chungcheongnam-do, 336-745, Korea
jkwak@sch.ac.kr
[3] INDI SYSTEM Co., Ltd.,
187-10 Guro-dong, Guro-gu, Seoul, 152-848, Korea
[4] Department of Computer and Media Engineering, Kangnam University,
Gugal-dong, Giheung-gu, Yongin-si, Gyeonggi-do, 446-702, Korea
hkyang@kangnam.ac.kr

Abstract. RFID systems have become vital technology for realizing
ubiquitous computing environments. However, features of RFID systems
present potential security and privacy problems. In an effort to resolve
these problems, many kinds of security and privacy enhancement technologies have been researched. However, solutions produced to date still
have flaws and are not sufficiently effective for real RFID systems such
as the EPCglobal network™. Therefore, in this paper, to make RFID systems more secure and efficient, improved technology based on password,
is proposed. The proposed technology combines an encryption algorithm
with a password-derived key, and can be applied to low-cost RFID systems for enhancing the security and privacy of these systems.

Keywords: RFID, EPCglobal network™, authentication, encryption algorithm, password, password-derived key.

1 Introduction

RFID (Radio Frequency Identification) is data carrier technology that transmits information via signals in the radio frequency portion of the electromag-
netic spectrum [5]. RFID systems are used for supply chain management, animal
identification, access control, passport security and so on.

---

* This research was supported by the MIC (Ministry of Information and Communication), Korea, under the ITRC (Information Technology Research Center) support
program supervised by the IITA (Institute of Information Technology Assessment).
** Corresponding author.

M.S. Szczuka et al. (Eds.): ICHIT 2006, LNAI 4413, pp. 607–616, 2007.
© Springer-Verlag Berlin Heidelberg 2007

However, RFID systems read the globally unique identity of an item without its owner's awareness, and present new security problems, such as the exposure of a corporation's marketing data, or privacy problems, such as the disclosure of a user's private information. Due to these security and privacy threats, a large-scale trial planned by Benetton Corporation to affix RFID tags to clothing, resulted in massive protest that eventually resulted in the firm's decision to back away from RFID deployment [2, 21]. This attitude is a serious obstacle in the popularization of RFID technology.

Generally applicable techniques have been researched to resolve security and privacy threats and accept RFID systems as a component of daily life. Representative techniques include, the "Kill command" technique [23], hash-lock protocol [20, 22, 23], randomized hash-lock protocol [22, 23], external re-encryption protocol [9], "Blocker tag" technique [10], hash based ID variation protocol [7], and challenge-response based protocol [17]. In addition to these technical efforts, various political measures have been proposed. In Korea, the Ministry of Information and Communication published "RFID Privacy Guideline" in July 2005.

These political measures can restrict useful RFID applications. In addition, current technical measures are insufficient to apply to real systems such as the EPCglobal network™, since most of the technical measures are based on an abstract model.

Therefore, in this paper, a password based RFID authentication protocol, secure against security and privacy threats in real RFID systems, is proposed. Section 2 analyzes the requirements and limitations in RFID systems. Section 3 proposes a new RFID authentication protocol using password techniques. Section 4 analyzes the security and effectiveness of the proposed protocol in real RFID systems. Finally, Section 5 presents the conclusion.

2   Requirements and Limitations in RFID Systems

As mentioned previously, many security and privacy threats exist in RFID systems. In addition, low-cost passive RFID tags have low computation power and storage.

2.1   Security and Privacy Requirements in RFID Systems

Simson L. Garfinkel et al. analyzed various security and privacy threat contexts in RFID deployment [4]. In order to prevent these threats, the RFID system satisfies the following security and privacy requirements.

[Security and Privacy Requirements]

- *Weak anonymity*: Uncheckability. In order to prevent exposure of the tag's identification information, the malicious recipient of the responses generated by the tag should not be able to discover which tag created the information.

The legitimate recipient can verify the identity of the tag from the response [12,17].

- *Strong anonymity*: Unlinkability. In order to prevent exposure of the tag's location information, the malicious recipient of the responses should not distinguish different responses generated by the same tag [12,17].
- *Anti-counterfeiting*: In order to prevent tag cloning, the data should be stored in secure memory in the tag, or the counterfeited tag should be unusable.
- *Recognizability*: In order to prevent a denial of service attack, the RFID system should be recognizable against denial of service attacks.

2.2 Computation Capability and Storage Limitation in RFID Systems

In order to promote the spread of RFID systems, the cost of RFID tags should be low. Most RFID systems use low-cost passive RFID tags. Currently, 5¢ is a reasonable price [19]. However, this kind of RFID tag suffer from two limitations. The first is the extremely low computation capability and the second is the small storage size. Though high-cost RFID tags can embed a resource efficient scheme such as NTRU [8,14], the 5¢ RFID tag cannot implement a symmetric-key cryptographic system, public-key cryptographic system, hash operation, random number generation and so on. The low-cost RFID tag can perform certain exclusive-OR and AND operations. In addition, the current commercial EPC-global Class-1 Gen-2 [3] RFID tag, Philips' UCODE EPC G2 tag, has 512 bits of on-chip memory [16].

However, most previous solutions for security and privacy in RFID systems demand more resources than these limitations. The hash-lock protocol [20,22,23], randomized hash-lock protocol [22, 23], hash based ID variation protocol [7], hash-chain protocol [15], and challenge-response based protocol [17] use hash function operations. The PRF based protocol [13] uses a pseudo random number generation function. The universal re-encryption protocol [6] and re-encryption with a check [18] protocol use exponential operations. In addition, the one-time pad based protocol [11] uses considerable memory for one-time pads.

3 Proposed Technical Solution

In this section, a new technical solution for the real RFID system is proposed in order to enhance security and privacy and improve efficiency in RFID systems.

The proposed protocol is similar to Manfred Aigner et al.'s mutual authentication [1] in that both protocols use an encryption algorithm instead of hash function. However, the proposed protocol is distinguishable from Manfred Aigner et al.'s mutual authentication in that the key distribution between the RFID tag and reader-side RFID system is not necessary since the proposed protocol uses an encryption algorithm with a password-derived key. In addition, in the proposed protocol, there is no generation of a random number in RFID tags.

In the proposed protocol, it is assumed that the communication channel between the RFID tag and reader is insecure since the computation resource of the RFID tag is limited and the communication channel is wireless. However, it is assumed that other communication channels are secure since the components of the RFID system have high computation ability, except for the RFID tag. In addition, the process of the user inputting a password is assumed to be secure.

[Notations and Parameters]

- $Query$: Request for the responses of the tags.
- ID: Secret identification information of the RFID tag such as EPC.
- PW: Password.
- $F(I)$: Key generation function. $F(I)$ means that a key is generated from input I.
- K: Encryption and decryption keys derived from password. $K = F(PW)$.
- $E_K(M)$: Encryption. $E_K(M)$ means that message M is encrypted using encryption key K.
- $D_K(C)$: Decryption. $D_K(C)$ means that ciphertext C is decrypted using decryption key K. If C is $E_K(M)$, $D_K(C)$ equals $D_K(E_K(M)) = M$.
- R: Random number.
- URL: Uniform Resource Locator.
- $Info.$: Information related with a specific tag ID such as product type, manufacturing date and so on.
- $\oplus$: Exclusive-OR (XOR).
- $\|$: Concatenate function.

3.1 Initial Set-Up Process

In order to authenticate an RFID tag, some data should be embedded in the tag when the tag is produced by the manufacturer. In addition, data related with the tag's data is also stored in the Information Services.

[Set-Up Process]

1. Manufacturer selects password, PW, and generates two different random numbers, R_0, and R_1.
2. Manufacturer concatenates the ID and random number R_0. $ID\|R_0$.
3. Manufacturer generates the encryption and decryption key K from PW. $K = F(PW)$.
4. Manufacturer encrypts $ID\|R_0$ with K. $E_K(ID\|R_0)$.
5. Manufacturer writes $E_K(ID\|R_0)$, R_0, and R_1 on the memory of the RFID tag.
6. Manufacturer stores ID, R_0, R_1 and other related information in the manufacturer's Information Services.

3.2   Authentication Process

After set-up, the RFID tag can be used in the authentication process. In the process from manufacturer to customer, the following authentication process is used. Fig. 1 presents the authentication process of the proposed protocol.

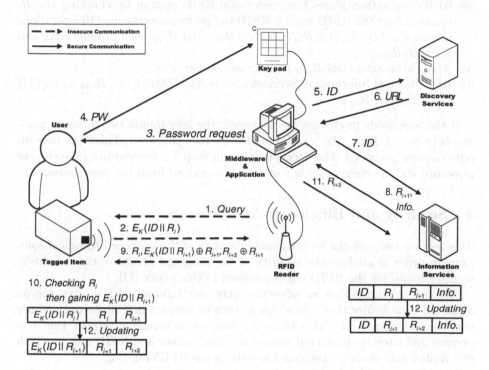

Fig. 1. Authentication process

[Authentication Process]

1. RFID reader broadcasts *Query* to RFID tags.
2. RFID tag transmits $E_K(ID\|R_i)$ as a response to the RFID reader.
3. RFID reader obtains the response and Application requests a *PW* from the RFID tag user.
4. User inputs the *PW*.
5. RFID Middleware decrypts the $E_K(ID\|R_i)$ received from the RFID tag using K, a decryption key derived from the *PW*. RFID Middleware obtains the *ID* and R_i. Application transmits the *ID* to Discovery Services.
6. Discovery Services search the *URL* of Information Services that contain information relating to the *ID*, then transmit the corresponding *URL* to the Application.
7. Application transmits the *ID* to the Information Services, where the address is the *URL*.

⑧ Information Services authenticate the RFID tag with the ID then transmit $Info.$ and R_{i+1} relating to the ID to the Application.

⑨ RFID Middleware concatenates ID and R_{i+1}, re-encrypts $ID\|R_{i+1}$ using K, and creates R_{i+2}. RFID reader transmits $E_K(ID\|R_{i+1})\oplus R_{i+1}$, $R_{i+2}\oplus R_{i+1}$, and R_i to the RFID tag.

⑩ RFID tag authenticates the reader-side RFID system by checking the R_i received from the RFID reader. RFID tag performs exclusive-OR operations among $E_K(ID\|R_{i+1})\oplus R_{i+1}$, $R_{i+2}\oplus R_{i+1}$ and R_{i+1} then obtains R_{i+2} and $E_K(ID\|R_{i+1})$.

⑪ Application transmits R_{i+2} to Information Services.

⑫ RFID tag and Information Services update $E_K(ID\|R_i)$, R_i, R_{i+1} as $E_K(ID\|R_{i+1})$, R_{i+1}, R_{i+2}.

If the user needs to change the password, the user inputs two different passwords in Step 4. First, the user inputs the current password. Then, the user inputs the new password. The decryption key in Step 5 is derived from the current password and the encryption key in Step 9 is derived from the new password.

4 Security and Efficiency Analysis

This section analyzes the security and efficiency of the proposed technical solution. In order to analyze the security of the proposed protocol, the Ari Juels' security model for the RFID tag is assumed in this paper [11].

Ari Juels assumed that an adversary may only interact with a given tag on a limited basis before that tag is able in turn to interact in a protected manner with a valid verifier [11]. This protected interaction means a 'refresh' that is a privacy and integrity protected session between verifier and RFID tag, in which the verifier may update identification data in the RFID tag [11].

In addition, the Ari Juels' security model imposes two restrictions on attacker interaction with RFID tags between refreshes. In the proposed protocol, these restrictions are defined differently.

[Restrictions]

- *Limited successive tag queries*: The probability that an attacker can successively transmit a *Query* to targeted RFID tags in different locations before updating the RFID tags' identification data is very low, since the attacking-devices of the attacker are much sparser than the legitimate RFID readers and the attacker should become nearer to the RFID tags in order to collect data from the RFID tags. In addition, the attacker may face difficulties in gathering data inside a monitored environment such as a shop [11].
- *Limited interleaving*: A sophisticated attacker has the potential to mount a full-blown man-in-the-middle attack [11]. However, the mobility of RFID tags [11] and password mechanism restrict the attacker's ability to perform such attack. The attacker must possess the correct password in order to complete the man-in-the-middle attack of the proposed protocol.

4.1   Security Analysis

This subsection analyzes that the proposed protocol satisfies the security and privacy requirements explained in section 2.

(1) Weak anonymity
The attacker can take the series of $Query$, $E_K(ID\|R_i)$, R_i, $E_K(ID\|R_{i+1}) \oplus R_{i+1}$, and $R_{i+2} \oplus R_{i+1}$ in an insecure communication channel. If the encryption algorithm is secure and the password is securely managed, the attacker cannot obtain ID from $E_K(ID\|R_i)$ and $E_K(ID\|R_{i+1}) \oplus R_{i+1}$. The proposed protocol satisfies the weak anonymity.

(2) Strong anonymity
The attacker can take the series of $Query$, $E_K(ID\|R_i)$, R_i, $E_K(ID\|R_{i+1}) \oplus R_{i+1}$, and $R_{i+2} \oplus R_{i+1}$ in an insecure communication channel. If the attacker obtains these messages between the targeted RFID tag and reader for more than two successive authentication sessions, the attacker can distinguish between different responses generated by the same tag. However, since the attacker has two restrictions, limited successive tag queries and limited interleaving, in the proposed protocol, the probability of a security breach is negligible and forward trace is limited. Therefore, the proposed protocol is strongly anonymous in this respect.

(3) Anti-counterfeiting
The attacker can counterfeit the RFID tag by reusing messages between the RFID tag and reader. In addition, the attacker can clone the RFID tag by physical attack. However, the attacker cannot use the clone without password since the user of the tag must input the correct password to be authenticated in the proposed protocol. Therefore, the proposed protocol satisfies anti-counterfeiting.

(4) Recognizability
The attacker can destroy RFID system infrastructure by destroying RFID system components. In addition, the attacker can cause interference in stable services by jamming or other attack tools such as "Blocker tag [10]". Generally, most RFID systems are not secure against these kinds of threats, including the proposed protocol. In addition, there is no technically secure solution. However, in the proposed protocol, if the attacker interferes with communications between the RFID tag and reader, the user can recognize a denial of service attack since the user cannot receive the password request in the authentication process. The proposed protocol can recognize a denial of service attack.

4.2   Efficiency Analysis

This subsection analyzes the efficiency of the proposed protocol into three factors, computational overhead, storage overhead, and frequency and traffic of insecure communications.

(1) Computational overhead
The previous technical solutions [6, 7, 11, 13, 15, 17, 18, 20, 22, 23] use hash function operations, random number generators or even encryption algorithms in the RFID tag. However, most RFID tags have extremely low computational ability. Therefore, previous protocols cannot be applied to low-cost passive tags. In the proposed technical solution, RFID tags only perform two exclusive-OR operations. The computational overhead of the exclusive-OR operation is extremely low. Therefore, the proposed technical solution can be comfortably applied to low-cost RFID tag.

(2) Storage overhead
In the proposed technical solution, if the length of the ID, $E_K(ID\|R_i)$, and R_i is 96 bits same as EPC-96, the storage overhead of the RFID tag is 288 bits. The storage overhead of the Information Services by the ID, $E_K(ID\|R_i)$, and R_i is light since the length of the $Info.$ is much longer than the ID, $E_K(ID\|R_i)$, and R_i.

(3) Frequency and traffic of the insecure communications
The frequency of insecure communications between RFID tag and reader is 3. In addition, the sum of the length of data between the RFID tag and reader is 384 bits without $Query$ if the $E_K(ID\|R_i)$ and R_i is 96 bits. In comparison with the previous technical solutions, this is an expected outcome.

Table 1 compares the efficiency of the proposed protocol with the hash-lock protocol (HLP) [19, 22], randomized hash-lock protocol (RHLP) [22, 23], hash based ID variation protocol (HIDVP) [7], hash-chain protocol (HCP) [15], and Challenge-response based protocol (CRP) [17].

Table 1. Comparison of the efficiency

	HLP	RHLP	HIDVP	HCP	CRP	Our Protocol
Computation	1H	1H, 1R	3H, 4X	2H	2H, 1R	2X
Storage	2L	L	2L	L	L	3L
Frequency	4	2	3	2	3	3
Traffic	3L	2L	4.5L	L	4L	4L

[Notations of Table]

- H: Hash function operation. $\{0,1\}^* \rightarrow \{0,1\}^L$.
- R: Random number generation. Random number r is L bits. $r \in_U \{0,1\}^L$.
- X: Exclusive-OR operation.
- L: Size of the data. It is assumed that the size of each data is L bits except TID, LST, DB-ID and $Query$.
- $Computation$: Number of computations on the RFID tag.
- $Storage$: Memory size of the RFID tag. It is assumed that the size of TID, LST is 0.5L bits and the size of DB-ID is not considered in HIDVP.

- *Frequency*: Number of insecure communications.
- *Traffic*: Size of the data transmitted through the insecure communication channel. The size of the *Query* is not considered. In addition, the size of DB-ID is not considered in HIDVP.

5   Conclusion

In order to resolve security and privacy threats in the RFID system, many kinds of solutions have been proposed. However, previous protocols are not suitable for real RFID systems such as the EPCgobal network™. Therefore, in this paper, a secure and efficient technical solution for the real RFID system, using an encryption algorithm with a password derived key to encrypt the RFID tag's ID, is proposed. Although the proposed protocol uses an encryption algorithm, the computational overhead of the RFID tag is extremely low. In addition, the user of the RFID tag is aware of the authentication process between the RFID tag and reader, since the user must input a password for authentication.

If the proposed protocol is combined with other political measures, the RFID system provides additional security.

References

1. Aigner, M., Feldhofer, M.: Secure Symmetric Authentication for RFID Tags. In: Telecommunications and Mobile Computing 2005 (TCMC2005), Graz, Austria (March 8-9, 2005)
2. The Boycott Benetton website, http://www.boycottbenetton.com/
3. EPCglobal, EPC Radio-Frequency Identity Protocols Class-1 Generation-2 UHF RFID Protocol for Communications at 860 MHz-960 MHz Version 1.0.9 (January 31, 2005), http://www.epcglobalinc.org
4. Garfinkel, S.L., Juels, A., Pappu, R.: RFID Privacy: An Overview of Problems and Proposed Solutions. Security and Privacy Magazine 3(3), 34–43 (2005)
5. GS1 US, Glossary Version 6.0 (May 2005)
6. Golle, P., Jakobsson, M., Juels, A., Syverson, P.: Universal Re-Encryption for Mixnets. In: Okamoto, T. (ed.) CT-RSA 2004. LNCS, vol. 2964, pp. 163–178. Springer, Heidelberg (2004)
7. Henrici, D., Müller, P.: Hash-based Enhancement of Location Privacy for Radio-Frequency Identification Devices using Varying Identifiers. In: PERCOMW 2004. Proceedings of the Second IEEE Annual Conference on Pervasive Computing and Communications Workshops, pp. 149–153. IEEE, Los Alamitos (2004)
8. Hoffstein, J., Pipher, J., Silverman, J.H.: NTRU: A Ring-Based Public Key Cryptosystem. In: Buhler, J.P. (ed.) Algorithmic Number Theory. LNCS, vol. 1423, pp. 267–283. Springer, Heidelberg (1998)
9. Juels, A., Pappu, R.: Squealing Euros: Privacy protection in RFID-enabled banknotes. In: Wright, R.N. (ed.) FC 2003. LNCS, vol. 2742, pp. 103–121. Springer, Heidelberg (2003)
10. Juels, A., Rivest, R.L., Szydlo, M.: The Blocker Tag: Selective Blocking of RFID Tags for consumer Privacy. In: Proceedings of 10th ACM Conference on Computer and Communications Security, pp. 103–111. ACM Press, New York (2003)

11. Jeuls, A.: Minimalist cryptography for Low-Cost RFID Tags. In: Blundo, C., Cimato, S. (eds.) SCN 2004. LNCS, vol. 3352, pp. 149–164. Springer, Heidelberg (2005)

12. Kwak, J., Rhee, K., Oh, S., Kim, S., Won, D.: RFID System with Fairness within the Framework of Security and Privacy. In: Molva, R., Tsudik, G., Westhoff, D. (eds.) ESAS 2005. LNCS, vol. 3813, pp. 142–152. Springer, Heidelberg (2005)

13. Molnar, D., Wagner, D.: Privacy and Security in Library RFID: Issues, Practices, and Architectures. In: ACM CCS. Proceedings of Conference on Computer and Communications Security, pp. 210–219. ACM Press, New York (2004)

14. NTRU Cryptosystems Inc., http://www.ntru.com/

15. Ohkubo, M., Suzuki, K., Kinoshita, S.: Hash-Chain Based Forward-Secure Privacy Protection Scheme for Low-Cost RFID. In: Proceedings of the SCIS 2004, pp. 719–724 (2004)

16. Philips, http://www.semiconductors.philips.com/

17. Rhee, K., Kwak, J., Kim, S., Won, D.: Challenge-Response based RFID Authentication Protocol for Distributed Database Environment. In: Hutter, D., Ullmann, M. (eds.) SPC 2005. LNCS, vol. 3450, pp. 70–84. Springer, Heidelberg (2005)

18. Saito, J., Ryou, J.-C., Sakurai, K.: Enhancing Privacy of Universal Re-encryption Scheme for RFID Tags. In: Yang, L.T., Guo, M., Gao, G.R., Jha, N.K. (eds.) EUC 2004. LNCS, vol. 3207, pp. 879–890. Springer, Heidelberg (2004)

19. Sarma, S.: Towards the Five-Cent Tag, White paper, MIT Auto-ID Center (2001)

20. Sarma, S., Weis, S., Engels, D.: RFID Systems and Security and Privacy Implications. In: Kaliski Jr., B.S., Koç, Ç.K., Paar, C. (eds.) CHES 2002. LNCS, vol. 2523, pp. 454–469. Springer, Heidelberg (2003)

21. Shepard, S.: RFID-Radio Frequency Identification. McGraw-Hill, New York (2005)

22. Weis, S.: Security and Privacy in Radio-Frequency Identification Devices. MS Thesis, MIT (May 2003)

23. Weis, S., Sarma, S., Rivest, R.L., Engels, D.: Security and Privacy Aspects of Low-Cost Radio Frequency Identification Systems. In: Hutter, D., Müller, G., Stephan, W., Ullmann, M. (eds.) Security in Pervasive Computing. LNCS, vol. 2802, pp. 201–212. Springer, Heidelberg (2004)

Design of WLAN Secure System Against Weaknesses of the IEEE 802.1x

Seong-pyo Hong[1], Jong-an Park[1], Seung-jo Han[1], Jae-young Pyun[1], and Joon Lee[2]

[1] Dept. of Information & Communication Engineering,
Chosun University, Gwangju, Korea
[2] Dept. of Computer Engineering, Chosun University, Gwangju, Korea
{hongsp,japark,sjbhan,jypyun,jlee}@chosun.ac.kr

Abstract. The IEEE 802.1x framework, what was known to have adjusted the IEEE 802.11b's weakness in client authentication is a port-based control mechanism that introduces the logical port idea and performs authentication through the AP or the bridge system. Unfortunately, there are two problems in existing access authentication scheme for wireless LAN, the IEEE 802.1x. One of the problems is that it is possible for a malicious user to disguise as a right authenticator because he/she does not take into account the authentication of authenticators. The other problem is that a malicious user can force an authentication Server to waste computational resource by continuously accessing requests. In this paper, we propose a Wireless LAN secure system that offers secure encrypted communication and user authentications. The purpose of the WLAN secure system that this study suggests is to improve the weakness in security of IEEE 802.1x and to guarantee a secure encrypted communication.

Keywords: Wireless LAN Security, Authentication, Privacy, IEEE 802.1x.

1 Introduction

Wireless Local Area Networks will facilitate ubiquitous communications and location independent computing in restricted spatial domains such as offices, factories, enterprise facilities, hospitals, and campuses. In such environments, WLANs will complement and expand the coverage areas of existing wired networks. The main attractions of WLANs includes of cost effectiveness, ease of installation, flexibility, tether-less access to the information infrastructure, and support for ubiquitous computing through station mobility.

Even now, for existing wireless networks, security is often cited as a major technical barrier that must be overcome before widespread adoption of mobile services can occur. Unlike wired networks, WLANs provide the transmitted data to anyone with a receiver that is in radio range[1][8]. As a result, one must consider WLAN traffic as being delivered to the adversary as well as the intended

M.S. Szczuka et al. (Eds.): ICHIT 2006, LNAI 4413, pp. 617–627, 2007.

party, and the adversary with a transmitter has the ability to inject or forge packets onto the network.

The IEEE 802.1x framework, what was known to have adjusted the IEEE 802.11b's weakness in client authentication, is a port-based control mechanism that introduces the logical port idea and performs authentication through the AP or the bridge system. The system makes use of various types of authentication mechanisms such as Challenge/Response, Kerberos, TLS, and OTP through EAP[2]. Unfortunately, there are two problems in existing access authentication scheme for wireless LAN, the IEEE 802.1x. One of the problem is that it is possible for a malicious user to disguise as a right authenticator because does not take into account the authentication of authenticators. The other problem is that a malicious user can forcing an authentication Server to waste computational resource by continuously access requests[9][11].

In this paper, we propose a Wireless LAN secure system that improve the weaknesses in security of the IEEE 802.1x. The proposed system does not allow any faking of the identity by performing a thorough mutual authentication to all associated objects. Furthermore, it provides an integrity service by encrypting *SUCCESS* and *EAP-SUCCESS* messages with distributed a new shared-key through the key distribution mechanism, when an authentication process is executed.

2   Weaknesses of the IEEE 802.1x

2.1   Denial of Service Attack

The IEEE 802.1x uses EAP as the transfer protocol for authentication data. The EAP authentication process begins when the client request for authentication to the authentication Server, via the AP, in return, responds by asking for identity of the client. The authentication protocol will be executed as soon as the client replies to authentication Server's request. Therefore, the legal clients may fail to connect to the service since having no countermeasures against malicious user's constant request for authentication, forcing the server to continuously allocate its resources. The reason this becomes a problem is because the authentication Server has to allocate resources to proceed with the authentication without even verifying the legality of the clients[6][10].

Such as EAP-TLS or EAP-TTLS[3] authentication protocols that are provided by IEEE 802.1x, often needs heavier loads of computation and resource allocation because their authentication method is based on Public Key Infrastructure. Hence, when a malicious user decides to continuously attempt to connect, the authentication Server will waste its resources to unnecessary computations.

2.2   Spoofing Attack

Although the IEEE 802.1x authentication mechanism defines the authentication of servers and clients, it does not take into account the authentication of authenticators. It also overlooks the confidentiality of the data being transferred.

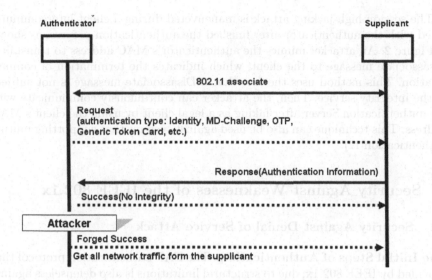

Fig. 1. Spoofing Attack by intercept an *EAP–SUCCESS* message

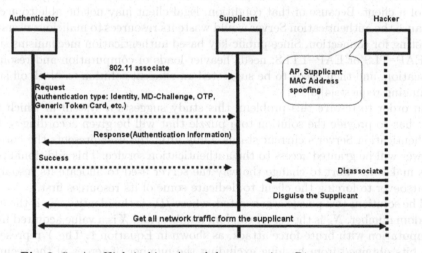

Fig. 2. Session High-jacking Attack by transmit a *Disassociate* message

Thus, the system is vulnerable to malicious user's various spoofing attacks such as faking as a legal authenticator to the client or in the reversely case[5].

Figure 1 shows, how the spoofing of *EAP–SUCCESS* messages can attack the system that one of many possible examples of spoofing. An *EAP–SUCCESS* message, which indicates the success of the client's authentication is not subject to the integrity service. Thus, the attacker can intercept an *EAP–SUCCESS* message previously initialized by another client's legitimate process and send that message back to the client after faking as a legal authenticator. This allows the attacker to identify and intercept all data being transferred.

The session high-jacking attack is maneuvered during a client intercommunicated with the authenticator after finished the authentication process, as shown in Figure 2. An attacker mimics the authenticator's MAC address to transmit a Disassociate message to the client, which indicates the termination of communication. This method uses the fact that a Disassociate message is not subject to the integrity service. Then, the attacker can continuously communicate with the authentication Server after faking as a legal client by using the client's MAC address. This technique can also be used against systems that supporting mutual authentication[12].

3   Security Against Weaknesses of the IEEE 802.1x

3.1   Security Against Denial of Service Attack

The Initial Steps of Authentication. The EAP authentication protocol that provided by IEEE 802.1x, due to structural limitations is also defenseless against Denial of Service attack. For example, the authentication Server has to allocate resources to proceed with the authentication without even verifying the legality of a client. Because of that condition, legal client may not be able to access because the authentication Server would waste its resources to malicious users requesting for connection. Since public-key based authentication mechanisms such as EAP–TLS or EAP–TTLS, needs heavier loads of computation and resource allocation, and then needs to be suggested an efficient solution to block off such damaging requests[4][5].

In order to resolve this problem, this study suggests a system in which the user has to provide the solution to a puzzle that will be given according to the authentication Server's current status. Only after the user presents the correct answer will be granted access to the authentication service. This additional process makes its effort to change the way the server used to allocate its resources whatsoever to forcing the client to dedicate some of its resources first.

The solutions to a puzzle is generated, where ID_C is the identity, N_C is the user random number, N_S is the server random number, and X is a value acquired from computation with brute-force attack, as shown in Equation 1. The Y represents the bits obtained from hashing excluding the number of zeroes of the security level parameter k.

$$h(ID_C, N_S, N_C, X) = 00.......00Y. \tag{1}$$

The degree of difficulty of puzzles depends on the security level parameter k. Only the client does not compute anything when the k is zero and, it is impossible to figure out the X value using the MD5 hash function when the parameter k is 128. This study will assign the k value to a number between 0 to 64. Therefore, the propose system can control the client's authentication request by arbitrarily assigning the security level parameter k according to authentication Server's current status.

Protocol of Initial Steps. The protocol of initial steps is executed, as shown in Figure 3. The authentication Server has generates a random number N_S, and then determine the security level parameter k to send the client according to it's current status. The client generates an random number N_C, and then compute the X value with brute-force by using the hash function according to the security level parameter k. Then the user random number N_C, server random number N_S, security level parameter k, and the user identity ID_C are sent to the authentication Server, which then compares the X value submitted by the user and the X' value computed by using the variables that the user submitted.

Only the authentication protocol will be executed, when the result of the comparison turns out to be acceptable. Moreover, the user will never be able to anticipate the X value because the server random number N_S and the security level parameter k will be newly regenerated each time to request for authentication.

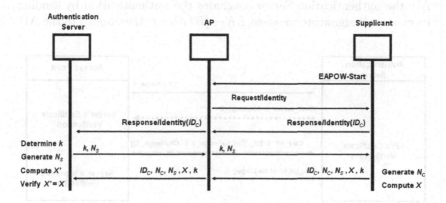

Fig. 3. Protocol of Initial Steps

3.2 Protection Against Spoofing Attack

Mutual Authentication. The AP is regarded as a solely reliable element in the IEEE 802.1x framework. Therefore, the IEEE 802.1x only performs the authentication about clients and authentication Servers while overlooking that of the AP. This fact, seduces malicious users to mimic the AP, allowing them to attack the clients by spoofing the unprotected $SUCCESS$ message or session high-jacking by spoofing the $Disassociate$ message[7].

The proposed system includes the AP authentication method, therefore providing the mutual authentication of all associated objects. Not only does this eliminate any attacks or spoofing that come from an unauthenticated user but it also guarantees security by the using the public-key based TLS. The following series of steps explain the authentication process, and the flow is illustrated in Figure 4.

1: Challenge_1 – The AP informs the connection request to the authentication Server and generates a random number(Challenge_1) and also sends it to the authentication Server.

2: Challenge_2, Cert_server – The authentication Server sends a random number(Challenge_2) and its Certificate to the AP for its own authentication.

3: Server Authentication – The AP inspects the authentication server's Certificate. The AP sends a failure message and terminates the connection if the Certificate is invalid and otherwise it sends an electronic signature message($EK_{AP}[H(Challenge\_1 \parallel Challenge\_2)]$) and Certificate of AP. This message is used to verify the Certificate of AP.

4: The AP Authentication – The authentication Server confirms the received electronic signature message and Certificate of AP.

5: Termination of Authentication – After confirming the Certificate of AP, the authentication Server concludes the authentication by sending its electronic signature message($EK_{SV}[H(All\ of\ Message)]$) to the AP.

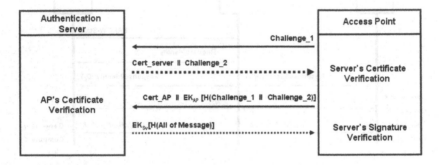

Fig. 4. Mutual Authentication between the AP and authentication Server

SUCCESS and EAP-SUCCESS Message Encryption. As soon as the mutual authentication process between a client and the authentication Server is completely finished, the authentication Server sends a *SUCCESS* message to the AP.

The *SUCCESS* message in the proposed system, is composed of a *FINISHED* message, which is generated with a shared–key(*WEP_Key*) and the authentication Server's electronic signature during the mutual authentication process between a client and the authentication Server, as expressed in Equation 2. The *FINISHED* message is indicates that the mutual authentication process between a client and the authentication Server is completely finished and the shared–key(*WEP_Key*) is what the client and the authentication Server shares throughout the key distribution mechanism. The shared–key*WEP_Key* is encrypted with the AP's public–key because only the AP should to be known.

$$SUCCESS = EK_{AP}[EK_{SERVER}[h(FINISHED \parallel WEP\_Key)] \parallel \\ FINISHED \parallel WEP\_Key]. \tag{2}$$

The objective of the *SUCCESS* message is to inform the AP that the client authentication was successful and to send the shared–key(*WEP_Key*) that will be used for the encrypted communication between the client and the AP. Although the shared–key(*WEP_Key*) is strictly contained between the client and the authentication Server, the AP should also be bale to identify this key in order to continue the WEP communication after terminate the authentication. Since the shared–key(*WEP_Key*) should never be exposed, it is encrypted using the AP's public–key and then sent.

The AP, receiving the *SUCCESS* message from the authentication Server decrypts the message and generates an *EAP-SUCCESS* message in the form of Equation 3, and then encrypts it back using the shared–key(*WEP_Key*) to send it to the clients. This shared–key(*WEP_Key*) is retrieved from decrypted the *SUCCESS* message that was previously received from the authentication Server.

$$EAP - SUCCESS = EK_{WEP\_Key}[EK_{SERVER}[h(FINISHED \parallel \\ WEP\_Key)] \parallel FINISHED \parallel WEP\_Key]. \tag{3}$$

The purpose of the *EAP-SUCCESS* message is for the AP to inform the client of its certified status. The AP that did not authenticated from authentication Server will not be able to decrypts the *EAP-SUCCESS* message because it wouldn't find out the private–key, and therefore can not sending the *EAP-SUCCESS* messages to the client by encrypted shared–key(*WEP_Key*).

Moreover, the authentication Server's electronic signature is included in the *EAP-SUCCESS* message. Even so if an attacker find out the shared–key(*WEP_K ey*), it would be impossible to generate an authentication Server's electronic signature without knowing the authentication Server's private–key. The client, receiving the *EAP-SUCCESS* message from the AP will disconnect the communication if the *EAP-SUCCESS* message has not been decrypted by the shared–key(*WEP_Key*), which was transferred during the mutual authentication between the client and authentication Server.

The fault of decryption indicates the failing of purchase a legitimate shared–key(*WEP_Key*) from the authentication Server. Even if the decryption is successful, the absence of an authentication Server's electronic signature or being unable to identify one will disconnect the communication because it means that the AP was failed to purchase an authentication from authentication Server and thus unable to acquire an authentication Server's electronic signature.

3.3 Security Assessment of the Proposed Mechanism

Protection against DoS Attack. Such as EAP–TLS and EAP–TTLS, examples of the IEEE 802.1x authentication mechanism, needs heavier loads of computation and resource allocation because based on the Public Key Infrastructure. The authentication Server would have to waste a large amount of its resources to unnecessary encryption when malicious user continuously request for connection. Therefore an initiative action needs to be taken, such as regulating the user's authentication request beforehand.

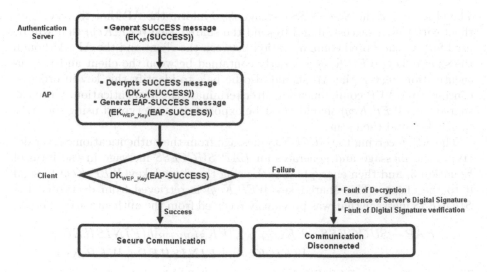

Fig. 5. Flows of Secure Encrypted Communication

This study suggests a solution in which a user must provide the correct answer to a puzzle that will be given according to the authentication Server's current status in order to access the authentication protocol. Thus, countless number of connection requests can also be regulated by forcing the clients to dedicate some of their resources and the authentication Server setting the security level parameter to control such requests according to its status. And then, the security level parameters can be arbitrarily set according to the authentication Server's current status, as shown in Figure 6.

Normally, when the k value is set to zero, the authentication process would be executed without computation of the X value. When the authentication Server detects a continuously authentication requests or in cases of resource shortage, the security level parameter k is incremented, and then the clients load of computation also increases, therefore regulating the malicious users from continuously trying to approach the authentication Server.

Protection against Spoofing Attack. The IEEE 802.1x authentication mechanism only defines the authentication of a client and an authentication Server. As a result, malicious users can use this fact to practice various types of spoofing attacks such as faking as a legal authenticator to the client or in the reversely case.

In effort to prevent from spoofing attacks, this study provides a solution in which the AP also has to be authenticated. The AP authentication mechanism can be protected against any spoofing attacks by utilizing the public–key based TLS, which has already been proved to be effective in security. Moreover, the *EAP–SUCCESS* message would also have to be encrypted to prevent from

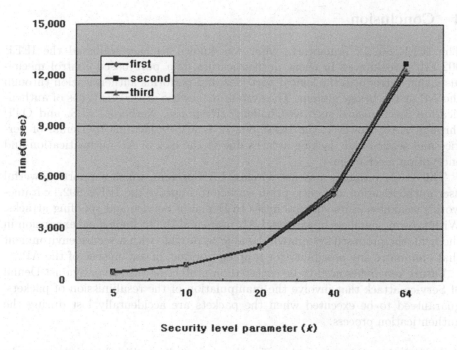

Fig. 6. Times to Computation of X according to the Security Level Parameter

Table 1. Comparison of Existing System with Proposal System

	Existing System	Proposal System
Authentication method	MD–5, EAP–TLS, EAP–TTLS(option)	EAP–TLS
AP authentication	N	Y
Integrity service	N	Y
MIM attack	Available(Not supporting mutual authentication)	N/A
Session Hijacking attack	Available(Not supporting mutual authentication and Integrity service)	N/A
DoS attack	Available	N/A

spoofing attacks that result from the lack of integrity service of *EAP–SUCCESS* messages. Therefore, not only will the attackers impossible to fake as a legal AP but they will also not be able to generate *EAP–SUCCESS* messages, thus successfully protecting the system.

4 Conclusion

The IEEE 802.1x framework, what was known to have adjusted the IEEE 802.11b's weaknesses in client authentication, is a port–based control mechanism that introduces the logical port idea and performs authentication through the AP or the bridge system. The system makes use of various types of authentication mechanisms, such as Challenge/Response, Kerberos, TLS, and OTP through EAP. However, the IEEE 802.1x is also vulnerable to Denial of Service and session high–jacking attacks due to the lack of AP authentication and encryption mechanism.

This study proposes a secure wireless LAN system, which supports powerful user authentication and encryption service to improve the IEEE 802.1x framework's weakness in its vulnerabilities to Denial of Service and spoofing attacks. With growing numbers of wireless LAN users and the technological expansion in the field, the proposed system will be able to provide with a secure environment that eliminates any possibility for tapping, forging, or any misuse of the AP.

Future researches need to be centered on protecting the server against Denial of Service attack that involve the manipulation of the resubmission of packets, guaranteed to be executed when the packets are accidentally lost during the authentication process.

Acknowledgments. A portion of authors who take part in this work were take funds by the Second stage of Brain Korea 21 Project.

References

1. Patiyoot, D., Shepherd, S.J.: Cryptographic security Techniques for wireless networks. In: ACM SIGOPS Operating Systems Review, vol. 33, pp. 36–50. ACM Press, New York (1999)
2. Faria, D.B., Cheriton, D.R.: DoS and authentication in wireless public access networks. In: WiSe 2002 Conference, pp. 47–56. ACM Press, New York (2002)
3. Funk, P., Blake–Wilson, S.: EAP Tunneled TLS Authentication Protocol (EAP–TTLS). IETF PPPEXT Working Group (2005)
4. Cam–Winget, N., Housley, R., Wagner, D., Walker, J.: Wireless networking security: Security flaws in 802.11 data link protocols. In: Communications of the ACM, pp. 35–39. ACM Press, New York (2003)
5. Mishra, A., Arbaugh, W.A.: An Initial Security Analysis of the IEEE 802.1X Standard, University of Maryland, pp. 1–12 (2002)
6. Wi–Fi Alliance: Wi–Fi Protected Access, Wi–Fi Alliance White Paper (2003)
7. Arbaugh, W.A. (ed.): 802.11 Security Vulnerabilities, University of Maryland (2003)
8. Stubblefield, A., Ioannidis, J., Rubin, A.D.: A key recovery attack on the 802.11b wired equivalent privacy protocol(WEP). In: ACM Transactions on Information and System Security (TISSEC), vol. 7, pp. 319–332 (2004)
9. Chen, J.-C., Jiang, M.-C., Liu, Y.-W.: Wireless LAN Security and IEEE 802.11i. IEEE Wireless Communications, 1–19 (2004)

10. Vollbrecht, J., Moskowitz, R.: Wireless LAN Access Control and Authentication, InterLink Networks White Paper (2002)
11. He, C., Mitchell, J.C.: Security analysis: Analysis of the 802.11i 4-way handshake. In: WiSe 2004 Conference, pp. 43–50. ACM Press, New York (2004)
12. Allen, J., Wilson, J.: Securing a wireless network. In: Proceedings of the 30th annual ACM SIGUCCS conference on User services, pp. 213–215. ACM Press, New York (2002)

A Sophisticated Base Station Centralized Simple Clustering Protocol for Sensor Networks

Giljae Lee, Yoonjoo Kwon, Woojin Seok, Jaiseung Kwak, and Okhwan Byeon

Korea Institute of Science and Technology Information,
52, Eoeun-dong, Yuseong, Daejeon, Korea
{giljael,yulli,wjseok,jskwak,ohbyeon}@kisti.re.kr

Abstract. In wireless sensor networks, energy efficiency has been a key factor. So far, many energy-efficient routing protocols have been proposed and much attention has been paid to cluster-based routing protocols due to their advantages. However, some cluster-based sensor network routing protocols need location information of the sensor nodes in the network to construct clusters efficiently. Owing to the cost, it is not feasible to know the locations of all sensor nodes in the sensor network. In this paper, we propose a sophisticated base station centralized simple clustering protocol (SBCSP). The proposed protocol utilizes the remaining energy of each sensor node, standard deviation of their energy consumed and the number of cluster heads changed depending on the number of sensor nodes alive in the sensor network. Throughout the performance experiments, we show that SBCSP has better performance than low-energy adaptive clustering hierarchy (LEACH).

1 Introduction

Recent advances in wireless communication and electronics have enabled to develop low cost, low power, and multi-functional sensor nodes, which are small and able to communicate with neighbor nodes within short distances. Wireless sensor network (WSN) usually consists of a base station and many sensor nodes. The sensor nodes are randomly distributed in the sensor network field and monitor temperature, air pressure, motion, and so on. After sensing, the sensor nodes send the sensing data to the base station. Here, the base station acts as a gateway to deliver the data from the sensor nodes to users who need it. WSN can have thousands of sensor nodes. Therefore, it could be difficult to initialize the sensor nodes and to manage the sensor network. This is why self-configuring the sensor nodes and saving energy via an efficient routing protocol [1] are desirable in WSN.

Conventionally two types of protocols have been used in WSN. One is direct communication protocol and the other is multi-hop communication protocol such as the Minimum Transmission Energy *Minimum Transmission Energy* (MTE) routing protocol. In the direct communication protocol, the sensor nodes distant from the base station dissipate their energy faster than others sending their data to the base station directly. In the MTE routing protocol, each node acts

M.S. Szczuka et al. (Eds.): ICHIT 2006, LNAI 4413, pp. 628–637, 2007.
© Springer-Verlag Berlin Heidelberg 2007

as a router for other sensor nodes. Therefore, energy is consumed faster if more sensor nodes are adjacent to the base station [4]. To overcome these problems of the conventional protocols, data centric protocols such as the Sensor Protocols for Information via Negotiation *the Sensor Protocols for Information via Negotiation* (SPIN) [5] and *the directed diffusion* (DD) [6] have been proposed. As far as data transmission and data collection are concerned, the data centric protocols are more efficient than the conventional transmission methods. However, the data centric protocols require many control messages and thus cause long latency to set routing paths. On the other hand, cluster-based routing protocols involve reduced control messages. In addition, the cluster-based routing protocols have bandwidth reusability, enhanced resource allocation, and improved power control. One disadvantage of the cluster-based routing protocols is that cluster heads are fixed, carry out more energy-intensive processes, and then their energy consumption is greater than non-cluster head nodes. As a result, network lifetime becomes shorter.

In this paper, we propose a *sophisticated base station centralized simple clustering protocol* (SBCSP). SBCSP enhances performance with using standard deviation of energy consumed per each cluster during the data communication phase. Some related works on cluster-based routing protocols for sensor networks are presented in Section 2. In Section 3, we explain SBCSP in detail. In Section 4, we present performance results of SBCSP and compare them with other protocols. Finally, we conclude with some conclusions and future works.

2   Related Works

Low-energy adaptive clustering hierarchy (LEACH) was proposed to solve the problems of the cluster-based protocols to adapt them for WSN. LEACH is based on hierarchical clustering structure model. In order to increase the lifetime of the sensor network, LEACH changes cluster heads periodically. LEACH involves two main steps: the setup phase and the steady state phase. The setup phase involves cluster head selection part and cluster construction part. After cluster heads are chosen among sensor nodes, the selected sensor nodes broadcast advertisement message. The message contains cluster head ID that is the sensor node ID. Based on the advertisement message, non-cluster head sensor nodes know which nodes are new cluster heads in the sensor network. The new cluster head nodes use *the carrier-sense multiple access* (CSMA) *medium access control* (MAC) protocol to transmit it. Non-cluster head sensor nodes then select a cluster head node according to the signal strength of the advertisement message from the cluster head nodes and then send join request message back to the selected cluster head node for registration. Once the selected cluster head nodes receive join message, they make a *time division multiple-access* (TDMA) schedule to exchange data with non-cluster sensor nodes without collision. The cluster head node broadcasts the schedule to its member nodes, and then the member sensor nodes start sending their data to the base station through their cluster head node during the steady state phase.

In LEACH, a cluster head node is selected with sensor node's probability value. As far as optimal energy consumption is concerned, it is not desirable to select cluster head nodes randomly and construct clusters with them. However, repeating round can improve total energy dissipation and performance in the sensor network. In spite of these advantages, LEACH has some shortcomings: no remaining energy of sensor nodes is considered to construct clusters, and the number of cluster head nodes is fixed in the sensor network. Moreover, LEACH does not guarantee the number of cluster head nodes and their distribution because the cluster head nodes are selected stochastically just by the value of probability [4].

LEACH-centralized (LEACH-C) is a base station controlled version of LEACH. It is the same as LEACH except the cluster construction phase. In the cluster construction step, all sensor nodes in the sensor network send their information, which includes their ID, remaining energy level, and position information to the base station. Then, the base station calculates average energy of the sensor network and chooses a candidate set of cluster head nodes that have more energy than the average in the sensor network. Here, the base station executes *annealing algorithm* to find the most optimized and energy-efficient number of clusters with the candidate set of cluster head nodes. After constructing clusters, the base station broadcasts information about cluster head nodes, their members, and their TDMA schedule to the sensor network. After, non-cluster head sensor nodes decide own TDMA slot and come in sleep state until their own data transmission turn [3].

In *the Base-Station Controlled Dynamic Clustering Protocol* (BCDCP), the base station constructs *cluster-to-cluster* (CH-to-CH) routing paths, and carries out other energy-intensive tasks such as data aggregation, compression, and fusion. BCDCP constructs balanced clusters where each cluster head has approximately equal number of member nodes to avoid overload, and cluster head nodes are uniformly distributed in the sensor field. It uses CH-to-CH routing to transfer data to the base station. Similar to LEACH-C, energy information sent by all sensor nodes is used to construct clusters in the setup phase. The base station uses *balanced clustering technique* for distributing load of cluster head nodes, and *iterative cluster splitting algorithm* to find optimal number of clusters. After the setup phase, the base station makes a multiple CH-to-CH routing paths, creates a schedule for each cluster, and then broadcasts schedule information to the sensor network. In the data communication phase, the cluster head nodes transfer data from sensor nodes to the base station through the CH-to-CH routing paths [7].

3   A Proposed Base Station Centralised Protocol

In this section, we explain SBCSP, the proposed protocol. SBCSP uses the remaining energy of each sensor node and the number of sensor nodes alive in the sensor network. Therefore, SBCSP do not need location information of the sensor nodes. LEACH-C and BCDCP is base station controlled and they assume a base station can know location information of the sensor nodes and use it to increase system lifetime of the sensor network. It is, however, difficult to know locations of all sensor nodes because of cost. There are also network overhead for all sensor nodes to send their information to the base station at the same time every setup phase. On the other hand, SBCSP surmise the base station primarily does not know location information of sensor nodes such as LEACH although they are base station centralized. Instead, their base stations change the number of the cluster head nodes depending on the number of sensor nodes alive in the sensor network. Moreover, SBCSP uses the variation of energy of each sensor node to increase efficiency of cluster construction. After choosing cluster head nodes, the base station broadcasts the list of the new cluster head nodes to the sensor network. In the data communication phase, sensor nodes transmit their energy information together with their own sensing data. Therefore, SBCSP can keep away from overhead originated by directly sending setup information of sensor nodes to the base station during the cluster construction phase.

In SBCSP, we assume a sensor network model, similar to those used in LEACH, with the following properties.

- Sensor nodes are energy-constrained.
- The base station is supplied by power outside, and has enough memory and computing capability for energy intensive tasks.
- Each sensor node always transmits data to the base station at a fixed rate.
- Each sensor node can directly communicate with other sensor nodes by varying their transmission power within the sensor network.
- The location of each sensor node is fixed.

Energy model of SBCSP is the same as LEACH [3]. In the energy model, we suppose that power attenuation depends on distance between a transmission node and a reception node. The propagation loss is assumed to be inversely proportional to d^2 while it is assumed to be inversely proportional to d^4 at long distances. To transmit a k-bit messages to a distance d, the radio expends the following energy:

$$
\begin{aligned}
E_{T_x}(k, d) &= E_{T_x-elec}(k) + E_{T_x-amp}(k, d) \\
&= \begin{cases} E_{elec} \times k + \epsilon_{friss-amp} \times k \times d^2 & \text{if } d < d_{crossover} \\ E_{elec} \times k + \epsilon_{two-way-amp} \times k \times d^4 & \text{if } d \geq d_{crossover} \end{cases}
\end{aligned} \quad (1)
$$

At reception, exhausted energy is given by

$$
E_{R_x}(k, d) = E_{R_x-elec}(k) = E_{elec} \times k. \quad (2)
$$

Here, E_{elec}, $\epsilon_{friss-amp}$ and $\epsilon_{two-way-amp}$ are identical to those of LEACH.

3.1   Cluster Construction Phase

In SBCSP, during the cluster construction phase, the base station broadcasts advertisement message, which includes a list of new cluster head nodes selected by the base station according to the amount of remaining energy of each sensor node. To choose cluster head nodes, first, SBCSP selects some candidate sensor nodes as cluster head nodes according to their remaining energy and then calculates the standard deviation of consumed energy of each sensor node during the data communication phase.

At the first cluster construction phase, the base station sends cluster information as zero because it cannot recognize energy level of each sensor node and the number of sensor nodes alive in the sensor network. Therefore, sensor nodes carry out cluster head selection algorithm (equation 3). If a sensor node knows to be a new cluster head node itself according to the advertisement message, the sensor node broadcasts new advertisement message that includes its own information. Otherwise, non-cluster head nodes discard advertisement message from the base station and wait new advertisement message from new cluster head nodes. Receiving messages from cluster head nodes, the non-cluster head nodes choose one cluster head node according to the signal strength, and then send join message to the cluster head node. The other operations in the cluster construction phase are identical to those of LEACH. As the cluster construction ends, each cluster head node makes a schedule for its sensor nodes and broadcast it.

N is the number of sensor nodes in the sensor network, and k is the desirable number of cluster head nodes. The $P_i(t)$ represents the probability that a sensor node i is a cluster head node. It is identical to the formation of LEACH, too.

$$P_i(t) = \begin{cases} \frac{k}{N-k\times(r \bmod \frac{N}{k})} & C_i(t)=1 \\ 0 & C_i(t)=0 \end{cases}. \tag{3}$$

In some cases, non-cluster head nodes sending data to the cluster head node do dissipate more energy than cluster head nodes. Therefore selecting cluster head nodes just using remaining energy may not be efficient. SBCSP considers that the cluster whose standard deviation of energy consumed is the smallest of all clusters is relatively well constructed. Therefore, the base station observes the difference of energy level between previous and present cluster construction phase, computes standard deviation of it and then selects the cluster head node whose standard deviation is the smallest of clusters. The cluster head node selected is going to be one of the new cluster head nodes together with sensor nodes having most amount of energy in the sensor network.

The following is the algorithm to select cluster head nodes in the cluster construction phase.

– Calculating standard deviation of energy consumed of sensor nodes in each cluster
– Selecting the cluster head node of the cluster which has the smallest standard deviation among clusters
– Making a list of remaining energy of sensor nodes in the sensor network
– Sorting the list with descending order
– Inserting the cluster head node which has the smallest standard deviation to the cluster head node list
– Selecting the cluster head nodes according to the desirable number of cluster head nodes
– Broadcasting the final cluster head node list into the sensor network

3.2   Data Communication Phase

In data communication phase, each sensor node transmits data to its own cluster head node according to the communication schedule. After cluster head nodes aggregate data sent by sensor nodes and do fusion and compression them, they transmit processed data, which include remaining energy level of each sensor node, to the base station.

The base station passes raw data through the Internet or other networks, and it stores information of remaining energy of each sensor node for the next cluster construction phase. Now, the base station recognizes not only the amount of energy of each sensor node but also the number of sensor nodes in the sensor network. The base station can use various kinds of algorithms to calculate the suitable number of cluster heads considering the sensor network size as well as the number of sensor nodes. In this paper, we use 5 percent of the number of sensor nodes as the optimum number of cluster head nodes like LEACH.

4   Performance Evaluation

In this section, we estimate the performance of SBCSP. We simulated with ns-2 leach extension [8] and implementation [9]. Simulation environment is equal to those of LEACH and Table 1 shows simulation parameters. We assume that sensor nodes initially have equal initial energy, $0.5J$. The extent of the sensor network was set to $100m \times 100m$. In addition, we presume that the base station is located at $(50m, 50m)$ inside the sensor network. In simulation, we define system lifetime as the time that sensor nodes send data to the base station. Performance was measured by total messages successfully delivered to the base station, system lifetime, average consumed energy in each simulation time, and so on. Then, we compared the simulation results of SBCSP with two protocols: LEACH and LEACH-C.

Table 1. Parameters for simulation

Parameter	Value
Network Grid	$(0, 0) \times (100, 100)$
BS	$(50, 50)$
$d_{crossover}$	$87m$
ϵ_{elec}	$50nJ/bit$
$\epsilon_{friss-amp}$	$10pJ/bit/m^2$
$\epsilon_{two-way-amp}$	$0.0013pJ/bit/m^4$
$\epsilon_{aggregation}$	$5nJ/bit$
Data packet size	$500bytes$
Packet header size	$25bytes$
Initial energy	$0.5J$
Number of nodes (N)	100
Number of clusters (k)	5 (5% of N)

Figure 1 shows the total number of received data at the base station during system lifetime. We can see that SBCSP sends more data to the base station than LEACH by 17 percent. Besides it is worthy of close attention that SBCSP elevates the number of data transmitted to the base station up to 92 percent of LEACH-C. Figure 2 presents average number of data transmitted per round. Expectedly LEACH-C has the biggest value.

Figure 3 shows the number of sensor nodes alive in simulation time. The figure shows that the system lifetime of SBCSP is the longest. The system lifetime of LEACH is the shortest among protocols. Throughout this result, we can see that using standard deviation of energy consumed in SBCSP is very important contribution. In result, SBCSP exceeds the system lifetime of LEACH-C by 86 percent and so far as LEACH by no less than 98 percent.

In Figure 4, we can see the average energy consumed in each simulation round. Throughout this figure, we can know LEACH to be the most variable in energy consumption, and the biggest standard deviation of the average of energy consumed in Figure 5. As LEACH chooses cluster head nodes randomly regardless of the amount of energy of each sensor node, the number of cluster head nodes of LEACH is so irregular that LEACH is the least efficient among three protocols. On the other hand, LEACH-C has the fixed number of the cluster head nodes by making use of the location and energy information of each sensor node in the sensor network. This is the reason why SBCSP, which flexibly changes the number of cluster head nodes according to the number of sensor nodes alive, has system lifetime much longer than LEACH-C.

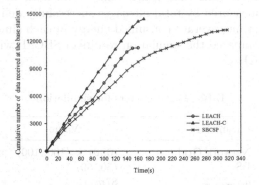

Fig. 1. The cumulative number of data received at the base station

Figure 5 presents that the analysis of energy consumption for simulating each protocol. When we regard the amount of data transmitted per $1J$ of LEACH as a basis, LEACH-C makes it appear that it is the most efficient followed by SBCSP. Although LEACH dissipates more energy than others, its average data sent to the base station per $1J$ is the smallest, for much of energy consumed of each sensor node is used to construct clusters as well as communicate with cluster head nodes. LEACH-C is the most efficient within the frame of energy but its average of energy consumption of LEACH-C has the biggest, which is

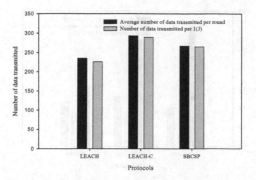

Fig. 2. The analysis of data transmitted in simulation time

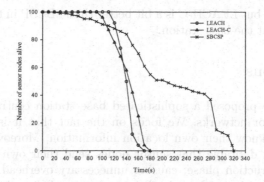

Fig. 3. The number of sensor nodes alive

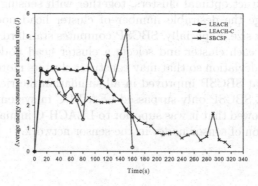

Fig. 4. Average consumed energy per simulation time

inefficient, because LEACH-C dissipates its energy to send the most data for the short time. On the other hand, the average of energy consumption of SBCSP is the smallest because its system lifetime is the longest.

Throughout the simulations, we can see that SBCSP is better than LEACH in all simulation metrics. Most of all, SBCSP is superior to LEACH-C in terms

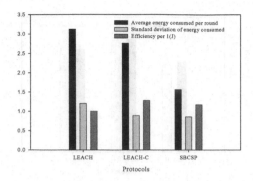

Fig. 5. The analysis of consumed energy in simulation

of system lifetime but LEACH-C is a bit better than SBCSP in the total number of data received at the base station.

5   Conclusions

In conclusion, we proposed a sophisticated base station centralized clustering protocol for sensor networks. We focus on the fact that it is expensive that all sensor nodes know their own location information. Moreover, to construct clusters, it is not desirable that sensor nodes send their own information every cluster construction phase, causing unnecessary overhead. SBCSP carries out cluster construction without inefficient sensor-node-broadcast to notify the base station of their information because sensor nodes send their information, necessary to construct optimal clusters, together with sensing data. Moreover, SBCSP can change the desirable number of cluster head nodes according to the sensor network size. Especially, SBCSP computes standard deviations of energy consumed of each cluster and selects a cluster head node, which has the smallest standard deviation so that may construct clusters more efficiently. Simulation showed that SBCSP improved in all simulation metrics compared with LEACH. However, SBCSP only surpassed LEACH-C in system lifetime. Nevertheless, SBCSP showed that it was superior to LEACH-C in many areas without location information of sensor nodes in the sensor network.

References

1. Akyildiz, I.F., et al.: Wireless Sensor Networks: A Survey. Elsevier Sci. B. V. Comp. Networks 38(4), 393–422 (2002)
2. Al-Karaki, J.N., Kamal, A.E.: Routing Techniques in Wireless Sensor Networks: A Survey. IEEE Wireless Communications 11(6), 6–28 (2004)
3. Heinzelman, W.B., Chandrakasan, A.P., Balakrishnan, H.: An Application-Specific Protocol Architecture for Wireless Microsensor Networks. IEEE Wireless Communication 1(4), 660–670 (2002)

4. Heinzelman, W.R., Chandrakasan, A.P., Balakrishnan, H.: Energy-Efficient Communication Protocol for Wireless Microsensor Networks. In: Proc. 33rd Hawaii International Conf. System Sciences (2000)
5. Heinzelman, W.R., Kulik, J., Balakrishnan, H.: Adaptive Protocols for Information Dissemination in Wireless Sensor Networks. In: Proc. 5th Annual international Conf. Mobile Computing and Networking, pp. 174–185 (1999)
6. Intanagonwiwat, C., Govindan, R., Estrin, D.: Directed Diffusion: A Scalable and Robust Communication Paradigm for Sensor Networks. In: Proc. 6th Annual International Conf. Mobile Computing and Networking, pp. 56–67 (2000)
7. Muruganathan, S.D., Ma, D.C.F., Bhasin, R.I., Fapojuwo, A.O.: A Centralized Energy-Efficient Routing Protocol for Wireless Sensor Networks. IEEE Communication Magazine 43(3), S8–S13 (2005)
8. The MIT uAMPS ns Code Extensions Version 1.0, http://www.mtl.mit.edu/research/icsystems/uamps/research/leach/leach_code.shtml
9. J. A. Pamplin, NS2 Leach Implementation, http://www.internetworkflow.com/resources/ns2_leach.pdf

Plus-Tree: A Routing Protocol
for Wireless Sensor Networks

Yongsuk Park and Eun-Sun Jung*

Networking Technology Laboratory
Samsung Advanced Institute of Technology
{victorious.park,eun-sun.jung}@samsung.com

Abstract. We study tree-based routing protocols for wireless sensor networks. The existing tree-based algorithm has a few shortcomings; (1) it may take a long path since sensor node must transmit packets via either its parent or its children nodes and (2) it is vulnerable to a link failure since the tree has to be reconstructed in case of a single link failure. We propose Plus-Tree, a routing protocol, which overcomes the shortcomings and improves the performance of the existing protocol. A node in Plus-Tree may have neighbor links other than tree links that can be used for alternative path. Simulation results show that Plus-Tree performs better than the existing tree-based protocol with respect to hop counts.

1 Introduction

With recent advances in processor and radio technology, wireless sensor networking has received significant attention in the last few decades. Wireless sensor network (WSN) consists of a large number of distributed sensor nodes which are small, low cost, and low-power devices. Due to the hardware constraints, sensor node has limited capabilities of sensing, computing, and communicating. Sensor node monitors some data such as sound and vibration, and then transmits the collected data to a gateway node that connects a WSN to wired networks. WSN can be used in many applications such as transportation, environment monitoring, military applications, and ubiquitous home network. In this paper, we study tree-based routing protocol for WSN.

In general, routing protocols for wireless networks are classified in three categories:

- *Proactive (or table driven) protocols*: Proactive protocols maintain routes to all nodes in the network by broadcasting periodic routing updates. The main drawback of these protocols is control overhead - that is, there are routing updates even when they are not needed. Examples are DSDV [1], OLSR [2], WRP [3], etc.

---

* Corresponding author.

M.S. Szczuka et al. (Eds.): ICHIT 2006, LNAI 4413, pp. 638–646, 2007.
© Springer-Verlag Berlin Heidelberg 2007

- *Reactive (or on-demand) protocols*: Reactive protocols create and maintain a route only when required. Thus, there is latency in packet delivery and in reacting to a link failure. Examples are AODV [4], DSR [5], etc.
- *Hybrid protocols*: Hybrid protocols combine the advantages of both proactive and reactive protocols. Examples are ZRP [6], VBR [7], etc.

Sensor nodes are typically not mobile and the network traffic is relatively low. Therefore, in WSN, link failure does not occur as frequently as in MANET (Mobile Ad-hoc Network). Another characteristic of WSN is that there is a gateway (or root) node to which rest sensor nodes transmit collected data. Based on these observations, tree-based routing protocols have been proposed for WSN. Tree-based routing algorithm is a proactive protocol which maintains routing information by means of tree structure. Tree-based protocols effectively keep routing information and provide rout-ing path in a simple manner. Also, it is suitable for data fusion (or aggregation) - that is, parent nodes in WSN can aggregate collected data from their children nodes. How-ever, it is well known that it may provide a long path to neighbor nodes. Since sensor node must transmit packets via either its parent or its children nodes, it usually creates a long path rather than an optimal route to a destination. Moreover, it is vulnerable to a link failure - the tree has to be reconstructed in case of a single link failure. In this paper, we propose a new routing protocol to reduce a routing path by using neighbor information.

The rest of the paper is organized as follows. Section 2 reviews related work and Section 3 provides an overview of existing tree-based routing protocols. Section 4 presents our proposed approach, Plus-Tree. Performance evaluation is given in Section 5 and Section 6 concludes the paper.

2 Related Work

In this section, we discuss relevant research on routing specially designed for WSN.

A classical routing protocol is flooding which is the simplest way of routing. How-ever, there are several disadvantages of flooding-based routing protocol. First, nodes often receive the same message from different neighbors (implosion). Second, nodes that are responsible for the same region may send the same message to their neighbors (overlap or duplication). Third, there is no way to know nodes' resource level, such as energy level (resource blindness). To overcome the drawbacks of flooding, many approaches have been proposed for WSN.

SPIN (Sensor Protocols for Information via negotiation) [8] is an event-driven protocol based on 3-way handshake: ADV (advertisement), REQ (request), and DATA. A source node advertises its data by sending ADV. Then, only nodes that wish to receive the data send REQ and receive it. Directed diffusion [9], gossiping, and rumor routing [10] are other variations to avoid flooding.

Location-based routing [11,12] is another approach to avoid flooding. In Location-based routing protocols, location information is utilized to limit the amount of flood-ing to a specific region. Disadvantage of location-based routing is that each node must know the locations of other nodes.

LEACH (Low-Energy Adaptive Clustering Hierarchy) [13] is a cluster-based rout-ing protocol, where self-elected cluster heads collect data from all nodes in their cluster. The cluster headers aggregate the collected data and transmit the data directly to the gateway node. LEACH minimizes global energy consumption by distributing the load to all the nodes. PEGASIS (Power-Efficient GAthering in Sensor Information Systems) [14] is another example of hierarchical protocol.

3 Tree-Based Routing

In this section, we present a basic tree routing algorithm [15]. Fig. 1 shows an example of WSN which will be used for the remainder of the paper. The network consists of 12 nodes and each node is shown as a circle. We assume that node 6 is the root of the entire network and the transmission range for a node is one-hop. We do not claim that this is a typical example for WSN but it is good enough to show how the wireless sensor network is formed into a tree structure in Fig. 2 and to see the problems of tree structure.

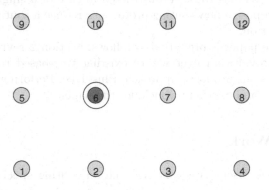

Fig. 1. Example of wireless sensor network: node 6 is the root

Before we describe Tree-based routing in detail, let us define and as follows:

– C_m: the maximum of children that a node can have in a tree
– L_m: the maximum height that a tree structure can have

Then, a tree structure is determined by C_m and L_m.

At the beginning of tree structure formation, the root node sends an associ-ation message to its neighbors. After that, the neighbor nodes are attached to the root node and then send association messages to their neighbors.

Every time a new node attempts to join a tree, a parent node checks if its number of children is less than C_m and its height is less than L_m. As long as the condition is met, the new node gets a connection to the tree structure and gets a new node ID.

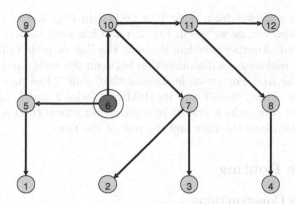

Fig. 2. WSN after the tree formation

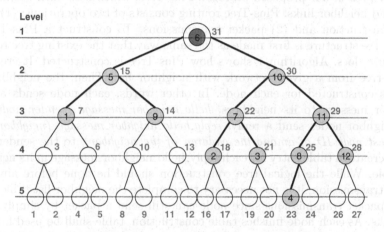

Fig. 3. Logical Tree Structure for Fig. 2

Fig. 2 shows the constructed tree for the wireless sensor network in Fig. 1. Each arrow between two nodes indicates a relationship between a parent and a child node – node initiating an arrow is the parent node for the other node. Node 6 is the root and nodes 1, 2, 3, 4, 9, and 12 become the leaves of the tree.

Fig. 3 shows the corresponding logical tree structure of the wireless sensor network shown in Fig. 2 after the tree formation where $C_m = 2$ and $L_m = 5$. The node number in a circle is the node ID given in association processing. The position ID, the node's position number from C_m and L_m, is shown outside a circle. The root gets the biggest number among its children. Since the root may have $2^5 - 2(C_m^{L_m} - 2)$ number of children, its position number is 31.

As mentioned earlier, this tree structure is considered to be a good structure to collect data information because the root is the only device connected to the main computer (or wired networks) and all the other wireless sensors in general send data packets to the root. On the other hand, this tree structure has a few problems. One is that a tree path may take longer than it could be. For example, to send a packet from node 1 to node 2 in Fig. 3, a tree path 1-5-6-10-7-2 is used. However, as we see in Fig. 2, node 1 is next to node 2 so a path 1-2 can be selected. Another problem is about the link or node failure. A single link/node failure may cause a disconnection between the node (and its children) and the rest of the tree. For example, assume that node 7 lost its connection to node 10 in Fig. 3. Then, node 7 and its children, nodes 2 and 3, also lose their connection to the root, which results in a situation where there is no way for them to communicate to the root and the rest of the tree.

4   Plus-Tree Routing

4.1   Plus-Tree Construction

We now present our proposed protocol, Plus-Tree, which uses both normal tree links and neighbor links. Plus-Tree routing consists of two operations: (1) Plus-Tree construction and (2) packet transmissions. To construct a Plus-Tree, a logical tree structure is first made as the same way that the existing tree routing procedure does. Algorithm 1 shows how Plus-Tree is constructed. It creates a logical tree from a given network with neighbor list. Then, the neighbor list table is constructed for each node. In other words, each node sends a hello neighbor message to its neighbors, *hello_neighbor_message (sender node ID)*, and neighbor nodes send a reply, *reply_hello_neighbor_message (neighbor node id, lowest node ID of among the children of the neighbor)*, to the sender. The sender creates a table entry for each reply_hello_neighbor_message in its neighbor list table. While the logical tree construction should be done before any data packet transmission, it is not necessary to complete the neighbor list table setup before packet transmissions since node can still send a packet by simply using tree links. As each node finishes table construction, table shall be used for path discovery and data transmissions.

Algorithm 1. Plus-Tree Construction

1: Given a WSN, construct a logical tree subject to C_m and L_m and assign a logical tree node ID to each node.
2: /* Each node constructs a neighbor list; entry of each table keeps "to whom" and "for whom" node IDs */
3: Each node sends hello message to its one-hop away neighbors and the neighbor node responds reply_hello message with its ID (to whom) and lowest node ID from its children (for whom)

Fig. 4 illustrates new links after neighbor discovery for node 6. Initially, node 6 has links only to nodes 5 and 10. After the neighbor discovery process, node 6 has new links, which are indicated as dot lines in the figure. The new links can be used for alternative paths in the event of link failure as well as for shorter paths to destination. Suppose that node 1 wants to send a packet to node 2 and the link between nodes 1 and 5 is lost. Node 1 can send data packets to node 6 by using the new link between nodes 1 and 6.

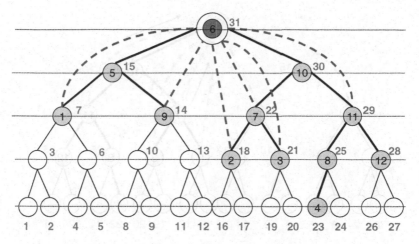

Fig. 4. Selecting a shorter path than tree path using neighbor list

Algorithm 2 describes how a node sends a packet to a destination. In Plus-Tree, the node will send a packet using a shorter path if available. Otherwise, it takes a tree path. For example, a packet from node 1 to node 3 takes a path 1-5-6-10-7-3 if the tree routing is used while it takes 1-6-3 if the Plus-tree routing is used.

Algorithm 2. Packet Transmission (On the packet arrival)

1: Node checks
2: **if** destination node is the current node **then**
3:     stop forwarding
4: **else if** the destination node is a child of "to whom" node in neighbor list **then**
5:     **if** the destination node is a child of multiple "to whom" nodes in neighbor **then**
6:         Forwards the packet to the closest ancestor node among "to whom" nodes.
7:     **end if**
8: **else**
9:     Forwards a packet to the node that tree routing tells.
10: **end if**

4.2 Link Failure

In Plus-Tree, a node may use neighbor links to transmit data packets in case of link failure. However, the data packets that attempt to use the broken tree link are lost because the nodes do not know how to reach the other node. For instance, suppose that the link between nodes 7 and 10 is lost as we see in Fig. 5. In this case, node 7 can transmit its packets to node 10 via its neighbor node 6. However, node 10 does not know how to reach node 7. Also, its descendants do not know the link failure and may send data packets to the node 10. For this reason, we propose algorithms that find a new path for the communication

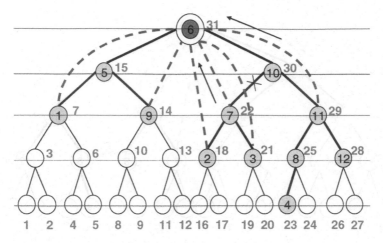

Fig. 5. Gateway set up and path finding for link failure

between two nodes of the link failure. In our example, a new path will be found for nodes 7 and 10.

The main idea of finding a new path is to let the missing child, called a leader node, to find a gateway to the root while the parent node reports the missing child information to its ancestors. A new path is built between the gateway and the root while the root notifies the new path information to the ancestors of the missing child. Now, the missing child and its descendants can send packets to the root and the rest of the tree. Also, the ancestors use the new path information to transmit packets to the missing child and its children. In our example in Fig. 5, node 7 becomes a leader node to select a gateway node and in this specific case, node 7 is selected as the gateway node because it has a connection to the root. Packets originated from node 7 and its descendants use the link between node 7 and the root. Also, node 10 (the parent node) sends a message to its ancestors (node 6 in Fig. 5) to notify that the link between nodes 7 and 10 is no longer available. The root records the notification and will not use the link.

Based on this principle, we present the algorithms for link failure. Algorithm 3 shows how the parent node transmits a message to its ancestors informing that it lost its child. Algorithm 4 shows how the child node sends a message to its descendants for electing a gateway to the root. In the last, Algorithm 5 shows how descendant nodes send messages to the leader node when they have a path to the root. Fig. 5 shows an example of a link failure (x marked) and the message sending (arrows) for discovery of a new gateway and for link failure notification to the root.

Algorithm 3. Parent Node

1: The parent node sends a lost_child (child_ID, parent_ID) message to the root.
2: The root and nodes between the parent and the root record the link failure between the parent node and the child node.

Algorithm 4. Child node (leader node)

1: Send a find_new_parent (child node ID) message to the root.
2: The root sends a reply_new_parent message to both the parent and the child.
3: If there is no reply_new_parent message, the child node sends a lost_parent (child_ID) message to its children.
4: It waits for reply_lost_parent message from its descendant.
5: The node that sends reply_lost_parent is chosen as a gateway node.

Algorithm 5. Descendant node

1: As a lost_parent (child_ID) message arrives, the descendant node sends a find_new_parent message to the root.
2: If it receives a reply_new_parent, it sends reply_lost_parent (descendant ID) message to the leader node.
3: The leader node sets the descendant as a gateway node.

---

5   Performance Evaluation

In this section, we present the simulation results to compare the performance between tree routing and plus-tree routing. Ten different cases are considered where for each case, five different pairs of source and destination are randomly selected. The performance metric is the average number of hops.

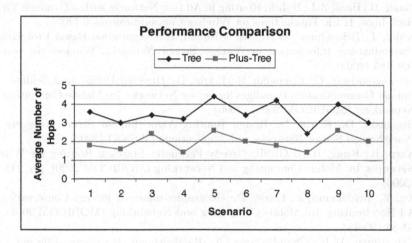

Fig. 6. Average number of hops for different cases

Fig. 6 depicts the average number of hop counts. In all cases, Plus-Tree protocol performs better than Tree protocol. Since Plus-Tree uses neighbor list, the source gets a chance to send a packet with a shorter path than a tree path.

6 Conclusion

In this paper, we have discussed tree-based routing protocol for WSN. We have presented routing algorithms that find shorter path than the existing tree algorithm. By simulation, we have shown that the average number of hops from the plus-tree algorithm is smaller than that from the tree algorithm. We have also proposed algorithms for link failure.

References

1. Perkins, C., Bhagwat, P.: Highly Dynamic Destination-Sequenced Distance-Vector Routing (DSDV) for Mobile Computers. Computer Communications Review, 234–244 (1994)
2. Clausen, T., Jacquet, P., Laouiti, A., Minet, P., Muhlethaler, P., Qayyum, A., Viennot, L.: (Op-timized Link State Routing Protocol (OLSR)) IETF MANET, Internet Draft
3. Murthy, S., Garcia-Luna-Aceves, J.J.: An Efficient Routing Protocol for Wireless Networks. MONET 1, 183–197 (1996)
4. Perkins, C., Royer, E.M., Das, S.R.: (Ad Hoc On-Demand Distance Vector Routing (AODV)) IETF 3561
5. Johnson, D.B., Maltz, D.: Dynamic Source Routing in Ad-Hoc Wireless Networking. Mobile Computing (1996)
6. Haas, Z.J., Pearlman, M.R.: Providing Ad-hoc Connectivity with the Reconfigurable Wireless Networks. In: ACM SIGCOMM (1998)
7. Liang, B., Haas, Z.J.: Hybrid Routing in Ad Hoc Networks with a Dynamic Virtual Back-bone. IEEE Transactions on Wireless Communicaitons 5 (2006)
8. Kulik, J., Heinzelman, W.R., Balakrishnan, H.: Negotiation-Based Protocols for Disseminat-ing Information in Wireless Sensor Networks. Wireless Networks 8, 169–185 (2002)
9. Intanagonwiwat, C., Govindan, R., Estrin, D.: Directed Diffusion: A Scalable and Robust Communication Paradigm for Sensor Networks. In: Mobile Computing and Networking (MOBICOM'2000) (2000)
10. Braginsky, D., Estrin, D.: Rumor Routing Algorithm for Sensor Networks. In: Workshop on Sensor Networks and Applications (WSNA) (2002)
11. Karp, B., Kung, H.T.: GPSR: Greedy Perimeter Stateless Routing for Wireless Networks. In: Mobile Computing and Networking (MOBICOM'2000), pp. 243–254 (2000)
12. Xu, Y., Heidemann, J., Estrin, D.: Geography-informed Energy Conservation for Ad Hoc Routing. In: Mobile Computing and Networking (MOBICOM'2001), pp. 70–84 (2001)
13. Heinzelman, W.R., Chandrakasan, A., Balakrishnan, H.: Energy-Efficient Communication Protocol for Wireless Microsensor Networks. In: IEEE Hawaii International Conference on System Sciences (2000)
14. Lindsey, S., Raghavendra, C.S.: PEGASIS: Power-Efficient Gathering in Sensor Information Systems. In: IEEE Aerospace Conference, vol. 3, pp. 1125–1130 (2002)
15. Hester, L., Huang, Y., Allen, A., Andric, O., Chen, P.: neuRFon Netform: A Self-Organizing Wireless Sensor Network. In: IEEE ICCCN Conference (2002)

Optimization and Routing Discovery for Ad Hoc Wireless Networks: A Cross Layer Approach

Reizel Casaquite and Won-Joo Hwang*

Department of Information and Communications Engineering, Inje University,
607 Obang-dong, Gimhae, Gyungnam, 621-749 South Korea
rzl_16@yahoo.com, ichwang@inje.ac.kr

Abstract. Wireless communication links are time-varying in nature, causing degradation of network's performance; this however, could be alleviated by allowing cross layer interaction in the protocol stack. Hence, optimization problems formulated in this paper considered parameters from the physical, MAC, and network layers where an optimal transmission power vector that minimizes the overall transmission power of the network, satisfying the SINR, required data rate, and maximum transmission power constraints was formulated to exist. Distributed algorithms for the joint scheduling, power control and routing were also derived and an energy-aware routing algorithm which maximizes network lifetime was proposed. The energy consumed by each node in the routing path and delays associated on each transmission were considered as link cost such that a route with minimum link cost is utilized. Basically, the routing algorithm searched an energy-efficient route, satisfying the energy constraint of each node in a routing path at a tolerable delay.

1 Introduction

Cross-layer design is an extensive research area in ad hoc wireless networks nowadays, particularly under energy constraints. It allows communication of a layer with any other possibly non-adjacent layer in the protocol stack to help improve network performance. According to Goldsmith et al. [1], a cross-layer protocol design that supports adaptivity and optimization across multiple layers of the protocol stack is needed since the inflexibility and sub-optimality of the layered architecture design usually result in poor performance of a network, especially when energy is a constraint or the application has high bandwidth needs, and/or stringent delay constraints. Many research works in routing assume a fixed underlying protocol for access control and most research on multiple access control assume fixed routes and flow requirements. The authors in [2] state that the choice of MAC protocol does, in fact, affect the relative performance of the routing protocol and a cross layer design should be implemented since the functionalities of the two layers interact. Likewise, coupling between power control in the physical layer and scheduling in the MAC layer should be investigated further.

---

* Corresponding author.

M.S. Szczuka et al. (Eds.): ICHIT 2006, LNAI 4413, pp. 647–658, 2007.
© Springer-Verlag Berlin Heidelberg 2007

Our contributions are as follows: an energy-aware routing algorithm with new link cost (metric) was proposed in this paper. This link cost considers node's energy and link delay associated on each transmission. Moreover, a mathematical model and optimization problems were formulated. The joint optimization problem is related to the works of Chiang [3,4] in which other power levels and TCP windows sizes were optimized. In our paper, we extend it to routing and allocating flows at the network layer, considering the energy of each node and link delay. The main technique used was dual decomposition for convex problems.

This paper is organized as follows: Section 2 presents the related works while Sect. 3 presents the mathematical analysis and problem formulation. The mathematical model and assumptions are presented in Sect. 3.1 and the linear programming (LP) problem that solves an optimal transmission power vector is presented in Sect. 3.2. Section 3.3 solves a utility maximization problem with elastic link capacities, and Sect. 4 discuss the link cost which is minimization of energy and delay in Sect. 4.1 and 4.2 respectively. The proposed routing algorithm on the other hand, is described in Sect. 4.3. Numerical examples of the optimization problems and relationships between parameters are given in Sect. 5 and Sect. 6 concludes the paper.

2 Related Works

The interoperability among the physical, MAC, and network layers is being studied since they take part in the energy expenditure and throughput performance of the system. The MAC layer controls the interference level at any time instance which may lead to transmit power adaptation in the physical layer. The physical layer, on the other hand, classifies the node's local neighborhood and defines the context in which access, routing, and other higher-layer protocols operate. Routing for ad hoc wireless networks is a significant design challenge under energy constraints since the exchange or routing of data consumes precious energy resources. Most research dealing with routing considered only the minimum energy path regardless of the delay associated when transporting data while some papers dealt only with end-to-end delay which leads to finding the shortest-hop path ([5] and references herein). In this way, one or more nodes are heavily loaded and nodes tend to have widely differing energy consumption resulting in an early death of some nodes. Hence, the path to be selected must consider the energy reserves of the nodes such that nodes with depleted energy reserves do not lie on many paths.

In Fig. 1, if node 1 has some packets to send to node 6 and node 2 has also some packets to be sent to node 6, and if we assume that $\{1, 3, 6\}$ and $\{2, 3, 6\}$ are the minimum energy paths, node 3 is selected as a route for packets going to every other node, thus, it will consume power at a faster rate and will die first compared to other nodes in the network. If nodes' energy consumption is not considered, a route consuming more energy may be selected which consumes more of the overall networks energy which could lead to a short network lifetime. Also, if all traffic is routed through the minimum energy path to the destination,

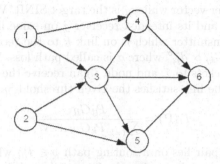

Fig. 1. A Simple Topology

the nodes in that path will run out of batteries quickly. This would result to a network partition [6]. Hence, traffic should use routes with sufficient remaining energy to maintain balance in the network.

To maximize network lifetime in a more balance manner, an intuitive technique called Minimal Battery Cost Routing (MBCR) was proposed by the authors in [7]. The algorithm utilized the node's remaining battery capacity as a cost function and the route that has a minimal battery cost is selected, thereby increasing the lifetime of the network. A Maximum System Lifetime (MSL) routing algorithm suggested by Tassiulas and Chang [6] also routes traffic to nodes with sufficient energy such that the energy consumption is balanced among nodes in proportion to their energy reserves. However, their routing algorithm uses shortest-path. According to [1], a route utilizing a small number of hops (low delay) may use significantly more energy (per node and/or total energy) than a route consisting of a larger number of hops. Hence, routing protocols under energy constraints must somehow balance delay constraints, battery lifetime, and routing efficiency. In our paper, we trade-off delay and energy such that our routing algorithm searches an energy-efficient route considering the energy link cost and end-to-end delay link cost.

3 Problem Formulation

3.1 System Model and Assumptions

An ad hoc network can be generally represented by a graph $G(V, L)$ where V denotes the set of all nodes and L is the set of links between those nodes. At this time, the approach considers quasi-stationary or fully stationary wireless ad hoc networks. We denote a link l as an active direct communication pair (i, j) and $(i, j) \in L$. We assume that there are M numbers of source-destination pair in the network and throughout this paper we define link (i, j) and link l to have same meaning. A source node i is denoted by $s(i)$ and a destination node is denoted by $d(i)$. We assume a symmetric hearing matrix among the nodes and the channel gain between two nodes is approximately same in both directions. We let $\Gamma_l(P)$ be the measured SINR at link (i, j) where $\mathbf{P} = (P_1, P_2, \ldots, P_M)^T$

is a transmission power vector while γ_l is the target SINR. We let G_{ll} be the link gain of transmitter i and its intended receiver j on same link l and G_{lk} is the link gain of other transmitter which is on link k to the receiver on link l. Gain can be computed as $G_{ij} \propto d_{ij}^{-\sigma}$ where σ is called path loss. We let node i be the node transmitting packets to j, and node j can receive the packets successfully only if the SINR at the link satisfies the given threshold γ_l.

$$\Gamma_l(\mathbf{P}) = \frac{P_{ll}G_{ll}}{\sum_{l \neq k} P_k G_{lk} + N_l} \tag{1}$$

A source-destination pair lies on a routing path $p \in P_R$ where P_R contains all possible routes or a set of directed paths from $s(i)$ to $d(i)$ containing no loops. We let $N_l = \eta_0 B_T$ as the thermal noise at the receiver node of link l, where η_0 and B_T denotes the noise density and bandwidth respectively. The Shannon link capacity (theoretical maximum) over a frequency bandwidth W, assuming presence of Gaussian noise and interference is given by (2). We denote that the link capacity is a function of SINR, which in turn is determined by the power levels at all transmitters.

$$c_l(\mathbf{P}) = W \log_2(1 + \Gamma_l(\mathbf{P})). \tag{2}$$

Scheduling decides which links will transmit and when to transmit. It is similar to choosing an independent set of flow contention graph to be active at each time slot. When link l is active, node i and node j cannot transmit to other nodes or receives from other nodes. Given a contention graph, we can identify its maximal cliques [3] where flows within same maximal cliques cannot transmit simultaneously but flows in different cliques may transmit simultaneously.

3.2   Minimize Overall Transmission Power

Our mathematical approach considers the entire network i.e. if there exist an optimal transmission power vector that satisfies the given constraints and the overall transmission power of the network is minimized, the maximum transmission power needed by the transmitters to have a successful transmission on each link is also minimized. Our problem is to find a transmission power vector that minimizes the overall transmission power of the network subject to minimum required data rate, maximum transmission power constraint, and SINR constraints. Our optimization problem could be stated as

$$\text{Minimize} \quad \sum_{l=1}^{M} \alpha_l P_l \tag{3}$$

$$\text{Subject to} \quad 0 \leq P_l \leq P_l max \tag{4}$$

$$\Gamma_l(\mathbf{P}) \geq \gamma_l \tag{5}$$

$$c_l(\mathbf{P}) \geq X_i \tag{6}$$

where $\alpha_l \geq 0$ is the cost or weight assigned for each transmission power and $\sum_{l=1}^{M} \alpha_l = 1$. In addition, X_l is the minimum required data rate of the receiving node on link l. The constraints (4), (5) and (6) denote the node's maximum transmission power, SINR constraint, and the minimum data rate constraint, respectively. As can be seen in (1) and (2) the constraints (5) and (6) are non linear. However, (5) can be written in the form of $\mathbf{AP} \geq \mathbf{b}$ where $\mathbf{A}$ (receiver by transmitter matrix) and $\mathbf{b}$ are given below.

$$\mathbf{A} = \begin{bmatrix} 1 & -\frac{\gamma_1 G_{12}}{G_{11}} & \cdots & -\frac{\gamma_1 G_{1M}}{G_{ll}} \\ -\frac{\gamma_2 G_{21}}{G_{22}} & 1 & \cdots & \frac{\gamma_2 G_{2M}}{G_{22}} \\ \vdots & \vdots & \ddots & \vdots \\ -\frac{\gamma_M G_{M1}}{G_{MM}} & -\frac{\gamma_M G_{M2}}{G_{MM}} & \cdots & 1 \end{bmatrix} \quad \mathbf{b} = \begin{bmatrix} \frac{\gamma_1 N_1}{G_{11}} \\ \frac{\gamma_2 N_2}{G_{22}} \\ \vdots \\ \frac{\gamma_M N_M}{G_{MM}} \end{bmatrix}$$

Likewise, from (2) and (6) and by letting $\varphi_l = 2^{\frac{x_l}{W}} - 1$ the constraint (6) can be written in the form of $\mathbf{YP} \geq \mathbf{z}$ given by

$$\mathbf{Y} = \begin{bmatrix} 1 & -\frac{\varphi_1 G_{12}}{G_{11}} & \cdots & -\frac{\varphi_1 G_{1M}}{G_{ll}} \\ -\frac{\varphi 2 G_{21}}{G_{22}} & 1 & \cdots & \frac{\varphi_2 G_{2M}}{G_{22}} \\ \vdots & \vdots & \ddots & \vdots \\ -\frac{\varphi_M G_{M1}}{G_{MM}} & -\frac{\varphi_M G_{M2}}{G_{MM}} & \cdots & 1 \end{bmatrix} \quad \mathbf{z} = \begin{bmatrix} \frac{\varphi_1 N_1}{G_{11}} \\ \frac{\varphi_2 N_2}{G_{22}} \\ \vdots \\ \frac{\varphi_M N_M}{G_{MM}} \end{bmatrix}$$

The optimization problem can be formally written as

$$\text{Minimize} \quad _{P_l \in [0, P_l max]} \alpha_l P_l \tag{7}$$

$$\text{Subject to} \quad AP \geq b \tag{8}$$

$$YP \geq z \tag{9}$$

An optimal solution to the LP problem in (7) exists if and only if there is a solution to the constraints given by equations (8)-(9) i.e. there is at least one set of transmission power $\mathbf{P}$ which ensures successful reception at all receiver nodes and at the same time satisfies maximum node's transmission power, SINR, and data rate constraints. Observe that, transmit power is bounded by $P_l max$. Hence, an optimal solution exists by virtue of Theorem 3.4 in [8].

3.3   Maximize Utility

We consider a network with L links, each with a capacity of c_l (bits per second) and with $s \in \mathbf{S}$ users, each transmitting at a source rate of x_s (bits per second). Each user emits one flow using a set of links $L(s)$ in its path, associated with a utility function for each flow, where U_s is continuously differentiable, increasing, strictly concave and unbounded as x_s approaches zero. We assume that all utility functions are logarithmic $U_s(x_s) = log(x_s)$ and is additive so that the aggregate traffic utility is $U(x) = \sum_s U_s(x_s)$. Our objective is to choose a source rate x_s so as to maximize the sum of utilities

$$\text{Maximize} \quad U(x) = \sum_{s \in S} U_s(x_s) \tag{10}$$

$$\text{Subject to} \quad \sum_{s:l \in L(s)} x_s \leq c_l(\mathbf{P}) \tag{11}$$

$$x_s, \mathbf{P} \geq 0 \quad \forall l \tag{12}$$

The constraint (11) says that the aggregate source rate at any link l does not exceed the effective link capacity while equation (12) ensures that the data rate and transmission power are non-negative. The optimization variables of problem (10) are both source rate x_s and transmit powers $\mathbf{P}$. As can be seen in (2), the link capacity c_l is a function of transmit powers $\mathbf{P}$ and is nonlinear. If we introduce a Lagrange multiplier β_l for the constraint (11) the relaxed function is given by

$$L(x, \mathbf{P}, \beta) = \sum_{s:l \in L(s)} U_s(x_s) + \sum_{l=1}^{L} \beta_l (c_l(\mathbf{P}) - \sum_{s \in L(s)} x_s). \tag{13}$$

By linearity of the differentiation operator, (13) could be decomposed into two sub-problems (14) and (15), which are functions of the network and MAC/Phy layer respectively.

$$L(x, \beta) = \sum_{s:l \in L(s)} U_s(x_s) + \sum_{l=1}^{L} \beta_l \sum_{s \in S, s \in L(s)} x_s \tag{14}$$

$$L(\mathbf{P}, \beta) = \sum_{l=1}^{L} \beta_l c_l(\mathbf{P}) \tag{15}$$

The Lagrange variable β_l can be interpreted as the price per unit bandwidth at link l. We would like to maximize (14) as given by

$$\text{Maximize} \quad \sum_{s:l \in L(s)} U_s(x_s) + \sum_{s:l \in L(s)} \beta_l x_s \tag{16}$$

which admits a unique maximizer that adjusts the source rate according to β_l at the source. The optimal data rate that maximizes (16) can be solved and readily obtain as given by (17) where U^{-1} is the derivative of the inverse of the utility function.

$$x_s(\beta) = U_s^{-1}(\beta_l) \tag{17}$$

The problem (16) defines both congestion control and routing behavior. Congestion control is based on (17) where the source node will adjust its rate according to the path price β_l while routing is based on the minimum cost path. Furthermore, the problem (15) is the scheduling problem for link layer flows according to β_l. Since the link capacity is a function of SINR, which in turn is determined by the power levels at all transmitters, the scheduling scheme can be determined by solving the problem (18) where at each link (i, j) the corresponding power assignment that achieves maximum capacity is to be chosen.

$$c_l(\beta) = argmax \sum_{l=1}^{L} c_l(\mathbf{P})\beta_l \tag{18}$$

Similar to the Jointly Optimal Congestion Control and Power Control (JOCP) algorithm in [3], each transmitter can update its power and maximize (18) by

$$P_l(t+1) = P_l(t) + \kappa \left(\frac{\beta_l(t)}{P_l} - \sum_{j \neq i} \frac{\beta_j(t)G_{jl}}{\sum_{k \neq j} P_k G_{jk} + N_j} \right) \tag{19}$$

where $\kappa > 0$ and is a constant that denotes step-size. The Lagrange multiplier can be updated using (20) where δ is a positive step-size. The convergence analysis of such algorithm is already well known [9].

$$\beta_l(t+1) = \left[\beta_l(t) + \delta \left(\sum_{s \in S, s \in L(s)} x_s - c_l(\beta) \right) \right]^{+} \tag{20}$$

4   Routing Mechanism

In this section, the proposed routing mechanism is presented. Our goal is to develop a routing algorithm considering node's energy and end-to-end delay as link cost. In particular, an energy-efficient route satisfying the energy constraint at each node in the routing path at a minimum end-to-end delay is utilized. The first two sections discuss the proposed link costs metrics while the last section is the routing discovery mechanism.

4.1   Energy Link Cost

To maximize the life of all nodes and the network itself, the path to be chosen must consider energy reserves such that nodes with depleted energy reserves do not lie on many paths. Hence, traffic should use routes with sufficient remaining energy to maintain balance in the network. We define the total cost of sending a packet along some path $p \in P_R$ as the sum of the weights of all nodes that lie along that path. The cost of sending a packet from $s(i)$ to $d(i)$, [5] via intermediate nodes is given by

$$F(z_k) = \sum_{k=1}^{d-1} f_k z_k. \tag{21}$$

We denote $f_k(z_k) = 1/z_k$ where z_k is the amount of remaining energy available or measured voltage at a node k which gives a good indication of the energy used so far. The lower the value of the cost function $F(z_k)$, the higher the remaining energy of the nodes lying in path p. Similarly, the higher the value of the cost function, the lower the remaining lifetime of the nodes. Our goal is to choose a path with minimum total energy-link cost of sending a packet from a source to a destination node, i.e. $F(z_k)$ subject to $z_k \geq E_T$, where E_T is the node's battery/voltage threshold. The remaining energy associated with each source destination pair in a path p is given by the energy of a node which has the minimum amount of remaining energy [10] among all other nodes lying in path p or in equation form, $E(p) = min\, z_k$. Thereby, we could define the maximum transmission power for each link as $P_l max = E(p)$.

4.2   Delay Link Cost

For any link in the network, we let f_{ij} denote the expected traffic on link (i, j) or the amount of flow traversing link (i, j) per unit of time and $0 \leq f_{ij} \leq c_{ij}$. Similar to [11], we assume $D_{ij}(\cdot)$ to depend only on flow rate f_{ij} and link characteristics such as link capacity c_{ij} and propagation delay. We further assume that the function is continuous and convex that approaches infinity as f_{ij} approaches c_{ij}. For simplicity of analysis, we considered an M/M/1 queue where the delay of a given link is computed by

$$D_{ij}(f_{ij}) = \left(\frac{1}{c_{ij} - f_{ij}} \right). \tag{22}$$

For a given path $p \in P_R$, the total expected delay per message on all links is given by

$$D(p) = \sum_{(i,j) \in L(s)} D_{ij} f_{ij} \tag{23}$$

and $D(p)$ should be minimized. In our case, it should satisfy the delay threshold D_T, i.e. $D(p) \leq D_T$. Equation (23) is a nonlinear programming problem but there are already existing algorithms that could solve this problem ([11] and references herein).

4.3   Routing Discovery

The joint algorithm consists of three major components: route discovery where an energy-aware path connecting a given source-destination pair is being searched for transmission; minimization of the link's transmission power, where the maximum transmit power needed for a successful transmission on each link is being minimized; and route maintenance, if a specific path breaks, a new route is searched for transmission. The route maintenance mechanism is similar to AODV (Ad hoc On-Demand Distance Vector Routing) protocol, hence we will only discuss route discovery in this section.

If a source node receives a request for a route to an intended destination and no route has been known yet, the source will initiate a route discovery process to acquire a possible route by broadcasting RREQ and starts a timer. The RREQ contains the value of the transmission power used (19) and the value of the end-to-end delay which is initially zero. After some time, still not receiving RREP from other nodes, the source node will broadcast an RREQ at a power level greater by a certain amount than the previous one until a path is found or until no path is found even after increasing the transmission power to P_lmax. An intermediate node only forwards the first RREQ it receives and as it forwards the RREQ, it calculates the energy link cost (21) and the accumulated delay link cost (23) to be added to the path cost in the header of the RREQ packet. The remaining energy of each node, z_k will be computed and see if $z_k \geq E_T$. If not, it will not be used in the next transmission until there are no other nodes available. The destination node may receive multiple possible paths for the transmission and is responsible for comparing path costs and selecting the minimum cost path. The destination node will compute the end-to-end delay (23) of that specific path, if it satisfies the delay requirement, D_T. Also, the energy link cost of the path (21) will be computed. If it satisfies the delay constraint, the destination node will reply an RREP packet back to the source node using the reverse path informing that an energy-efficient path has been discovered and all nodes in the said path satisfies the battery's threshold voltage. However, if the destination node will receive the same RREQ with same energy link cost but with smaller delay than the previously received RREQ, it will reply with a new RREP to the source node again. Basically, a routing path with minimum energy link cost (21) from a source to a destination node and at the same time meets the delay requirement will be used for transmission. Once the source receives the RREP, it makes the destination's route table entry and starts sending data packets to the destination. If the source receives the same RREP with smaller delay than the previously received RREP, it will update the destination's route table entry. The data is transmitted to the destination node at a transmission power given by (19) and at a transmission rate given by (17).

5   Numerical Analysis

We consider a simple asymmetric diamond topology as shown in Fig. 2. Nodes in this topology are equipped with omni directional antennas. Node 1 is the source node and node 4 is the sink node. The link gain is given by $G_{ij} = 1/d_{ij}^4$ and we assume that the peak transmission power of each node is 1Watt. We use a channel frequency of 916MHz, a thermal noise n= -104dBm, and a SINR threshold of 6dB for all nodes. We split the path into $\{1, 2, 4\}$ and $\{1, 3, 4\}$ transmissions. Since a node is unable to transmit and receive packets at the same time, we only allowed transmissions $\{(1, 2), (3, 4)\}$ and $\{(1, 3), (2, 4)\}$ alternately.

Using MATLAB, we solved the optimization problem in (3) i.e. there exists a power vector that minimizes the total transmission power while satisfying the identified constraints. We first consider (1,2) and (3,4) transmissions and plot the optimal/minimum transmission power with increasing required data rate demands at the receiver node (Fig. 3). We vary the data rate where same target SINR is assumed for all links. As observed, as the required data rate of the receiving node increases, the minimum link's transmission power needed

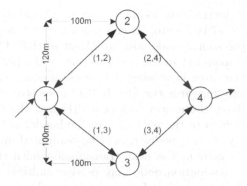

Fig. 2. Asymmetric Diamond Topology

for successful transmission also increases. Hence, high data rate demands also require high transmission power. Still considering the transmissions (1,2) and (3,4), we plot the optimal transmission power with increasing SINR requirement in Fig. 4. We assume a required data rate of 3Mbps for all receiving nodes. The transmission power of the nodes increases with increasing SINR.

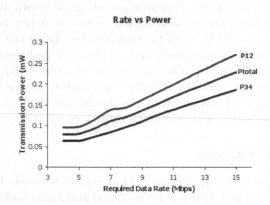

Fig. 3. Transmit power with increasing required data rate in (1,2), (3,4) transmissions

Using same topology, we determine the SINR of the links (1,3) and (3,4) given the computed optimal transmission power on each link i.e. the minimum transmission power needed by the transmitter to have a successful transmission to the receiver. To examine the trade-off between power and delay, we assume first that collision and contention issues are negligible. We plotted the computed end-to-end delay of the path $\{1,3,4\}$ and its relationship with power in Fig. 5. We used equation (22) in computing the delay on each transmission where we set $f_{ij} = 2000s$. As can be seen in the graph, delay decreases as power per transmission increases. This also means that the faster the service rate or transmission capacity on each link, the lesser will be the end-to-end delay. Thereby,

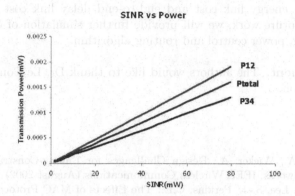

Fig. 4. Transmission power with increasing SINR demands in (1,2),(3,4) transmissions

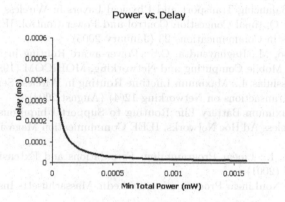

Fig. 5. Trade-off of power and delay in transmission path $\{1, 3, 4\}$

nodes have limited initial amount of energy which is consumed in different rates depending on the transmit power level and the distance from intended receiver. The relationships between the parameters obtained are applicable when each user has a respective target SINR and users will access different services with different transmission rates.

6 Conclusion and Future Works

We have formulated and solved optimization problems for minimizing transmission power, and maximizing utility of the network involving parameters from the physical, MAC, and network layers, which presents that interaction between different layers of the protocol stack is necessary to achieve performance gains. We also proposed a routing algorithm which maximizes network lifetime and at the same time trade off the energy consumed of each node, and the delay associated for every transmission. The algorithm searches an energy-efficient route considering the energy link cost and end-to-end delay link cost of a routing path. For our future work, we will provide further simulation of our proposed joint scheduling, power control and routing algorithm.

Acknowledgment. The authors would like to thank Dr. Le Cong Loi for the helpful discussion.

References

1. Goldsmith, A., Wicker, A.: Design Challenges for Energy-Constrained Ad Hoc Wireless Networks. IEEE Wireless Communications (August 2002)
2. Royer, E.M., Lee, S.-J., Perkins, C.E.: The Effects of MAC Protocols on Ad Hoc Network Communication. In: Proc. WCNC (2000)
3. Chiang, M., Man, R.: Jointly Optimal Congestion Control and Power Control in Wireless Multihop Networks. In: GLOBECOM 2003, vol. 1 (December 2003)
4. Chiang, M.: Balancing Transport and Physical Layers in Wireless Multihop Networks: Jointly Optimal Congestion Control and Power Control. IEEE Journal on Selected Areas in Communications 23 (January 2005)
5. Singh, S., Woo, M., Raghavendra, C.S.: Power-aware Routing in Mobile Ad hoc Networks. In: Mobile Computing and Networking, MOBICOM (1998)
6. Chang, J., Tassiulas, L.: Maximum Lifetime Routing in Wireless Sensor Networks. IEEE/ACM Transactions on Networking 12(4) (August 2004)
7. Toh, C.K: Maximum Battery Life Routing to Support Ubiquitous Mobile Computing in Wireless Ad Hoc Networks. IEEE Communication Magazine 39, 138–147 (2001)
8. Vandenberghe, L.: Linear Programming: Foundations and Extensions, 2nd edn., Princeton, NJ (2001)
9. Bertsekas, D.: NonLinear Programming, 2nd edn. Massachusetts Institute of Technology (1999)
10. Zhang, B., Mouftah, H.: Energy-aware on-demand routing protocols for wireless ad hoc networks. Wireless Networks 12(4) (July 2006)
11. Vutukury, S., Garcia-Luna-Aceves, J.J.: A Traffic Engineering Approach based on Minimum-delay Routing. In: International Conference on Computer Communications and Networks (October 2000)

Analysis of the Characteristics of Rain Attenuation in the 12.25GHz Band for Wireless Networking

Dong You Choi

#375 Seosuk-dong, Dong-gu, Gwangju 501-759 Korea
Deparment of Information Communication, Chosun University
dy_choi@hanmail.net

Abstract. Quantitative analysis and prediction of radio attenuation is required in order to improve reliability of satellite-earth communication links and for economically efficient design. For this reason, many countries have made efforts to develop their own rain attenuation prediction models which are fit for their rain environment. In this study the slant path length adjustment factor and rain height proposed in Korea and Japan was applied to ITU-R model (P.618-5, P.618-8), which is most widely used in the world. Their results were compared to measured data of rain attenuation and their effectiveness and validity were examined through evaluating the Pearson correlation coefficient.

1   Introduction

Attenuation rain is mainly due to scattering: The electric field of the radio wave polarizes the water molecules of the raindrop, and then the rain drop acts like a small electric dipole radiating over a large solid angle. A heavy rain makes long radio hope impossible at frequency above $10GHz$. In a heavy rain the drops are large and their shape is ellipsoidal. Then a horizontally polarized wave attenuates more than a vertically polarized wave. This phenomenon, depending on the wind speed, also causes depolarization of the wave, if the electric field is not along either axis of the raindrop. For this reason, in order to predict the effect of rain on radio attenuation quantitatively, developed countries in the field of satellites such as America, Europe, and Japan have developed their own rain attenuation prediction models since early 1960s. Particularly, the International Telecommunication Union Radio Communication Sector (ITU-R) recommends methods to Rec.P.618 and this prediction method is widely used.

However, as the characteristics of rain vary with time and space, the characteristics of rain attenuation appearing in Korean unique rain environment may be different from the prediction results of these models. By comparing and analyzing prediction data and measured data by several rain attenuation prediction models, I could see difficulty of direct application there of in [1].

M.S. Szczuka et al. (Eds.): ICHIT 2006, LNAI 4413, pp. 659–668, 2007.

Accordingly, in this study, I applied the rain height [2] proposed by K.Satoh in Japan having locally similar environment and the slant path length adjustment factor [3,4] proposed by Joo-hwan Lee, considering the Korean rain environment, to ITU-R model (P.618-5, P.618-8). Also, by comparing and analyzing prediction results with measured rain attenuation, the feasibility and validity of the application of ITU-R model in Korea will be described.

2   Rain Attenuation 1 Minute Rain Rate Measurements

In general, attenuation at frequencies over $10GHz$ is mainly due to absorption by oxygen and water vapor, snow, and atmospheric gases. Attenuation due to oxygen is influenced by antenna height, air pressure, and temperature, and it is not regionally different. However, rain attenuation is highly influenced by rainfall intensity and it varies over time and space. Thus it is required to obtain spatial and temporal data of rainfall intensity for precise prediction of rain attenuation. Therefore, in this study used rain rate data [5] measured by the Electronic and Telecommunications Research Institute (ETRI) from 1984 to 1993. Figure 1 shows the log-normal approximated time percentage curves of rain rate derived from measured rain rate data.

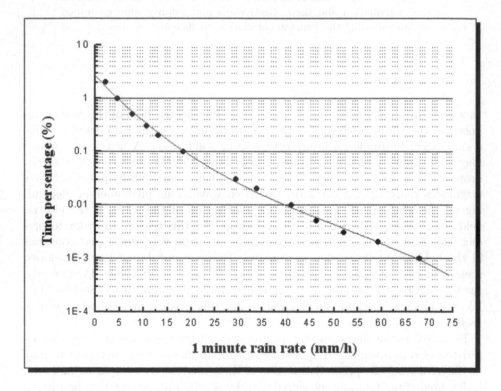

Fig. 1. Conditional cumulative distribution of 1 minute rain rate at $12.25GHz$

In this study, I selected Koreasat-3, which uses the $12.25GHz$ band frequency, and analyzed the beacon signal level data according to the rainy wet season in 2001 at Yong-in, Korea (ITU-R rain zone K). To measure the rain rate, I used the rain measurement system, which was installed when the Yong-in Satellite Control Office was established. To measure the beacon signal level, which is always received at a certain level from a satellite, the controlling equipment of Koreasat-3 was used [1]. The accumulated rain rate data was first collected in a data collector and was then saved in a computer. The rain rate data may be saved in either 1every 10 minute or at 10 second intervals. Since the 10 minute interval data was not sufficiently accurate for use with the satellite beacon signal level, the 10 second interval data collection was used. Table 1 shows the measurement condition of rain attenuation in Yong-in Satellite Control Office.

Table 1. Measurement condition of rain attenuation

	Latitude(Yong-in)	37.43°N
	Longitude(Koreasat-3)	116°E
System location	Elevation angle	45.20°
	Azimuth angle	198.1°
	Sea level	0.142km
Climate zone	ITU-R model	K-zone
	Polarization	Dual liner
Down link	EIRP	34dBW
	Frequency	12.25GHz
Antenna	Type	Cassegrain
	Diameter	7.2m
	Type	Tipping bucket
	Size	Diameter 200mm
Rain gage	Resolution	0.5mm
	Accuracy	Less than 5% (in rain rate 10mm/hr)
	Operative temperature	−40°C ∼ +50°C
Observation period	June - August 2001	

3 Comparison of Correlation of Predicted and Measured Rain Attenuation

The ITU-R Rec.P.618 describes a method of predicting rain attenuation for earth-space telecommunication systems in its section 2.2.1.1. The latest version is P.618-8 [6], which has been enforced since 2003. The P.618-8's prediction model is consideration different from the old version of P.618-5 [7] especially in computing slant path length adjustment factors for the horizontal and vertical direction whereas the P.618-5 employs only one adjustment factor [8].

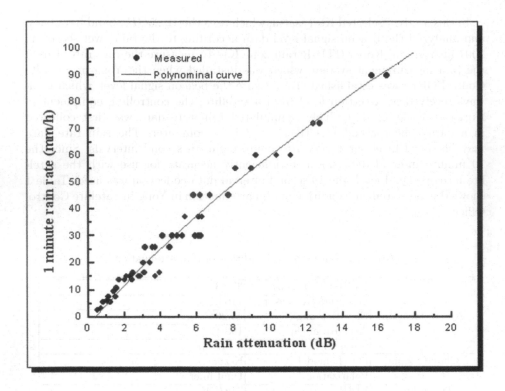

Fig. 2. Measured rain attenuation

Comparisons are based on the Pearson correlation coefficient [9,10] defined as follows:

The correlation $\rho_{X,Y}$ between two random variables X and Y with expected values μ_X and μ_Y and standard deviations σ_X and σ_Y is defined as,

$$\rho_{X,Y} = \frac{Cov(X,Y)}{\sigma_X \sigma_Y} = \frac{E((X-\mu_X)(Y-\mu_X))}{\sigma_X \sigma_Y} \tag{1}$$

where E is the expected value of the variable and Cov means covariance. Since $\mu_X = E(X)$, $\sigma_X^2 = E(X^2 - E^2(X))$ and likewise for Y, it may also write

$$\rho_{X,Y} = \frac{E(XT) - E(X)E(Y)}{\sqrt{E(X^2) - E^2(X)}\sqrt{E(Y^2) - E^2(Y)}} \tag{2}$$

The correlation is defined only if both standard deviations are finite and both of them are nonzero. It is a corollary of the Cauchy-Schwarz inequality that the correlation cannot exceed 1 in absolute value. The correlation is 1 in the case of an increasing linear relationship, -1 in the case of a decreasing linear relationship, and some value in between in all other cases, indicating the degree of linear dependence between the variables. The closer the coefficient is to either -1 or 1, the stronger the correlation between the variables.

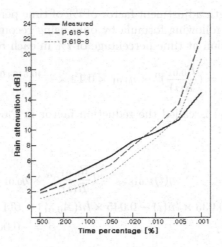

Fig. 3. Comparison of correlation of predictions by ITU-R model for measured rain attenuation

Figure 3 shows correlation of rain attenuation in the $12.25GHz$ band. In more than 0.02% of time percentage, the increasing relationship of P.618-5 were bigger than P.618-8. When time percentage was 0.02% or lower, the increasing relationship of P.618-8 were bigger than P.618-5. From this figure, it is found that the P.618-5 and P.618-8 gives a Pearson correlation coefficient of same about 0.97.

4   Analysis of Slant Path Length Adjustment Factor and Rain Height Attenuation on the Predictions

4.1   Effect of Slant Path Length Adjustment Factor

To predict rain attenuation at i% of the time percentage $A_p(i)$, the ITU-R prediction model uses the basic formula,

$$A_p(i) = k \times R(i)_\alpha \times L_s \times \eta(i)[dB] \tag{3}$$

where k and α are frequency dependent parameters that can be computed by ITU-R Recommendation P.838-2 [11] and $R(i)$ is the estimated one minute rain rate i yearly time percentage. L_s is the "actual" slant path length determined by elevation angle θ and rain height (or freezing height) h_r and height above mean sea level of the earth station, h_s in km. For $\theta \geq 5°$, L_s is computed by the following formula:

$$L_s = \frac{h_r - h_s}{sin\theta}[Km] \tag{4}$$

In this formula (4), h_r is calculated by P.839-3 [12] when applying P.618-8, and h_r is obtained by a formula in P.618-5 when applying P.618-5. $\eta(i)$ is defined as the slant path length adjustment factor of $i\%$ of time percentage and $\eta(i)$ can be ex-pressed as the following formula by only using recommended formulas to predict rain attenuation at time percentage of $i\%$ in each recommendation.

$$\eta(i)_{P.618-5} = (\frac{R_{0.01}}{R(i)})^\alpha \times \eta_{0.01} \times 0.12 \times i^{-(0.546+0.043 \times log(i))} \qquad (5)$$

Here, $\eta_{0.01}(= \gamma_{0.01})$ is called the reduction factor in P.618-5 and its formula can be referred in [7].

For P.618-8,

$$\eta(i)_{P.618-8} = (\frac{R_{0.01}}{R(i)})^\alpha \times \eta_{0.01} \times (\frac{i}{0.01})^{-\lambda}$$

$$\lambda = 0.655 + 0.033 \times ln(i) - 0.045 \times ln(A_{0.01}) - \beta(1 - i) \times sin\theta$$

$$\beta = -0.005(|\varphi| - 36) \qquad (6)$$

where $\eta_{0.01}(= \gamma_{0.01} \times v_{0.01})$ is for P.618-8. The $\gamma_{0.01}$ is called horizontal reduction factor and the $v_{0.01}$ is called vertical adjustment factor [13] and its formula can be referred in [6]. φ is latitude of the reception point $(< 36°)$ and is the elevation angle $(\geq 25°)$. This factor $\eta_{0.01}$ is multiplied by an actual slant path length to calculate an effective slant path length for rain attenuation at time percentage of $i\%$.

Joo-Hwan Lee divided the slant path length adjustment factor F_s into horizontal and vertical path components and suggested the resultant divided components. The F_h is called horizontal adjustment factor and the F_v is called vertical adjustment factor [3,4].

$$F_s = F_h cos^2\theta \times F_v sin^2\theta \qquad (7)$$

$$F_h = \frac{1}{1 + 0.025857 \times L_g \times e^{0.015R}} \qquad (8)$$

$$F_v = 0.703 + 0.24e^{-0.039R} \qquad (for \; R \leq 60mm/h)$$
$$F_v = 0.714 + 0.13e^{0.04(R-60)} \qquad (for \; R > 60mm/h) \qquad (9)$$

where R is the one minute rain rate and L_g is the horizontal projection [13].

As shown in Figure 4, the slant path length adjustment factor by the measured data was less than one in all experiment sections and decreased as the rain intensity increased.

It is seen that, for P.618-5 and P.618-8, the slant path length adjustment factor de-creased when the rain intensity was 50 mm/h or less, but increased when it was more than 50 mm/h. While, in case of 55 mm/h or less of rain intensity, P.618-5 relatively approximated the measured data, P.618-8 did so in more than 55 mm/h. Both of the prediction models, however, showed a large difference from the measured data.

The slant path length adjustment factor suggested by Joo-hwan Lee had a value less than one in all experiment sections and decreased as the rain intensity

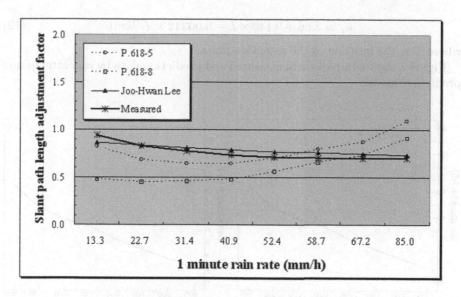

Fig. 4. Comparison of slant path length adjustment factor

increased, which value best conforms to the measured data on the whole. It is considered that it results from well reflecting the Korean rain environment and shows its feasibility and validity.

4.2   Effect of Rain Height

In ITU-R Rec.P.618, several different methods to obtain the mean rain height are used. For P.618-8, P.839-8 is used and some formula are defined in P.618-5 [14]. Several different methods including old methods to calculate the mean rain height in P.618 and P.839 are evaluated but the accurate result in terms of the best method for the rain height is not obtained [14].

The following formula are used to calculate the rain height h_r.

$$h_r = 5[km]   (for 0° \leq \phi < 23°)$$
$$h_r - 5 - 0.075(\phi - 23)[km]   (for \phi \geq 23°) \tag{10}$$

where ϕ is the latitude of the station.

For P.618-8, P.839-3 is used to obtain h_r. In P.839-3, h_r can be calculated by next formula using the 0 isotherm height data h_o given by ITU-R [15].

$$h_r = h_o + 0.36[km] \tag{11}$$

K. Satoh in Japan locally adjacent to Korea suggested a unique rain height as shown in Formula (11) [2].

$$h_r = 3.66 + 0.140 \times L - 0.00342 \times L^2 [km] \tag{12}$$

where L is the latitude of the reception point.

Figure 5 show comparison of measured and prediction data by rain attenuation prediction models.

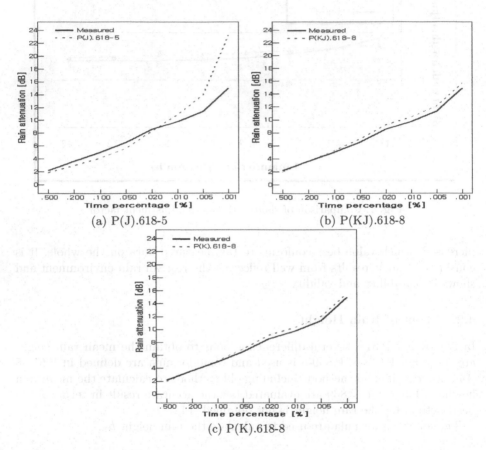

(a) P(J).618-5   (b) P(KJ).618-8

(c) P(K).618-8

Fig. 5. Comparison of correlation of prediction model for measured rain attenuation

For P(J).618-5, only the rain height of formula (12) suggested by K. Satoh without applying adjustment factors of formulae (7) to (9) suggested by Joo-hwan Lee was applied because P.618-5 had only one slant path length adjustment factor which only considers a horizontal characteristic. As a result, P(J).618-5 show the same about 0.97 (P.618-5, P.618-8) in a Pearson correlation coefficient with respect to measured data. Accordingly, although Japan is locally similar to Korea, it is decided that the rain height of formula (12) suggested by K. Satoh has very little effect to Korean rain environment.

For P(KJ).618-8, both adjustment factors of formulae (7) to (9) divided into vertical and horizontal components and suggested by Joo-hwan Lee and the rain

Table 2. Comparison of Pearson correlation coefficient

	Measured	P.618-5	P.618-8	P(J).618-5	P(KJ).618-8	P(K).618-8
Measured	1	0.97	0.97	0.97	0.99	0.99

height of formula (12) suggested by K. Satoh were applied because P.618-8 had two slant path length adjustment factors which consider horizontal and vertical characteristics.

For P(K).618-8 only adjustment factors of formulae (7) to (9) which are divided into horizontal and vertical components suggested by Joo-hwan Lee were applied to P.618.8. As a result, P(KJ).618-8, P(K).618-8 significantly increased from about 0.97 (P.618-5, P.618-8, P(J).618-5) to about 0.99 in a Pearson correlation coefficient with respect to measured data.

It suggests that the slant path length adjustment factor suggested by Joo-hwan Lee well reflects the Korean rain environment and shows its feasibility and effect.

5   Conclusions

In this study, to P.618-5, the most internationally used rain attenuation prediction model currently in the world, and P.618-8 improving P.618-5, respectively, the slant path length adjustment factor proposed by Joo-hwan Lee was applied that consider the annual average rain height suggested by K. Satoh in Japan that is locally similar to Korea. We calculated the Pearson correlation coefficient with respect to measured data. Key results are as follows:

- P(KJ).618-8, P(K).618-8 show the same about 0.99 in a pearson correlation coefficient with respect to measured data. Therefore, in order to further predict Korean rain attenuation, P(KJ).618-8 or P(K).618-8 including the slant path length adjustment factor suggested by Joo-hwan Lee should be greatly considered.
- As compared to the slant path length adjustment factor used in P.618-5 and P.618-8, respectively, that of Joo-hwan Lee conformed very well to measured data on the whole. It is considered that it results from well reflecting Korean rain environment and shows its feasibility and effect (Figure 4, Figure 5, and Table 2).
- The rain height of K. Satoh was suggested in consideration of the annual average rain height of Japan. Although Japan is locally similar to Korea, it was very little effect to apply it directly to Korean environment without modification (Figure 5 and Table 2).

References

1. Choi, D.Y.: Measurement of Rain Attenuation of Microwaves at 12.25GHz in Korea. In: Boutaba, R., Almeroth, K.C., Puigjaner, R., Shen, S., Black, J.P. (eds.) NETWORKING 2005. LNCS, vol. 3462, pp. 1353–1356. Springer, Heidelberg (2005)

2. Satoh, K.: Studies on spatial correlation of rain rate and raindrop layer height. IEICE Trans. Commun. J66-B(4), 493–500 (1983) (Japanese Edition)
3. Lee, J.-H.: Prediction of Rain Effective Path Length Using Satellite Beacon Measure-ments Data, Chungnam National University, A thesis of master (July 1999)
4. Lee, J.-H., Choi, Y.-S., Park, D.-C.: An Empirical Model of Effective Path Length for Rain Attemuation Prediction. KEES 11(5), 813–821 (2000)
5. ETRI, Data Implementation of Rainfall Intensity over South Korea, ETRI (December 1996)
6. Recommendation ITU-R P.618-8 (2003)
7. Recommendation ITU-R P.618-5 (2001)
8. Minematsu, F., Suzuki, Y., Kamei, M., Shogen, K.: Comparison of Measured Rain Attenuation in the 12 GHz Band with Predictions by ITU-R Methods. IEICE Trans. Commun. E88-B(6), 2419–2426 (2005)
9. Hogg, R.V., Craig, A.T.: Introduction to mathematical statistics-Fourth Edition. Collier Macmilan, 73–78
10. Bain, L.J., Engelhardt, M.: Introduction to Probability and Mathematical Statistics, pp. 187–190. PWS Publishers (1987)
11. Recommendation ITU-R P.838-2 (2003)
12. Morita, K., Higuchi, I.: Statistical studies on electro magnetic wave attenuation due to rain. NTT Communication Laboratory Technical Journal 19(1), 97–150 (1970)
13. Dissanayake, A., Allnutt, J., Haidara, F.: A Prediction Model that Com-bines Rain Attenuation and Other Propagation Impairments Along Earth-Satellite Paths. IEEE Transactions on Antennas and Propagation 45(10), 1546–1558 (1997)
14. Ito, C., Hosoya, Y., Kashiwa, T.: A study on rain height of ITU-R rain attenuation prediction method on earth-space links. IEICE Trans. Commun. J82-B(4), 687–690 (1999) (Japanese Edition)
15. Recommendation ITU-R P.839-3 (2001)

Author Index

Lecture Notes in Artificial Intelligence (LNAI)